ENCYCLOPEDIA OF WIRELESS TELECOMMUNICATIONS

McGRAW-HILL TELECOMMUNICATIONS

Bates	*Broadband Telecommunications Handbook*
Bates	*GPRS*
Bates	*Optical Switching and Networking Handbook*
Bates	*Wireless Broadband Handbook*
Bedell	*Wireless Crash Course*
Benner	*Fiber Channel for SANs*
Camarillo	*SIP Demystified*
Chernock	*Data Broadcasting*
Clayton	*McGraw-Hill Illustrated Telecom Dictionary,* 3/e
Collins	*Carrier Grade Voice over IP*
Faigen	*Wireless Data for the Enterprise*
Guthery	*Mobile Application Development with SMS and the SIM Toolkit*
Harte	*3G Wireless Demystified*
Harte	*Delivering xDSL*
Held	*Deploying Optical Networking Components*
Kobb	*Wireless Spectrum Finder*
Lee	*Mobile Cellular Telecommunications,* 2/e
Lee	*Lee's Essentials of Wireless*
Louis	*Broadband Crash Course*
Louis	*M-Commerce Crash Course*
Louis	*Telecommunications Internetworking*
Minoli	*Ethernet-Based Metro Area Networks*
Mueller	*Convergence APIs: Parlay and JAIN*
Muller	*Bluetooth Demystified*
Muller	*Desktop Encyclopedia of Telecommunications,* 3/e
Nichols	*Wireless Security*
OSA	*Fiber Optics Handbook*
Pecar	*Telecommunications Factbook,* 2/e
Radcom	*Telecom Protocol Finder*
Richard	*Service Discovery: Protocols and Programming*
Roddy	*Satellite Communications,* 3/e
Rohde/Whitaker	*Communications Receivers,* 3/e
Russell	*Signaling System #7,* 3/e
Russell	*Telecommunications Pocket Reference*
Russell	*Telecommunications Protocols,* 2/e
Sayre	*Complete Wireless Design*
Shepard	*Optical Networking Demystified*
Shepard	*SONET/SDH Demystified*
Shepard	*Telecom Convergence,* 2/e
Shepard	*Telecom Crash Cource*
Simon	*Spread Spectrum Communications Hanbook*
Smith	*Wireless Telecom FAQs*
Smith/Gervelis	*Cellular System Design and Optimization*
Snyder	*Wireless Telecommunications Networking with ANSI-41,* 2/e
Sulkin	*Implementing the IP-PBX*
Vacca	*I-Mode Crash Course*
Wetteroth	*OSI Reference Model for Telecommunications*
Whitaker	*Interactive Television Demystified*

Encyclopedia of Wireless Telecommunications

Francis Botto

McGraw-Hill
New York Chicago San Francisco
Lisbon London Madrid Mexico City Milan
New Delhi San Juan Seoul Singapore
Sydney Toronto

Library of Congress Cataloging-in-Publication Data

Botto, Francis.
Encyclopedia of wireless telecommunications / Francis Botto.
p. cm.
ISBN 0-07-139025-1 (alk. paper)
1. Wireless communication systems—Encyclopedias. I. Title.

TK5103.2.B68 2002
621.382′03—dc21 2002021273

McGraw-Hill

*A Division of The **McGraw-Hill** Companies*

1 2 3 4 5 6 7 8 9 0 DOC/DOC 0 8 7 6 5 4 3 2

ISBN 0-07-139025-1

The sponsoring editor for this book was Marjorie Spencer, the editing supervisor was Stephen M. Smith, and the production supervisor was Sherri Souffrance. It was set in Century Schoolbook by TechBooks.

Printed and bound by R. R. Donnelley & Sons Company.

This book is printed on recycled, acid-free paper containing a minimum of 50% recycled, de-inked fiber.

CONTENTS

Contents

PREFACE

This encyclopedia provides a comprehensive reference about wireless telecommunications, including the intricate network infrastructures, third-generation technologies and concepts, evolving software architectures, security perimeters, video services, multimedia applications, and the handheld devices themselves. It is intended as an industry guide for individuals currently working in mobile telecommunications development, engineering, and management, and for those on the paths that lead to these destinations.

The previously meandering path of development of wireless technology has become linear in contemporary times, where all the services and applications are now visualized clearly and of course implemented using commercially available technologies and development tools. The publication of this encyclopedia is therefore timely as wireless telecommunications is emerged fully in the mass market.

FRANCIS BOTTO

ABOUT THE AUTHOR

Francis Botto is a consultant and researcher in the field of wireless applications and computer/Internet technologies. He has authored or co-authored a number of books and has published numerous articles about computer technology and the Internet.

Numerals

Num-2

1G (First-Generation) Networks

The 1G category of public mobile network is now consigned to history in first-world countries. 1G networks were analog and offered mainly telephony services, including AMPS (Advanced Mobile Phone System) that operated in the 800MHz cellular band.

By the late 1970s early cellular networks began to emerge, and formed the basis for the wireless communications used today. In 1977 Illinois Bell introduced a cellular network in Chicago. Called AMPS (it was developed by AT&T's Bell Laboratories and operated between the 800MHz and 900MHz bands, and was the most used service in the United States until the early 1990s. In 1982 the development of an international standard called GSM (Global System for Mobile Communications) was begun, and by 1993 a rapid rise in compatible networks had emerged globally. This marked the beginning of wireless telecommunications for the mass market and drove the many growing cell phone services available today.

The capacity of 1G networks was naturally small, because authentic 1G networks were not cellular. Later cellular networks provided greater geographic coverage through cells using the same frequencies. An early internationally agreed standard analog network did not exist, and so countries and continents had disparate network systems. These included Nordic Mobile Telephone (NMT–Scandinavia), Total Access Communications System (TACS), C-Netz (West Germany), Radiocomm 2000 (France), and, of course, AMPS.

NMT was also used in central and southern Europe, and was introduced in Eastern Europe during the late 1990s. NMT had two variants, namely, NMT-450 and a later NMT-900 system using the 900Mhz frequency band. Like some of the later networks, NMT offered the option of international roaming—although this was not as seamless as it is today.

The United Kingdom adopted TACS, which was based on AMPS, but used the 900MHz frequency band; TACS became successful in the Middle East and in Southern Europe. The American AMPS standard used the 800MHz frequency band and was also used in South America, Far East, and in the Asia Pacific region, including Australia and New Zealand. In the Asia Pacific country of Japan was NTT's MCS system that was the first commercial delivery of a mobile 1G Japanese network. While most first-world countries are closing 1G networks, many Less Developed Countries (LDCs) are actively investing in and also upgrading them.

Telecommunications networks remain a foundation for the World Wide Web, and connectivity between tier 1 and tier 3/4 devices and servers and hosts, but it was the advent of the TCP/IP (Transmission Control Protocol/ Internet Protocol), which emerged from the early DARPA and ARPANET

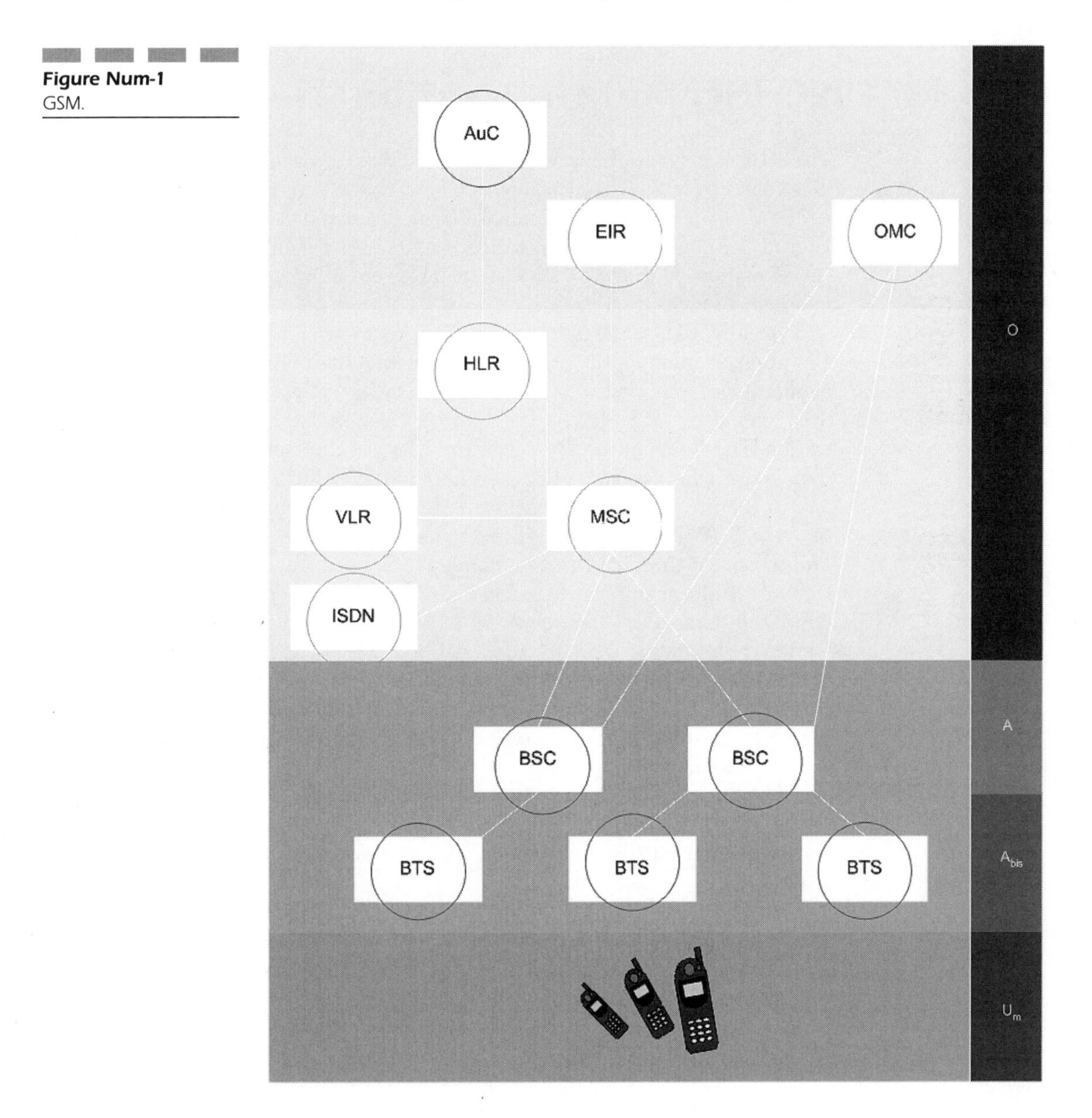

Figure Num-1
GSM.

networks, that introduced a common transport protocol for communications. In 1983 all ARPANET networks were running TCP/IP, and later 32bit IP addressing meant that networks could include millions of addressable hosts. Some 18 years later, the growing global network is now driving the development of IPv6, that is, a 128bit addressing scheme specified by the IETF. See Fig. Num-1.

(*See* 2G Networks, 2.5G Networks, 3G, all below; cdmaOne under letter C, GPRS under letter G, UMTS under letter U.)

2G (Second-Generation) Networks

The 2G category of public mobile network offers the earliest digital telecommunications and includes variations of Global System for Mobile (GSM), Digital-AMPS (D-AMPS), Code Division Multiple Access (CDMA–IS95), and Personal Digital Cellular (PDC). GSM is the most popular globally, and in August 2000 it was estimated that 372 GSM networks were in operation with a collective mobile user base of 361.7 million.

Typically a 2G GSM network provides users with data rates of 9.6kbps or 14.4kbps. As many as 20 to 25 MSCs (Mobile Switching Centers) or control elements are included in a network, and each will cover a given geographic area served by BTSs (Base Transmitter Stations) each with a 5km to 8km radius. The BTSs naturally determine the cellular network coverage where mobile users may be served by a number of BTSs that switch seamlessly from one to the next. It may be assumed that GSM operates in the uplink band between 890MHz and 915MHz and usually offers a user data rate of 9.6kbps, and the speed threshold on the mobile station is around 250km/h.

Two GSM variants include Digital Cellular Systems 1800 (DCS-1800 or GSM-1800) and PCS-1900 or GSM-1900 that is used in North America and Chile. The different frequency is used because of the lack of capacity in the 900MHz band. The 1800MHz band accommodates a larger number of mobile users particularly in densely populated areas. The coverage area of 1800MHz networks is often smaller, and therefore dual-band phones are used, able to roam between either network.

ETSI has also published GSM-400 and GSM-800 specifications, with the former suited to large geographic area coverage, and can therefore be used in conjunction with higher frequency band networks in sparsely populated regions.

Comparable to GSM, DCS-1800 (Digital LPC Cellular System) is used in the United Kingdom. It operates in the uplink radio band between 1710MHz and 1785MHz and can be assumed to provide a user data rate of 9.6kbps, and exhibits a 250 km/h speed threshold on the mobile station.

(*See* 2.5G Networks below, 3G below, GSM under letter G, UMTS under letter U.)

GSM Network Operation

1. When a mobile phone is switched on, it registers its presence with the nearest MSC that is then informed of the location of the mobile user.
2. If the user is outside the geographic area of the home MSC, the nearest MSC will implement a registration procedure. This procedure uses the home MSC to acquire information about the mobile device. This information is held by the home MSC in a database called the home location register (HLR) that holds mapping information necessary so that calls can be made to the user from the PSTN (Public Switched Telephone Network). The local MSC duplicates part of this information in the VLR (Visitor Location Register) for as long as the caller is in the MSC area.
3. Normally one HLR and one VLR are associated with each MSC that provides switching and a gateway to other mobile and fixed networks.
4. Mobile devices have SIM chips holding user identification and configuration data. SIM chips permit an authorization procedure to be implemented between MSCs and EIRs (Equipment Identification Registers). The EIR has a blacklist of barred equipment, a gray list of faulty equipment or for devices that are registered for no services, and a white list for registered users and their service subscriptions.

When either voice or data traffic originates at the subscriber terminal, it goes over the air interface to the BTS, from where it goes to the BSC.

2G Origins

In 1982 CEPT (Conference Europeene des Postes et Telecommunications) assembled the Groupe Special Mobile (GSM) committee so as to specify a pan-European cellular radio system that would increase the capacity of the analog systems like the Nordic Mobile Telephone system, NMT. A pan-European bandwidth of 890 to 915MHz and 935 to 960MHz was agreed.

Eight system proposals included:

- Bosch proposed the S900-D system that used four-level frequency shift keying (FSK) modulation.
- ELAB's (Norway) proposed system-employed adaptive digital phase modulation (AD PM) and a Viterbi equalizer to combat the effects of intersymbol interference (ISI).
- Ericsson proposed the DMS90 system that used frequency hopping, GMSK modulation, and an adaptive decision feedback equalizer (DFE).
- Televerket proposed the Mobira system and the MAX II system that were similar to the DMS90 system.

- SEL proposed the CD900 wideband TDMA system that is also used in conjunction with spectral spreading.
- TEKADE proposed the MATS-D system that incorporated three different multiple access schemes, namely, code division multiple access (CDMA), frequency division multiple access (FDMA), and time division multiple access (TDMA).
- LCT proposed the SFH900 system that used frequency hopping in combination with Gaussian minimum shift keying (GMSK) modulation, Viterbi equalization, and Reed-Solomon channel coding.

The proposed systems were piloted in Paris in 1986, when ELAB's offering was chosen. By June 1987 a narrow-band TDMA system based on ELAB's was agreed, and would support eight (and eventually 16) channels per carrier.

Six different codecs at 16kbps were considered, where a residual excited linear prediction (RELP) codec and a multipulse excitation linear prediction codec (MPE-LPC) proved best. Eventually these were merged to produce a regular pulse excitation LPC (RPE-LPC) with a net bit rate of 13kbps.

GMSK was chosen because of its improved spectral efficiency, and the initial drafts of the GSM specifications were published in mid-1988 when it emerged that it would not be possible to fully specify every feature before the 1991 launch. So the system specification was given two phases: The most common services (including call forwarding and call barring) were in Phase 1; and remaining services (including supplementary services and facsimile) were in Phase 2.

At the request of the United Kingdom, a version of GSM operating in the 1800MHz dual band was included in the specification for Personal Communications Networks (PCN) that became Digital LPC Cellular System at 1800MHz (DCS 1800 or GSM 1800). In Phase 2 of the specifications (June 1993), GSM900 and DCS 1800 were merged in the same documents.

Because of the complexity of processes required for a third (Phase 3) revision, it was decided that an incremental advancement strategy would be adopted as new features surfaced; this is known as Phase 2+. Significant Phase 2+ proposals include an increase in the maximum mobile speed and the half-rate speech coder.

1988 saw GSM become a Technical Committee within the European Telecommunications Standards Institute (ETSI). In 1991 the GSM Technical Committee was renamed Special Mobile Group (SMG) and given the task of specifying a successor to GSM. The group SMG5 was assigned the task of specifying the Universal Mobile Telecommunication System (UMTS), but has since been discontinued and the task of specifying UMTS is given to other committees.

(*See* 2.5G Networks below, 3G below, GSM under letter G, UMTS under letter U.)

2.5G Networks

The 2.5G category of public mobile network offers improved data rates by adding an overlay such as GPRS (General Packet Radio Service) to a 2G network like GSM. Other 2.5G solutions include HSCSD (High-Speed Circuit-Switched Data) and EDGE (Enhanced Data Rates for Global Evolution). These technologies seek to increase the user data rates of 2G networks like GSM that typically exhibit 9.6 and 14.4kbps.

2.5G GPRS

The GPRS (General Packet Radio Service) overlay increases the GSM and TDMA user data rate to a maximum of about 171kbps, and includes the following physical network overlay core solutions:

- IP-based GPRS backbone
- GGSN (Gateway GPRS Serving Node)
- SGSN (Server GPRS Serving Node)

Principal features of GPRS include:

- Packet-based transmission
- Packet-based charging as opposed to time-based charging
- Always-on—so a VPN connection, for example, does not require constant logging on to the remote/enterprise network/intranet
- Eight time slots
- A maximum user data rate of about 171kbps when all time slots are used
- Alleviates capacity impacts by sharing radio resources among all mobile stations in cells
- Lends itself to bursty traffic

See Fig. Num-2.
GPRS applications include:

- E-mail, fax, messaging, intranet/Internet browsing
- Value-added services (VAS), information services, games
- m-commerce: retail, banking, financial trading, advertising
- Geographic: GPS, navigation, traffic routing, airline and/or rail schedules
- Vertical applications: freight delivery, fleet management, sales-force automation

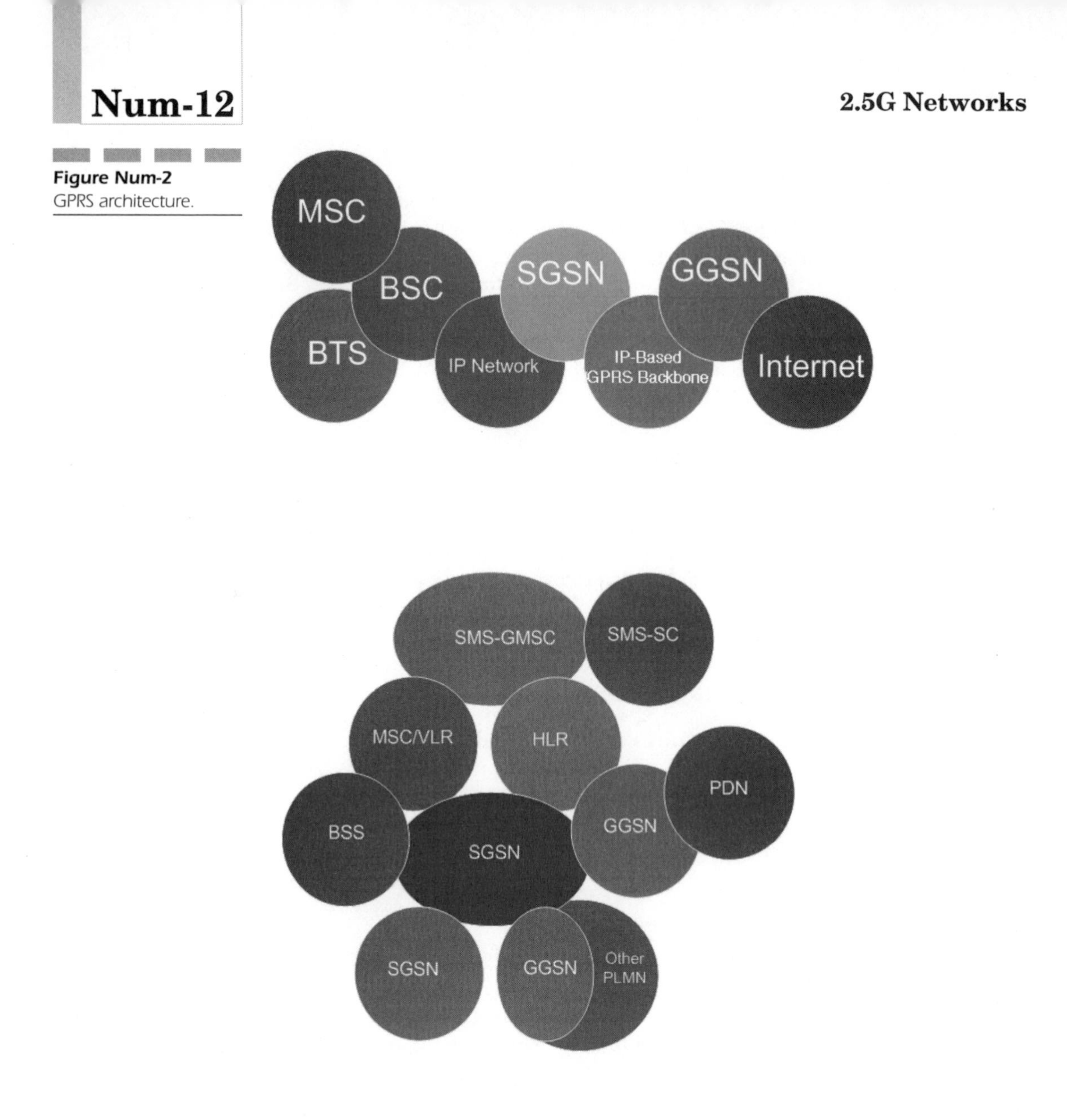

Figure Num-2
GPRS architecture.

GPRS Terminal Classes

A GPRS terminal phone may be:

- Class A: Supports GPRS, GSM, and SMS, and may make or receive calls simultaneously over two services. In the case of CS services, GPRS virtual circuits are held or placed on busy as opposed to closed down.
- Class B: Supports either GSM or GPRS services sequentially, but monitors both simultaneously. Like Class A, the GPRS virtual circuits

will not be closed down with CS traffic, but switched to busy or held mode.

- Class C: Make or receive calls using the manually selected service, with SMS an optional feature.

GPRS Architecture

The GPRS architecture is an overlay network for 2G GSM networks, and provides packet data for user data rates between 9.6 and 171kbps. GPRS's air-interfaces are shared among multiple users.

BSC BSCs require software upgrades and packet control unit (PCU) hardware that directs data to the GPRS network.

Core Network Nodes and gateways that are essentially packet-switched MSCs are required, including:

- Serving GPRS Support Node (SGSN)
- Gateway GPRS Support Node (GGSN)

Databases (HLR and VLR) Databases require software upgrades to handle GPRS functions and call models.

Subscriber Terminal New subscriber terminals are required to access GPRS services.

GPRS BSS

To become a GPRS BSC, each GSM BSC requires the installation of one or more PCUs that give interfaces for packet data out of the BSS. Like GSM when either voice or data traffic originates at the subscriber terminal, it goes over the air interface to the BTS, from where it goes to the BSC. Unlike GSM, however, the traffic is separated; voice is sent to the MSC (like GSM) and data is sent to the SGSN, through the PCU using a Frame Relay interface.

GPRS Network

2G GSM MSCs are circuit switched (CS) and cannot therefore handle packet data, so Nodes are introduced:

- Serving GPRS Support Node (SGSN)
- Gateway GPRS Support Node (GGSN)

The SGSN acts like a packet router or packet-switched MSC that directs packets to MSs.

SGSN

A 2.5G GPRS SGSN performs the following functions:

- Queries home location registers (HLRs) to get subscriber profiles
- Detects new GPRS MSs
- Processes registration of new subscribers
- Keeps records of subscribers' locations
- Performs mobility management functions such as mobile subscriber attach/detach and location management
- Connects to the base-station subsystem using a Frame Relay connection to the BSC's PCU

GGSN

A 2.5G GPRS GGSN performs the following functions:

- Interfaces with IP networks like the Internet and enterprise intranets, and with other mobile service providers' GPRS provisions
- Maintains routing information used to tunnel protocol data units (PDUs) to SGSNs that serve MSs
- Networks and handles subscriber screening
- Addresses mapping
- Supports multiple SGSNs

GPRS Mobility Management

- GPRS mobility management functions track the location of an MS as it moves within a given area.
- Visitor location registers (VLRs) store the MS profiles that are accessed by SGSNs via the local GSM MSC.
- The MS and the SGSN are connected via a logical link in each mobile network.
- When transmission ends, or when an MS moves out of an SGSN's area, the logical link is released.

2.5G HSCSD HSCSD (High-Speed Circuit-Switched Data) usually uses a maximum of four time slots (that may be 9.6kbps or 14.4kbps) for data connections. Because HSCSD is circuit switched, used time slots are constantly allocated even when there is no transmission. This disadvantage makes HSCSD appropriate for real-time applications with short latencies. HSCSD also requires appropriate handsets that are not as widespread as GPRS, for example, but at the same time HSCSD is a less expensive network upgrade than GPRS for operators.

2.5G EDGE EDGE is an overlay solution for existing ANSI-136/TDMA networks, and may use the existing ANSI-136 30kHz air-interface. EDGE is on the migration path to UMTS, and may even coexist with it so as to provide services for wide-area coverage. EDGE standards support mobile services in ANSI-136/TDMA systems with data rates of up to 473kbps.

A significant change in the ANSI-136/TDMA standards to support higher data rates is the use of modulation schemes, including 8-PSK (Phase Shift Keying) and GMSK (Gaussian Minimum Shift Keying). GMSK provides for wide-area coverage, while 8-PSK provides higher data rates but with reduced coverage.

EDGE provides high data rates over a 200kHz carrier, giving up to 60kbps per timeslot that may equate to 473kbps. EDGE is adaptive to radio conditions, giving the highest data rates where there is sufficient propagation.

(*See* EDGE under letter E.)

EGPRS (Enhanced General Packet Radio Service) The dominant data networking protocol, on which most data network applications are running, is TCP/IP, the Internet protocol. All Web applications are run on some form of TCP/IP, which is by nature a protocol family for packet-switched networks. This means that EGPRS is an ideal bearer for any packet-switched application, such as an Internet connection. From the end user's point of view, the EGPRS network is an Internet subnetwork that has wireless access. Internet addressing is used, and Internet services can be accessed. A new number, the IP.

(*See* 3G below, UMTS under letter U.)

8-PSK (8-Level Phase Shift Keying) An enhanced modulation method that is used in the EDGE radio interface. In this context 8-PSK:

- Has three bits per symbol
- Gives a gross bit rate per slot of 69.2kbps (including overhead, and given that the symbol rate is 271ksymbols/s).

All rates per time slot include 22.8, 34.3, 41.25, 51.6, 57.35, and 69.2kbps for code rates of 0.33, 0.50, 0.60, 0.75, 0.83, and 1.

(*See* EDGE under letter E.)

3G (Third Generation)

The 3G category of public mobile network is capable of offering user data rates that may extend to Mbps, and describes the Universal Mobile Telecommunication System (UMTS) that was shaped in part by the Third-Generation Partnership Project (3GPP). 3G cellular technology greatly surpasses the multiple impacts of 2G services and, to a lesser extent, those of 2.5G. The promise of sophisticated video and multimedia wireless applications able to deliver MPEG video and high-quality audio is attracting investment in the WASP and Mobile Telecommunications sectors, as the potentially profitable 3G services are foreseen, including:

- Video applications using MPEG standards
- Video telephony
- Videoconferencing
- Video on demand (VoD)
- Telepresence
- Surrogate services such as exploration
- Client for remote services
- Web/Internet browsing
- Client for VPN
- Client connection for teleworkers
- mcommerce—retailing, online banking, etc.
- Point of Information (POI) in real estate sector, etc.
- Point of Sale (POS) for secure purchasing
- CCTV (Closed Circuit Television) security
- UAV (Unmanned Aerial Vehicle) video communication and navigation

A UMTS provides global roaming and is architected using orbiting satellites that may integrate BTSs (Base Transmitter Stations) and BSCs (Base Switching Centers). To create this type of network, satellites may orbit at altitudes between 780 and 1414km so as to minimize signal transmission latency. One example of a satellite network is Teledesic whose consortium is led by Bill Gates. Other mobile satellite services include Motorola's Iridium (with 66 satellites), Loran/Qualcom's Globalstar (with 48 satellites), and TRW-Matra's Odyssey (with 10 satellites orbiting at high altitude).

UMTSs offer broadband user data rates and operate in the K-band (10.9 to 36GHz) and L-band (1.6 to 2.1GHz). Aeronautical and maritime telecommunications were catalysts in the development and deployment of satellite mobile telephone services with the first maritime satellite launched in 1976. Called MARISAT, it consisted of three geostationary satellites and was used by the

U.S. Navy. This later evolved into the INMARISAT (International Maritime Satellite Organization) that provides public telecommunications services to airliners.

3G Layers

A 3G network consists of layers dedicated to:

- Transport: Carries data (bits) over the IP backbone and wireless access network that may be ATM, SONET, or an alternative.
- Control: Controls calls, authenticates calls, manages mobility, manages sessions, and is accommodated in the network nodes that include RNC/BSC, MSC, SGSN, and GGSN.
- Applications/services: Hosts applications and services and is otherwise known as the *service network*.

See Fig. Num-3.

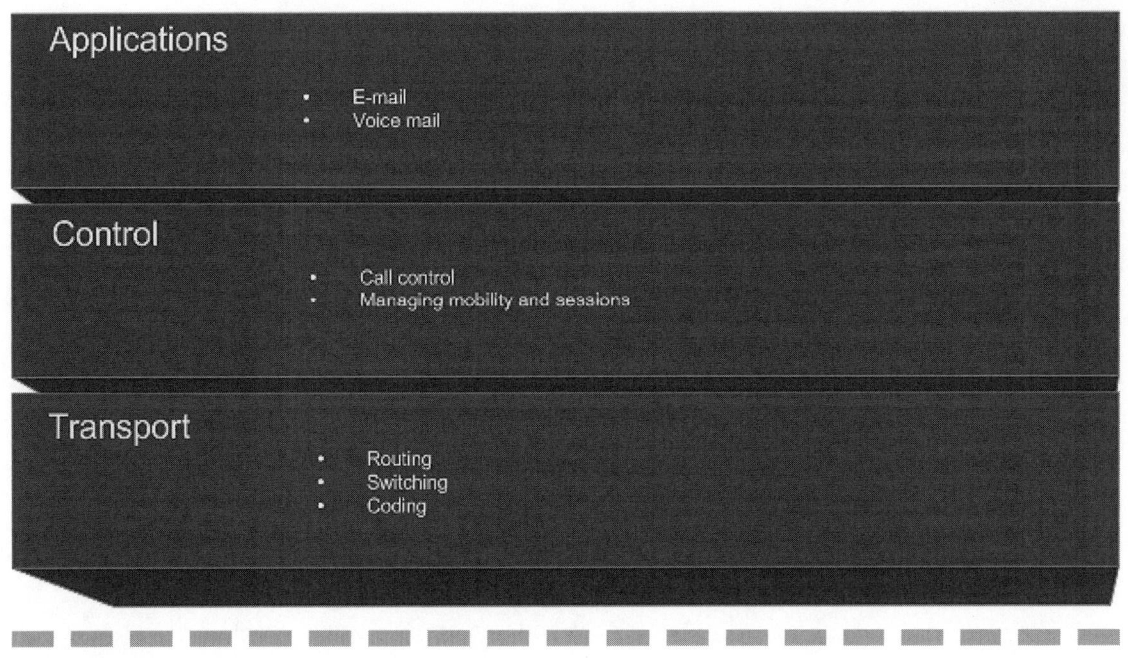

Figure Num-3 3G layers.

3G Origins

WAP and I-mode have been key to mass-market mobile wireless applications that converge on the Internet, giving limited access to Web content. The precursor to WAP is, of course, SMS (Short Message Service), where text is sent to mobile users' handsets, and even though this is considered the trailing edge of mobile applications, such solutions remain a practical industry for many WASPs globally. WASPs such as these are (generally) positioned close to the trailing edge, as opposed to being leading-edge enterprises that may be engaged in developing core software solutions.

The European RACE 1043 project began with the aim of identifying services that Y2K 3G services would deliver, and evaluating how the mobile network infrastructure would evolve in the mass market telecommunications sector. The project's forecasts include a displacement theory where TACS (Total Access Communications System) would be displaced by GSM that would then be displaced by UMTS (Universal Mobile Telecommunications System)—a term that was coined by the project.

However, the intermediate transitions from 2G GSM to 3G were not foreseen, including the incremental advancements of GPRS (General Packet Radio Service) and EDGE.

The 3GPP were assigned the task of specifying a 3G system based on the underlying GSM network.

UMTS Acronyms

ACLR	Adjacent channel leakage power ratio
AI	Acquisition indicator
AICH	Acquisition indication channel
BCH	Broadcast control channel
CCPCH	Common control physical channels
CPCH	Common packet channel
CPICH	Common pilot channel
DCH	Dedicated channel
DPCCH	Dedicated physical control channel
DPDCH	Dedicated physical data channel
FACH	Forward access channel
FBI	Feedback information
FDD	Frequency division duplex
GGSN	Gateway GPRS support node
GMSC	Gateway MSC
I_{uCS}	Interface between an RNC and an MSC
I_{uPS}	Interface between an RNC or BSC and an SGSN
I_{ur}	Interface between RNCs MAC Medium access control

MSC	Mobile switching center
MUD	Multiuser detection
Node B	Base station transceiver
OVSF	Orthogonal variable spreading factor
P-CCPCH	Primary common physical channel
PCH	Paging channel
PCPCH	Physical common packet channel
P-CPICH	Primary CPICH
PI	Paging indicator
PICH	Pilot channel
PRACH	Physical random access channel
PSC	Primary synchronization code
QPSK	Quadrature phase shift keying
RACH	Random access channel
RNC	Radio network controller
RNS	Radio network subsystem
S-CCPCH	Secondary common control physical channel
S-CPICH	Secondary CPICH
SCH	Synchronization channel
SF	Spreading factor
SGSN	Serving GPRS support node
SSC	Secondary synchronization code
TFCI	Transport format combination indicator
TPC	Transmit power control
UARFCN	UTRA absolute radio frequency channel number
UE	User equipment
UMTS	Universal mobile telecommunication system
USIM	Universal subscriber identity module

UMTS Network Architecture

The UMTS network shown in Fig. Num-4 is essentially a GSM Phase 2+ core network that is optimized for higher bit rates, and includes:

- Mobile switching center (MSC) and gateway MSC (GMSC) for circuit-switched GSM networks
- Serving GPRS support node (SGSN) and a gateway GPRS support node (GGSN) because GSM Phase 2+ accommodates GPRS packet data
- GSM base station subsystems (BSSs)
- UMTS radio network subsystems (RNSs)
- A-interface between a base station controller (BSC) and a mobile switching center (MSC)

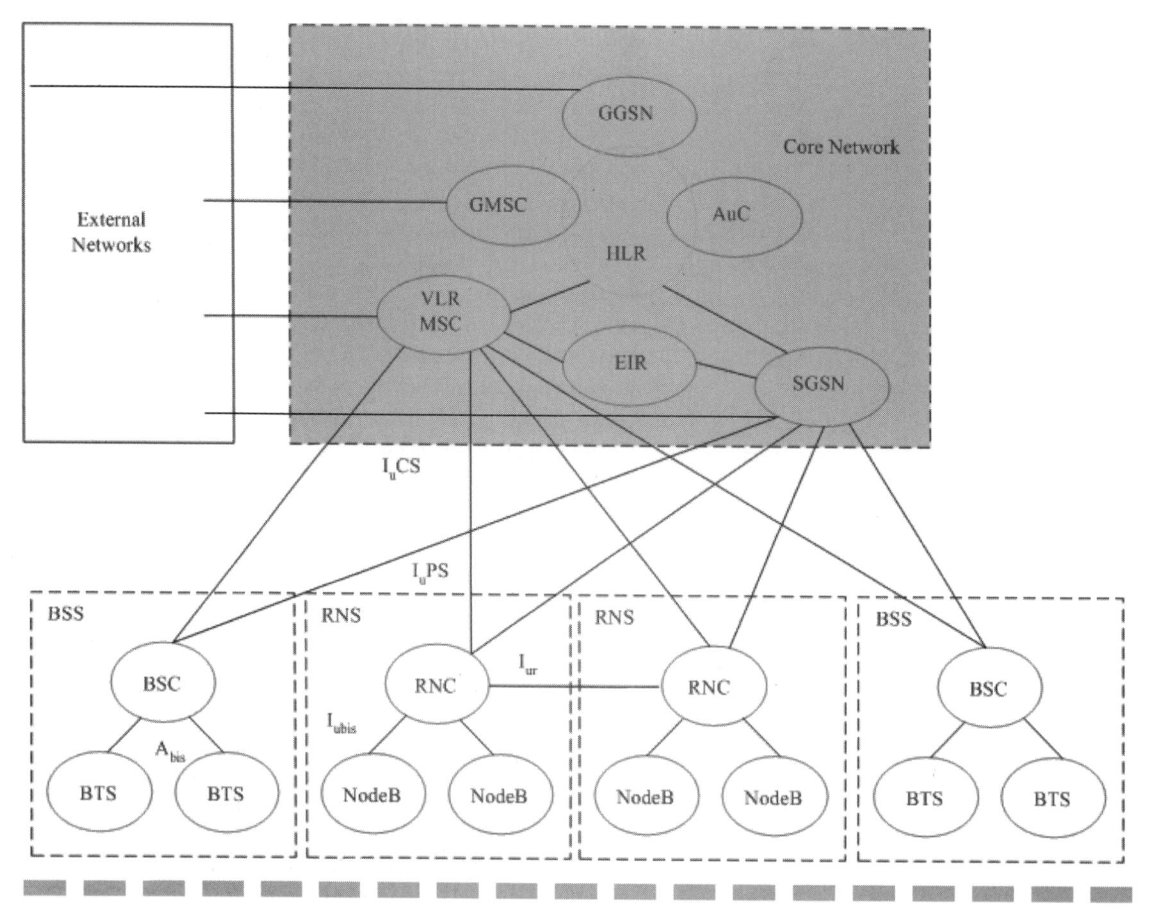

Figure Num-4 3G UMTS architecture.

- I_{uPS} between a BSC and an SGSN, where the subscript "uPS" signifies a packet-switched interface
- Abis interface between a BTS and a BSC
- Interfaces between the RNC and MSC, SGSN, and RNC of I_{uCS} (circuit-switched), I_{uPS}, and I_{ur}, respectively

The physical channels in UMTS transfer information across the radio interface. A physical channel is defined by its:

- Code and carrier frequency in an FDD version
- Code, carrier frequency, and timeslot in TDD

Operations of a UMTS Transmitter at the Physical Layer The transport channel data from layer 2 and above are arranged in blocks depending on the type of data. The blocks are cyclically redundancy coded (CRC) for error detection at the receiver. The data are segmented into blocks and channel coding ensues. The coding may be convolutional or turbo. Sometimes channel coding is not used. Data is interleaved to decrease the memory of the radio channel and thereby render the channel more Gaussian-like. The interleaved data are then segmented into frames compatible with the requirements of the UTRA interface.

Rate matching is performed next. This uses code-puncturing and call data repetition, where appropriate, so that after transport channel multiplexing the data rate is matched to the channel rate of the dedicated physical channels. A second stage of bit FI interleaving is executed, and the data are then mapped to the radio interface frame structure.

Suffice it to say that at this point there are different types of physical channels, namely:

- Pilot channels that provide a demodulation reference for other channels
- Synchronization channels that provide synchronization to all UEs within a cell
- Common channels that carry 0.2MHz of information to and from any user equipment (UE)
- Dedicated channels that carry information to and from specific UEs.

The physical layer procedures include:

- Cell search for the initial synchronization of a UE with a nearby cell
- Cell reselection, which involves UE changing cells
- Access procedure that allows a UE to initially access a cell
- Power control to ensure that a UE and a BS transmit at optimum power levels
- Handover, the mechanism that switches a serving cell to another cell during a call

UMTS Terrestrial Radio Interface

The UMTS terrestrial radio interface (UTRA) frequency duplex (FDD) mode is the W-CDMA radio interface of the UMTS, and is designated by the ITU as IMT DS. The UTRA FDD mode uses segment 3 for up-link transmission, and segment 6 for down-link transmission. These segments are described in Table Num-1.

The nominal spacing between radio carriers is 5MHz, with a channel raster of 0.2MHz. This means that the carrier separation may be adjusted in steps of 0.2MHz, e.g., the carrier spacing may be 4.8MHz.

TABLE Num-1

IMT-2000 Spectrum and Segments

Segment	Frequency	Comment Number Band (MHz)	Uses
1	1885–1900	Unpaired	Currently used for DECT in Europe, and for PHS, PCS, and DECT in other parts of the world Suitable for time division duplex (TDD) operation
2	1900–1920	Unpaired	Segment 2 is used at present for PCS and PHS in the United States and Japan, respectively
3	1920–1980	Paired with 6	Forms 60MHz frequency division duplex (POD)
4	1980–2010	MSS (mobile satellite services) paired with 7	Supports the earth-to-space links Used for PCS in the United States. Mobile satellite services (MSS) are in Segments 4 and 7, providing 30MHz POD bands
5	2010–2025	Unpaired	Segment 5 may be used in the United States for earth-to-space MSS services Suitable for time division duplex (TDD) operation
6	2110–2170	Paired with 3	Forms 60MHz frequency division duplex (POD)
7	2170–2200	MSS (mobile satellite services) paired with 4	Provides the space-to-earth links

The carrier frequency is defined by the UTRA absolute radio frequency channel number (UARFCN). This number is defined over a frequency band from 0 to 3.7GHz, and is the transmission frequency multiplexed by five.

The UARFCN (Nu-uplink and Nd-downlink) will always be an integer because of the raster frequency of 0.2MHz. It cannot be assumed that radio channels in the UTRA FDD are paired as in GSM.

(*See* 2G Networks above, UMTS under letter U.)

3G GGSN

A Gateway serving node for a 3G public mobile network such as UMTS.

(*See* GPRS under letter G, UMTS under letter U.)

3G QoS

A threshold or series of thresholds that determines the overall standard of service provided by the operator, and includes among other things:

- Latencies or delays
- Minimum bit rate guarantee that may be within a specified coverage area

3G SGSN

A Server serving node for a 3G public mobile network such as UMTS. (*See* UMTS under letter U.)

3GPP (Third-Generation Partnership Project)

An international group of telecommunications representatives and /or entities that shaped the UMTS.

The 3GPP consisted of:

- Standards organizations: ARIB (Japan), CWTS (China), ETSI (Europe), TI (USA), TTA (Korea), and TTC (Japan)
- Market representation partners: Global Mobile Suppliers Association (OSA), the OSM Association, the UMTS Forum, the Universal Wireless Communications Consortium (UWCC), and the IPv6 Forum
- Observers: TIA (USA) and TSACC (Canada)

2-Tier

A client/server architecture where application logic, data, and presentation are distributed between client systems (at tier 1) and one or more servers (at tier 2). Now consigned to history, early versions were based on “dumb terminals” (or client systems) that did little more than send and receive messages to and from a server that was invariably a mainframe.

The World Wide Web of the early 1990s was a 2-tier client/server model where the Web/HTTP server simply published HTML documents to the client via a largely unidirectional path. The early static Web and the many intranets were 2-tier, where the user simply received published information (or Web pages) from the Web server. There was no feedback from the client system, and the application elements were partitioned so that data and logic were on the server-side.

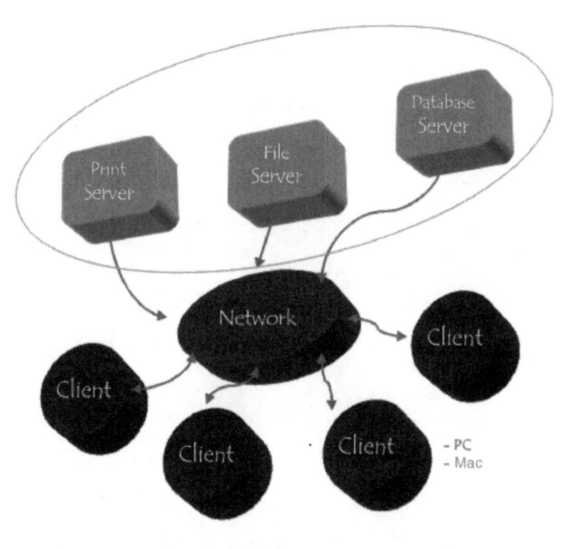

Figure Num-5
2-tier client/server.

This changed in 1995 with the introduction of CGI (Common Gateway Interface). The dynamic or active Web model that was initially driven by CGI impacted the partitioning of the application elements of data, logic, and presentation. See Fig. Num-5.

File servers, print servers, and database servers may also be integrated in the design architecture so as to distribute processing and optimize performance. The connection or access technology between servers and clients is provided by a LAN variant.

(*See* 3-Tier below, CGI under letter C, Client/Server under letter C.)

3-Tier

A client/server architecture where the elements presentation, application logic, and data may be perceived as distributed across different platforms. The three tiers are separate and independent, and interact via appropriate glues or middleware and include:

- Tier 1: presentation is the front end and may be composed of view objects.
- Tier 2 is the application logic that is the middle-tier.
- Tier 3 is data that is the back end.
- Tier 0 devices are those that connect with clients (at tier 1) and include printers, palmtops, Palm PCs, etc.

The partition that separates these three entities, in terms of those that reside on the client and those that reside on the server, is a function of the client/server implementation, and the clients may be PCs, Macintosh

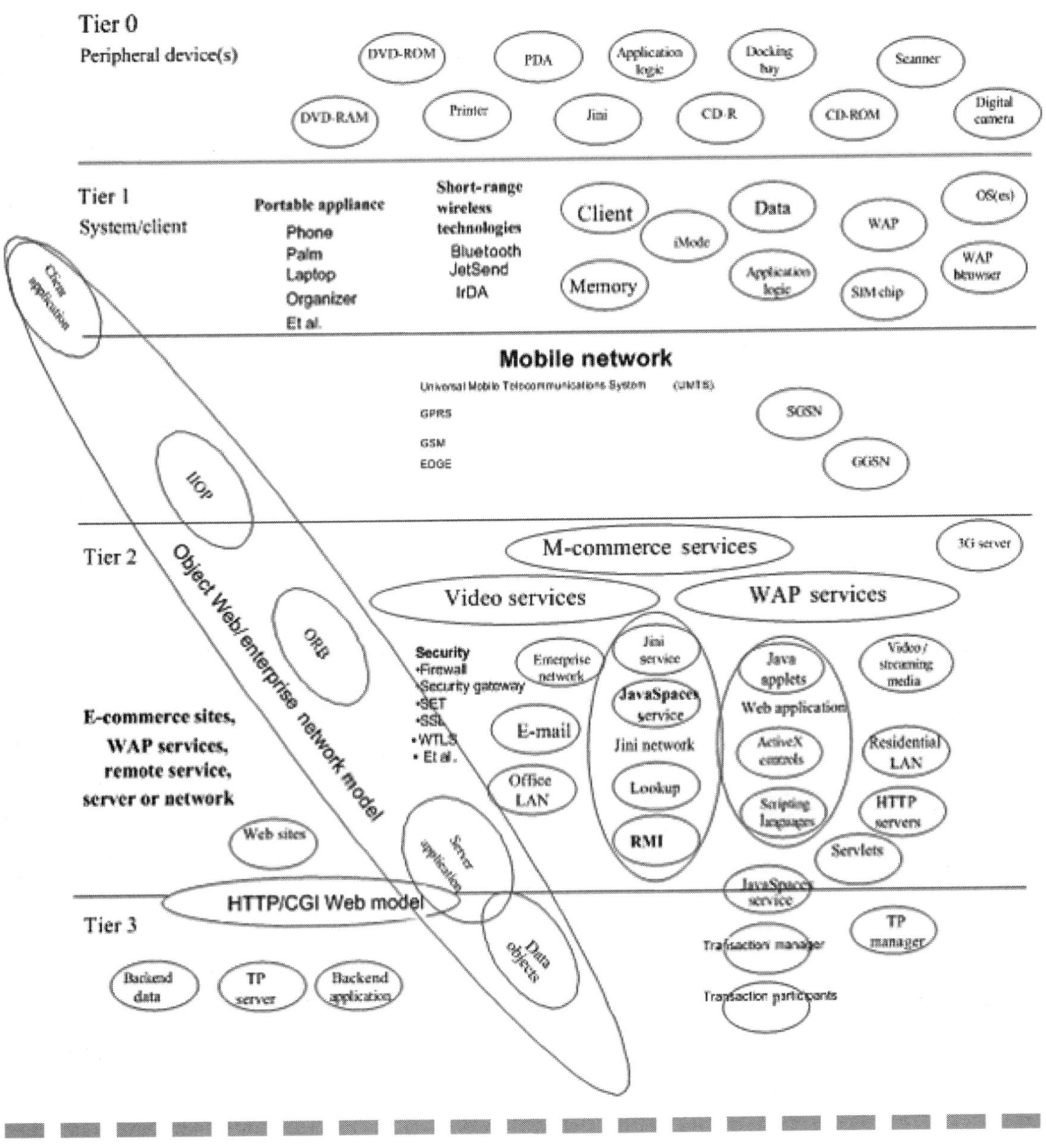

Figure Num-6 3-tier.

computers, or NCs. If middleware is included, it may be based on an interface definition language (IDL) like CORBA. See Fig. Num-6.

(*See* 2G Networks earlier in this chapter, 2.5G Networks earlier in this chapter, Client/Server under letter C, CORBA under letter C, IDL under letter I, UMTS under letter U, WAP under letter W.)

Data Rates and Capacities

1.5Mbps

A maximum user data transfer rate for which MPEG-1 was designed. This figure was chosen because it was the maximum user data transfer rate of a single-speed CD-ROM drive that is calculated as follows:

Average user data transfer rate = blocks/sec × user data/block

- Mode 1: Average user data transfer rate = 75 × 2048 = 153,600bytes/s = 150kbytes/s
- Mode 2: Average user data transfer rate = 75 × 2336 = 175,200bytes/s = 171kbytes/s

(*See* MPEG-1 under letter M.)

1.544Mbps

1. A data transfer rate offered by a single T1 line. (*See* T1 under letter T.)
2. A data transfer rate of a primary rate multiplex of 255 channels of 64kbps ISDN channels. (*See* ISDN under letter I.)

2Mbps (2Mbits/s)

A threshold bandwidth beyond which a network or access technology is described as broadband. 2Mbits/s = 2,000,000 bits per second.
(*See* Access Technology under letter A, B-ISDN under letter B.)

2.048Mbps

The bandwidth offered by an E1 four-wire digital trunk.

9.6kbps

An initial user data rate offered by a single physical channel or single timeslot per TDMA frame over a GSM network. This was increased in early 2000 to 14.4kbps by reducing the power of the channel coding using code symbol puncturing.

The data rates may be increased further by allowing the MS to access multiple timeslots per TDMA, or to use a higher-level modulation scheme such as QAM (Quadrature Amplitude Modulation).

(*See* TDMA under letter T.)

9.6kbps

A mobile network data rate associated with early GSM networks.

10Mbps

A user data transfer rate that the MPEG-2 video standard was originally created for. The Motion Pictures Experts Group (MPEG), and its many subgroups, were given the task of creating MPEG-2. This second phase of MPEG work began as long ago as 1990.

(*See* CD-ROM under letter C, DCT under letter D, DVD under letter D, JPEG under letter J, MPEG under letter M, Video under letter V.)

11Mbps

A maximum data rate delivered by an IEEE 802.11b wireless LAN.

(*See* 802.11 later in this chapter, WLAN under letter W.)

16bit

1. A data path that offers 16 parallel lines or data bits.
2. A sample size, which modern sound cards commonly use for recording and playing wave audio. When the sampling frequency is set at 44.1kHz, the resulting quality is that of audio CD.
3. A 16bit digital video or computer-generated image or animation is generated and stored using 16bits of color information for each pixel (or dot). This results in a maximum of about 65K (or 2^{16}) colors.

(*See* 24bit Image Depth later in this chapter.)

32bit

1. A program or operating system that uses 32bit instructions. 32bit operating systems include Windows 98, Windows NT, and OS/2 Warp. Windows 95 is not a pure 32bit operating system due to certain 16bit instructions, but is generally regarded as 32bit OS.

32bit software is able to access memory more efficiently than 16bit variants. It is capable of flat memory addressing in which 4GBytes (2^{32}) memory segments can be addressed. A 32bit segment register is used to point to addresses within a 4GByte range. (*See* Operating Systems under letter O, Windows under letter W.)

2. A 32bit processor uses 32bit instructions. The earliest Intel 32bit processor was the third-generation 80386.
3. A data bus width (in terms of the number of its lines) connected to a device such as a processor, hard disk controller, memory card, or graphics card.
4. An extension of the 24bit image depth, an additional Byte (or Alpha channel) provides control over the transparency of pixels. Red, green, and blue each represented by eight bits, giving 256 tones of each, which in turn leads to over 16.7 million (256 × 256 × 256) colors. The additional eight bits (the Alpha Channel in Apple parlance) are used to control transparency. 32bit graphics make photographic-quality images possible. The Apple Macintosh is remembered as the first platform on which the 32bit graphics capability became commercially available.

53

The number of bytes in cells, which are used in ATM networks, and includes a five-byte header.

(*See* ATM under letter A, Frame Relay under letter F.)

64bit

1. A program or operating system that uses 64bit instructions. (*See* Operating Systems under letter O, Windows under letter W.)
2. A 64bit processor uses 64bit instructions.
3. A data bus width (in terms of the number of its lines) connected to a device such as a processor, hard disk controller, memory card, or graphics card.
4. An image depth.

64kbits/s

A bandwidth of an ISDN (Integrated Services Digital Network) line. ISDN is used widely for videoconferencing, and high-speed Internet access.

ISDN services were defined by the CCITT in 1971 and published in 1984 in the Red Book. ISDN is based on PCM (Pulse Code Modulation) that was conceived by A. H. Reeves, and experimented with in the Second World War, and was used in American telecommunications in the 1960s so as to increase network capacity.

In 1986 a pre-ISDN service named Victoria was offered by Pacific Bell in Danville, California, offering RS232C ports that were configurable from 50bps to 9.6kbps. In the same year, official ISDN systems were introduced in Oak Brook, Illinios, and by 1988 some 40 similar pilot schemes were installed.

ISDN digital networks eventually developed into B-ISDN where multiple lines could be used to provide data rates in increments of 64kbps, and videoconferencing and high-speed Internet access were made possible. B-ISDN implementations could even be used to implement the lower data rates of 1.544Mbits/s offered by modern T1 digital links, which arrived some time later. By the late 1990s digital networks had risen to levels where real estate agents were being asked questions such as, "Does the apartment have ISDN or T1?"

(*See* B-ISDN under letter B, ISDN under letter I, Videoconferencing under letter V.)

150kbytes/s

1. The average user data transfer rate of a pure single-speed CD-ROM drive operating in mode 1. The data transfer rate of a CD-ROM drive broadly increases in multiples of 150kbps:
   - 10 × speed—approximately 1500kbps
   - 20 × speed—approximately 3000kbps
   - 24 × speed—approximately 3600kbps

   In practical tests, the data transfer rate rarely increases in precise multiples of 150kbps. (*See* CD-ROM under letter C, DVD under letter D.)
2. The average user data transfer rate of a CD-I form 1 track when read using a pure single-speed player.

171kbps

The highest user data rate offered by a GPRS (General Packet Radio Service) network.

(*See* GPRS under letter G.)

171kbytes/s

The average user data transfer rate of a pure single-speed CD-ROM drive operating in mode 2.

(*See* CD-ROM under letter C.)

171.1kbytes/s

A data transfer rate of an audio CD encoded according to the CD-DA or Red Book audio standard.

527.3Mbytes

The user data capacity of 1 h, Mode 1 CD-ROM disc and Form 1 CDI disc.
(*See* CD-ROM under letter C.)

602Mbytes

The user data capacity of 1 h, Mode 2 CD-ROM disc.
(*See* CD-ROM under letter C.)

1000

The number of bits transferred in 1s, using the unit kbps.

1024

1. A kilobyte (kB) has 1024 bytes.
2. A megabyte (MB) has 1024 kilobytes.
3. A gigabyte (GB) has 1024 megabytes.
4. A terabyte (TB) has 1024 gigabytes.

2048bytes

The user data capacity of a CD-ROM mode 1 data block.

2336bytes

1. The user data capacity of a CD-ROM mode 2 data block.
2. The user data capacity of an audio CD sector.

14,400

1. A standard modem speed measured in bps.
2. A user data rate offered by a single physical channel or single timeslot per TDMA frame. The data rate may be increased further by allowing the MS to access multiple timeslots per TDMA, or to use a higher-level modulation scheme such as QAM (Quadrature amplitude modulation).

(*See* Access Technology under letter A, Modem under letter M.)

33,600

A standard modem speed measured in bps. It was superseded by the V.90 56.6kbps analog modem standard.

(*See* 56,600 below, Access Technology under letter A, Modem under letter M.)

56,600

A standard modem speed measured in bps. Two initial 56K standards include: x2 Technology and Rockwell K56flex.

A standard analog modem speed. It exceeds the proven bandwidth limit calculated using Shannon's theorem. The higher speed is achieved using PCM and a digital link between the telephone company and the ISP.

56.6kbps modems are asymmetrical, offering wider downstream bandwidths, thus downloading times are shorter than those of uploading.

The ITU has attempted to amalgamate two of the industry standards: X2 and K56flex.

The resulting V.90 standard was specified provisionally and finally released in 1998.

(*See* Access Technology under letter A, Modem under letter M.)

1,000,000

The number of bits transferred in 1s using a 1Mbps data transfer rate.

Display Technology

12.1in

A standard TFT/DSTN display size used in modern laptop systems.

13.3in

A standard TFT/DSTN display size used in modern laptop systems.

15in

A standard display size, where its CRT (Cathode Ray Tube) is measured diagonally. The measurement cannot always be equated with the screen image size, which may or may not be the same.

16.7 Million

A 24bit digital video, animation, or color graphic may have up to 16.7 (2^{24}) million colors.

1600-by-1200 Pixels

A standard graphics resolution used on many PCs, and its delivery requires an appropriate graphics card and display.

16bit

1. A data path that offers 16 parallel lines or data bits.
2. A sample size, which modern sound cards commonly use for recording and playing wave audio. When the sampling frequency is set at 44.1kHz, the resulting quality is that of audio CD.
3. A 16bit digital video, or computer-generated image or animation is generated and stored using 16bits of color information for each pixel (or dot). This results in a maximum of about 65K (or 2^{16}) colors.

(*See* 24bit Image Depth below, 32bit below.)

24

A playback frame rate of a movie recording.

24bit Image Depth

A 24bit digital video, computer-generated image or animation is generated and stored using 24 bits of color information for each pixel (or dot). This results in a maximum of over 16.7 million (2^{24}) colors. 24bit digital videos, animations, and images are described as truecolor.

Red, green, and blue are each represented by eight bits, giving 256 tones of each, which in turn leads to over 16.7 million (256 × 256 × 256) colors. 24bit graphics make possible near-photographic-quality images.

(*See* Computer Graphics under letter C.)

25

A playback frame rate of a PAL or SECAM broadcast television/video signal; this frame rate prevails in most countries except the United States and Japan.

(*See* MPEG under letter M.)

30

A playback frame rate of an NTSC broadcast television/video signal. It is used in the United States and in Japan.

(*See* MPEG under letter M.)

30bit Image Depth

A 30bit digital video or computer-generated image or animation that is generated and stored using 30 bits of color information for each pixel (or dot). This results in a maximum of about 1 billion (or 2^{30}) colors.

(*See* 24bit Image Depth above.)

32bit

An extension of the 24bit image depth, an additional Byte (or Alpha channel) provides control over the transparency of pixels. Red, green, and blue are each represented by eight bits, giving 256 tones of each, which in turn leads to over 16.7 million (256 × 256 × 256) colors. The additional eight bits (the Alpha Channel in Apple parlance) are used to control transparency. 32bit graphics make possible photographic-quality images. The Apple Macintosh is remembered as the first platform upon which the 32bit graphics capability became commercially available.

352-by-240 Pixels

A frame resolution that is described as the SPA (Significant Pel Area) for an MPEG-1 video sequence encoded using an NTSC broadcast television/video source. The playback frame rate is standardized at 30 frames/s.

(*See* MPEG under letter M.)

352-by-288 Pixels

A frame resolution that is described as the SPA (Significant Pel Area) for an MPEG-1 video sequence encoded using a PAL broadcast television/video source.

(*See* MPEG under letter M.)

360-by-240 Pixels

A frame resolution that may be used as an SIF (Source Input Format) for an MPEG-1 video sequence encoded using an NTSC broadcast television/video source. The playback frame rate is standardized at 30 frames/s.

(*See* MPEG under letter M.)

360-by-288 Pixels

A frame resolution that is described as the SIF (Source Input Format) for an MPEG-1 video sequence encoded using a PAL broadcast television/video source.

(*See* MPEG under letter M.)

36bit

An image depth.

(*See* 24bit Image Depth, 30bit Image Depth, 32bit, all above.)

3D (Three-Dimensional)

A 3D computer image or animation stored and generated using absolute or relative coordinates that include X (horizontal), Y (vertical), and Z (depth) dimensions.

Standard file formats and standard languages for developing 3D animations for multimedia and virtual reality (VR) have emerged. The VRML

(Virtual Reality Modeling Language) is suitable for the development of 3D World Wide Web (www) pages.

Web content development tools may be used to create 3D graphics and animations for Web pages, and often do not require knowledge of VRML.

Chips aimed at the acceleration of 3D graphics include the Glint family that was developed by 3DLabs. Creative Labs licensed Glint technology from 3DLabs in 1994 after which they collaborated to develop the GLINT 3D processor.

This is used in the Creative 3D Blaster that was first shown at Creativity '95 in San Francisco—a milestone in the development of 3D graphics cards.

3D engines and APIs that can be used to generate 3D visuals and animations include:

- OpenGL (accommodated by Mac OS X)
- Microsoft Direct3D
- Apple QuickDraw3D

Authentic 3D animations depend on matrix multiplication where sets of coordinates are multiplied by a transformation matrix. 3D vectors, or ordinary 3D coordinates, [X Y Z], may be exchanged for homogeneous vector coordinates [X Y Z H].

The homogeneous dimension (H) is added to accommodate a four-row transformation matrix, increasing the number of possible 3D transformations. The transformation of homogeneous coordinates is given by:

$$[X \quad Y \quad Z \quad H] = [x \quad y \quad z \quad 1]T$$

The resulting transformed coordinates can be normalized to become ordinary coordinates:

$$[x^* \quad y^* \quad z^* \quad 1] = [X/H \quad Y/H \quad Z/H \quad 1]$$

Consider the 4 × 4 transformation matrix:

$$\begin{matrix} a & b & c & p \\ d & e & f & q \\ h & i & j & r \\ l & m & n & a \end{matrix} = T$$

Scaling, shearing, and rotation are achieved using the 3 × 3 matrix sector:

$$\begin{matrix} a & b & c \\ d & e & f \\ h & i & j \end{matrix}$$

The transformation matrix:

$$\begin{array}{cccc} 1 & 0 & 0 & 0 \\ 0 & \cos 0 & \sin 0 & 0 \\ 0 & -\sin 0 & \cos 0 & 0 \\ 0 & 0 & 0 & 1 \end{array}$$

is used to rotate a 3D object by the angle 0 around the X axis.

A rotation of an angle 0 about the y axis is achieved using the transformation matrix:

$$\begin{array}{cccc} \cos 0 & 0 & -\sin 0 & 0 \\ 0 & 1 & 0 & 0 \\ \sin 0 & 0 & \cos 0 & 0 \\ 0 & 0 & 0 & 1 \end{array}$$

A rotation of an angle 0 about the z axis is achieved using the transformation matrix:

$$\begin{array}{cccc} \cos 0 & \sin 0 & 0 & 0 \\ -\sin 0 & \cos 0 & 0 & 0 \\ 0 & 0 & 1 & 0 \\ 0 & 0 & 0 & 1 \end{array}$$

It is possible to concatenate the rotational transformation matrices so as to perform two rotations concurrently through one matrix multiplication. However, the rotations are noncommutative, so attention must be paid to the order of the transformation matrices during multiplication.

To perform a rotation about the x axis and the y axis, the transformation matrix can be achieved as follows:

$$\begin{array}{cccc} 1 & 0 & 0 & 0 \\ 0 & \cos 0 & \sin 0 & 0 \\ 0 & -\sin 0 & \cos 0 & 0 \\ 0 & 0 & 0 & 1 \end{array} * \begin{array}{cccc} \cos 0 & 0 & -\sin 0 & 0 \\ 0 & 1 & 0 & 0 \\ \sin 0 & 0 & \cos 0 & 0 \\ 0 & 0 & 0 & 1 \end{array}$$

$$= \begin{array}{cccc} \cos 0 & 0 & -\sin 0 & 0 \\ \sin 20 & \cos 0 & \cos 0 \sin 0 & 0 \\ \cos 0 \sin 0 & -\sin 0 & \cos 20 & 0 \\ 0 & 0 & 0 & 1 \end{array}$$

Translation is achieved through the 1×3 matrix sector:

$$[1 \quad m \quad n]$$

Perspective transformation is achieved using the 3×1 matrix sector:

$$\begin{matrix} p \\ q \\ r \end{matrix}$$

The remaining element "a" produces overall scaling. For instance, overall scaling is achieved using the transformation matrix:

$$\begin{matrix} 1 & 0 & 0 & 0 \\ 0 & 1 & 0 & 0 \\ 0 & 0 & 1 & 0 \\ 0 & 0 & 0 & s \end{matrix}$$

Normalizing the transformed coordinates drives the scaling effect:

$$[x^* \quad y^* \quad z^* \quad 1] = [x/s \quad y/s \quad z/s \quad 1]$$

It is important to note that 3D images can also be stored using 2D vector matrices that include X and Y dimensions only.

Graphics transformation algorithms can be written in appropriate high-level languages such as C++, Java, and Visual Basic, and even in machine code or assembly language. Any high-level programming language that supports arrays may be used to develop graphics transformation software.

However, APIs for popular 3D engines such Microsoft Direct3D or Apple QuickDraw3D provide the necessary high-level programming statements, in order to bypass the underlying mathematical elements.

Intel MMX technology gives improved delivery of 3D graphics and animations.

(*See* VRML under letter V.)

3D Curves

A curve or space curve that exists in three dimensions.

Algorithms that include the necessary mathematical elements drive the generation of 3D curves. APIs for popular 3D engines include OpenGL, Microsoft Direct3D, Apple QuickDraw3D. Equally, Web-content development tools may be used to create 3D graphics and animations.

3D Surface

A surface that exists in three dimensions. APIs for popular 3D engines such as Microsoft Direct3D or Apple QuickDraw3D provide the necessary high-level programming statements.

3D Vector Coordinate

Authentic 3D animations depend on matrix multiplication where sets of coordinates are multiplied by a transformation matrix. 3D vectors, or ordinary 3D coordinates are represented by [X Y Z].

(*See* 3D above.)

4:3

A standard aspect ratio adopted in broadcast television, video, and graphics display technology. The IBM VGA graphics standard, and the MPEG-1/2/3/4 video standards offer resolutions that have 3:4 aspect ratio.

(*See* MPEG under letter M.)

640-by-480 Pixels

The standard resolution of SVGA.

1024-by-768

A standard display resolution sometimes referred to as XGA (Extended Graphics Array).

64bit

An image depth.

(*See* 32bit above.)

8bit Image Depth

An 8bit image depth gives a maximum of 256 colors for digital video and computer-generated animations and images. The color information for each pixel (or dot) is stored using eight bits giving a maximum of 256 (2^8) colors.

The 8bit color information can be edited using a palette editor such as Microsoft PalEdit so as to:

- Alter the order of color cells in a palette
- Reduce the number of colors in a palette by deleting unwanted color cells

- Alter brightness
- Alter color contrast
- Fade and tint colors
- Copy color cells from one palette to another
- Merge two or more palettes into one
- Develop common color palettes that can be used with a number of different 8bit video sequences so as to reduce any flicker that may occur as a result of palette switching, which occurs when one image, animation, or video sequence is exchanged for another. This operation may also be implemented using a palette optimizer.

Palettes can be pasted into 8bit video sequences using a video-editing program such as Adobe Premier, Asymetrix Digital Video Producer, and Microsoft VidEdit (which is part of the full implementation of Microsoft Video for Windows).

Palettes can be applied to a complete video sequence or a preselected portion of a video sequence, or even to a single frame.

They can be pasted in still 8bit images using an editing program such as Microsoft BitEdit, which is supplied with Microsoft Video for Windows.

(*See* AVI under letter A, MPEG under letter M, Streaming under letter S, Video under letter V.)

Frequencies

See Fig. Num-7.

0–130kHz

Very low frequency/low frequency (VLF/LF).

1GHz

A frequency band allocated by ITU-R for public network cellular mobile radio.

10GHz

A frequency band allocated by ITU-R for cellular base station backhaul.

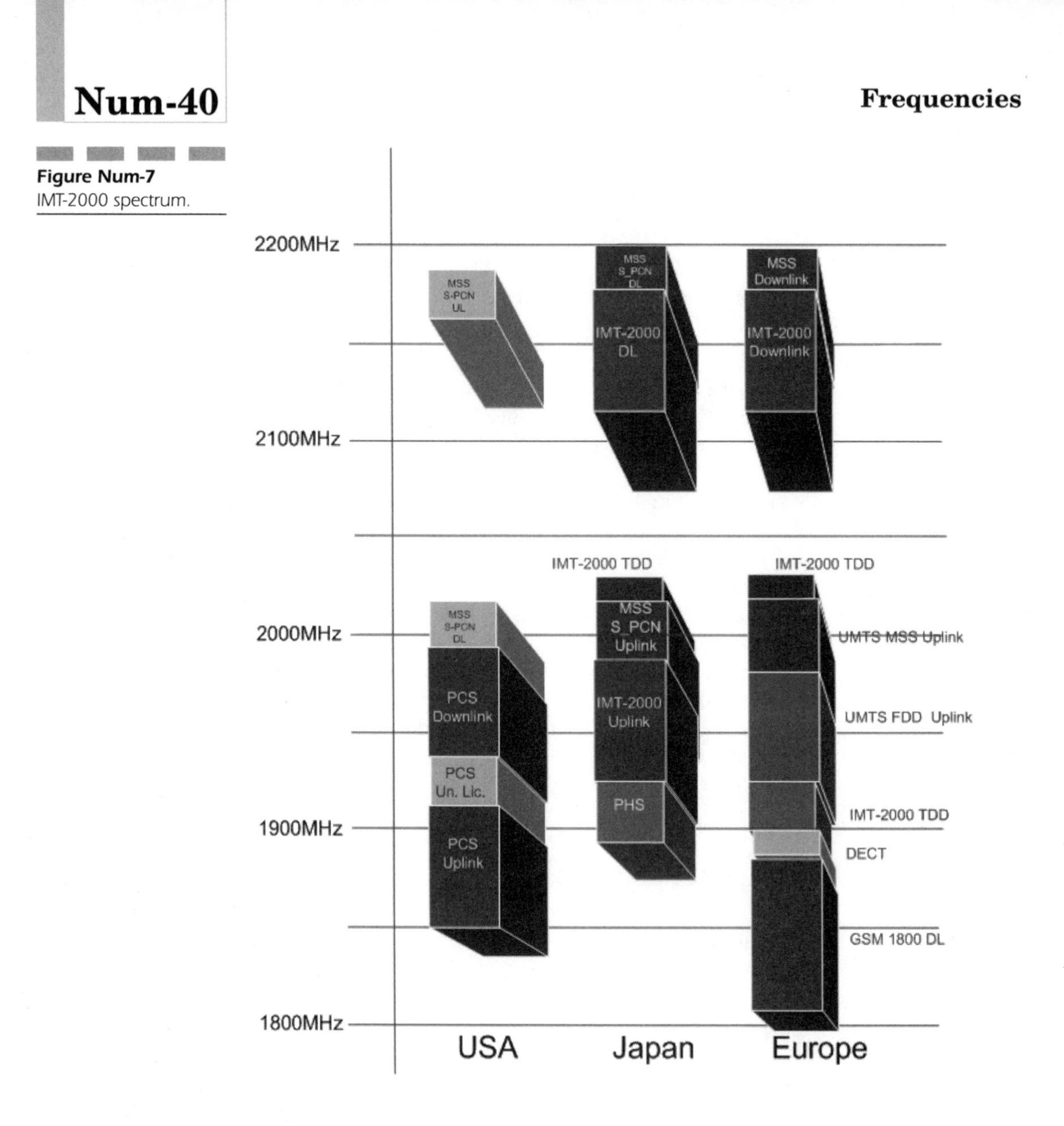

Figure Num-7
IMT-2000 spectrum.

11GHz

A frequency band allocated by ITU-R for medium capacity (E3, T3).

11.025kHz

A standard sampling rate featured by virtually all sound cards. (*See* Wave Audio under letter W.)

11.7–12.7GHz, 28–29GHz

Small-dish satellite TV frequencies.

12, 14GHz

A frequency band allocated by ITU-R for satellite communication direct broadcast by satellite.

13, 15GHz

A frequency band allocated by ITU-R for short to medium haul point to point (25km).

130–505kHz

Low frequency/medium frequency (LF/MF).

16kHz

A standard sampling rate featured by many sound cards that results in 16,000 samples per second during the sound recording process.

(*See* ADC under letter A, ISDN under letter I.)

1610–1626.5MHz

Satellite telephone uplink frequency.

162.0125–322MHz

Very high frequency/ultra high frequency (VHF/UHF).

174–216MHz

Television frequency.

18 to 30GHz

An approximate frequency range called the Ka band.

18, 23, 26GHz

A frequency band allocated by ITU-R for short haul point to point (10–15km).

1850–1990MHz

Personal communications frequency.

1885–1900MHz

IMT-2000 spectrum: Currently used for DECT in Europe, and for PHS, PCS, and DECT in other parts of the world. Suitable for time division duplex (TDD) operation.

1900–1920MHz

IMT-2000 spectrum: Segment 2 is used at present for PCS and PHS in the United States and Japan, respectively. Suitable for time division duplex (TDD) operation.

1900MHz

The North American operating frequency for GSM, as opposed to the 900MHz frequency used in Europe.

1920–1980MHz

IMT-2000 spectrum: Forms 60 MHz frequency division duplex (POD).

1980–2010MHz

IMT-2000 spectrum: Supports the earth-to-space links. Used for PCS in the United States. Mobile satellite services (MSS) are in Segments 4 and 7, providing 30MHz POD bands.

2GHz

A frequency band allocated by ITU-R for cellular mobile radio.

2.5–2.7GHz

1. A 6MHz frequency band allocated by the FCC for TV broadcasting.
2. Used for a wireless broadband technology known as MMDS (Multichannel Multipoint Distribution System).

2010–2025MHz

IMT-2000 spectrum: Segment 5 may be used in the United States for earth-to-space MSS services. Suitable for time division duplex (TDD) operation.

2107–3230kHz

Medium frequency/high frequency (MF/HF).

2110–2170MHz

IMT-2000 spectrum: Forms 60 MHz frequency division duplex (POD).

2170–2200MHz

IMT-2000 spectrum: Provides the space-to-earth links.

22.05kHz

A standard sampling rate featured by many sound cards, resulting in 22,050 samples per second during the sound recording process.
(*See* Wave Audio under letter W.)

2483.5–2500MHz

Satellite telephone downlinks frequencies.

25MHz

A radio band used by GSM. Uplink channels occupy the band 890MHz–915MHz, and the downlink channels from the base station channels are between 935MHz and 960MHz.

(*See* 2G Networks earlier in this chapter, 3G earlier in this chapter, GSM under letter G.)

2655–3700MHz

Ultra high frequency/super high frequency (UHF/SHF).

27.5–32GHz

Super high frequency/extremely high frequency (SHF/EHF).

28GHz

A frequency band allocated by ITU-R for point-to-multipoint, multimedia applications).

28 to 31GHz

A high frequency band known as LMDS (Local Multipoint Distribution Service) that is used for line-of-sight broadband communications that include Internet access. In Canada LMDS is known as LMCS (Local Multipoint Communications Service).

3GHz

A frequency band allocated by ITU-R for rural radio.

32–400GHz

Extremely high frequency (EHF).

322–2655MHz

Ultra high frequency (UHF).

3230–28,000kHz

High frequency (HF).

33–162.0125MHz

High frequency/very high frequency (HF/VHF).

3700MHz–27.5GHz

Super high frequency (SHF).

38GHz

A frequency band allocated by ITU-R for short haul (5–7km).

4–6GHz

Large-dish satellite TV frequency.

4GHz

A frequency band allocated by ITU-R for high capacity point-to-point satellite communications.

4kHz

The bandwidth of POTs (Plain Old Telephone services).

40GHz and above

A frequency band allocated by ITU-R for short haul (1–3km).

44.1kHz

A sampling frequency used to record CD quality audio. All MPC-2– and MPC-3–compliant sound cards can record in stereo at 44.1kHz. The incoming analog signal is digitized at least 44,100 times per second. (*See* Wave Audio under letter W.)

The resulting digital audio may be stored using media such as:

- CD
- CD-ROM
- Hard disk
- Zip disk
- Jaz disk
- Mini Disc
- CD-R
- DVD-ROM
- DVD-RAM

44.49MHz

Analog cordless telephone frequency.

470–806MHz

Television frequency.

48kHz

A sampling rate required to produce high-quality digital audio that may be stored on Mini Disc, DAT, hard disk, or another DSM that offers an appropriate data transfer rate.

(*See* Digital Audio under letter D.)

505–2107kHz

Medium frequency (MF).

535–1635kHz

Wireless AM radio frequency.

54–88MHz

Television frequency.

6GHz

A frequency band allocated by ITU-R for satellite communications.

7–8GHz

A frequency band allocated by ITU-R for long-haul point-to-point microwave (of up to 50km).

700MHz

Wireless data frequency.

800MHz

RF wireless modem frequency.

806–890MHz

Cellular.

88–108MHz

FM radio frequency.

890MHz–915MHz

The uplink channel from mobile to base station using GSM.

900–929MHz

Personal communications frequency.

900MHz

Digital cordless frequency.

900MHz

The European operating frequency for GSM, as opposed to the 1900MHz frequency used in North America.

929–932MHz

Nationwide paging frequency.

935MHz and 960MHz

The downlink channel from the base station to mobile using GSM.

Miscellaneous

0.2dB

The minimum core losses of a silica optic fiber that make possible practical lightwave communications.

(*See* 1979 below.)

100BaseT

A network technology, which yields a data transfer rate of 100Mbits/s. Its implementation requires structured cabling and compatible network interface cards (NICs) on network systems.

(*See* Ethernet under letter E, LAN under letter L.)

1043

The RACE 1043 project was launched in 1988 by the Europeans with the intention of exploring and discovering the potential of 3G networks and services. It is a milestone in the evolution of the 3G services industry, and marked the birth of a new vernacular including terms like UMTS (Universal Mobile Telephone Service.)

(*See* 3G earlier in this chapter, UMTS under letter U.)

10base2

An industry name for thin-Ethernet or *cheapernet* LAN technology. It uses inexpensive coaxial cable, and is popular for small networks. Compliant network computers and/or devices are fitted with Ethernet cards (or chipsets) and are connected using coaxial cables.

(*See* Ethernet under letter E, LAN under letter L.)

10base5

An industry name for basic Ethernet LAN technology. Network computers and/or devices are fitted with Ethernet cards (or chipsets) and are connected using coaxial cables. It provides 10Mbits/s data rates up to a distance of 500 m.

(*See* Ethernet under letter E, LAN under letter L.)

10baseT

An industry name for larger Ethernet LANs that are based on structured cabling. Unshielded twisted-pair telephone cabling and LAN hubs are included in the structured cabling system, which is built around a star LAN topology. It delivers data to connected workstations at a rate of 10Mbits/s.

(*See* Ethernet under letter E, LAN under letter L.)

95 (IS-95)

An official designation and internationally agreed standard for cdmaOne.

(*See* cdmaOne under letter C.)

95

An IS standard for dual mode, wireless spread spectrum communications in the 800 to 1900MHz frequency band.

1394

An IEEE designation for Firewire.

9660

An abbreviation for the ISO9660 standard, which is the official designation for a refined version of the High Sierra Group (HSG) industry standard for storing data on CD-ROM.

(*See* CD-ROM under letter C, DVD under letter D.)

120mm

The diameter of a CD, CD-ROM, or DVD disc.

1325 L Street

The address of Alexander Graham Bell's laboratory in Washington, D.C.

136 (UWC-136)

A family of compatible TDMA standards advocated by the Universe ** Wireless Consortium (UWC).

136 (IS-136)

An official designation and internationally agreed specification for a 2G system. IS-136 is the U.S. TDMA system that is an evolution of AMPS.

(*See* 2G Networks earlier in this chapter, TDMA under letter T.)

213

The distance in meters over which the first wireless telecommunications took place in 1880 in Washington, D.C., using a system invented by Alexander Graham Bell.

3270

A family of industry-standard client/server products from IBM, which includes dumb terminals.

802.11

An IEEE standard addressing wireless LANs.
(*See* WLAN under letter W.)

Significant Years/Dates

1792

The year when the French semaphore system was invented.

1837

The year in which Cooke and Wheatstone invented their five-wire telegraph, and when Samuel Morse put forward his well-known communications system.

1880 (February 19)

The day when wireless telecommunications were invented by Alexander Graham Bell, as he and his assistant Sumner Tainter transmitted a beam of light that was modulated with a voice signal and was successfully decoded. The transmission over 213 m took place in Washington, D.C., from the roof of Franklin School House to Bell's Laboratory at 1325 L Street. Bell concluded that it was his most important invention.

This momentous occasion was, of course, also the invention of lightwave communication. This invention was mocked in the *Washington Post* where an anonymous person wrote: "Does Professor Bell intend to connect Boston and Cambridge with a line of sunbeams hung on telegraph posts, and if so, what diameter are the sunbeams to be, and how is he to obtain the required size?" This cynicism is ironic because it has proved to be an accurate and prophetic statement, when considering modern optic-fiber networks.

(*See* Optic Fiber under letter O.)

1979

The year when Japan's Nippon Telegraph and Telephone Public Corporation developed a silica fiber with losses of just 0.2dB per kilometer. Until this point losses in the fiber's core that were caused by an unacceptable density of impurities meant that optic fibers were only practical over short distances because of the attenuation.

The resulting low fiber core losses meant the optic fibers became a practical lightwave communications medium, and a viable replacement for copper and aluminum conductors. The advantages of silica-based optic fiber became apparent, and included:

- Ease of installation due to their light weight, compactness, and flexible construction
- Wide bandwidth, offering large-scale multiplexing
- Immunity to corrosion
- Immunity to electrical and electromagnetic interference
- Cost effective—in certain instances the replaced copper scrap value helped cover the cost of optic-fiber installation
- Durable, reliable, offering longevity
- Inexpensive

(*See* Optic Fiber under letter O.)

1984

The year in which the CCITT published recommendations for ISDN services.

(*See* ISDN under letter I.)

1986

The year in which Philips (Netherlands) published a preliminary draft of the Green Book or CD-I (Compact Disc—Interactive) specification. This marked the real technical beginnings of multimedia, and clearly showed that the CD could become a multimedia distribution medium.

(*See* Multimedia under letter M.)

1987

A significant year in the development of the PC as a multimedia appliance. The launch of IBM's PS/2 range of computers saw the emergence of the VGA (Video Graphics Array) graphics controller standard. An important milestone in the evolution of the PC as a multimedia device, because VGA delivered color graphics using an analog port.

The VGA standard initially included just 16 colors, but third-party manufacturers increased this to include 256 colors. The other graphics standard introduced through PS/2 range was MCGA (Multicolor Graphics Array), which included 256 colors, but it failed to become the industry's chosen graphics standard.

(*See* VGA under letter V.)

1988

The year of the formation of the MPEG group.

1989

In circa 1989 MPEG had architected an industry standard video compression algorithm simply called MPEG-1 or Phase 1. This was for narrow bandwidth DSM (Digital Storage Media), namely, the single-speed CD/CD-ROM of the time.

1990

A year which saw the launch of the first consumer multimedia appliance in the United States. The appliance was the CDTV (Commodore Dynamic Total Vision).

(*See* CD-ROM under letter C, JPEG under letter J, MPEG under letter M, Multimedia under letter M, Streaming under letter S, Video under letter V.)

1990 (April)

The year that ISO IEC/JTC1/SC29 WG11 became responsible for MPEG.

1992

The year in which the World Wide Web became interwoven with the global network of computer networks that was to become the worldwide Web.

(*See* ActiveX Control under letter A, Browser under letter B, HTML under letter H, Java under letter J, JavaScript under letter J, Web under letter W.)

1992 (March)

A time when WARC (World Administration Radio Congress) assigned 200MHz to IMT-2000 for worldwide use, giving 1885–2025MHz and 2110MHz–2200MHz. Parts of these bands, however, are used by other services.

(*See* Frequencies earlier in this chapter.)

1995

The year in which Internet Telephony became a reality.

(*See* Internet Telephony under letter I.)

1995

The year when the World Wide Web became a 3-tier client/server architecture based on the HTTP/CGI model.

(*See* 3-Tier earlier in this chapter, CGI under letter C, HTTP under letter H.)

1995

The year when SunSoft announced the Java programming language.

(*See* Java under letter J.)

1998

The year that saw the release of the V.90 56.6kbps analog modem speed.

1999

The year when SunSoft launched its Jini technology and JavaSpaces.

(*See* Jini under letter J.)

Symbols and Common Syntax

&

An ampersand symbol used as a prefix in the hexadecimal counting system.

?

1. A part of a URL address that marks the beginning of data used by a CGI program that may be executed using a GET method. The URL defines the CGI program (such as credit.cgi, for example, as well as the accompanying data used by the server that follows the question mark: http://www.FrancisBotto.com/cgi-bin/credit.cgi?subject=transaction
2. A wildcard that may be used as a substitute for a single undefined character in a search string.

(*See* CGI Environment Variables under letter C.)

<! DOCTYPE wml PUBLIC "-//WAPFORUM//DTD WML 1.2//EN"

A WML document prologue that all WML scripts contain. In the example above, the first line declares that the WML deck consists of XML statements.

The second line defines the document using the DTD (Document Type Definition) mnemonic as adhering to the WAP Forum WML 1.2 specification.

(*See* WML under letter G.)

<APPLET>

An HTML tag that encloses a Java applet.

(*See* Applet under letter A, Java under letter J.)

<EMBED>

Browsers harness plug-ins using the <EMBED> tag that includes the SRC attribute that points to the file used. The following form plays a sound file called mozart.wav using a plug-in:

```
<EMBED SRC="fleetwoodmac.wav" HEIGHT=40 WIDTH=100>
```

(*See* Plug-In under letter P.)

/etc/password

A Unix file used to store passwords.

(*See* Linux under letter L, Unix under letter U.)

<FORM>

An HTML tag for creating forms:

```
<FORM> NAME="Customer"
ACTION="http://botto.com/cgibin/form/cgi
METHOD=get>
```

The <FORM> tag may have the attributes:

- NAME: the form's name.
- ACTION: indicates the URL where the form is sent to.
- METHOD: indicates the submission method that may be POST or GET.
- TARGET: indicates the windows or frame where the output from the CGI program is shown.

<INPUT TYPE>

An HTML tag used to define input components such as radio buttons. For example, using HTML, you may add radio buttons using the following form that merely displays four radio buttons labeled $30, $40, $50, and $60:

```
<FORM>     NAME="Customer"
ACTION="http://botto.com/cgibin/form/cgi
METHOD=get>
<INPUT TYPE="radio" NAME="rad" VALUE="1">
$30
<INPUT TYPE="radio" NAME="rad" VALUE="2">
$40
<INPUT TYPE="radio" NAME="rad" VALUE="3">
$50
<INPUT TYPE="radio" NAME="rad" VALUE="4">
$60
</FORM>
```

(*See* <META> below, <TITLE> below, HTML under letter H, Search Engine under letter S, Web Page Description under letter W.)

<META>

An HTML tag that may be used to enclose descriptive meta data used by search engines as an alternative to the 200 characters that follow the <BODY> tag.

```
<HEAD>
  <TITLE>Francis Botto home page</TITLE>
  <META name="description" content="IT
Research">
  </HEAD>
Francis Botto
IT Research
```

The <META> tag may also be used to add keywords of up to 1000 characters to a Web page, and may be retrieved through appropriate search phrases, for example:

```
  <META     name="keywords"
content="Multimedia, MPEG, DVD">
```

(*See* <TITLE> below, HTML under letter H, Search Engine under letter S.)

<tag/>

A WML structure that identifies elements without content.
(*See* <tag> content </tag> below, WML under letter G.)

<tag attr = "wxyz"/>

A WML attribute that specifies additional information about an element.
(*See* WML under letter G.)

<tag> content </tag>

A WML expression that specifies elements holding content in the WML deck. These may be:

- Tasks performed in response to events
- Character entities
- Card delimiters

(*See* WML under letter G.)

<TITLE>

An HTML tag that encloses the Web page title that is used as meta data by popular search engines when retrieving Web documents, displaying it as the document's title. Such data is collected by search engines periodically, but may remain transparent to some if your ISP uses a robots.txt file to stop Web robots from indexing Web pages. It is possible to determine if a server has a robots.txt file by entering the Web page's URL (including its domain name and domain category) and including robots.txt as a suffix:

```
http://www.FrancisBotto.com/robots.txt
```

Sending Web page URLs to search engines may cause them to be categorized as available via additional search words and phrases other than those contained in the Web pages themselves.

(*See* <META> above, HTML under letter H, Search Engine under letter S, Web Page Description under letter W.)

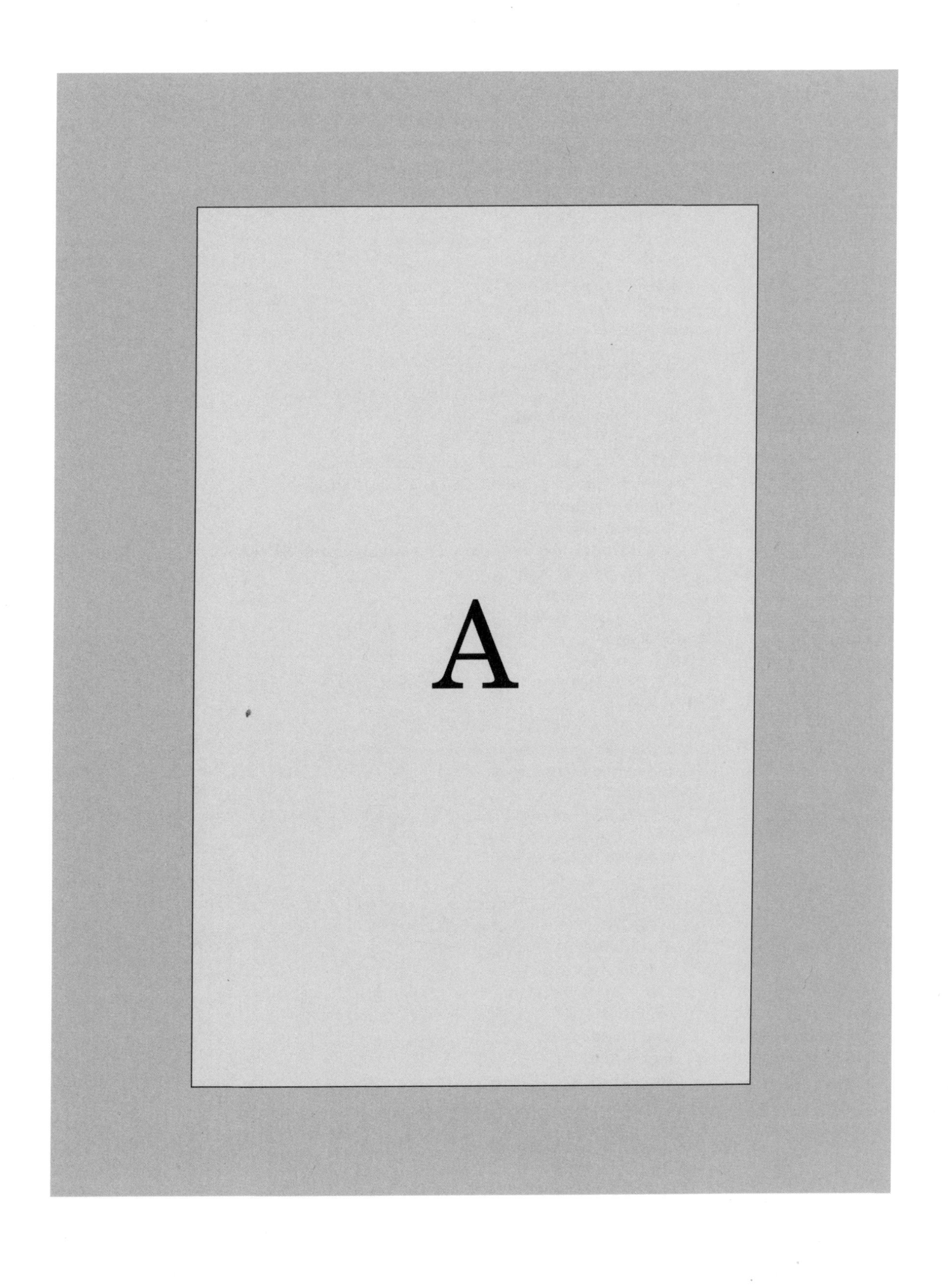

A

A-2

Access

Access Technology

Access technologies include mobile networks and data links used to connect users, and also to link with gateways to remote services that may be hosted on IP networks and include WML sites, I-mode sites, or to services hosted at secure data centers that may provide streaming video.

Wireless access technologies are as follows:

- 1G
- 2.5G—GPRS, EDGE
- 2G—GSM, DCS
- 2.5G—GPRS, EDGE
- 3G—UMTS
- DBS (Direct Broadcast Satellite)
- Wireless ISDN
- Wireless LAN

Physical access technologies are as follows:

- ADSL
- ATM
- B-ISDN (multiples of 64kbps)
- Cable
- DSL
- IP data links
- ISDN (64kbps)
- Kilostream
- POTS (56.6kbps)
- T1
- T2

(*See* 1G Networks, 2G Networks, 2.5G Networks, 3G, all in Numerals chapter; ADSL later in this chapter, ATM later in this chapter, B-ISDN under letter B, Cable Modem under letter C, ISDN under letter I, Modem under letter M.)

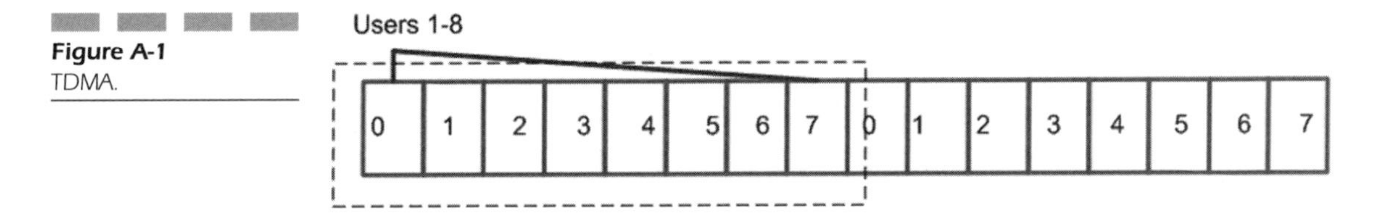

Figure A-1
TDMA.

Access is the process that allows users to gain the rights to operate a local or remote system, service, application, or program. The user may be required to enter an ID and password. Using mobile applications users are allocated isolated channels of communication that may be achieved through frequency division, time division, and code division.

- *Frequency Division Multiple Access* (FDMA) provides each user with a different frequency and was used on 1G networks.
- *Time Division Multiple Access* (TDMA) separates users by allocating different time slots for each channel. GSM has eight time slots per frame, allowing each user to send every eighth time slot.
- *Code Division Multiple Access* (CDMA) is used in 3G networks and in cdmaOne, and separates users by using different codes.

When access to a network is obtained, the mobile user has an uplink from the handset as well as a downlink to the handset. This bidirectional data flow requires separation that may be achieved by using a duplex. For example, Time Division Duplex (TDD) separates the uplink and downlink channels in time, and is used by Bluetooth.

Frequency Division Duplex (FDD) achieves separation through the allocation of different frequencies for uplink and downlink channels; a popular example includes WCDMA FDD. This method relates to the terms *uplink* and *downlink frequencies.*

(*See* 2G Networks, 2.5G Networks, 3G, all in Numerals chapter.)

Attaching to GPRS Networks A GPRS Attach is a logical link between an SGSN and an MS (or an MS-SGSN link). A GPRS detach cancels this link and usually occurs when the phone is switched off, and removes the GPRS terminal from the network. A combined GPRS/IMSI Attach is also possible, making the phone available for voice and packet data. An MS-SGSN link is implemented when:

1. An Attach message is sent from the MS to the SGSN.
2. The SGSN attempts to find a unique IMSI (International Mobile Security Identity) number for the MS.
3. If the SGSN is unfamiliar with the MS, then it requests the old SGSN for an IMSI and for authentication triplets.

4. If the old SGSN is unfamiliar with the MS, it sends an error message. The new SGSN requests the MS to send its IMSI.
5. An authentication of the MS is performed.
6. The HLR is updated if the MS is in a new service area.
7. The VLR is updated if the MS is in a new location area.
8. The SGSN passes the MS details of the TLLI (Temporary Location Link Identifier) that is used to identify the MS-SGSN link.

When an MS-SGSN link is established, the mobile requires an IP address and connection parameters that are acquired using the PDP (Packet Data Protocol) context activation. These include IP or X.25 protocols, IP address, compression option, and QoS profile, and may be set by the application using AT commands (*see* Miscellaneous later in this chapter). PDP context activation reveals the mobile phone to the GGSN, making external connections possible, and requires the following stages:

1. A PDP context request is sent from the MS to the SGSN.
2. Security procedures between the MS and SGSN are implemented.
3. The SGSN examines the QoS.
4. Sets up a logical link to the GGSN by setting up a tunnel that exists using the GTP (GPRS Tunneling Protocol).
5. Relays information to the GGSN concerning connections with the MS.
6. The GGSN contacts a RADIUS in the network to obtain the MS's IP address.
7. The IP address is returned to the MS.

With both the Attach and Context Activation complete, the GPRS device may send and receive packets.

An MS may obtain its IP address using other methods that include:

- Dynamic IP address where the GGSN assigns a dynamic IP address from its own pool.
- Static IP address where the SGSN obtains the IP address from the HLR. This is rarely used, however, because IPv4 places constraints on the supply of public IP addresses.

(*See* 2.5G Networks in Numerals chapter, GPRS under letter G.)

ALCAP (Access Link Control Application Protocol) The ALCAP is used by the Iu interface and is responsible for the data bearer setup, and holds all data about the user plane. ALCAP is made redundant at times when the Iu-PS interface is used, because this uses preconfigured data bearers. The signaling used by the ALCAP may be the same as that used by the application protocol. When the signaling bearers are present, the application protocol in the radio network layer may request that data bearers are set up. This request is sent to the ALCAP using the transport network layer.

See Fig. A-2.

1G (First-Generation) Networks The 1G category of public mobile network is now consigned to history in first-world countries. 1G networks were analog and offered mainly telephony services, including AMPS (Advanced Mobile Phone System) that operated in the 800MHz cellular band.

By the late 1970s early cellular networks began to emerge, and formed the basis for the wireless communications used today. In 1977 Illinois Bell introduced a cellular network in Chicago. Called AMPS (it was developed by AT&T's Bell Laboratories and operated between the 800MHz and 900MHz bands, and was the most used service in the United States until the early

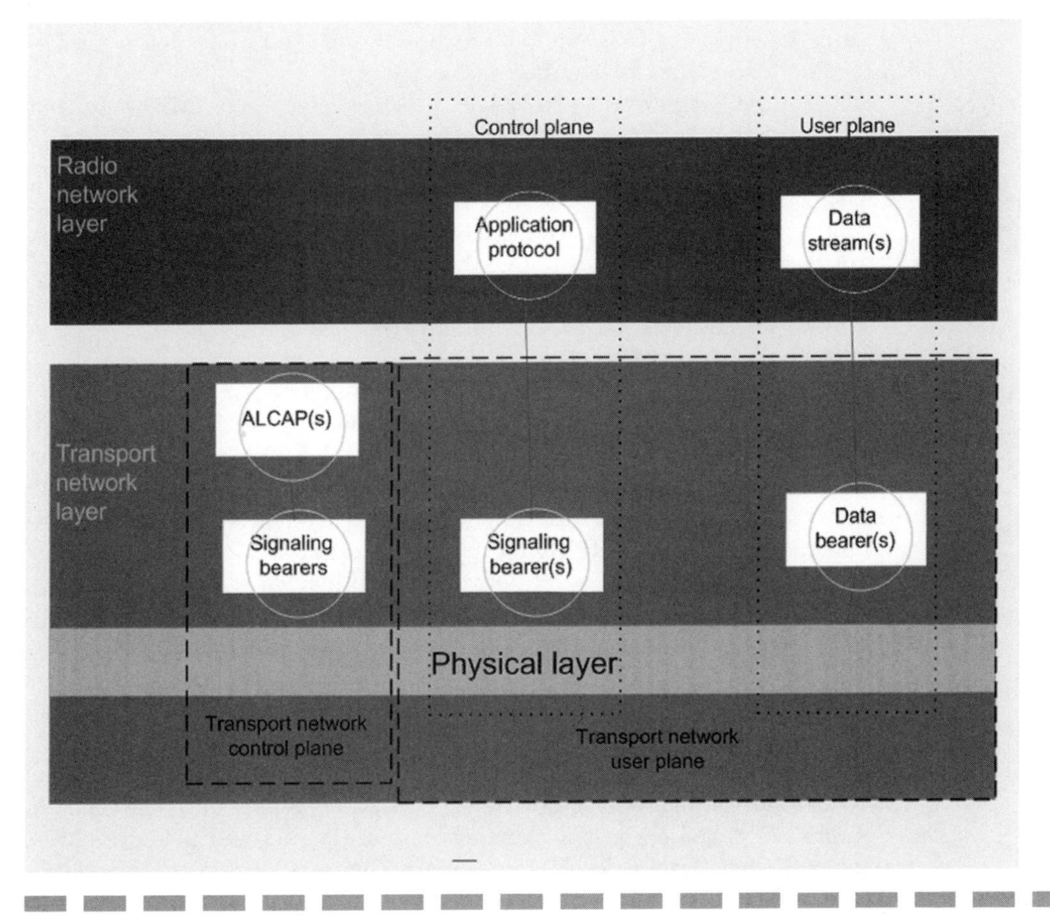

Figure A-2 ALCAP used in UTRAN.

1990s. In 1982 the development of an international standard called GSM (Global System for Mobile Communications) was begun, and by 1993 a rapid rise in compatible networks had emerged globally. This marked the beginning of wireless telecommunications for the mass market and drove the many growing cell phone services available today.

The capacity of 1G networks was naturally small, because authentic 1G networks were not cellular. Later cellular networks provided greater geographic coverage through cells using the same frequencies. An early internationally agreed standard analog network did not exist, and so countries and continents had disparate network systems. These included Nordic Mobile Telephone (NMT; Scandinavia), Total Access Communications System (TACS), C-Netz (West Germany), Radiocomm 2000 (France), and, of course, AMPS.

NMT was also used in central and southern Europe, and was introduced in Eastern Europe during the late 1990s. NMT had two variants, namely, NMT-450 and a later NMT-900 system using the 900Mhz frequency band. Like some of the later networks, NMT offered the option of international roaming–although this was not as seamless as it is today.

The United Kingdom adopted TACS, which was based on AMPS, but used the 900MHz frequency band; TACS became successful in the Middle East and in Southern Europe. The American AMPS standard used the 800MHz frequency band and was also used in South America, Far East, and in the Asia Pacific region, including Australia and New Zealand. In the Asia Pacific country of Japan was NTT's MCS system that was the first commercial delivery of a mobile 1G Japanese network. While most first-world countries are closing 1G networks, many Less Developed Countries (LDCs) are actively investing in and also upgrading them.

Telecommunications networks remain a foundation for the World Wide Web, and connectivity between tier 1 and tier 3/4 devices and servers and hosts, but it was the advent of the TCP/IP (Transmission Control Protocol/Internet Protocol), which emerged from the early DARPA and ARPANET networks, that introduced a common transport protocol for communications. In 1983 all ARPANET networks were running TCP/IP, and later 32bit IP addressing meant that networks could include millions of addressable hosts. Some 18 years later, the growing global network is now driving the development of IPv6, that is, a 128bit addressing scheme specified by the IETF.

(*See* 2G Networks, 2.5G Networks, 3G, all in Numerals chapter; cdmaOne under letter C, GPRS under letter G, UMTS under letter U.)

2G (Second-Generation) Networks The 2G category of public mobile network offers the earliest digital telecommunications and includes variations of Global System for Mobile (GSM), Digital-AMPS (D-AMPS), Code Division Multiple Access (CDMA—IS95), and Personal Digital Cellular (PDC). GSM is the most popular globally, and in August 2000 it was estimated that

372 GSM networks were in operation with a collective mobile user base of 361.7 million.

Typically a 2G GSM network provides users with data rates of 9.6kbps or 14.4kbps. As many as 20 to 25 MSCs (Mobile Switching Centers) or control elements are included in a network, and each will cover a given geographic area served by BTSs (Base Transmitter Stations) each with a 5 km to 8 km radius. The BTSs naturally determine the cellular network coverage where mobile users may be served by a number of BTSs that switch seamlessly from one to the next. It may be assumed that GSM operates in the uplink band between 890MHz and 915MHz and usually offers a user data rate of 9.6kbps, and the speed threshold on the mobile station is around 250 km/h.

Two GSM variants include Digital Cellular Systems 1800 (DCS-1800 or GSM-1800) and PCS-1900 or GSM-1900 that is used in North America and Chile. The different frequency is used because of the lack of capacity in the 900MHz band. The 1800MHz band accommodates a larger number of mobile users particularly in densely populated areas. The coverage area of 1800MHz networks is often smaller, and therefore dual-band phones are used, able to roam between either network.

ETSI has also published GSM-400 and GSM-800 specifications, with the former suited to large geographic area coverage, and can therefore be used in conjunction with higher frequency band networks in sparsely populated regions.

Comparable to GSM, DCS-1800 (Digital LPC Cellular System) is used in the United Kingdom. It operates in the uplink radio band between 1710MHz and 1785MHz and can be assumed to provide a user data rate of 9.6kbps, and exhibits a 250 km/h speed threshold on the mobile station.

(*See* 2G Networks, 2.5G Networks, 3G, all in Numerals chapter; GSM under letter G, UMTS under letter U.)

2.5G The 2.5G category of public mobile network offers improved data rates by adding an overlay such as GPRS (General Packet Radio Service) to a 2G network like GSM. Other 2.5G solutions include HSCSD (High-Speed Circuit-Switched Data) and EDGE (Enhanced Data Rates for Global Evolution). These technologies seek to increase the user data rates of 2G networks like GSM that typically exhibit 9.6 and 14.4kbps.

2.5G GPRS The GPRS (General Packet Radio Service) overlay increases the GSM and TDMA user data rate to a maximum of about 171kbps, and includes the following physical network overlay core solutions:

- IP-based GPRS backbone
- GGSN (Gateway GPRS Serving Node)
- SGSN (Server GPRS Serving Node)

Principal features of GPRS include:

- Packet based transmission
- Packet-based charging as opposed to time-based charging
- Always-on—so a VPN connection, for example, does not require constant logging on to the remote/enterprise network/intranet
- Eight time slots
- A maximum user data rate of about 171kbps when all time slots are used
- Alleviates capacity impacts by sharing radio resources among all mobile stations in cells
- Lends itself to bursty traffic

GPRS applications include:

- E-mail, fax, messaging, intranet/Internet browsing
- Value-added services (VAS), information services, games
- m-commerce: retail, banking, financial trading, advertising
- Geographic: GPS, navigation, traffic routing, airline and/or rail schedules
- Vertical applications: freight delivery, fleet management, sales-force automation

GPRS Terminal Classes A GPRS terminal phone may be:

- Class A: Supports GPRS, GSM, and SMS, and may make or receive calls simultaneously over two services. In the case of CS services, GPRS virtual circuits are held or placed on busy as opposed to closed down.
- Class B: Supports either GSM or GPRS services sequentially, but monitors both simultaneously. Like Class A, the GPRS virtual circuits will not be closed down with CS traffic, but switched to busy or held mode.
- Class C: Make or receive calls using the manually selected service, with SMS an optional feature.

IP Address A physical IP address consists of 32 bits that identifies hosts, servers, networks, and connected computers. The syntax for such addresses consists of four bytes, each written in decimal form, and separated by a full stop: 118.234.165.124. This physical address is obtained by a DNS server when it is given a Web address.

The three types of IP address are

- *IP Address Class A:* A networks may have between 2^{16} (65,536) and 2^{24} (16.7 million) hosts. See Fig. A-3.

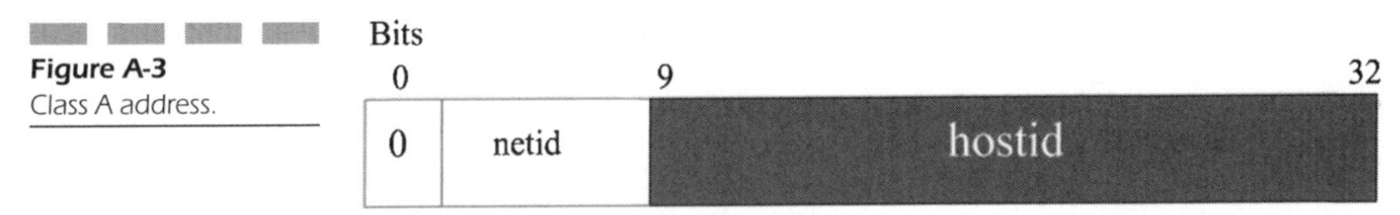

Figure A-3
Class A address.

- *IP Address Class B:* B networks may have between 2^8 (256) and 2^{16} (65,536) hosts. See Fig. A-4.
- *IP Address Class C:* C networks may have up to 253 hosts, but not 255 because two values are reserved. See Fig. A-5.

The addresses consist of a network address (netID) and a host address (hostID). The leftmost digits represent the netID address. This is set to zero when addressing hosts within the network.

(*See* IPv6 under letter I.)

2.5G HSCSD HSCSD (High-Speed Circuit-Switched Data) usually uses a maximum of four time slots (that may be 9.6 or 14.4kbps) for data connections. Because HSCSD is circuit switched, used time slots are constantly allocated even when there is no transmission. This disadvantage makes HSCSD appropriate for real-time applications with short latencies. HSCSD also requires appropriate handsets that are not as widespread as GPRS, for example, but at the same time HSCSD is less expensive for operators to upgrade than GPRS.

2.5G EDGE EDGE is an overlay solution for existing ANSI-136/TDMA networks, and may use the existing ANSI-136 30kHz air-interface. EDGE is on the migration path to UMTS, and may even coexist with it so as to provide services for wide-area coverage. EDGE standards support mobile services in ANSI-136/TDMA systems with data rates of up to 473kbps.

A significant change in the ANSI-136/TDMA standards to support higher data rates is the use of modulation schemes, including 8-PSK (Phase Shift Keying) and GMSK (Gaussian Minimum Shift Keying). GMSK provides for

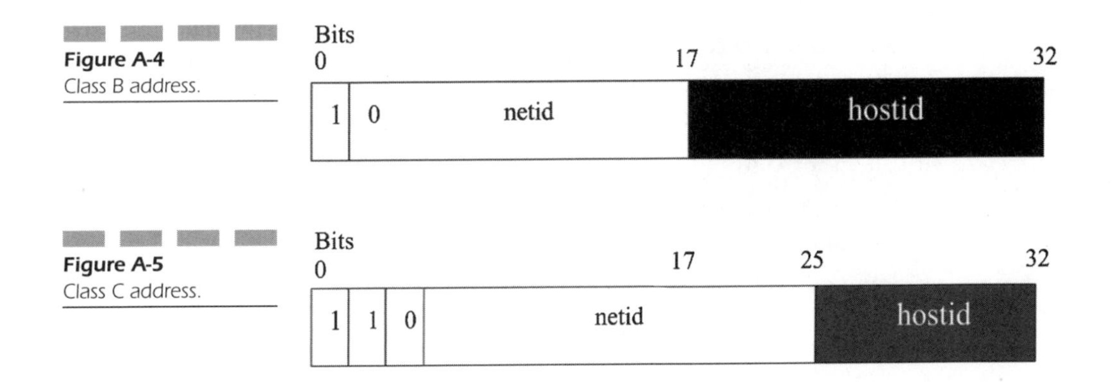

Figure A-4
Class B address.

Figure A-5
Class C address.

wide area coverage, while 8-PSK provides higher data rates but with reduced coverage.

EDGE provides high data rates over a 200kHz carrier, giving up to 60kbps per timeslot that may equate to 473kbps. EDGE is adaptive to radio conditions, giving the highest data rates where there is sufficient propagation.

(*See* EDGE under letter E.)

EGPRS (Enhanced General Packet Radio Service) The dominant data networking protocol, on which most data network applications are running, is TCP/IP, the Internet Protocol. All Web applications are run on some form of TCP/IP, which is by nature a protocol family for packet-switched networks. This means that EGPRS is an ideal bearer for any packet-switched application, such as an Internet connection. From the end user's point of view, the EGPRS network is an Internet subnetwork that has wireless access. Internet addressing is used, and Internet services can be accessed. A new number, the IP.

(*See* 3G in Numerals chapter, UMTS under letter U.)

3G (Third Generation) The 3G category of public mobile network is capable of offering user data rates that may extend to Mbps, and describes the universal mobile telecommunication system (UMTS) that was shaped in part by the Third Generation Partnership Project (3GPP). 3G cellular technology greatly surpasses the multiple impacts of 2G services and, to a lesser extent, those of 2.5G. The promise of sophisticated video and multimedia wireless applications able to deliver MPEG video, and high-quality audio is attracting investment in the WASP and Mobile Telecommunications sectors, as the potentially profitable 3G services are foreseen, including:

- Video applications using MPEG standards
- Video telephony
- Videoconferencing
- Video on demand (VoD)
- Telepresence
- Surrogate services such as exploration
- Client for remote services
- Web/Internet browsing
- Client for VPN
- Client connection for teleworkers
- m-commerce—retailing, online banking, etc.
- Point of Information (POI) in real estate sector, etc.
- Point of Sale (POS) for secure purchasing

- CCTV (Closed-Circuit Television) security
- UAV (Unmanned Aerial Vehicle) video communication and navigation

A UMTS provides global roaming and is architected using orbiting satellites that may integrate BTSs (Base Transmitter Stations) and BSCs (Base Switching Centers). To create this type of network, satellites may orbit at altitudes between 780 and 1414 km so as to minimize signal transmission latency. One example of a satellite network is Teledesic whose consortium is led by Bill Gates. Other mobile satellite services include Motorola's Iridium (with 66 satellites), Loran/Qualcom's Globalstar (with 48 satellites), and TRW-Matra's Odyssey (with 10 satellites orbiting at high altitude).

UMTSs offer broadband user data rates and operate in the K-band (10.9-36GHz) and L-band (1.6 to 2.1GHz). Aeronautical and maritime telecommunications were catalysts in the development and deployment of satellite mobile telephone services with the first maritime satellite launched in 1976. Called MARISAT, it consisted of three geostationary satellites and was used by the U.S. Navy. This later evolved into the INMARISAT (International Maritime Satellite Organization) that provides public telecommunications services to airliners.

ATM ATM is an internationally agreed telecommunications standard that supports transmission speeds of up to 622Mbits/s. Other line speeds include 2Mbits/s, 12Mbits/s, 25Mbits/s, 34Mbits/s, 45Mbits/s, 52Mbits/s, and 155 Mbits/s. ATM is used in UMTS where it provides the core network transport, and is based on asynchronous time division multiplexing.

The CCITT accepted ATM in 1990 as an internationally agreed standard for data, voice, and multimedia networks.

ATM bases itself on cell relay, which is a form of statistical multiplexing, and is similar to packet switching. The data transmission consists of cells, which have 53 octets or bytes, including a 5-octet header.

Using 52Mbits/s line speed, a single cell can be transmitted in:

$$53 \times 8/52\text{Mbits} = 8.15 \times 10^{-6}$$
$$= 8.15\ \mu\text{s}$$

The cells from different signals are interleaved, and the signal propagation delay, or jitter, is a function of the transmission line speed. It is sufficiently low to give a stream of contiguous cells, which is acceptable for real-time data, voice, audio, and video transmission. Similar to packet headers, cell headers also contain destination addresses.

Above the ATM layer in UMTS is AAL (ATM Adaptation Layer) that processes data from higher levels for ATM transmission. This involves arranging the data into 48-byte chunks and reassembling the data at the receiver. AAL exists at levels 0 to 5, and level 0 means that no adaptation is required, and the remaining adaptation layers properties are based on:

- Real-time requirements
- Constant or variable bit rate
- Connection-oriented or connectionless data transfer

(*See* Frame Relay under letter F.)

ADSL (Asymmetrical Digital Subscriber Line) ADSL is an access technology that uses the existing copper wire networks that are synonymous with POTS (Plain Old Telephone Services), though these may include fiberoptics also. Its downstream bandwidth is considerably wider than its upstream bandwidth:

- Downstream bandwidth of between 1.5Mbits/s and 52Mbits/s. Typically it is 1.5Mbits/s.
- Upstream bandwidth of the order of 784kbps to 2Mbits/s, which can be a function of the line length. Typically it is 1.5Mbits/s.

Applications include:

- High-speed Internet access
- VHS quality videoconferencing
- VoD (Video-on-demand)
- Multimedia networks

Discrete multitone (DMT) modulation according to ANSI T1.413 separates upstream data from downstream data by separating the signal into separate 43kHz carrier bands. See Fig. A-6.

See Table A-1.

(*See* Access Technology earlier in this chapter.)

Asynchronous Connectionless Link (ACL) The ACL link is used by the Bluetooth air interface for data transfer and asynchronous services, and incorporates packet switching, and grants transmission slots using a polling access scheme. When more than one Bluetooth unit is communicating, it results in a piconet where one is a master unit that tells slaves when it wants to send, and the slave then receives. Slaves send using slots when in agreement with masters.

(*See* Bluetooth under letter B.)

A-bis Interface The interface between a BSC and a BTS is defined by an open or public specification that allows operators to acquire BSC and BTS core

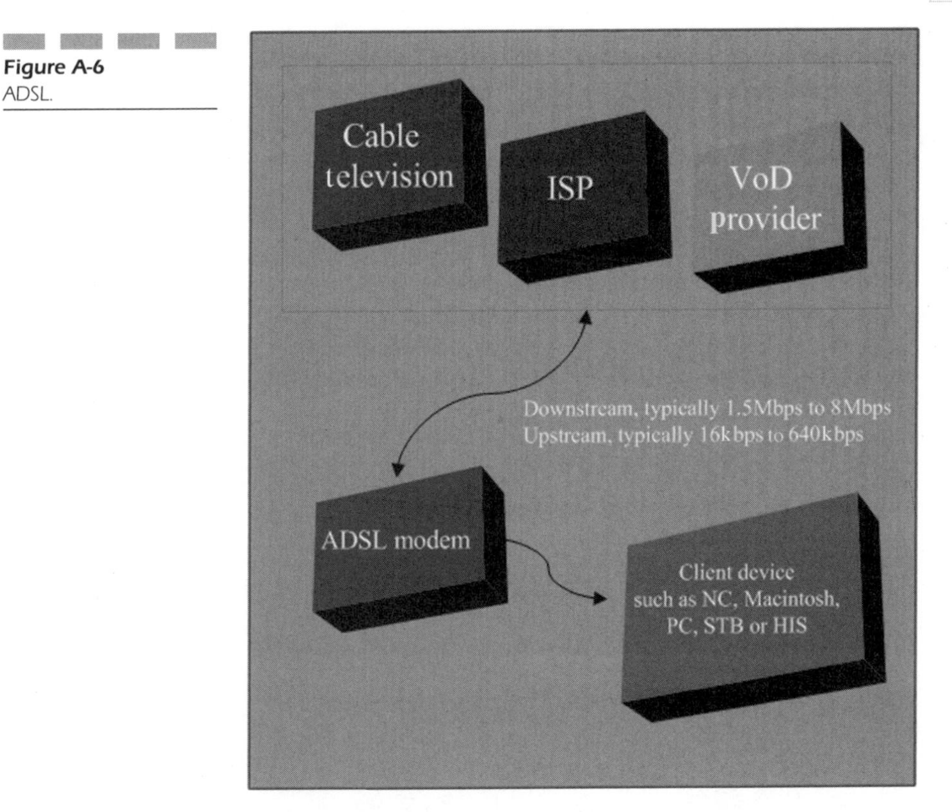

Figure A-6
ADSL.

elements from different sources. The A-bis interface is not a multivendor interface, however, but has proprietary solutions for each manufacturer. Essentially the BSC controls the BTS using the A-bis interface, with the collective partnership referred to as a base station subsystem (BSS), or as the GSM radio access network. A BSS has one BSC and at least one base transceiver station (BTS). See Fig. A-7.

The BTS has the transmitter and management equipment, while the BSC has:

- Radio resource management for BTSs
- Functions for intercell handovers
- Frequency allocation to BTSs

(*See* 2G Networks in Numerals chapter, 3G in Numerals chapter, A Interface later in this chapter, Base Station Controller under letter B, GSM under letter G.)

Access Burst The access burst used by an MSC gains initial access to a network, and is the first up-link burst demodulated by the BTS.

TABLE A-1

DSL Variants

DSL Categories	Maximum Data Rate Upstream	Maximum Data Rate Downstream
Asymmetrical Digital Subscriber Line (ADSL) allocates bandwidth asymmetrically in the frequency spectrum.	1Mbps	8Mbps
High Bit Rate Digital Subscriber Line (HDSL) allocates bandwidth symmetrically and may replace T1 or E1 four-wire links. HDSL permits users to be located greater distances from a central office without the use of repeaters.	1.544/2.048Mbps	1.544/2.048Mbps
Rate Adaptive Digital Subscriber Line (RDSL) is one of the most recent evolutions of ADSL and adapts to the highest data rate possible on the local loop.	784kbps	4Mbps
Symmetric Digital Subscriber Line (SDSL) allocates bandwidth symmetrically and operates over a single pair link.	2Mbps	2Mbps
Very-high-bit-rate Digital Subscriber Line (VDSL) gives the highest downstream data speed but may not operate successfully over a maximum distance more than 1000 ft.	1.5Mbps	52Mbps

Like the synchronization burst, tail bits at the beginning of the burst are extended to eight:

b0	b1	b2	b3	b4	b5	b6	b7
0	0	1	1	1	0	1	0

A training sequence ensues:

b8	B9	B10	B11	B12	B13	B14	B15	B16	B17 to B24	B25	B26	B27	B28	B29	B30	B31	B32	B33	B34	B35	B36	B37	B38	B39	B40	B41	B42	B43	B44	B45	B46	B47	B48
0	1	0	0	1	0	1	1	0	1	0	0	1	1	0	0	1	1	0	1	0	1	0	1	0	0	0	1	1	1	1	0	0	0

Because access bursts are shorter than other GSM bursts, a guard period of 68.25bit compensates for propagation delays between MSs and BTSs.

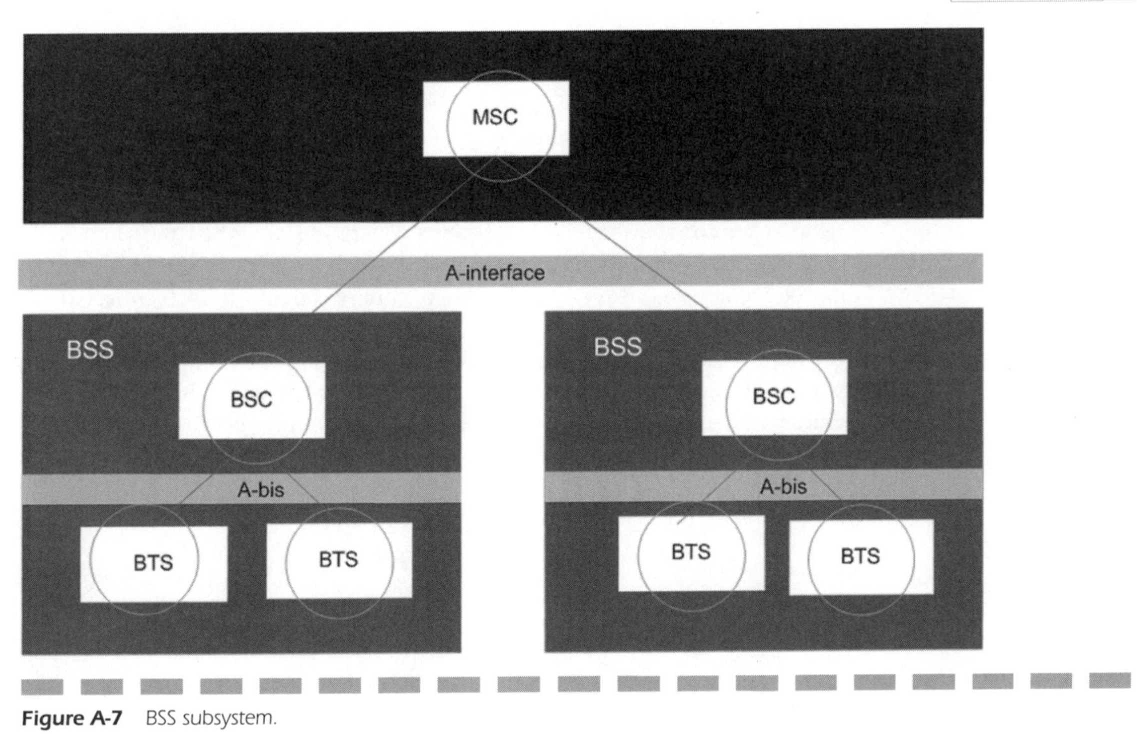

Figure A-7 BSS subsystem.

Absolute Radio Frequency Channel Number (ARFC) An ARFC is a number assigned to each RF carrier frequency pair using the GSM radio interface.

In the specifications:

$Fl(n)$ describes the frequency of the carrier in the lower up-link frequency band with an ARFCN of n.

$Fu(n)$ is used for the upper down-link frequency band.

See Table A-2.

The down- and up-link bursts of a duplex link are separated by three timeslots, and the frequency separation between the duplex carriers is 45MHz

TABLE A-2 Absolute Radio Frequency Channel Numbers

Band	Frequency	Channel Numbers
P-GSM900	$Fl(n) = 890 + 0.2n \quad Fu(n) = Fl(n) + 45$	$1 \leq n \leq 124$
E-GSM900	$Fl(n) = 890 + 0.2n \quad Fu(n) = Fl(n) + 45$	$0 \leq n \leq 124$
	$Fl(n) = 890 + 0.2(n - 1024)$	$975 \leq n \leq 1023$
DCS1800	$Fl(n) = 1710.2 + 0.2(n - 512) \quad Fu(n) = Fl(n) + 95$	$512 \leq n \leq 885$

for GSM900 and 95MHz for DCS1800. The MS, therefore, is not required to transmit and receive simultaneously.

The MS receives a down-link burst from the BTS, and then transmits an up-link burst three timeslots later.

The timing schedule is such that each duplex carrier supports a number of timeslots that are 15/26 ms (~0.577 ms) in duration.

These are arranged into TDMA frames consisting of eight time slots with a duration of 60/13 ms (~4.615 ms). Each timeslot within a TDMA frame is numbered from zero to seven, and these numbers repeat for each consecutive frame.

The timeslot and frame durations are derived from the fact that 26 TDMA frames are transmitted in 120 ms. The reasons for choosing these particular numbers will become clear when we examine GSM's complex frame structure.

The TDMA frame duration is:

$$\frac{120}{26} \text{ ms} = \frac{60}{13} \text{ ms}$$

The timeslot duration is:

$$\frac{120}{26 \times 8} \text{ ms} = \frac{15}{26} \text{ ms}$$

(*See* TDMA under letter T.)

Access Service Class The UTRAN MAC (Medium Access Control) obtains a set of access service classes (ASCs) from the RRC and chooses one to define the parameters used in an RACH transmission. The parameters include access slots and preamble signatures, and the algorithm uses two variables: MLP (MAC Logical Channel Priorities) and NumASC, the maximum number of ASCs.

See Table A-3.

(*See* Reverse Access Channel under letter R.)

TABLE A-3

Obtaining an ASC Number

Condition	ASC
All transport blocks in a transport block set have the same MLP.	ASC = min(NumASC, MLP)
If transport blocks in a transport block have differing MLPs.	ASC = min(NumASC, MinMLP)
The ASC enumeration corresponds to prioritization.	ASC0—highest priority
	ASC7—lowest priority

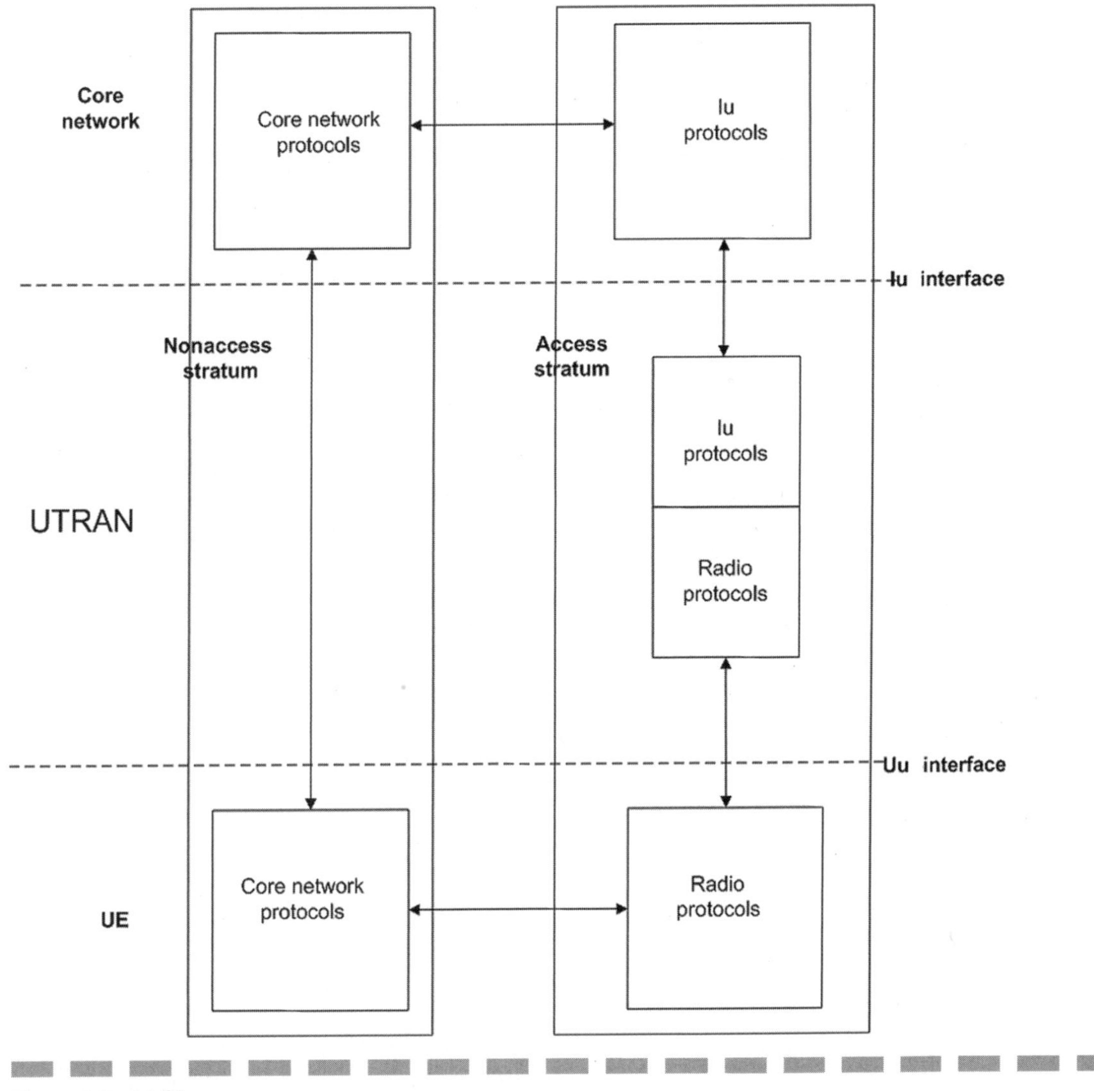

Figure A-8 UMTS.

Access Stratum In the UMTS architecture, UTRAN sees the air interface isolated from the MM (mobility management) and the connection management layers. This notion takes the form of the access stratum (AS) and the nonaccess stratum (NAS). The AS includes radio access protocols that terminate in the UTRAN, while the NAS has core network protocols between the UE and CN that are not terminated in the UTRAN that is transparent to the NAS.

The NAS attempts to be independent of the radio interface, as do the protocols MM, CM, GMM, and SM. These protocols are theoretically constant

TABLE A-4

AS Layers

Layer	Sublayers
Physical layer (L1)	
Data link layer (L2)	Medium Access Control (MAC)
	Radio link control (RLC)
	Broadcast/multicast control (BMC)
	Packet data convergence protocol (PDCP)
Network layer (L3)	Radio resource control (RRC)

in terms of the radio access specification that carries them, so any 3G radio access network (RAN) should connect to any 3G CN. The GSM's MM and CM layers are almost the same as 3G NAS, and NAS layers will compare with future GSM MM and CM layers according to the phase 2+ specification.

Lower layers from the AS are different from GSM where the radio access technology (RAT) is TDMA, as opposed to CDMA for UTRAN. The protocols used therefore are also radically different, as is the packet-based GPRS protocol stack. So this reveals an obstacle in the so-called smooth migration path from 2.5G GPRS to 3G. However, the GPRS CN components can be reused when renovating the system to become a 3G solution. See Fig. A-8.

The AS has three protocol layers that include sublayers (see Table A-4). The two vertical planes include the control (C-) plane and the user (U-) plane, and both have the MAC and RLC layers, while the RRC layer is present only in the C-plane. See Fig. A-9.

(*See* UMTS under letter U.)

Admission Control Using 3G CDMA systems, users have different subscriptions and QoS levels, so admission control is used to prioritize users when making connections over the network. Admission control monitors make requests for connections and for changes to existing connections. If levels of network traffic reach the saturation point, low-priority users will not be admitted, and high-priority users may be admitted while reducing the bandwidths of already connected users.

Applications

Application

Applications are hosted services that ideally should be available to mobile users irrespective of the network or handset that they use. Historically this

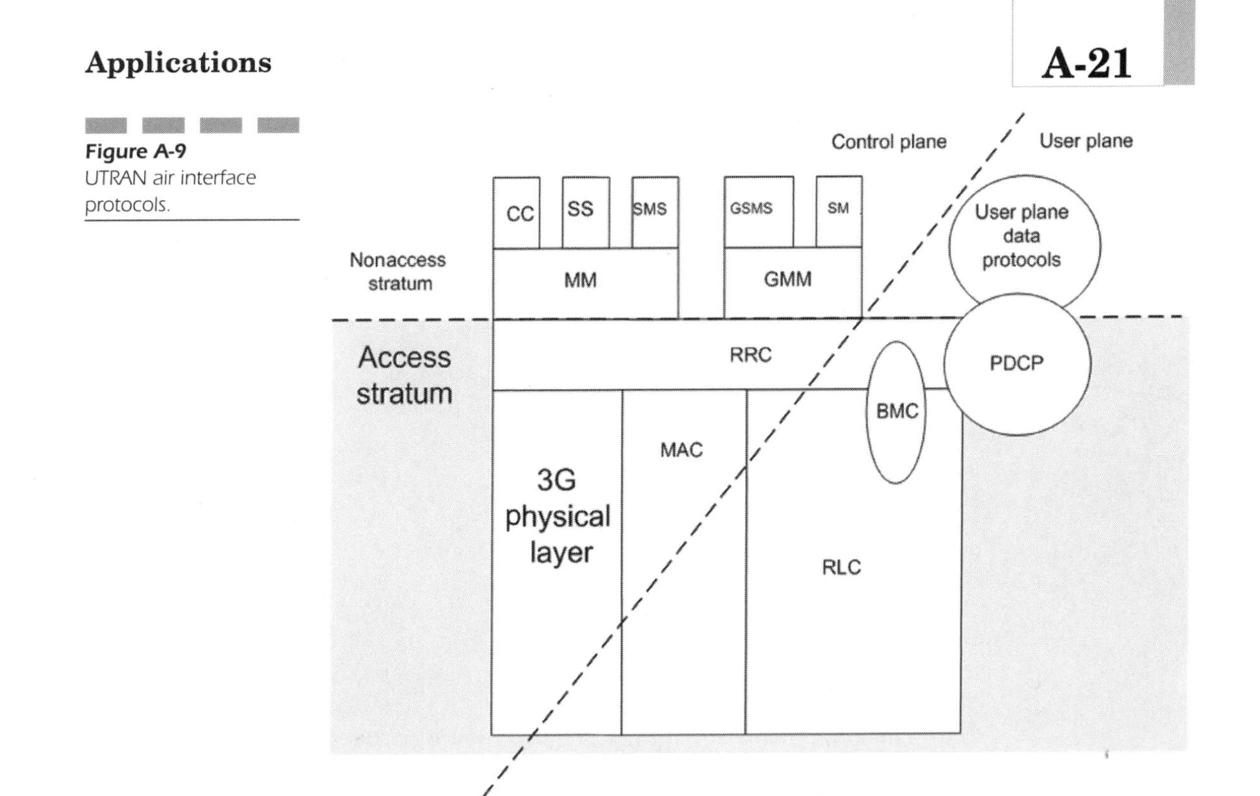

Figure A-9
UTRAN air interface protocols.

has not been the case because of the lack of international standards regarding the networks and accompanying mobile devices. 3G wireless systems remedy this problem using a horizontally layered and logically separated architecture including Applications/Services, Control, and Transport planes.

- The Application/Services plane holds services such as e-mail, voice mail, and real-time information.
- The Control plane sets up calls, tracks mobiles, and manages billing information.
- The Transport plane transports calls set up in the Control plane, including routing, coding, and switching.

Applications or services can be hosted on a LAN sometimes referred to as a DC (data center) that may hold various services, including 3G multimedia services such as VOD. However the data center may also hold old-fashioned SMS services that may be hosted using a core solution that provides backend data and messaging. This type of data center would typically have a real-time data feed from perhaps a weather bureau, traffic information center, news bureau, or a business that provides shared data. This data can then be relayed from the data center to subscribers. The data feed may consist of an ISDN connection and may not be real time. See Figs. A-10 and A-11.

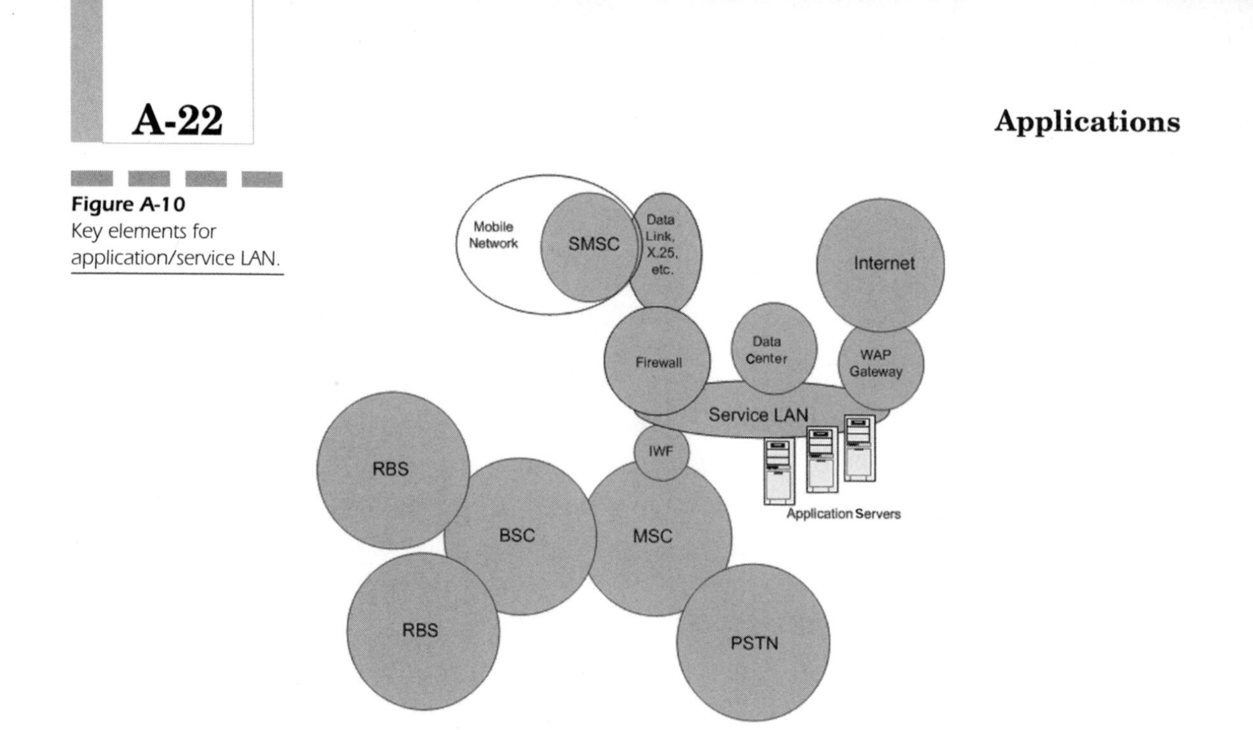

Figure A-10
Key elements for application/service LAN.

The LAN should be tightly coupled, using perhaps the fastest Ethernet technologies, and the hosted core application should be easily scaled to grow with usage. One solution to this is to use distributed computing applications/services running on the LAN, where it is possible to include an infinite number of servers—in theory, of course. For example, an SMS solution written

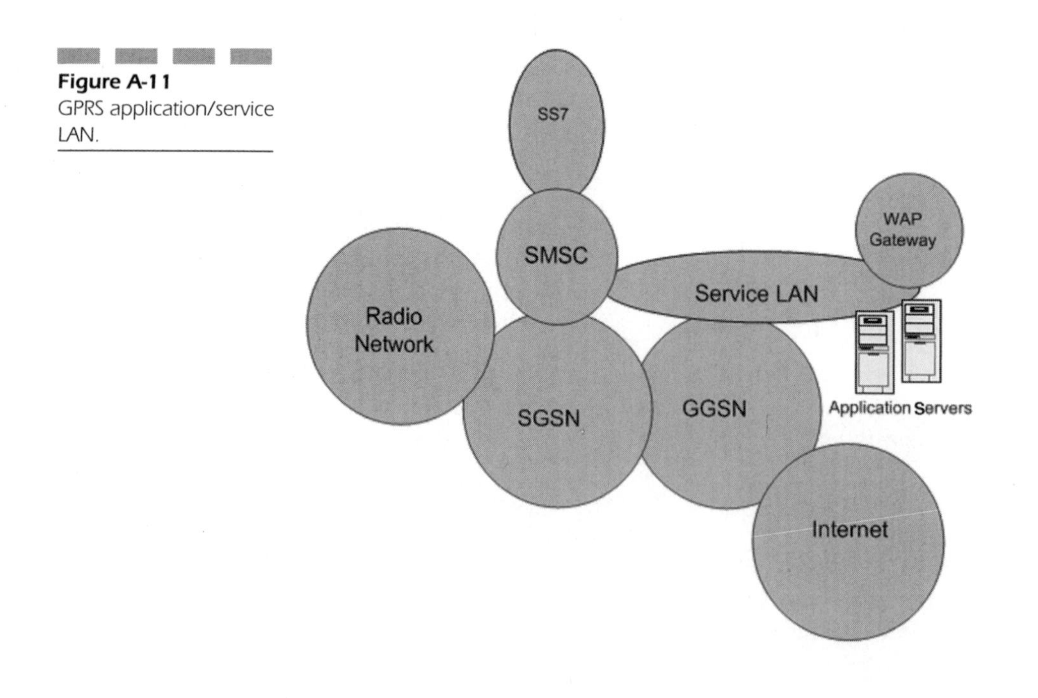

Figure A-11
GPRS application/service LAN.

in Objective-C using software servers and software clients can deliver such a scalable solution. Services such as these may be architected so as to provide interfaces with different types of fixed and mobile networks. Data centers may be linked, perhaps using IP data links.

(*See* 3G in Numerals chapter.)

WAP (Wireless Application Protocol)

WAP permits compatible devices to browse compliant Internet sites using an appropriate browser. A core part of the WAP architecture is the WAP gateway that unites dissimilar networks providing a bridge between an IP network (namely the Internet) and a Mobile Operator network.

Figure A-12 presents a "multiple scenarios" high-level representation of a WAP gateway with all its surrounding software architectures and physical infrastructures. It does not actually use every component part of the shown architecture, as the diagram is intended to show the collective client/server

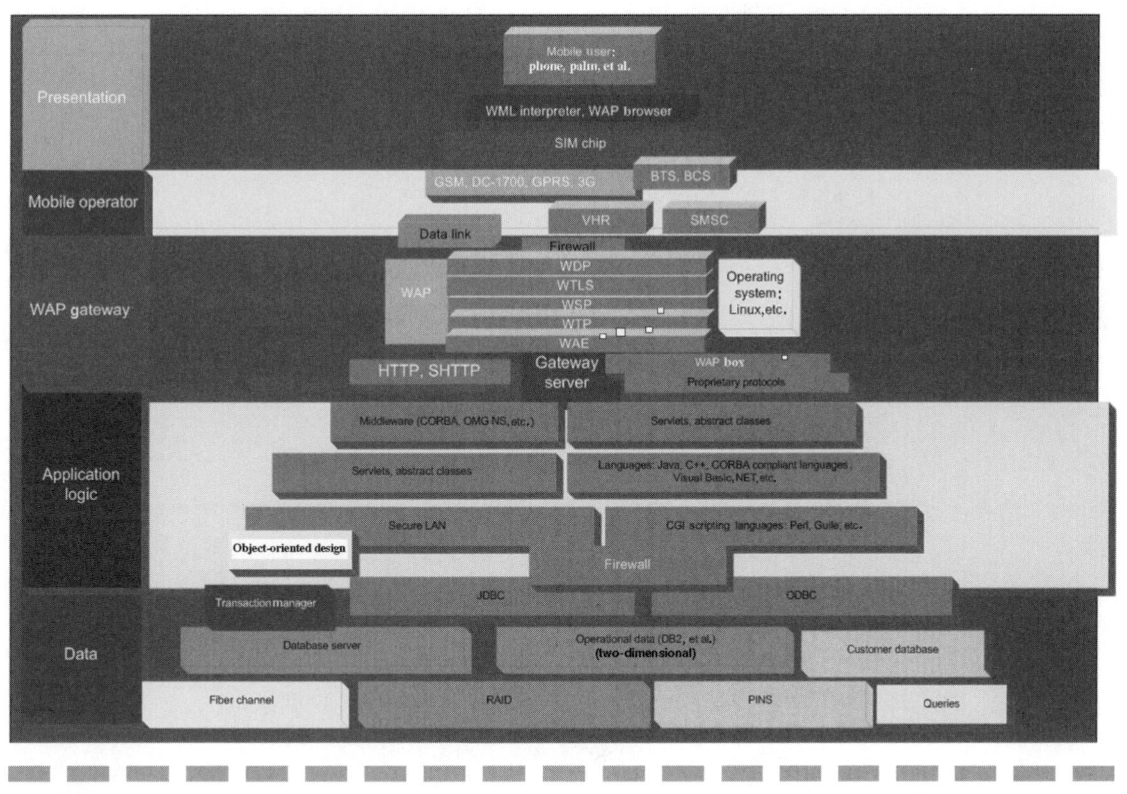

Figure A-12 Client/server architecture, including WAP.

architecture. The illustration attempts to explain the basic strata beginning with the presentation layer through to the backend.

A practical implementation of WAP would obviously focus on a single descent through the shown layers and tiers (see Table A-5). A WAP gateway core solution exhibits external interfaces, that is, what other systems it talks to, and how. Nokia and Ericsson have conventionally marketed and sold WAP gateway solutions; open source implementations for the Linux environment include Kannel (www.kannel.org). See Fig. A-13.

The Kannel WAP gateway has up to six external interfaces:

- SMS centers: The SMS centers use a variety of mostly proprietary protocols (CIMD, EMI, SMPP) over TCP/IP, modem lines, or various other carriers. The gateway needs to support as many SMS center protocols as possible, and to make it easy to add new ones.

TABLE A-5

WAP Tiers

Presentation
A handset such as GSM WAP phone, GPRS phone, or Palm
An OS such as Palm OS, Windows CE, or even a Jini browser
Mobile Network
GSM—Vodafone, Cellnet, Orange, One2One
GPRS
WAP Gateway
WAP Gateway core solution—Nokia, Kannel, etc.
OS—Windows NT, Unix, Linux
Application Logic
Middleware CORBA, OMG NS, proprietary classes that may be written in Java, C++, and on rare occasions Objective-C
Servlets, Java classes
CGI scripts
Data
Database glues—ODBC, JDBC
Operational data

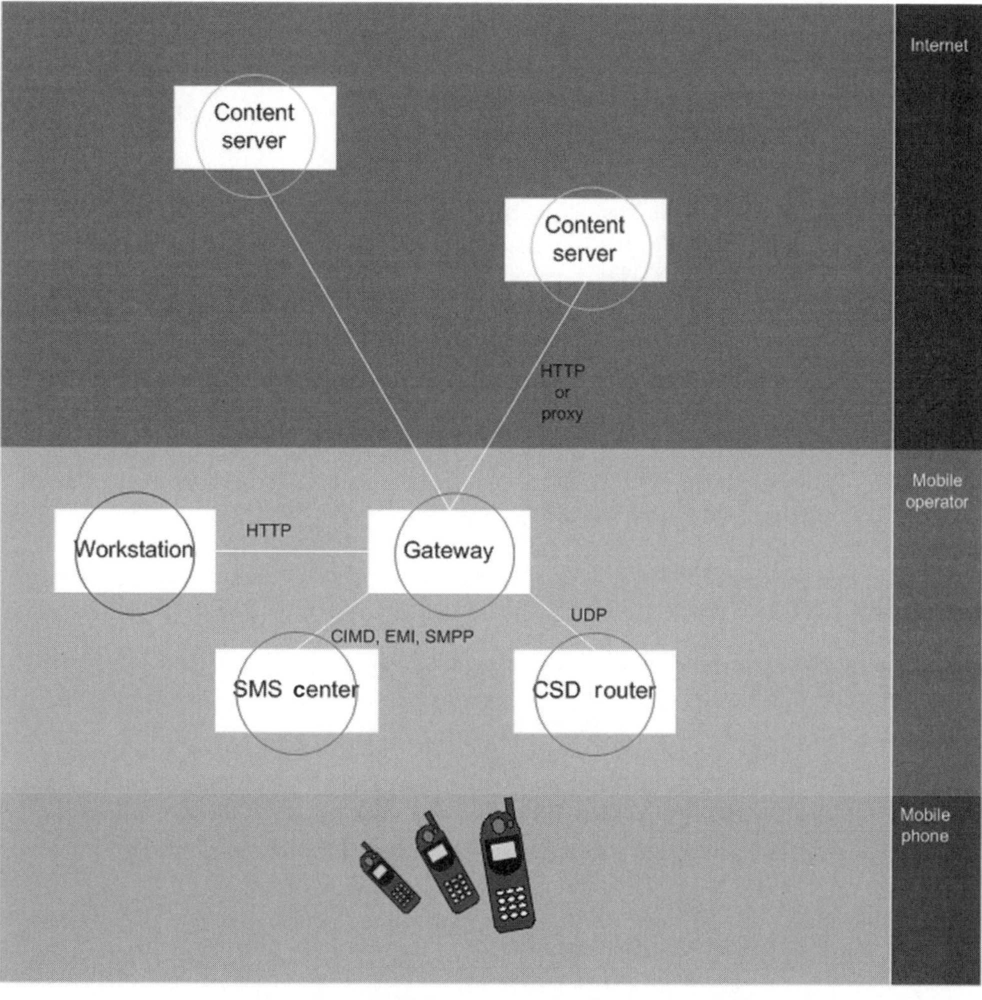

Figure A-13 External interfaces of the Kannel WAP gateway.

- CSD routers: Communication with phones via CSD routers is plain UDP (of the TCP/IP stack), i.e., there is no special CSD protocol.
- Configure/monitor/control workstation (which could also be considered part of the gateway, but isn't): The c/m/c workstation uses HTTP, where the gateway works as an HTTP server, and similarly for those sending SMS via HTTP.
- Content servers: The content servers also use HTTP, but with them the gateway is a client. HTTP proxies also use HTTP (but a slightly different kind), and the gateway is then also a client.

- Clients sending SMS messages via HTTP.
- HTTP proxy.

The WAP gateway divides the processing load on several hosts, which are of three different types:

- Bearer box: This host connects to the SMS centers and CSD routers, and provides a unified interface to them for the other boxes. It implements the WDP layer of the WAP stack.
- WAP box: These hosts run the upper layers of the WAP stack. Each session and the transactions that belong to that session are handled by the same WAP box. Sessions and transactions are not migrated between WAP boxes.
- SMS box: These hosts run the SMS gateway. They can't connect directly to the SMS centers, because the same SMS center connection can be used for both SMS services and WAP.

(*See* WAP under letter W.)

I-mode

I-mode was introduced in Japan by NTT DoCoMo and is used primarily in 2G PDC networks but 3G handsets in Japan retain I-mode functionality. I-mode is an alternative to WAP or, to be more precise, I-mode is an equivalent to WAP over GPRS, and features packet data and packet volume charging. I-mode is used to access Internet sites using a scripting language based on HTML, and is very successful in Japan.

(*See* I-mode under letter I.)

Electronic Payment

Mobile e-commerce or m-commerce is driving the deployment of secure transaction technologies using protocols like WAP's WTLS, for example. Three secure electronic payment services include Radicchio launched by Sonera SmartTrust, Gemplus, and EDs. These services use a SIM card that also acts as a credit card; the credit card reader is the mobile handset itself. The mobile device may have two SIM cards: one to act as a normal SIM card, and another to act as a debit or credit card that would be given to the user by an issuer such as a bank or credit card company. Ericsson, Motorola, and Nokia have developed MeT (Mobile electronic Transactions) that relies on the mobile user possessing a SIM-like card to prevent tampering.

Multimedia and Video

Multimedia services, consisting of mixed media such as voice, audio, and video, fit naturally into the 3G services/applications area given the higher user data rates offered by the networks. Video delivery over mobile networks requires compression using MPEG-1/2/3/4 or other standard algorithms that the targeted mobile devices can decode and play using an appropriate codec. Typically the data rate required to deliver streaming video to mobile devices is high:

$$\text{data\_rate} = \text{pixels} \times \text{bits\_per\_pixel} \times \text{playback\_rate}$$

The MPEG solution to this is basically the use of a *lossy* algorithm that omits redundant picture information from video when compressing or encoding it using an MPEG encoding solution. This encoding may occur in real-time as video is captured from analog or digital sources such as D1.

The MPEG video stream consists of intermittent full frames (or I-frames) that are followed by partial screen frames called P-frames that hold only changes to the previous frame. This is, of course, a fairly simple description of MPEG and video compression; for more detailed information you should refer to MPEG under letter M.

(*See* Multimedia under letter M, Video under letter V.)

Videoconferencing Videoconferencing allows users in remote locations to communicate in real time both visually and verbally.

Systems may be divided into the categories of:

- Mobile videoconferencing
- Desktop videoconferencing using conventional desktop or notebook computers
- Conference room videoconferencing that typically includes appropriately large displays.

Videoconferencing systems include a camera, microphone, video compression-decompression hardware and/or software, and an interface device that connects the system to an access technology.

The interface device might be:

- A wireless 3G interface
- A conventional modem used to connect with an ISP or intranet server, and thereafter use an Internet-based videoconferencing solution such as CU See-Me
- A cable modem that might provide high-speed Internet access via cable
- An ISDN interface that provides connection to the Internet or appropriate IP network
- A Network Interface Card (NIC) that connects to a LAN

- A wireless interface that provides connection over GSM or other mobile communications network

Access technologies for videoconferencing include 3G, PSTN, ISDN, ADSL, cable, GSM, ATM, T1 frame relay, and proprietary wireless technologies.

Point-to-point videoconferencing involves communication between two sites, while multi-point videoconferencing involves interaction between three or more sites. The latter might require a chairperson to conduct proceedings. Also the collective system might be voice-activated, switching sites into a broadcasting state when the respective participant begins speaking.

SIF An SIF frame has an MPEG-1 frame resolution of 352-by-288 pixels. This resolution equates to the MPEG Source Input File (SIF), which is achieved by omitting odd or even lines from a standard interlaced PAL (Phase Alternating Line) signal.

This is an exceptionally "lossy" procedure that omits a great deal of picture information and loses video quality. It is this single operation that limits the quality of video that can be achieved using MPEG-1, although it has to be implemented in order to confine the video stream to the narrow bandwidth of about 1.5Mbps. The MPEG claim that this is VHS quality is an area of debate.

(*See* MPEG under letter M.)

MPEG Video Production MPEG-1 video production has become increasingly popular using comparatively inexpensive video capture hardware and encoding software.

MPEG video production provides a gateway to producing Video CD or White Book compatible MPEG-1 video that can be integrated into multimedia productions; embedded into applications in the form of OLE objects; or used as noninteractive pop videos, movies, and documentaries.

Using a White Book compatible CD-R recorder, you can produce a Video CD with more than 70 minutes of linear MPEG-1 video.

This is a bridge format that can be played on a variety of different systems and appliances that includes appropriately specified PCs and Macs, Multimedia PC-3s, CD-I players fitted with DV (Digital Video) cartridges, Karaoke players, and 3DO players fitted with MPEG decoders.

An alternative to such video production is to use the services of a fully equipped bureau.

The decision as to whether or not to use the services of a bureau is driven by a number of obvious key factors, including the amount of encoding you require, for which you will be charged on a per-minute basis.

Other factors include the person-hours and hardware/software costs required to encode the MPEG-1 video yourself, and the resultant video quality, which will relate not only to the way in which the MPEG-1 encoding parameters are applied but also to the video source recording, the video source format, and captured digital video file.

It comes as a surprise to many to learn that bureaus are not bound to produce MPEG-1 video noticeably superior to that produced using low-cost MPEG encoding software. A bureau's use of expensive, real-time MPEG encoder cards is driven by the need to produce MPEG video as quickly as possible, as opposed to achieving optimum quality. Generally, however, it can be assumed that a reputable bureau will produce Video CD—compatible MPEG-1 video that is of an acceptably high quality and will certainly meet commercial standards.

The production of MPEG-1 video using encoding software begins with the capture of a video sequence from a source recording that might be in the VHS or S-VHS formats. Film studios and production companies might rely on professional- and broadcast-quality formats such as Digital Betacam or D1 for their source recordings.

The capture process can be carried out using an appropriate video capture card such as Spea's Crunch It, used in conjunction with a video capture program. You can use almost any Windows video capture software, including Asymetrix Digital Video Producer, Microsoft Video for Windows VidCap, or the VideoMaestro Video Capture program.

All these programs provide adequate control over capture parameters, allowing you to set a capture frame rate of 25 frames/s, a truecolor image depth of 24 bits providing 16.7 million colors per captured pixel, and an acceptable frame resolution of 352-by-288 pixels.

This resolution equates to the MPEG Source Input Format (SIF), which is achieved by omitting odd or even lines from a standard interlaced PAL (Phase Alternating Line) signal.

This is an exceptionally "lossy" procedure that omits a great deal of picture information and loses video quality.

It is this single operation that limits the quality of video that can be achieved using MPEG-1, although it has to be implemented in order to confine the video stream to the narrow bandwidth of about 1.5Mb/s.

If you are unable to capture video at 25 frames/s, then you can increase the frame rate following video capture using Video for Windows VidEdit or an equivalent digital video editing program. The increased frame rate is achieved merely by duplicating frames, but it does mean that the finally encoded MPEG video stream will at least be an authentic one.

If you capture video with a sound track, it is necessary to separate the video and audio components into different files. Using VidEdit this can be achieved by first making the audio track active, and selecting the whole video sequence using Selection from the Edit menu. Then by selecting Extract from the File menu, the sound track can be saved as a Windows wave file or in any other format that is compatible with the MPEG encoding software. If necessary, delete the audio sound track, and then save the video sequence. This general sequence of events is echoed using other digital video editing programs.

An alternative to capturing audio and video simultaneously, is to record them separately. A wave audio sound track can be recorded using a video capture program or a Windows wave audio recorder/editor such as Creative Wave Studio.

This may result in higher-quality video capture than might otherwise be possible, although it will probably lead to video and sound tracks that have different play times. Consequently when mixing or "multiplexing" them together using MPEG encoding software, the audio and video information will not be synchronized correctly, which will be most noticeable with sequences that are lip-synched or during "hits" where sounds and visual events correspond.

This may be remedied by opening the video file using a video editor such as VidEdit, and then inserting the wave audio file as its sound track.

A video editing program can be used to synchronize audio and video streams, usually by introducing a time offset for the audio track. By then separating the file into video and wave audio files, once again using the video editing program, their play times should become equal.

This will help ensure that they are synchronized after multiplexing using the MPEG encoding software. Although some MPEG encoders will automatically alter the length of the input audio file so it matches the length of the input video file, this does not guarantee that the audio and video material are synchronized correctly when multiplexed. It should be added that the synchronization of audio and video information can also be carried out at the decoding stage.

Having obtained audio and video input files in an appropriate format, they can be compressed or encoded separately into MPEG audio and MPEG video streams. This may take some time using MPEG encoding software, particularly when creating the MPEG video stream. It is therefore advantageous to have a decent Pentium-class system, which may reduce the encoding time to around two to three times the period required for real-time MPEG encoding.

The final stage of encoding is to mix or multiplex the MPEG audio and video streams into an MPEG system stream. Using PixelShrink, encoding and multiplexing jobs can be implemented in batches, relieving you of a degree of tedium. To test and evaluate the quality of the compressed system stream, you will need a fully specified MPEG player. You can use software-only players, but these will not allow you to evaluate the video adequately, as they may play video only, and are more likely to play the file at an incorrect speed, usually too slow.

Video Capture Video capture is a process of acquiring video in appropriate digital form, either compressed or uncompressed.

The video source recording might be analog or digital. The latter requires the video capture card to incorporate an appropriate input. The three general types of video capture include:

- *The real-time video capture* technique, which involves digitizing the incoming video source signal on the fly, and the video source device is not stopped or paused at any time during capture.

- *Automatic step-frame capture* requires that the source device is stopped, paused, and even rewound so as to digitize a greater amount of the source recording. It offers certain advantages; namely, it is possible to achieve a greater number of colors (or greater image depth), higher capture frame rates, and larger capture frame resolutions than would normally be possible using the same video capture hardware and software configuration to record video in real time.
- *Manual step-frame capture* usually depends on the operator clicking a button on screen in order to capture selected video frames.

Before video capture can begin, it is necessary to prime the capture program by choosing a number of different options that include color depth, video capture frame rate, frame dimensions, audio sample frequency, and audio sample size. If you want to capture video using an eight-bit color depth (or 256 colors), and you wish to use a color palette (which is a predefined set of colors), you also have to paste an appropriate palette into the video capture program or open an appropriate palette file. This, however, is a comparatively rare requirement.

A capture file has to be set up so as to optimize the rate at which digitized video can be written to the hard disk so as to improve video capture performance. If necessary, the target hard disk should be defragmented so that video data is written to a contiguous series of blocks, thus optimizing the target hard disk performance.

Available color depths using fully specified video capture card and capture program partnerships include 8bit, 16bit, and 24bit. The 8bit format gives a maximum of 256 colors stored in the form of a color palette that can be edited using programs such as PalEdit. Sixteen-bit and 24bit formats are described as truecolor, giving a maximum of 65K (2^{16}) and 16,777,216 (2^{24}) colors, respectively; and when using appropriately specified video capture hardware and software they can produce impressive results.

Using many video capture systems, the data throughput required to capture 16bit and particularly 24bit video in real time, limits both the capture frame rate and frame size. One solution to this problem is automatic step-frame capture where an MCI video source device is operated automatically.

The frame dimensions chosen hinge largely on the specification of the capture card, although the image depth chosen is also influential as is the capture frame rate.

Although the video frame dimensions can be scaled using video editing programs and even multimedia authoring tools, enlargement can result in a blocking effect as the individual pixels are enlarged. However, certain graphics cards, particularly those that enlarge Video for Windows video sequences, will apply a smoothing algorithm during playback in an attempt to minimize the blocking effect.

Video editing techniques also can be used to increase the playback frame rate (through frame duplication). Digital video editing techniques and hardware/software features of the playback system can help improve the quality of video playback.

However, capturing and compressing optimum quality digital video relevant to the intended playback platform remain the most important processes. There are limitations in what can be achieved through digital video editing, and through playback hardware that enhances digital video playback.

The original video sequence may be enhanced, even enlarged through duplication, but it cannot be used to play video information present in the source recording that it simply does not contain. Even though numerous algorithms can enhance digital video, and numerous others will emerge, it is reasonable to assume that if the video file does not contain a particular frame then that frame cannot be played.

The quality levels available using wave audio recorders together with mainstream sound cards also can be achieved through fully specified Windows video capture programs. Eight-bit or 16bit sample sizes are available, recorded at frequencies of 11.025kHz, 22.05kHz and 44.1kHz in mono or in stereo. The size of the sound track, which increases in relation to the recording quality chosen, can be monitored by selecting the Statistics command (or something similar on many video editing programs) using VidEdit.

Video Capture Card A video capture card can be used to capture and sometimes compress motion video, converting it into digital form. The majority of video capture cards are sold for the production of video for the Windows environment, and are often supplied with the full implementation of Video for Windows.

Manufacturers of mainstream video capture cards include Creative Labs, Fast Electronics, Intel, VideoLogic, and Spea. Increasingly as MPEG is entering into mainstream computing, video capture packages are able to capture and compress video according to MPEG. Important points to consider when purchasing a video capture card include:

- The image depths supported: All fully specified versions should support 8bit, 16bit, and 24bit color depths.
- The maximum capture frame rate and capture frame resolution supported at a given image depth.
- The video editing tool supplied with the package: Examples include VidEdit, Adobe Premier, and Asymetrix Digital Video Producer.
- The video capture program supplied: This will normally be Microsoft VidCap although other variants are available.
- Video sources supported: It can be assumed that all modern implementation will support both PAL and NTSC.
- The video formats and compression schemes supported: These include Intel Indeo, M-JPEG, and MPEG.

- The sound feature capabilities: With the exception of Video Spigot, all video capture cards feature a built-in sound facility able to record sound tracks of audio qualities that are equivalent to those available using an MPC-2—compliant sound card.
- The presence of a VL channel connector: This indicates that the card can be connected to graphics cards that also feature a VL channel connector. Occasionally such cards can be slightly less expensive, as well as be expandable to incorporate additional functionality such as M-JPEG video capture facility.

MPEG Frames An MPEG video sequence consists of partial frames in the form of Predicted (P) frames and Bi-directional (B) frames, and full frames or Intra (I) frames.

I frames are compressed in a similar way to JPEG (Joint Photographic Experts Group) images and do not rely on image data from other frames. They exist intermittently, perhaps between 9 and 30 frames, and provide nonlinear entry points.

Increasing the frequency of I frames provides a greater number of valid entry points, but the compression ratio of the overall file diminishes proportionally. Realistically, the compression ratios achieved using MPEG may be assumed to be around 50:1. Higher compression ratios lead to an unacceptable loss of quality, and it is wise to forget the 200:1 ratio that MPEG is supposedly capable of producing. Normally this is achieved through a pretreatment process that dramatically reduces the number of frame pixels.

I frames and P and B frames described below are termed Groups of Pictures (GOPs), and the occurrence of each frame might be predefined through the careful adjustment of MPEG parameters prior to encoding. However this fine level of control over compression parameters may not be provided by low-cost MPEG encoding programs.

MPEG-1 (Moving Picture Experts Group Algorithm) MPEG-1 is an internationally agreed digital video compression standard. It is used widely for local playback, for streaming multimedia over the Internet, and for other IP and multimedia networks. (*See* MPEG-2 under letter M.)

The early days of digital video were plagued by the problem of just how digital video data should be compressed, thus establishing the need for international standards for the digital storage and retrieval of video data.

Sponsored by the then ISO (International Standards Organization) and CCITT (Committee Consultitif International Telegraphique et Telephonique), the Motion Picture Experts Group (MPEG) was given the task of developing a standard coding technique for moving pictures and associated audio.

The group was separated into six specialist subgroups including Video Group, Audio Group, Systems Group, VLSI Group, Subjective Tests Group and DSM (Digital Storage Media) Group.

The first phase of MPEG work (MPEG-1) dealt with DSMs with up to 1.5 Mbits/s transfer rates, for storage and retrieval, advanced Videotex and Teletext, and telecommunications.

The second phase (MPEG-2) of work addressed DSMs with up to 10 Mbits/s transfer rates for digital television broadcasting and telecommunications networks. This phase would cling to the existing CCIR 601 digital video resolution, with audio transfer rates up to 128 kbits/s. MPEG-1 was finally agreed on, developed, and announced in December 1991.

MPEG participants included leaders in: computer manufacture (Apple Computer, DEC, IBM, Sun, and Commodore); consumer electronics; audio visual equipment manufacture; professional equipment manufacture; telecom equipment manufacture; broadcasting; telecommunications; and VLSI manufacture. Universities and research establishments also play an important role.

It provided a basis for the development of Video CD, which was specified publicly by Philips in late 1993. This is an interchangeable format that may be played using both PCs fitted with appropriate MPEG video cards and compatible CD-ROM drives, as well as Philips CD-I players fitted with Digital Video cartridges. Its development is constant so as to accommodate the increasing data transfer rates of both DSMs and other video distribution transports.

MPEG-1 compression is optimized for DSMs with data transfer rates of up to 1.5Mbits/s. MPEG-2 accommodates DSMs and video distribution transports capable of supporting higher data transfer rates of up to 10Mbits/s. MPEG-4 video compression is designed to transmit video over standard telephone lines.

An MPEG video stream generally consists of three frame types:

- Intra
- Predicted
- Bidirectional

Central to MPEG encoding is the use of reference or intra (I) frames that are complete frames and exist intermittently in an MPEG video sequence.

The video information sandwiched between intra frames consists of that which does not exist in the intra frames. Information that is found to exist in the intra frames is discarded or "lossed." Intra frames can act as key frames when editing or playing MPEG video as they consist of a complete frame.

Generally compressed MPEG video is difficult owing to the paucity of authentic access points. However, editable MPEG files do exist, one of which is backed by Microsoft. Additionally an MPEG video stream composed entirely of I frames lends itself to nonlinear editing.

The quality of MPEG video depends on a number of factors ranging from the source video recording quality to the use of important MPEG parameters that affect the overall compression ratio achieved.

Contrary to popular belief, the logical operations that provide a basis for obtaining high-quality MPEG video are by no means the exclusive property of expensive video production bureaus.

Equipped with a reasonably specified PC and a basic understanding of MPEG video, there is nothing to stop you producing good-quality White Book–compatible video on your desktop.

Probably the most obvious elements that influence MPEG video quality include the analog or digital source recording, the video source recording format, and the video source device specification.

It may be assumed that the higher resolution S-VHS format will provide slightly better results than VHS, but there will not be a dramatic improvement in resolution because the MPEG SIF is standardized at 352-by-288 pixels for PAL.

If you are digitizing the sound track of the source video recording also, then you will probably obtain the best results with camcorders and VCRs that offer hi-fi–quality stereo sound.

When capturing a video file so that it may eventually be compressed, it is important to choose an appropriate capture frame rate, capture frame size, and image depth.

The capture frame rate should be set for 25 frames/s for PAL and 30 frames/s for NTSC. Frame rates that differ from these will cause the MPEG video sequence to run at the wrong speed, and it will not be White Book compliant.

The capture frame size should correspond with the MPEG-1 SIF, which is 352-by-288 pixels for PAL and 352-by-240 pixels for NTSC. Authentic MPEG requires a truecolor image depth of 24 bits per pixel, giving a total of over 16.7 million colors, which are generated by combining 256 shades of red, green, and blue.

The quality of captured audio that is used as an input audio stream obviously depends upon the sample size, recording frequency, and whether mono or stereo is chosen.

You can assume that your wave audio recorder or video capture program will provide sampling rates of 11kHz, 22kHz, and 44.1kHz, and sample sizes of 8bit and 16bit. While higher sampling rates and larger sample sizes yield improved audio quality, the resultant audio stream can consume an unacceptably large portion of the available MPEG-1 bandwidth.

With regard to careful adjustment of the MPEG compression parameters there is not much you can do if the MPEG encoding software provides no control over them. If it does, then it may be assumed that a greater number of I frames can improve the quality slightly, although this will introduce an overhead in terms of lowering the compression ratio.

MPEG-2 (Moving Pictures Experts Group) MPEG-2 is an improved version of MPEG-1 video compression, and is supported by DVD technology. It was developed for media and networks able to deliver 10Mbits/s data transfer rates.

MPEG-1 was developed for narrow-bandwidth media, such as the original single-speed CD drive variants that offered average data transfer rates of approximately 150Kbytes/s or 1.2Mbytes/s.

MPEG-2 video may contain considerably more audio and video information than MPEG-1. The most noticeable improvement is the higher playback screen resolutions that are possible, making possible D1 or CCIR 601 quality.

DCT is key to MPEG-2, as it is to MPEG-1 and JPEG (or even M-JPEG). As is the case with MPEG-1, MPEG-2 requires decoding solutions that may be hardware based such as set-top boxes (STBs), or equivalent hardware implementations integrated in computers.

Applications of MPEG-2 video include, for example, Video-on-Demand, multimedia, videoconferencing. It may also be stored and delivered using DVD variants.

Computers/Software

Accumulator

A register within a processor architecture that can be used to store the results of arithmetic operations. It consists of one or more registers, and its overall size often indicates the size of instructions that can be processed.

ACID (Atomicity, Consistency, Isolation, Durability)

A series of properties, which define the real-world requirements for transaction processing (TP).

Atomicity A process of ensuring that each transaction is a single workload unit. If any subaction fails, the entire transaction is halted and rolled back.

Consistency A process of ensuring that the system is left in a stable state. If this is not possible the system is rolled back to the pre-transaction state.

Isolation A process of ensuring that system state changes invoked by one running transaction do not influence another running transaction. Such changes must only affect other transactions, when they result from completed transactions.

Durability A process of guaranteeing the system state changes of a transaction are involatile, and impervious to total or partial system failures.

Acrobat

An Adobe file format that permits formatted documents to be deployed efficiently over the Web.

Adobe Acrobat Reader is required to read Acrobat files (which have the .PDF extension).

Using Netscape Navigator, the Acrobat Reader requires a plug-in. Microsoft Internet Explorer uses an ActiveX control.

(*See* ASP later in this chapter.)

Active Template Library (ATL)

A development tool used to develop Active Server Components, which may be in-process or out-process.

(*See* ASP later in this chapter.)

Active Web Architecture

An architecture which provides bidirectional information flow between the HTTP server and HTTP client. The resulting interactivity on the client side permits data entry and the editing of HTML documents.

Active Web Architecture uses the Common Gateway Interface (CGI) between the HTTP server and its applications and databases. CGI is a protocol that provides the necessary communications. CGI scripts are created using a scripting language or programming tool.

ActiveX Control

An object or component that adds functionality to an application, which may be standalone, or deployed over the Web or network.

Microsoft ActiveX is an object architecture based on OLE 2.0, and intended for deployment over the Internet and compatible IP networks. More accurately, ActiveX is a reincarnation of OCX and may use COM and DCOM as glues.

ActiveX provides cost-effective functionality gains for Web browsers. An ActiveX control might take the form of a streaming video player, or a streaming audio player that might be added to the Microsoft Internet Explorer (which is a Web browser).

ActiveX controls may be created using Visual C++, Visual Basic 5 Control Creation Edition, and Java.

Guidelines for creating ActiveX controls:

- Refer to existing Active controls in the public domain, to those that are shareware, and to those that might be conventionally marketed and sold. The economics of recreating that which has already been created might prove undesirable. Study the functionality of the ActiveX Controls, and try to obtain real-world reviews of them, in order to gain an understanding of what may be expected from them.

- Use the latest editions of development tools such as Visual Basic Control Edition, Visual Studio, and others.
- Supply detailed design, architecture, implementation, and functionality documentation. If the ActiveX Control may be modified at the code-level, provide adequate comments in the source listing. Also include an impact statement of how the ActiveX Control changes targeted applications, together with a strategy of useful code segments or algorithms designed.
- Do not intentionally integrate patented algorithms in your ActiveX Control. It is accepted that such infringements can be implemented unwittingly by the developer and/or programmer.
- Test the ActiveX Control.
- Provide case scenarios giving real-world examples of their integration in Web applications.
- Refer to Microsoft Web sites for the latest ActiveX specification and development tools.
- Integrate configuration features, which may be used from within the application where the ActiveX Control is embedded.
- State the development environments and/or tools with which the ActiveX Control has been tested.
- Sign the ActiveX control.

ActiveX components running on the same system may interact using the COM protocol as a glue. Industrywide support beyond Microsoft, in the ActiveX compatible development tools include:

- Borland Delphi
- Powersoft PowerBuilder (see www.powersoft.com)
- Powersoft Optima++ (see powersoft.com)
- Symantec C++ (see www.symantec.com)
- MetroWerks Code Warrior (see www.metrowerks.com)

ActiveX Scripting

A process by which ActiveX Controls and Java Applets may be integrated into the underlying HTML code of an interactive Web application. Such scripting is generally used with Web applications, although standalone applications may also be built using the same scripting.

The scripting languages JScript and VBScript are used widely. Using them, a basic HTML listing may be given functionality, and responses to events through:

- JScript code
- VBScript code

- ActiveX Controls such as Shockwave and multimedia streaming components
- Java applets

Such a development strategy can be used to give the client-side a level of intelligence. Validations of user data and interactions distributes processing away from the server-side. This lessens the volume of data traffic, and serves to optimize application performance.

Because ActiveX Scripting may also be applied to the server-side, it is possible to create Active Server Pages (ASPs).

Adaptive Data Compression

A proprietary data compression technique integrated into the design of many Hayes modems. The algorithm adapts itself so as to optimize compression.

ADC (Analog to Digital Converter)

A device or electronic assembly used to convert continuously varying analog signals into digital form. The accuracy achieved depends largely on the size of samples and on the sampling rate.

Video capture boards and sound cards include analog-to-digital (ADC) converters. Standard PC and Macintosh sound cards tend to record using 8bit or 16bit samples at sampling rates of 11.25kHz, 22.05kHz, or 44.1kHz. Highly specified sound cards may record using sampling rates of up to 48kHz, which equates to DAT quality.

Video capture cards generally play a dual ADC role, converting audio as well as video into digital form. Normally audio is digitized using the same sample sizes and sampling frequencies available on most fully specified sound cards.

Whether capturing from a VHS or S-VHS video source recording, the process of digitizing a video signal requires a great deal more computation than that of an analog audio signal.

The maximum frame capture rate of a video capture card is a function of its maximum sampling rate, which is linked to the maximum data rate at which it can operate.

ADPCM (Adaptive Differential Pulse Code Modulation)

A process by which an analog signal is converted into digital form. It is a development of Pulse Code Modulation (PCM).

The sampling rate influences how accurately sharply varying analog signals are digitized. It is used in telecommunications, digital audio, video, and multimedia technologies.

Agent

1. An agent/manager architecture used for system management in client/server systems. The *agents* represent managed subjects, which are communicated with and manipulated by *managers*.
2. A *triggered* agent is a program that responds to events with appropriate actions. The actions might be little more than answering a telephone call. More sophisticated agents might modify software, or build databases or even data warehouses, or add items to a cache, in response to usage habits. Events such as changes to files or directories might also be used as triggers.
3. A *habitual* agent can be programmed to implement tasks at a precise frequency, such as hourly or daily.
4. A Microsoft ActiveX control intended to enhance the UI of local and Web applications.
5. In a telecommunications network, an agent interprets various commands, and responds to them appropriately.

AI (Artificial Intelligence)

A term used to describe the use of a system to emulate human decision making and learning abilities. The founding father of artificial intelligence is Alan Mathison Turing, through his writings, which include *Computing Machinery and Intelligence* (1950).

Alan Mathison Turing OBE, an English mathematician, World War II codebreaker and computer scientist and inventor, also described the "Turing machine," and how it could theoretically implement logical processes.

Expert systems, or Knowledge Based Systems (KBSs), neural networks, and genetic algorithms are perceived as part of AI.

A KBS includes a knowledge base of rules or heuristics, each comprising fact(s) and conclusion(s), that is, IF disk drive light off THEN check ribbon cable. Such conclusions and deductions may be weighted appropriately.

The rule-base is chained by an inference engine, which chains:

- Backward, comparing an inputted question with conclusions in the rule-base, and may compare subsequently located facts with conclusions of other rules
- Forward, comparing facts

A KBS can offer informed decision-making skills, the effectiveness of which is a function of the accuracy and comprehensiveness of its rule- or knowledge-base. A knowledge engineer is responsible for generating rule-base. Numerous KBS applications exist, including medicine, business, stock market, maintenance, etc. Web-based KBS solutions exist.

KBS can offer informed decision-making skills, the effectiveness of which is a function of accuracy and comprehensiveness of rule- or knowledge-base. Knowledge engineers are responsible for generating rule-base, which has proved to be a complex and arduous process. Numerous applications include medicine, business, stock market, maintenance, etc.

Neural Nets (NNs) may have one or more layers of neurons or simple processors. Preceptron NNs depend on weighted neuron connections that can be learned when given example data.

NNs are a subsymbol of expert knowledge because there is no rule-base that may be interpreted by an expert. Because it is difficult to gauge whether the NN has learned knowledge in an intelligent form, it is best considered a black box.

An NN is a network of neurons that function as processing units.

Their implementation is an attempt to reconstruct the operation of the human brain, which has some *ten thousand million* neurons.

The neuron connections have weights, which determine network behavior. Given an example, the weights may be learned.

Neural network variants include the

- Perceptron
- Multilayer perceptron

The perceptron neuron was proposed in 1962 by Frank Rosenblatt.

A computational neuron has input connections, each of which may have a different weight. The neuron is preprogrammed with a threshold value, which if equaled or exceeded by the total weight of inputs, will respond accordingly. Typically, the response is to output a specific value.

They differ from multiple input logic gates (such as AND, NAND, OR, or NOR) in that the inputs may not be one of two logical values.

The inputs i are assigned the weights w, and a positive output is yielded should a predefined threshold be exceeded:

$$\text{if } i_1w_1 + i_2w_2 + 1_3w_3 \ldots + i_nw_n > t,$$
$$\text{then output } = 1$$
$$\text{else}$$
$$\text{output } = 0$$

See Fig. A-14.

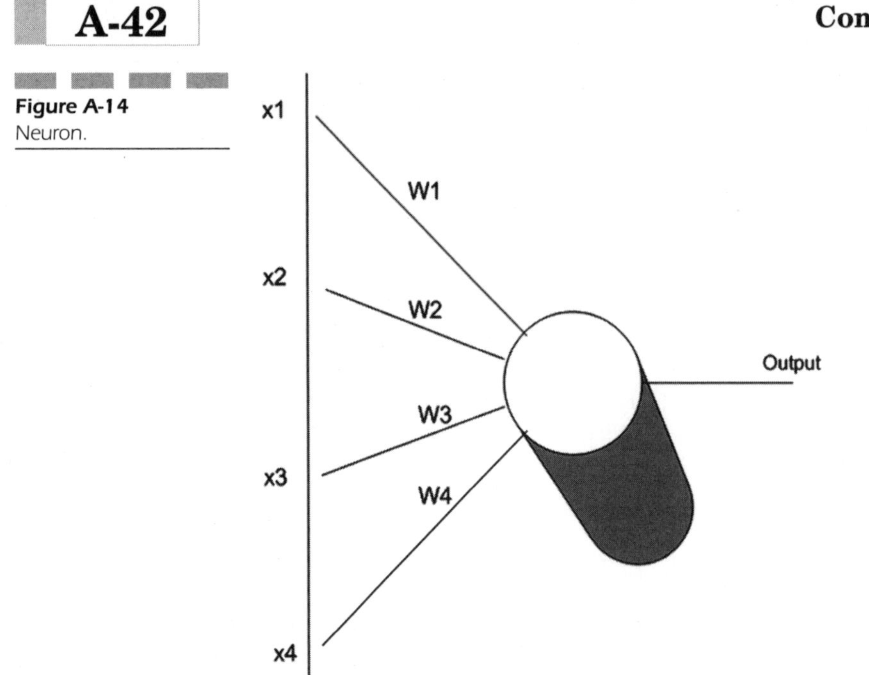

Figure A-14
Neuron.

Genetic algorithms are based on evolutionary theory and natural selection, where populations of Genomes combine to produce offspring that must reach a specified score if they are to survive. See Fig. A-15.

(*See* MPP under letter M.)

Algorithm

1. "An algorithm is a set of rules for getting a specific output from a specific input. Each step must be so precisely defined that it can be translated into computer language and executed by machine."—Donald E. Knuth
2. A collective name describing the components of the problem-solving process. It can be a program or series of steps defining a modus operandi, which yields what is regarded to be an acceptable solution.
3. A term loosely used to describe a program, or program segment. Algorithms for compression, and those that perform other operations, are often patented.

Allen, Paul

A cofounder of Microsoft, and sole founder of Asymetrix.

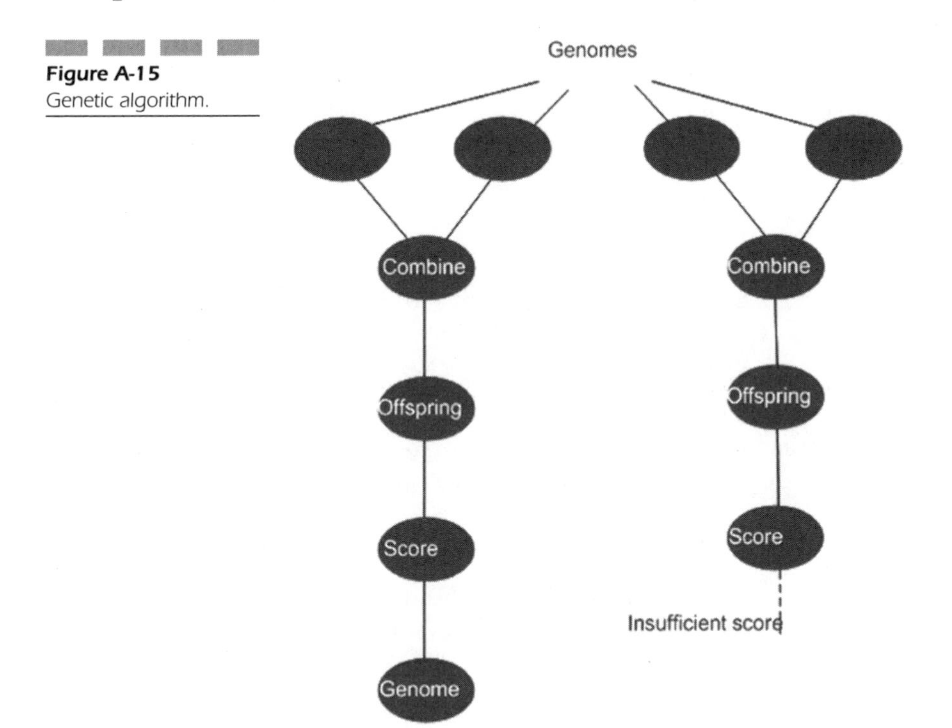

Figure A-15
Genetic algorithm.

Alpha

1. A family of RISC processors manufactured by Digital.
2. A prerelease copy of an application, which is distributed and tested in-house. It is the penultimate development stage that precedes beta testing.
3. An 8bit data channel on 32bit color systems that provides control over the transparency of pixels, thus facilitating numerous video effects.

AMD (Advanced Micro Devices)

A chip manufacturer that produces PC processors. AMD came to prominence when it reverse-engineered Intel's third generation 80386 processor, and won the legal right to market and sell it. More modern AMD offerings include the K6 MMX processor.

(*See* Pentium under letter P.)

AMIS

Audio Messaging Interchange Specification

Animation

A series of frames used to create the illusion of movement. Animation types include:

- Morphing, which dissolves one image into another, and may be created using dedicated morphing programs, or equivalent such features in animation programs
- Sprite, where one or more screen objects are moved
- Cell-based, where entire frames are updated fully or partially to give the illusion of movement
- Micons, where a continuous series of frames is repeated conditionally. The condition might be a mouse-click event.

Anonymous FTP (File Transfer Protocol)

An FTP server, to which users may connect, browse its files, download files, and possibly upload files.

ANSI (American National Standards Institute)

A highly influential standards institution. The array of ANSI standards covers everything from character sets to programming languages such as C++.

Anti–Money Laundering

A software solution that detects money laundering activity, and may provide automated decision making when it alerts users.

AOL (America Online)

A large, international ISP that has POPs (points of presence) in many major cities. The Compuserve ISP is part of the AOL Corporation.

Apple Computer

A computer manufacturer cofounded by Steve Jobs and Steve Wozniak. The launch of the highly successful Macintosh (Mac) computer in 1984 represented a significant point in the history of computers. The Macintosh has since been refined into several versions and is an excellent platform for multimedia delivery and development. HyperCard is synonymous with the Macintosh and marked the beginning of a deserved reputation for suitability to multimedia. HyperCard was the birth of hypertext/hypermedia in mainstream computing.

Applet

A program that resides on a server, and when requested, it is downloaded and executed by the client Browser. Such applets deployed on the Web require machine independence, and a virtual processor such as the Java Virtual Machine installed on the client. The applet concept is not new, predating Java by a considerable margin.

AppleTalk™

A network capability built into Macintosh computers, and permits integration into heterogeneous environments. It may be used with LocalTalk™ cabling, Token-Ring, and Ethernet.

Application Development

A process by which an application is created. In terms of authoring multimedia, the development life cycle might include various standard stages that include:

- Project planning
- Design
- Scripting
- Prototyping a storyboard design
- Multimedia production
- Production
- Coding in a multimedia language such as OpenScript or Lingo
- Coding in an Internet-related language such as WML, Java, VBScript, JScript, HTML, VRML, and Perl

- Coding in general-purpose languages such as Visual Basic and Visual C++
- Alpha testing
- Beta testing
- Packaging the application for distribution on a CD variant, or for deployment on a network.

In the perspective of the Web or Internet, application development may require the use of:

- Content authoring programs, such as those that permit the generation of animations, and multimedia production tasks.
- Web site development tools that permit production tasks such as integrating media with navigation schemes, etc.

Application-Level Gateway

An application-level gateway is able to process store-and-forward traffic, and provide security features. They may be programmed to maintain logs of application usage. Users must log in to the application gateway machine.

Application Message Queue

A buffer used in Microsoft Windows to store messages posted by an application using the PostMessage routine. The size of the queue can be set using SetMessageQueue.

Alternatively, e-mail messages with appropriate instructions can be sent to the relevant sites.

Application Programming Interface

An interface that provides programmers with the necessary high-level instructions to implement what would otherwise require a great deal of coding and specialist skills.

For example, an API might provide access to the complex functions of a 3-D graphics engine, through simple instructions. The underlying mathematical elements are transparent to the programmer, and need not be understood or coded. Multimedia-related APIs are released constantly.

Application Software

A program or suite of programs designed to perform a particular task, or set of tasks. Mainstream business applications include word processors, spreadsheets, relational databases, and contact managers. These are generally included in integrated packages.

Examples include:

- Integrated packages, such as Microsoft Office, Microsoft Works, and ClarisWorks
- Word processors, such as Microsoft Word, WordPerfect, and Lotus Word Pro
- Spreadsheets, such as Microsoft Excel, Lotus 123, and Quattro Pro
- Databases, such as Microsoft Access, Paradox, and DataEase
- Contact managers, such as Outlook, Goldmine, and those supplied with many integrated packages.

The three staple elements of an application are:

- Presentation that provides images, media, and text
- Programming logic used to process and manipulate data
- Data forms that may be of many different types

The physical, or perceived, location of the three functional elements depends upon a series of logical topologies devised by the Gartner Group. This is explained under the entry Client/Server in letter C.

Archive

A method of storing files for backup or long-term storage. Removable media that might be used for archiving purpose include 100MB Zip disks and 1GB Jaz disks from Iomega, as well as media devices from SyQuest. Other media include conventional hard disk, CD-R discs, and DVD-RAM discs. Various file compression utilities can be used for backup purposes, including the popular WinZip program.

ARP (Address Resolution Protocol)

An IP protocol that can be used to convert logical IP addresses (such as 18.170.103.34) into physical addresses. An ARP request results in a node's physical address that might be used by Ethernet networks, Token Ring, and FDDI, which may have a bandwidth of up to 100Mbps.

ARPA (Advanced Research Projects Agency)

A U.S. government agency formerly called DARPA (Defense Advanced Research Projects Agency).

ARPANET (Advanced Research Projects Agency Network)

An early network developed by the then DARPA (Defense Advanced Research Projects Agency) for researchers.

Originally coined DARPANET, its development was commissioned in 1969, resulting in a working network of four computers by 1970, and growing to 37 computers by 1972 at which time it became ARPANET.

Some assert that DARPANET was the technical, and possibly, conceptual birth of the *information superhighway*, or Internet. The key development resulting from ARPANET is the TCP/IP family of protocols. ARPANET ceased to exist in 1990.

Array

1. A two- or three-dimensional matrix of data values. The values might be character, numeric, or even binary objects. All modern high-level programming languages support arrays. The concept is similar to the use of tables in databases and data warehouses. (*See* Data Warehouse under letter D.)
2. An uncommitted logic array (ULA) is an electronic package that has electronic devices (or gates) that are unconnected. By adding the connections in the form of a metallization layer, the ULA is given a specific functionality.
3. A transformation array is used to manipulate a 2-D or 3-D set of coordinates.

ASCII (American Standard Code for Information Interchange)

A standard set of codes introduced to promote compatibility in terms of characters and symbols. Originally it consisted of 127 ASCII characters derived from seven bits. Eight bits were not used in order to preserve the sign bit. ASCII has since been extended into a larger highly standardized character set.

ASF (Advanced Streaming Format)

A storage container data format (and standard) for streaming multimedia and local media files. The contents of the container are not defined, and neither is the communications protocol which may be:

- HTTP
- TCP
- RTP
- UDP

The ASF container file contents are read by an appropriate media server, and transmitted to the client, where it may be stored or played.

ASP (Active Server Page)

A server-side software architecture that is browser neutral. ASPs may be normal HTML that glue together scripts written in Jscript, VBScript, JavaScript, or other ActiveScript-compliant language. The following ASP is compiled by the server and then the resulting HTML code is downloaded to the browser

```
<% for a = 1 to 2 %>
<font size= <% = a %> > Hello World </font> <br>
<% next %>
```

Downloaded HTML:

```
<font size= 1 > Hello World </font> <br>
<font size= 2 > Hello World </font> <br>
```

ASP applications may use ODBC databases , VBScript objects, and ActiveX DLLs. ASP objects include Request that may be used to retrieve cookies using request.cookies, and to request information from forms using request.form. ASP pages also have a response object that is used to write to the HTML file using Response.write, and to write to a cookie using response.cookies.

A technology included with IIS 3.0, which may be used to develop scalable, browser-independent Internet applications. DCOM and COM may be used as glues within ASP-based applications.

Resulting Web applications can acquire information about the client browser and act accordingly. This enables compatible HTML pages to be served to the browser without error messages.

Web browsers that are not ActiveX compliant, for instance, can be served appropriate Web pages. This intelligence facet is integrated into the architecture using server-side scripting.

The Microsoft Active Server provides the component parts to implement the aforementioned functionality, and includes the components:

- Browser Capabilities, which acquires the connected browser's key features

- ActiveX Data Object (ADO), which provides access to backend data (irrespective of its location), and is not limited to ODBC-compliant data sources
- TextStream, which is used to create and open files

Third-party and bespoke components can be integrated into an Active Web site. Such components are devoid of user-interfaces (UIs), and can be developed using an ActiveX control developer's workbench, including:

- Visual Basic
- Visual C++
- Visual C++ ControlWizard, which is used for OLE development
- ActiveX Software Development Kit

Active Server components fall into the categories of:

- *In-process,* which cannot execute in its own address space, and is a DLL.
- *Out-process,* which can execute in its own address space, and is an executable (.EXE).

Active Server Pages (ASPs) are seen as an alternative to CGI, and offer the advantages of:

- Shorter Web application development life cycle, particularly with developers and/or development teams that have little or no CGI programming experience
- Optimized server-side processing, because calls to CGI programs may invoke new processes on the server

Assembler

1. A compiler that converts assembly language mnemonics into machine code.
2. A device that assembles received packets in a packet-switched network.

Assembly Language

A low-level language used to program processors directly. It consists of mnemonics (such as LDA, DEC A, and INC A), as opposed to the more readable statements associated with high-level languages such as Java and C++.

Assembly language is loosely referred to as machine code, when in actual fact it requires an assembler to compile it into machine code. Of all the generations of program languages, it is the closest to machine code.

Asymmetrical Compression

A compression/decompression algorithm in which the processes that constitute compression are not reflected in decompression.

Asynchronous Messaging

A mode of communications between running threads, where a call from one thread to another does not require a response before it may continue processing. Rather it proceeds processing and receiving and sending messages.

ATA-2

A disk interface technology, which includes a controller. Like all others, it is an evolving disk controller specification.

Attachment

A file that is sent and received along with an e-mail message. The file may be binary or text, and is opened using an appropriate application.

Authenticode

A technology supported by the Microsoft Internet Explorer, which permits components such as ActiveX Controls and Java applets to be digitally signed. When such a signed component is encountered, Explorer checks its signature status. An unsigned component causes Explorer to display an appropriate prompt, while a signed component causes Explorer to display a certificate. The certificate includes information about the component and its author. The user is given the option to download the component.

Miscellaneous

A Interface

The Interface between an MSC and BSS and is defined in internationally agreed specifications.

(*See* 2G Networks in Numerals chapter, 3G in Numerals chapter, A-bis Interface earlier in this chapter, Base Station Controller under letter B, BSS Protocols under letter B, GSM under letter G.)

Access Channel (of cdmaOne)

A channel used by a mobile station (MS) initially to access the network when a call is initiated or as a response to a paging message.

Adjacent-Channel Interference

An interference caused by two adjacent channels.

Advice of Charge (AoC)

A GSM service that provides subscribers with monthly billing information.
(*See* Billing under letter B.)

ARIB (Association of Radio Industries and Businesses)

ARIB addresses Japanese 3G radio interface standards, and 3GPP UTRA FDD interface is based on proposals from ETSI and from ARIB. Japanese core networks are dealt with by the TTC.

Asynchronous

A data transmission technique where the sending device and receiver are not synchronized in real time. Each transmitted item, or packet, is encoded with start and stop bits, so the receiving device can decode it without ever receiving a timing signal from the sending device.

Because the asynchronous data transmission technique makes good use of available bandwidth or data rates, it is particularly suitable for networked multimedia.

AT&T (American Telephone and Telegraph)

A telecommunications giant (or telco).

Audio Compression

A general term used to describe the process of reducing audio data. Compressed audio data may be decompressed and played using streaming audio technologies.

In uncompressed form the large size of wave audio files occasionally place unreasonable demands on distribution media in terms of data capacity and/or bandwidth.

Wave audio compression operates on the actual audio data, compacting it in order to give smaller file sizes. It is decompressed on playback using either dedicated hardware and software, or software alone such as an appropriate driver in the Windows environment.

Standard wave audio compression algorithms include MP3, A-Law, IMA (Interactive Multimedia Association), ADPCM, and MPEG-1.

The latter can compress wave audio files, and record and compress audio from an analog source in real time. It can also be used to perform standard editing operations on MPEG-1 wave audio files, including cut and paste, and so on.

Whatever compression standard is used, the resultant file sizes, or the compression ratios achieved, depend on the compressor parameter settings chosen. As the compression ratio is increased, the resultant playback quality diminishes.

High-quality wave audio, therefore, tends to be compressed by a great deal less than a dialogue recording, for example.

Auto-Answer

A feature that permits a modem to respond appropriately to an incoming call.

Auto-Dial

A feature integral to all fully specified communications (comms) programs that permits stored telephone numbers to be dialed automatically.

Automatic Speed-Fallback

A modem that matches its data transfer rate with that of a communicating device or network.

AVI (Audio Video Interleave)

A Microsoft file format for storing interleaved audio and video. When introduced in 1990 it quickly became an industry standard. Using many video editing and video capture tools, the interleave ratio can be varied.

The ratio can be specified as a single figure where, for instance, an interleave ratio of 7 indicates that seven video frames separate each audio chunk. Using Microsoft VidEdit, the statistics of a video file can be shown where the interleave ratio is displayed next to the phrase Interleave Every.

The interleave ratio is expressed as the number of video blocks that separate audio blocks. Generally high interleave ratios are applicable to video stored on hard disk, whereas .AVI video stored on a CD variant is optimized using lower interleave ratios, which often equate to one video frame for every audio chunk.

Video stored in the AVI format can be compressed using schemes including Intel Indeo, Microsoft RLE, Cinepak, and Microsoft 1. Sound track quality commonly found in AVI files ranges from mono 8bit recordings digitized at 11kHz, to 16bit stereo recordings digitized at 44.1kHz.

Acronyms

AAL	ATM Adaptation Layer
AC	Authentication Center
ACCH	Associated Control Channel
ACIR	Adjacent Channel Interference Ratio
ACK	Acknowledgment
ACLR	Adjacent Channel Leakage Power Ratio
ACS	Adjacent Channel Selectivity
ACTS	Advanced Communication Technologies and Services
ADAC	Automatically Detected and Automatically Cleared
ADMC	Automatically Detected and Manually Cleared
AI	Acquisition Indicator
AICH	Access Link Control Application Part
AM	Acknowledge Mode
AMPS	Advanced Mobile Phone Service
AMR	Adaptive Multirate—speech codec
AOA	Angle of Arrival

AoC	Advice of Charge
AOL	America Online
AP	Access Preamble
ARIB	Association of Radio Industries and Businesses
ARQ	Automatic Repeat Request
AS	Access Stratum
ASC	Access Service Class
ASN.1	Abstract Syntax Notation One
ASP	Application Service Provider
ASuS	Application Support Servers
ATM	Asynchronous Transfer Mode
AuC	Authentication Center
AWGN	Additive White Gaussian Noise

B

B-2

Base Station Controller

A BSC (Base Station Controller) controls multiple Base Transceiver Stations (BTSs) using the A interface whose functions may be summarized as:

- Creates the operation and maintenance (O&M) interface
- Frequency management that is used to allocate frequencies to BTSs
- Inter-cell handovers
- Managing frequency hopping
- Power management
- Radio resource management for BTSs
- Time-delay measurements of uplinks relative to the BTS clock
- Traffic control that rationalizes the number of lines to BTSs and MSC

Base Transceiver Station

A Base Transceiver Station (BTS) has multiple transceivers (TRXs) that each support one carrier or eight time slots that represent eight physical channels. These also include control channels, so not all time slots can be used as traffic channels. Ordinarily a BTS serves a single cell, but it is possible to transmit to several cells. The radius of the BTS varies dramatically from a few meters (for indoor deployments) to over 30km usually in rural areas, and a radius up to 70km (and greater) is possible using modified BTSs. The spectral efficiency of a BTS is degraded as the radius increases, and it therefore follows that a very wide radius is applicable only in areas where there is a low traffic volume. The collective functions of the BTS include:

- Broadcast and control channel scheduling
- Channel coding, including error protection and encryption and/or decryption
- Detection of access bursts from mobile stations
- Frequency hopping
- LADPm protocol (layer 2)
- Transcoding and rate adaptation

The GSM specification indicates that the BTS integrates a TRAU (transcoder/rate adapter unit), but this may also be part of the MSC. The transcoder converts digital speech [FR (Full Rate) or EFR (Enhanced Full Rate) coded at 13kbps] from the air interface into 64kbps PCM for transmission over the network. The 13kbps channel is padded, making it a 16kbps channel, and making four such channels available over a 64kbps channel.

Small Base Transceiver Stations

Low-tier environments such as homes and offices could conceivably have a small GSM BSS that could in the future provide an alternative to 3G. Such indoor networks could be based on several technologies, including the FDD and TDD modes of WCDMA. A home base station (HBS) could be used for the residential application of GSM handsets perhaps with PDA features. This technology in the current climate would deliver economies when compared to cordless technology like DECT and CT2.

The HBS is a rather streamlined base station so as to make it affordable, and it may be a *cordless* or *base station* approach. The latter requires the HBS to be connected with the GSM PSTN, and requires an appropriate interface for that connection. This method is preferred by the operators because their infrastructure is not bypassed and there is therefore an opportunity bill for services. The cordless approach (or what ETSI has coined CTS—cordless telephone system), on the other hand, bypasses the GSM operator that cannot bill for calls made via the cordless HBSs. This is because the HBS serves only as an access point to the PSTN, and GSM services cannot be used while the mobile is using the HBS.

Residential HBS and office base station solutions are radically different in that office implementations are deployed to cope with traffic hotspots. A single-cell solution will have one cell for the entire office, and this yields a simple solution because there are no intraoffice handovers. The negative side, however, is poor spectral efficiency, and capacity is increased by allocating more carriers. A multicell solution requires that the office is divided into multiple cells, and capacity can therefore be increased merely by adding more cells. This is naturally a more technical and more expensive implementation than a single-cell solution, and caters to intraoffice cell handovers.

A hybrid IBS system combines the cordless and base station approach, and holds a single cell that is divided into multiple radio subcells. The cells use simplified BTSs in the form of RF-heads, and TRXs are kept in a centralized HUB. The RR protocol in the centralized HUB requires modification because the IBS requires location information as well as channel or frequency/timeslot pair.

Billing

Numerous revenue streams can be obtained in mobile applications or services with the most apparent being the actual billing of the customer that may be time charged, packet data charged, fixed charged, and/or subscription invoiced. A user's bill could contain monthly subscriptions for services that include pay-per-view movies, online games, along with call charges that may have been used for voice, multimedia, or for connection to remote enterprise networks over VPN links. The billing system creates this information in the form of CDRs (Charging Data Records) from the network, and calculates the cost of services, and dispatches statements to customers via the operator or service provider. See Fig. B-1.

Billing—GPRS

A GPRS billing system relies on the SGSN and GGSN to register usage and provide appropriate information using CDR (Charging Data Records) that are routed to the billing gateway. GPRS packet-based charging may be based on:

- Volume or bytes transferred
- Duration of a PDP context session

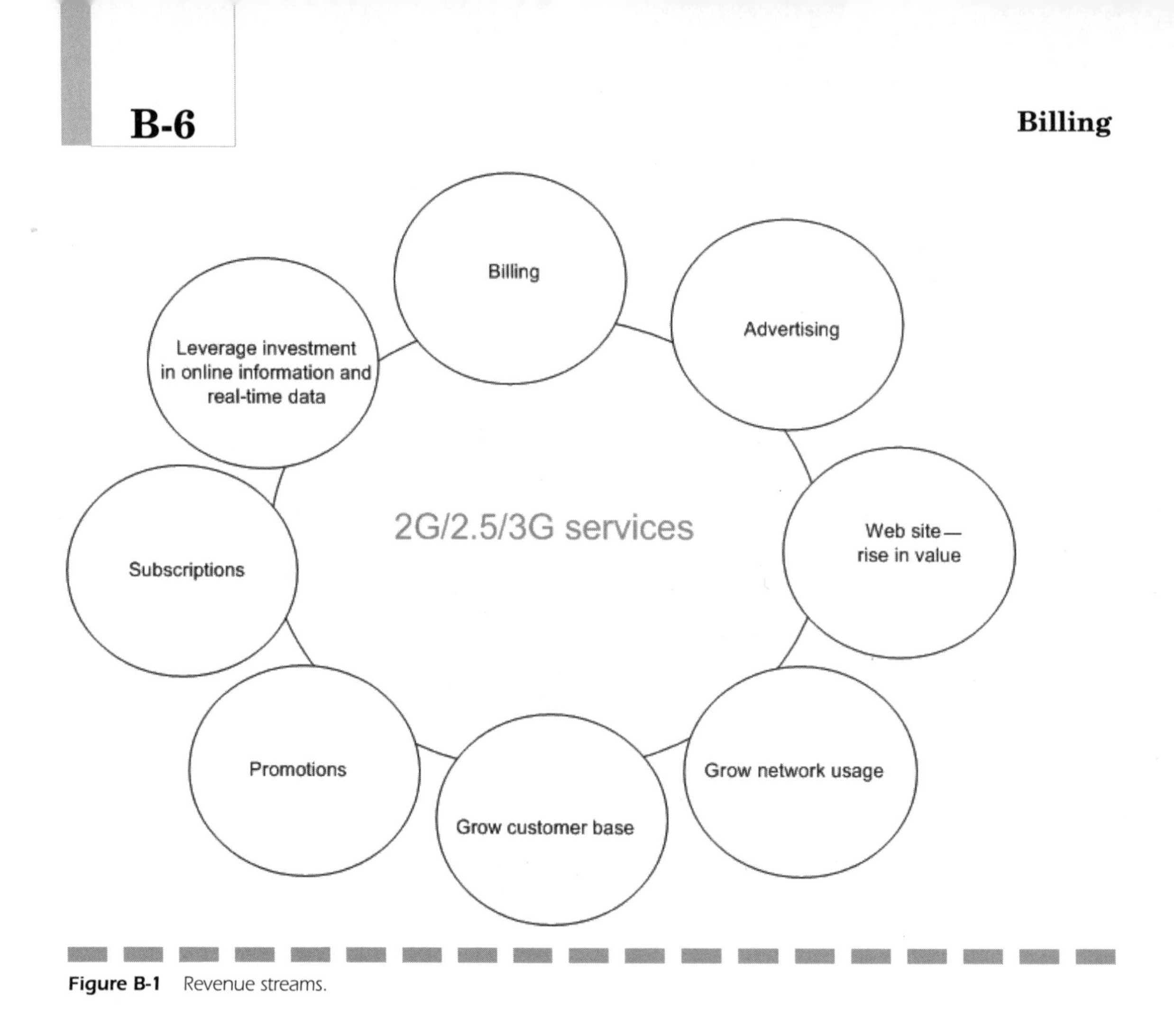

Figure B-1 Revenue streams.

- Time, including day and date so as to provide lower tariffs at off-peak periods
- Destination address of the network or proxy server holding the service that the subscriber has been accessing
- Quality of service assigned to the subscriber, giving predefined priority and bandwidth rights
- SMS usage where the SGSN produces specific CDRs.
- Served IMSI/subscriber classes that have different tariffs for business users, private users, and frequent users
- Reverse charging where the sender is billed
- "Free of charge" where the data or received service is not billed to the user
- Flat rate where a fixed fee (perhaps monthly) is billed to the user, and is appropriate for mass market telecommunications

- Bearer service used where the operator is offering multiple network types that may include GSM900, GSM1800, and perhaps wireless LAN for areas where the operation of other mobile networks is considerably less viable.

Current GPRS networks may not be equipped to bill users based on applications used, but the network may be upgraded to do so through the addition of a payment server on the service network.
(*See* Applications under letter A.)

Billing—Prepaid

Unlike postpaid billing, prepaid billing requires users to buy airtime for a SIM card, and does not require that the user have a subscription with an operator. Comparative advantages of prepaid services include:

- Credit checks are redundant because payment is made in advance.
- The phone and SIM can be purchased together and activated immediately.
- The user is unknown to the operator.
- The user has no contract with the operator, and may switch to another compatible operator merely by purchasing an appropriate SIM.
- There is no monthly subscription charge.
- If the phone is unused, the user is not charged.
- Prepaid lends itself to international travel by changing SIMs so as to use overseas (that are perhaps European) mobile networks.
- Prepaid is ideal for younger users.
- The operator requires no billing system.
- The user may purchase prepaid SIMs to receive calls only.

Disadvantages of prepaid billing include:

- Operators may not allow international roaming because they do not have real-time credit checks on prepaid-card credit status.
- New airtime vouchers must be physically purchased by the user.
- Anonymity of the customer (from the operator's perspective) creates security problems as it is difficult to trace fraudulent usage.
- Airtime vouchers can be counterfeited, stolen, and duplicated.

Prepaid mobile phones are popular throughout the world; 80 percent of mobile phone users in Italy and Portugal are on prepaid billing. Finland, however, with its high level of mobile phone usage has an unusually small percentage of its mobile users on prepaid billing schemes, because handset subsidies in Finland are illegal, and buyers must pay the market price. As a result, network usage is relatively inexpensive because the operator does not subsidize the cost of handsets, and it follows, therefore, that prepaid billing would make little or no difference to usage costs.

Billing—Service Providers

Service providers are essentially third-party companies, occasionally WASPs, that host a given mobile service or wireless application service that is secure and offers an agreed QoS and an agreed mobile device standard such as WAP, GPPRS, EDGE, or 3G. The service provider can, therefore, charge subscribers for use of its services on a monthly basis, or on a packet data or volume basis, or on a time charged basis. The service provider may provide various types of free and conventional marketed and sold services including:

- Lifestyle
- Traffic
- Travel
- Finance
- Weather
- City
- News items—news flashes and updates
- Text to speech
- Speech to text

The physical installation of a service provider may include:

- LAN
- Firewall
- Router
- Data link such as ISDN
- Real-time data feed from the information provider—if this is not generated on-site
- Non-real-time data feed—that may be free of charge
- Storage—usually RAID
- Secure backup servers
- Application server
- Software server to interface with the mobile network
- Core software solution that may be a software client/server architecture designed to process information that may be traditional SMS messages
- WAP gateway

In countries where service providers represent a large market share, they typically partner with established telcos or operators where they collaborate on marketing and selling mobile application services. And, to a lesser extent, there is collaborative research and development involving both the operator and the service provider.

In order to exit, the service provider has a number of options for billing or for mining revenues. These include:

- The service provider may charge a flat rate directly to the operator— or charge a flat payment (usually settled in agreed phases) for a service that is then available to the user of the operator's network. In this instance the service provider has no real-time billing system applied to the service. And the operator may not itself bill for the use of the service, but merely use the service as a value add, as a promotion used to grow its user base. The service may be a fully fledged wireless application used by the operator's entire user base, or it may be a pilot scheme where the service is being tested for technical and commercial viability.
- The service provider may have its own billing system, or charging gateway, that monitors usage and generates billing information, or it may simply sell flat-rate subscriptions, perhaps over the Internet, that eliminates the need for time or packet volume billing systems.

Bluetooth

Bluetooth is a short-range wireless technology that operates in the 2.45GHz radio band and is the namesake of the tenth-century Danish King Harald Bluetooth. It provides a data transmission rate of 722kbps and three channels that are used for voice. Bluetooth uses frequency hopping that changes the transmission or reception frequency after every packet is transmitted or received.

Forward error correction (FEC) is used to reduce the retransmission of packets that become corrupt usually because of the interference in noisy environments. Automatic request repeat is also included where CRC detects erroneous data packets that are then retransmitted as a result.

Devices and mobile phones use Bluetooth as a background task, and each connected Bluetooth device has a 12byte address, and device authentication and communications encryption take place in order to prevent users from infringing on other services. (See www.bluetooth.com.)

Bluetooth permits devices to communicate even when the line of sight is impeded by nonmetallic objects and provides omnidirectional communication, as opposed to directional communication offered by IR technologies like IrDA. Day-to-day applications of Bluetooth include a laptop and mobile phone

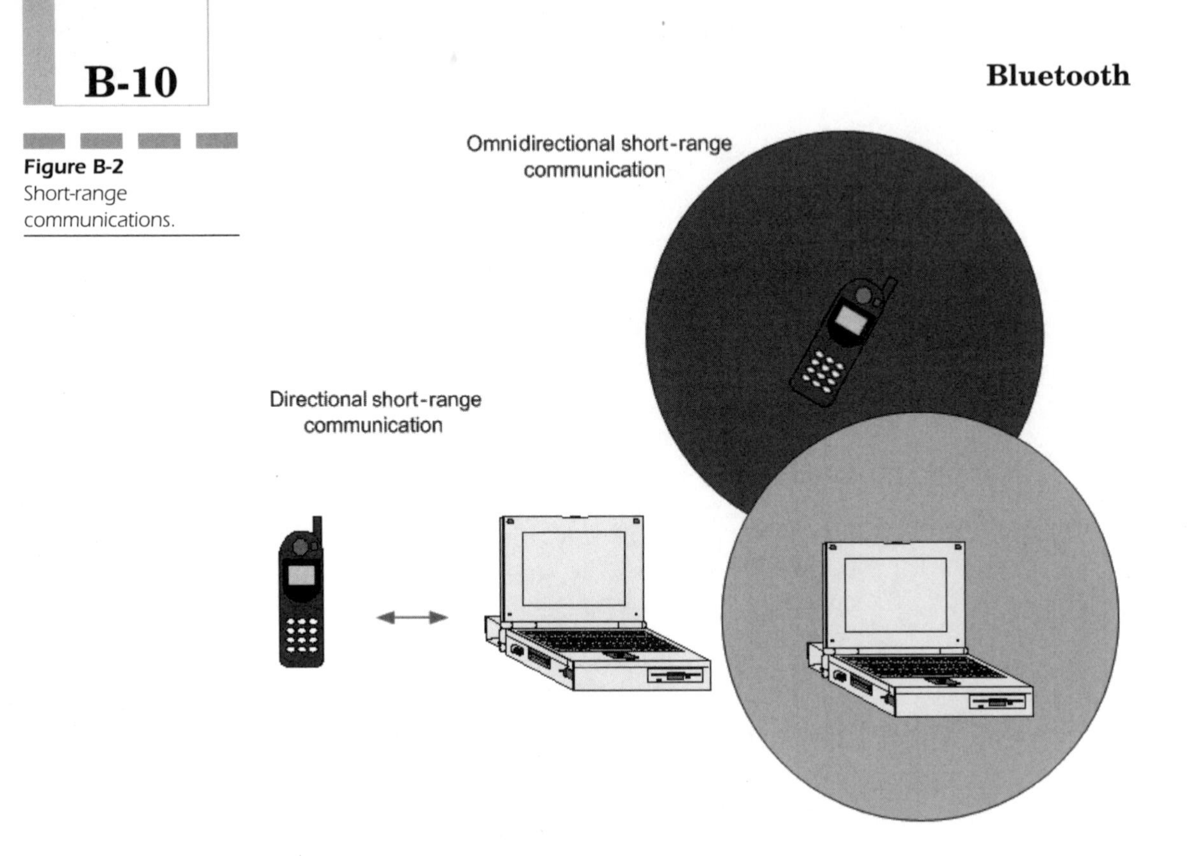

Figure B-2
Short-range communications.

connection to access a mail server, the Internet, or an intranet via a VPN, and a headset that can be freely interchanged between mobile phones, PDAs, and laptops. See Fig. B-2.

Bluetooth may also be used to set up LANs or piconets (or even scatternets), and a Bluetooth device can also be used as an Internet bridge that can be used to connect multiple users. The range is approximately 10m to 100m, though higher power is required to obtain the latter distance. The maximum data rate obtainable is 1Mbps, but in reality this drops to 722kbps for asynchronous transfer, and to 433kbps for symmetric transfer. Two or more Bluetooth devices communicating creates a piconet that may have up to eight units, and one is a master. The master controls the transmission and hopping scheme.

Origins

Bluetooth developed as a standard short-range wireless technology that would replace cabling and connectors for various appliances; it is similar to the IrDA interface that was less successful, however. In 1994 Ericsson began the development of a low-power radio technology for cable replacement and endeavored to satisfy the following requirements:

- Internationally agreed—in order to be accepted globally
- Low cost—so that the mass market could be targeted

- Low power—so that the power drain from the radio chip was close to zero
- Small footprint—so it could be integrated in small appliances
- Speech and data transmission—for simultaneous transfer

By 1998 Ericsson, Intel, IBM, Toshiba, and Nokia formed the Bluetooth Special Interest Group (SIG) that represented the Americas, Europe, and Asia. In 1999 SIG grew significantly and eventually members included 3Com, IBM, Intel, Lucent, Microsoft, Motorola, Nokia, Toshiba, and over 2000 associate member companies.

Bluetooth Air Interface

Bluetooth uses the Industrial Scientific Medical (ISM) frequency band (2402Hz to 2480Hz), and operates at 2.4GHz, using 79 channels that are 1MHz wide. The 2.4GHz band is unlicensed so other devices may operate in the same vicinity. Spread spectrum technologies are used to remedy interference between radio technologies, and frequency hopping means that a Bluetooth device changes frequency in pseudo-random at 1600 changes per second. If interference does cause errors, the packet-based solution allows erroneous packets to be detected and corrected by retransmitting them. See Fig. B-3.

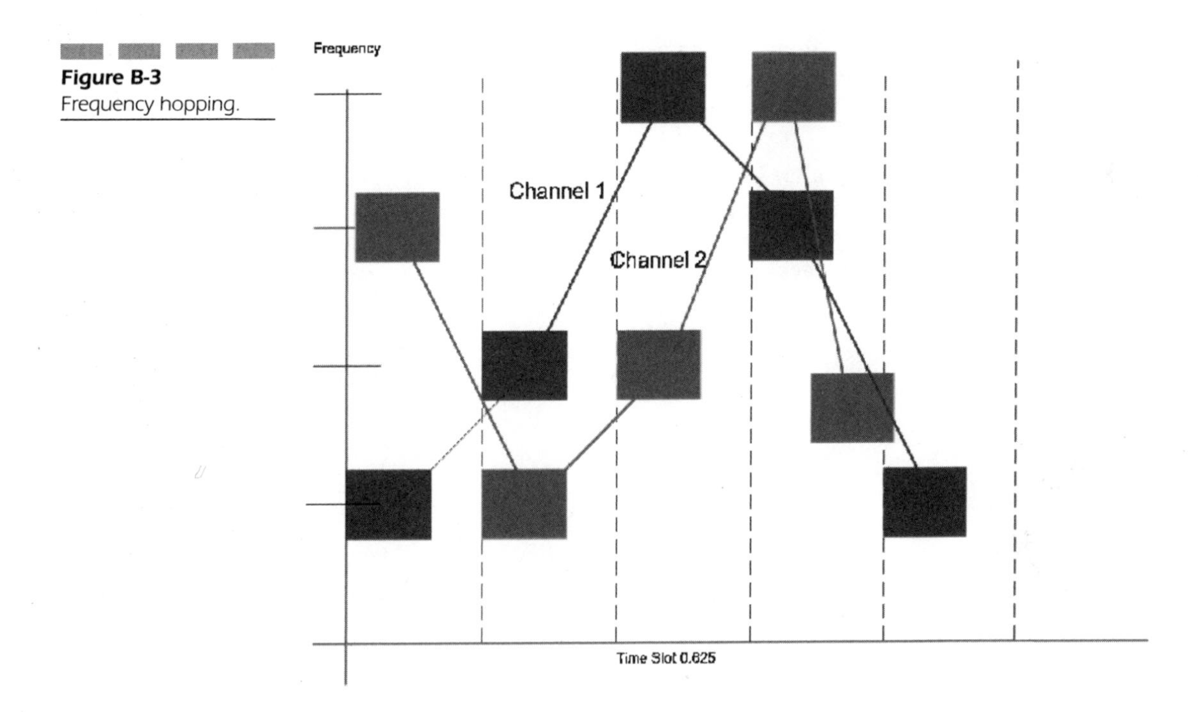

Figure B-3
Frequency hopping.

A Bluetooth channel has a master and one or more slaves; the master usually initiates the connection and selects a hopping scheme that is consistent with its internal clock. The slaves calculate the difference between the master and slave clock, and then a hopping frequency is determined. As a result, a master and slaves will hop in unison to the same frequencies. The uplink and downlink channels are separated using TDM (Time Division Multiplexing), and both channels use the same frequency hopping scheme.

Link Types

The delivery of different traffic like voice and bursty traffic is accommodated by two link types:

- *SCO (Synchronous Connection-Oriented)* is applicable using CS services with short latencies and high levels of QoS. This offers symmetric channels that are synchronous, and is used mainly for voice, though data is possible over the 64kbps channel.
- *ACL (Asynchronous Connectionless)* is used for data transfer and other asynchronous services, and offers PS channels. Transmission slots are not reserved but are granted using a polling access scheme.

A connection may have several links of ACL and/or SCO, and when using a piconet a master may have SCO and ACL links with several slaves, but there is a three-voice call limit.

Using Bluetooth analog voice may be digitally encoded using PCM (Pulse Code Modulation) or CSVD (Continuously Variable Slope Delta). PCM uses eight-bit samples at 8kHz, and so requires 64kbps in order to transmit the digitally encoded signal. CSVD uses 1-bit samples that indicate whether the slope of the voice signal is increasing or decreasing. Bluetooth voice calls are ISDN quality using 64kbps, and this is better than many 2G handsets that offer 8kbps.

Bluetooth packets can be sent one slot at a time, or via multislot where a large packet is sent over several slots and the frequency remains the

TABLE B-1

Bluetooth Bit Rates

Type	Symmetric, kbps	Asymmetric Downlink, kbps	Asymmetric Uplink, kbps
DM1	108.8	108.8	108.8
DH1	172.8	172.8	172.8
DM3	258.1	387.2	54.4
DH3	390.4	585.6	86.4
DM5	286.7	477.8	36.3
DH5	433.9	721	57.6

same. Multislot bit rates may reach 721kbps but only without protective coding.

See Table B-1.

Bluetooth Core Protocols

Bluetooth is architected for interchangeability among many consumer, household, and office appliances that have radically different protocol stacks. This is achieved using a chameleon-like set of core protocols that are described in the following sections. See Fig. B-4.

Baseband The baseband and link control layer holds the RF link between Bluetooth devices, and therefore it also has the air interface with its frequency hopping and synchronization. This layer is able to multiplex different link types, including SCO and ACL, that were described above in the section Link Types.

See Table B-2.

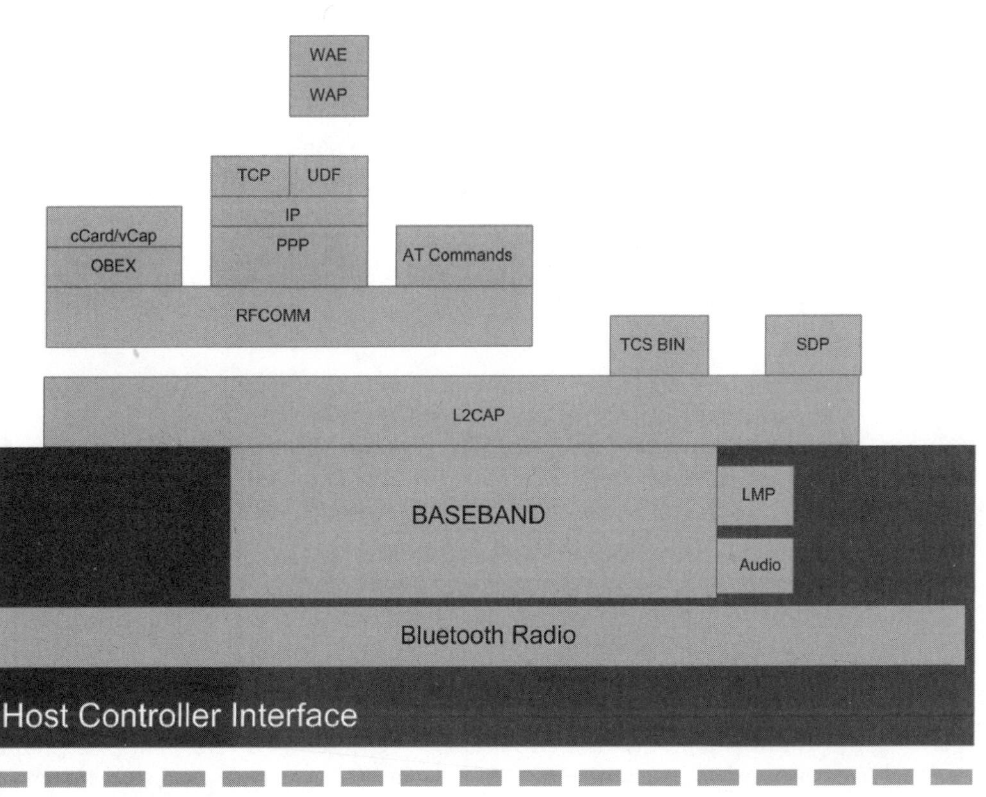

Figure B-4 Bluetooth core protocols.

TABLE B-2

Bluetooth Protocol Usage

Bluetooth Protocol Layer	Protocols
Adopted protocols	IP (Internet Protocol), PPP (Point-to-Point Protocol), TCP/UDP, OBEX (Object Exchange Protocol), WAP (Wireless Application Protocol), WAE (Wireless Application Environment), vCard, vCal, and IrMC (Ir Mobile Communications)
Bluetooth core protocols	Baseband, LMP (Link Manager Protocol), L2CAP (Logical Link and Control Adaptation Protocol), SDP
Cable replacement protocols	RFCOMM (Serial Cable Emulation Protocol)
Telephone control protocols	TCS-Bin (Telephone Control Specification—Binary), AT Commands

Link Manager Protocol The Link Manager Protocol (LMP) establishes links between Bluetooth units, negotiates packet sizes, and controls these negotiations during transmission. It also coordinates security issues such as keys. The baseband is also a conduit for audio over Bluetooth, bypassing the upper layers of the Bluetooth protocol. The baseband, or the lower layers, are accessed using the HCI (Host Controller Interface) that permits access to physical hardware capabilities. The HCI may be perceived as existing between the baseband and the L2CAP.

Logical Link Control and Adaptation Layer (L2CAP) L2CAP connects the HCI with the upper layers and provides connectionless and connection-oriented services, allowing for easy access to lower-layer operations and features that include:

- QoS management
- Multiplexing
- Segmentation and reassembly
- Abstraction of piconet groups

Service Discovery Protocol (SDP) The SDP is part of LMP and permits the discovery of services supported by a device. This first stage in the discovery process primes the connection between devices.

Cable Replacement Protocol (RFCOMM) The RFCOMM protocol is a serial line emulation protocol because it replaces serial cables.

Telephone Control Protocols Telephone control protocols take the form of the Telephone Control Specification—Binary (TCS-Bin) and AT Commands. TCS-Bin is a bit-oriented protocol that represents the call control signaling to establish data and voice calls between Bluetooth devices, and to control mobility management for groups of Bluetooth TCS devices. The AT Command set is applied to control the behavior of modems and mobile phones.

Bluetooth Security

Three modes of security are supported for devices:

- Mode 1 is a nonsecure mode of operation and devices do not implement security procedures.
- Mode 2, or Service-Level Enforced Security, allows services to have access policies, but does not require devices to implement security procedures before establishing channels.
- Mode 3, or Link Level Enforced Security, ensures that devices initiate security procedures prior to setting up links.

Bluetooth has methodologies for authenticating (the device and not the user) and for encrypting data, and allows non-Bluetooth protocols transported using Bluetooth to have different authentication and encryption. The Bluetooth authentication takes the form of a challenge-response system with connections requiring one-way, two-way, or no authentication. Authentication is based on a stored link key, and pairing is also permitted using PINs. A trust relationship between devices is adopted, and a trusted device is previously authenticated, while an untrusted device is not.

Network Connection

A network connection requires that the device is energized and in the standby mode of operation. The standby mode means that the device listens for appropriate messages every 1.28s, while addressing 32 hop frequencies. If communicating with another Bluetooth device, the methods followed depend on whether or not the device has been communicated with previously. In this instance, the device's address is known; otherwise an INQUIRY message will be transmitted. If it has communicated with the device previously, it sends a PAGE message.

Piconet The simplest network, or piconet, is where point-to-point communications occur between two Bluetooth devices, where one is master and the other slave. When there is more than one slave, point-to-multipoint communications are required with a shared channel. Each slave has a 3bit address, and parked slaves have 8bit addresses, and there can therefore be a maximum of 256 parked slaves per piconet. Each parked device can be activated and resume communications with the piconet that has 32 hop frequencies. These can be learned by any other Bluetooth device.

When an INQUIRY or PAGE message is sent, the device that initiates connection becomes the master unit, and the targeted device becomes the slave. APAGE message is sent by transmitting a sequence of 16 identical messages on 16 hop frequencies. If no response is received, the master retransmits on the remaining 16 hop frequencies. When using the INQUIRY message, a training period is required to gain a response from the slave. This period may range from 2.56 to 5.12s.

Error Correction Error correction is required for noisy environments that may extend to the extremities of industrial sites. The Bluetooth error correction schemes include one-third and two-thirds rate forward error correction (FEC) and automatic retransmission request (ARQ). ARQ is used while transmitting data, and a CRC field is created and appended to packets. The receiver uses the same computation to create a CRC that is compared to the received packet's CRC.

Scatternet A piconet may hold one master and up to seven active slaves; up to 256 parked slaves may be included. Two piconets may exist in the same wireless coverage area to create a scatternet. The piconets use their frequency hopping channel and are not synchronized. Devices may communicate with devices in another piconet when participating in a scatternet.

Broker

The broker concept is a useful mobile application service that may be implemented using numerous technologies including Jini that is described in the following sections.

JavaSpaces may form the basis of a bidding or auction system where a mobile device could be used to specify the amount of a particular fuel a gas station owner wanted to buy.

First of all, the user specifies the amount and type of fuel required, and using a mobile device this information is sent to the public JavaSpaces service.

JavaSpaces and Writing Entries

The following code segment uses the `write` method to copy the fuel requirement to a space:

```
public void writeFuel(CarFuel fuel) {
  Try {
      space.write(fuel, null, Lease.FOREVER);
```

```
        } catch (Exception e) {
            e.printStackTrace();
        }
    }
```

The write method as defined in the JavaSpaces interface is of the form:

```
Lease write (Entry e, Transaction txn, long lease)
  throws RemoteException, TransactionException;
```

An entry is manipulated using the write method, which may use the arguments *transaction* and *lease time*. If given a null value the transaction is a singular operation and is detached from other transactions. The *lease time* argument simply specifies the entry's longevity in the new space. When the lease expires, the space removes it, but if a Lease.FOREVER value is used the entry exists in the space indefinitely, or until it is removed by a transaction process or operation. The lease time may also be expressed in milliseconds like the three minute lease shown below:

```
  Long time = 1000 * 60 * 3 ; // three minute lease
  Lease lease = space.write(entry, null, time);
```

Leases may also be renewed using the *renew* method:

```
  void renew (long time)
  , ,
```

The time value is added to the lease time remaining. The renew method may raise the exceptions: LeaseDeniedException, UnknownLeaseException, and RemoteException. The former is raised should the space be unable to renew the lease, which may be caused by lack of storage resources. The UnknownLeaseException is thrown if the entry or object is unknown to the space, perhaps because its lease has already expired.

In real-world scenarios the use of Lease.FOREVER is considered an uneconomical use of resources. An intelligent solution to leasing is to use Lease.ANY, which leaves it up to the space to decide when expiry occurs. But when building JavaSpaces services, the use of Lease.FOREVER may compress the development life cycle, because it can reduce the probability of thrown exceptions.

The write method also throws two types of exceptions, including RemoteException and TransactionException. The former is raised when a communication breakdown occurs between the source process and the remote space, or when an exception occurs in the remote space while an entry is written to it. The RemoteException returns the detail field that holds the exception type. If the space is unable to grant a lease, a RemoteException is thrown. A `TransactionException` is thrown when the transaction is invalid or cannot be rolled forward or committed.

The broker then simply retrieves this using a server, and sends the information to a JavaSpace service for each supplier.

JavaSpaces and Getting Entries

The JavaSpace interface defines methods to *read* or copy entries from a space, and to *take* or copy entries from a space while removing them from spaces also. Locating entries in spaces is carried out by using associative lookup, and a template is used to match entry contents. For example, you may use a template like:

```
CarFuel anyFuelTemplate = new CarFuel();
AnyFuelTemplate.name = null;
AnyFuelTemplate.cost = null;
```

The *null* fields operate like wildcards and any fuel entry will be matched regardless of its name or cost. *Null* may also be used for the template so all entries in a space are matched. The match can be narrowed merely by adding values and name strings like:

```
AnyFuelTemplate.name = "Shell";
```

The JavaSpaces interface has two *read* methods:

```
Entry read (Entry tmpl, Transaction txn, long timeout)
```

And,

```
Entry readIfExists (Entry tmpl, Transaction txn,
  long timeout)
```

Both require three parameters: a *template*, *transaction*, and a *timeout* value that may be expressed in milliseconds.

When a matching entry is found, a copy of it is returned, or it is merely read. Where multiple matches are found, the space returns a single arbitrary entry. This clearly must be borne in mind when developing JavaSpaces services. The Read method waits for the *timeout* if no matching entries are found in a space, or until a matching entry is found. If no matching entry is found, a `null` value is returned by the Read method. The Long.MAX_VALUE may be used as a *timeout* value, causing the *read* operation to block until a matching entry is found. Alternatively, using the JavaSpaces.NO_WAIT value for the *timeout* causes the Read operation to return immediately.

The `readIfExists` method is nonblocking, and returns immediately when no matching entry is found, irrespective of the *timeout* parameter value. The timeout parameter is relevant only when the *read* operation takes place under a transaction. So with a `null` *transaction* the timeout value is equivalent to NO_WAIT.

Both *read* operations may throw the:

`InterruptedException` when the thread implementing the *read* operation is interrupted.

`RemoteException` when a failure occurs on the network or in the remote space.

`TransactionException` when the supplied transaction is invalid.

`UnusableEntryException` when an entry retrieved cannot be deserialized.

The `take` method defined in the JavaSpaces interface is of the same form as the *read* operation:

Entry take(Entry tmpl, Transaction txn, long timeout)

The *take* method removes entries from spaces, provided no exception is thrown. The aforementioned exceptions for *read* and `readIfExists` may also be thrown by the *take* method. These include `InterruptedException`, `RemoteException`, `TransactionException`, and `UnusableEntryException`.

A server at each supplier removes the information from its JavaSpaces service, and a quote is prepared for the fuel, which is written into a space specified by the buyer.

After a preset period, the user takes all bids from the space and naturally buys the cheapest fuel. See Fig. B-5.

The JavaSpaces system could just as easily be applied to customers, where they could be informed verbally or textually about the best garage to buy their fuel. An advanced car computer could even inform the driver of the quality of fuel, based on recorded information or data issued by an organization or petroleum company. This scenario would play out something like this:

1. The motorist, or a customer, is driving toward a garage, and with geographical information of the destination and fuel consumption patterns of the vehicle, the car computer indicates that fuel is required. It specifies an appropriate type and amount of fuel, and sends this information wirelessly to a public JavaSpaces service.
2. A broker then automatically retrieves this information using a server, and sends it to a JavaSpaces service for garages in the vicinity of the customer.
3. A server at the approaching garage or at a centralized facility removes the information from its JavaSpaces service, and writes the fuel cost into another space.

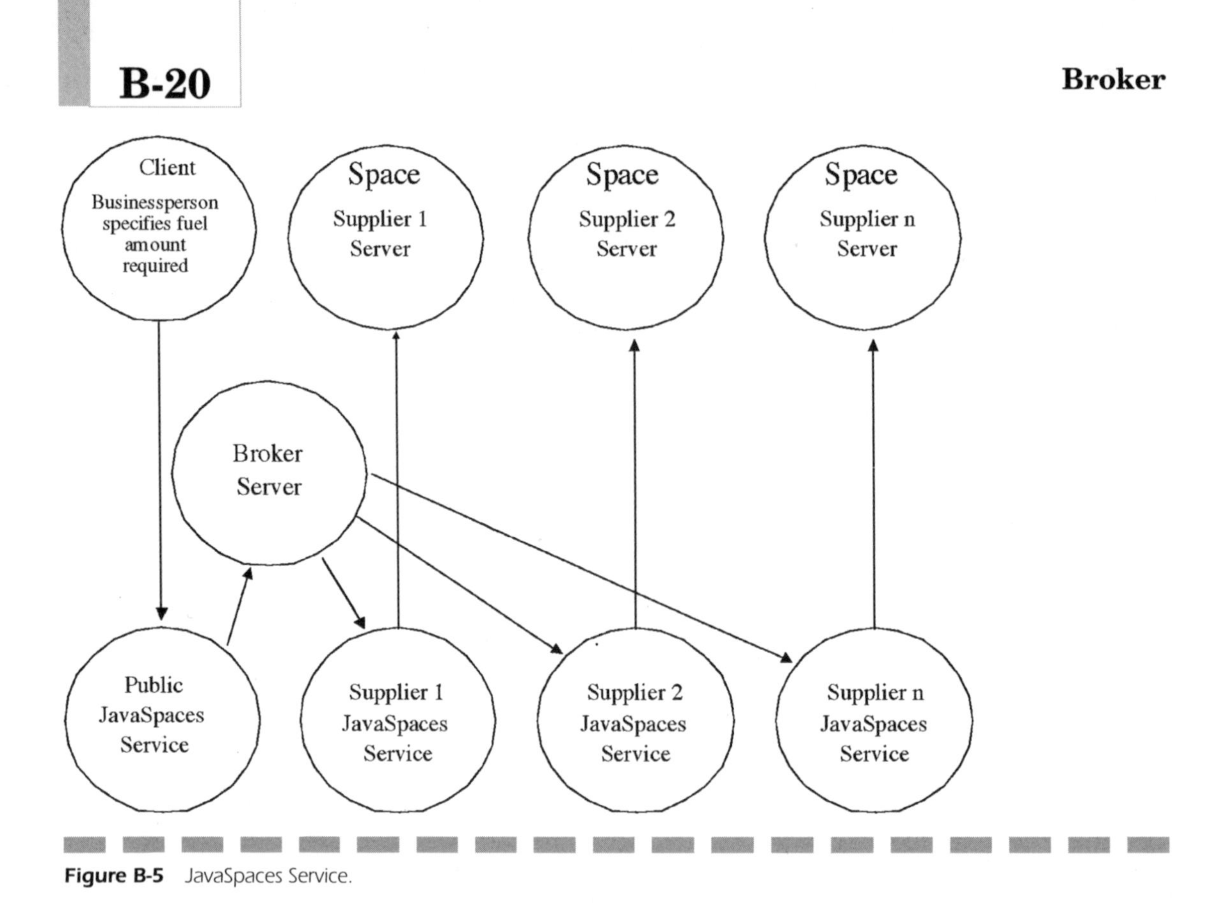

Figure B-5 JavaSpaces Service.

4. The car computer then removes the fuel cost and stores it in a space. It does the same perhaps for the next few garages the customer is about to pass, and then advises the motorist as to where to buy fuel.

The service station:

1. The car computer automatically sets up a wireless conversation with the network when the driver pulls up to the pump.
2. The car then finds the *lookup services* that provide a repository of registered Jini services on the network, which includes the gas pumps. It does this by sending a "looking for lookup services" message to the network, and all the lookup services respond with proxies of themselves which may be searched by the client. The lookup service may then return matching proxy objects, and, if necessary, download their code to the car.
3. The driver at some point would be required to type a PIN into a device in the car, rather than having to type one in at the pump, or alternatively a credit card or bankcard could be swiped from within the car. The car then indicates to the fuel pump the type and quantity of

fuel required, and the driver will then no doubt fill the gas tank manually.

The transaction scenario above naturally adheres to the ACID properties, and the two-phase commit protocol, and the deal is done without the need for the driver to physically attend the cashier's desk.

This technology could process transactions for:

1. Parking
2. Car servicing and repairs
3. Drive-in services of all kinds
4. Hotels and motels
5. Car ferries
6. Parking tickets
7. Tickets for road traffic offenses
8. Road tolls

Jini Transaction Processing—Two-Phase Commit

The two-phase commit protocol is implemented using the primary types:

1. `TransactionManager` that creates and coordinates transactions.
2. `NestableTransactionManager` that accommodates nested transactions or sub-transactions.
3. `TransactionParticpiant` that allows transactions to be joined by participants.

The two-phase commit protocol coordinates the changes made to system resources, which result from transactions. It tests for their successful implementation, in which case they are committed. If not, and any one fails, they are each rolled back. In transaction processing (TP), this is left to a transaction coordinator whose function is integrated in the `TransactionManager` using Jini.

The `TransactionManager` is key to the two-phase commit protocol. This requires that all `TransactionParticipants` vote in order to indicate their state. The vote may be *prepared* (or ready to *commit*), *not changed* (or read only), or *aborted* (when it is necessary to *abort* the transaction). Having received information of the readiness to commit through a *prepared* vote, the `TransactionManager` signals `TransactionParticipants` to roll forward and commit the changes resulting from the transaction. `TransactionParticipants` that vote *aborted* are signaled to *roll back* by the `TransactionManager`.

Jini Transaction Creation

To create a transaction it is necessary for the client to use a lookup service (or a similar *directory services* metaphor) in order to reference a Transaction-Manager object. A new transaction may be initiated using the create() method and by specifying a leaseFor period in milliseconds. The leaseFor duration is adequate for the transaction to complete, and the TransactionManager may forbid the lease request by throwing the LeaseDeniedException. Expiration of a lease before a TransactionParticipant's vote with a commit or abort leads the TransactionManager to abort the transaction.

Constants provide the currency of the described communications between TransactionParticipants and TransactionManagers and are defined in the TransactionConstants interface:

```
package net.jini.core.transaction.server;

/** Constants common to transaction managers and
    participants. */
public interface TransactionConstants {
  /** Transaction is currently active */
  final int ACTIVE = 1;
  /** Transaction is determining if it can be committed */
  final int VOTING = 2;
  /** Transaction has been prepared but not yet committed */
  final int PREPARED = 3;
  /** Transaction has been prepared with nothing to commit */
  final int NOTCHANGED = 4;
  /** Transaction has been committed */
  final int COMMITTED = 5;
  /** Transaction has been aborted */
  final int ABORTED = 6;
}
```

Jini Nested Transactions

At times a transaction requires the implementation of subtransactions for its successful completion. These nested transactions provide a logical hierarchy where a *parent* transaction is dependent on a *nested* transaction. Nested transactions are implemented using TransactionManagers that use the NestableTransactionManager interface:

```
Package net.jini.core.transaction.server;

import net.jini.core.transaction.*;

import net.jini.core.lease.LeaseDeniedException;

import java.rmi.RemoteException;

/**

 * The interface used for managers of the two-phase
   commit protocol for

 * nestable transactions. All nestable transactions must
   have a

 * transaction manager that runs this protocol.

 *

 * @see NestableServerTransaction

 * @see TransactionParticipant

 */

 public interface NestableTransactionManager extends
   TransactionManager {

  /**

   * Begin a nested transaction, with the specified
     transaction as parent.

   *

   * @param parentMgr the manager of the parent
     transaction

   * @param parentID the id of the parent transaction

   * @param lease the requested lease time for the
     transaction

   */

  TransactionManager.Created
    create(Nestable TransactionManager parentMgr,

      long parentID, long lease)

    throws UnknownTransactionException,
    CannotJoinException,

      LeaseDeniedException, RemoteException;

  /**

   * Promote the listed participants into the specified
     transaction.

   * This method is for use by the manager of a
     subtransaction when the
```

```
* subtransaction commits. At this point, all
  participants of the
* subtransaction must become participants in the
  parent transaction.
* Prior to this point, the subtransaction's
  manager was a participant
* of the parent transaction, but after a successful
  promotion it need
* no longer be one (if it was not itself a
  participant of the
* subtransaction), and so it may specify itself as a
  participant to
* drop from the transaction. Otherwise,
  participants should not be
* dropped out of transactions. For each promoted
  participant, the
* participant's crash count is stored in the
  corresponding element of
* the <code>crashCounts</code> array.
*
* @param id the id of the parent transaction
* @param parts the participants being promoted to
  the parent
* @param crashCounts the crash counts of the
  participants
* @param drop the manager to drop out, if any
*
* @throws CrashCountException the crash count
  of some (at least one)
* participant is different from the crash
  count the manager already
* knows about for that participant
*
* @see TransactionManager#join
*/
 void promote(long id, TransactionParticipant[] parts,
   long[] crashCounts, TransactionParticipant drop)
 throws UnknownTransactionException,
```

```
        CannotJoinException,
          CrashCountException, RemoteException;
  }
```

When a nested transaction is created, the originating manager joins the parent transaction using the join method when the managers are different. The create method may raise the UnknownTransactionException if the parent transaction manger does not recognize the transaction because of an incorrect ID, or has become inactive, or has been discarded by the manager.

Jini Transaction—Join

To join a transaction a participant invokes the join method held by the transaction manager using an object implementation of the TransactionParticipant interface. The TransactionManager uses the object to communicate with the TransactionParticipant. If the join method results in the RemoteException being thrown, the TransactionParticipant relays the same (or another exception) to its client. The join method has a *crash count* parameter that holds the TransactionParticipant's storage version that contains the transaction state. The *crash count* is changed following each loss or corruption of its storage. When a TransactionManager receives a join request, it reads the *crash count* to determine if the TransactionParticipant is already joined. If the crash count is the same as that passed in the original join, the join method is not executed. If it is different, however, the TransactionManager causes the transaction to abort by throwing the CrashCountException.

Jini TransactionServer

Jini is, of course, made of numerous Jini packages that combine to make the collective API that implements the many required interfaces. These include the methods and exceptions that make light work of developing Jini applications and include the shown TransactionManager TransactionManager interface.

```
package net.jini.core.transaction.server;

import net.jini.core.transaction.*;

import net.jini.core.lease.Lease;
```

```
import net.jini.core.lease.LeaseDeniedException;
import java.rmi.Remote;
import java.rmi.RemoteException;
public interface TransactionManager
extends Remote, TransactionConstants {
  /** Class that holds return values from create
      methods. */
  public static class Created implements
      java.io.Serializable {
      static final long serialVersionUID = -
      4233846033773471113L;
      public final long id;
      public final Lease lease;
      public Created(long id, Lease lease) {
        this.id = id; this.lease = lease;
    }
  }

  Created create(long lease) throws
    LeaseDeniedException, RemoteException;
  void join(long id, TransactionParticipant part, long
    crashCount)
    throws UnknownTransactionException,
    CannotJoinException,
      CrashCountException, RemoteException;
  int getState(long id) throws
    UnknownTransactionException, RemoteException;
  void commit(long id)
    throws UnknownTransactionException,
    CannotCommitException,
      RemoteException;
  void commit(long id, long waitFor)
    throws UnknownTransactionException,
    CannotCommitException,
      TimeoutExpiredException, RemoteException;
  void abort(long id)
    throws UnknownTransactionException,
```

```
      CannotAbortException,
        RemoteException;
    void abort(long id, long waitFor)
      throws UnknownTransactionException,
      CannotAbortException,
        TimeoutExpiredException, RemoteException;
}
```

Transaction Processing ACID Properties

Atomicity, *Consistency*, *Isolation*, and *Durability* (ACID) properties define the real-world requirements for transaction processing (TP):

1. *Atomicity* ensures that each transaction is a single workload unit. If any subaction fails, the entire transaction is halted, and rolled back.
2. *Consistency* ensures that the system is left in a stable state. If this is not possible, the system is rolled back to the pretransaction state.
3. *Isolation* ensures that system state changes invoked by one running transaction do not influence another running transaction. The changes must only affect other transactions, when they result from completed transactions.
4. *Durability* guarantees that the system state changes of a transaction are involatile, and impervious to total or partial system failures.

BSS Protocols

The logical BSS has an MSC and a BSC that are united, using a multivendor A-interface that in theory allows the operator to buy off-the-shelf and turnkey solutions. The interface is defined in the 08-series GSM specifications and though it is beginning to be consigned to history, it may unite a BSS with 3G MSC. The GSM network and the packet-based core network are linked by the Gb-interface that operates when GSM mobile stations use GPRS services. The GPRS data rate can be increased by EGPRS and that can raise the data rate speed in the GSM air interface to rates as high as 200kbps. Theoretically, the high-end GPRS variant data rates should provide access to certain 3G services. See Figs. B-6 and B-7.

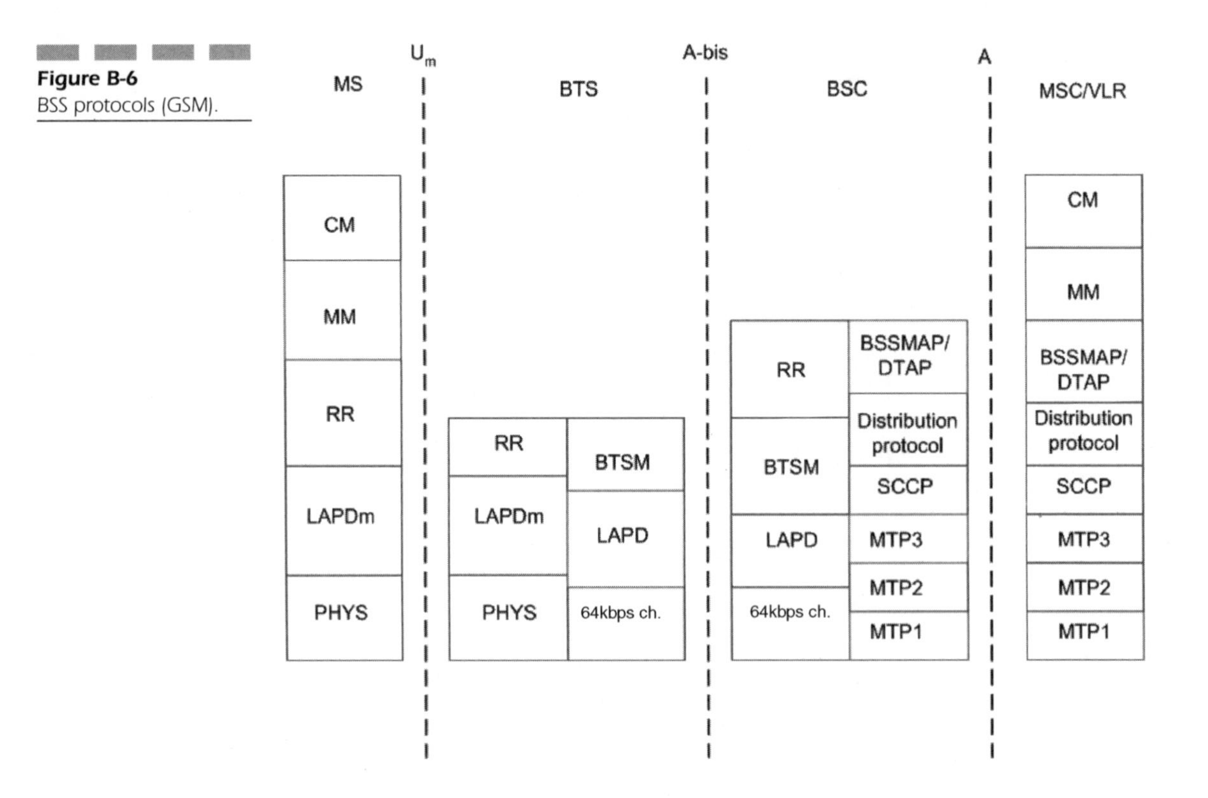

Figure B-6
BSS protocols (GSM).

Computers/Software

Backdrop

A background image or color in a multimedia presentation or title that remains constant for a given number of frames.

Backend

Another name for tier 3.

Background Task

A task that takes place in a multitasking operating environment and is allocated a specific and changing portion of processing time. The amount of time is less than that allocated to tasks that are in the foreground.

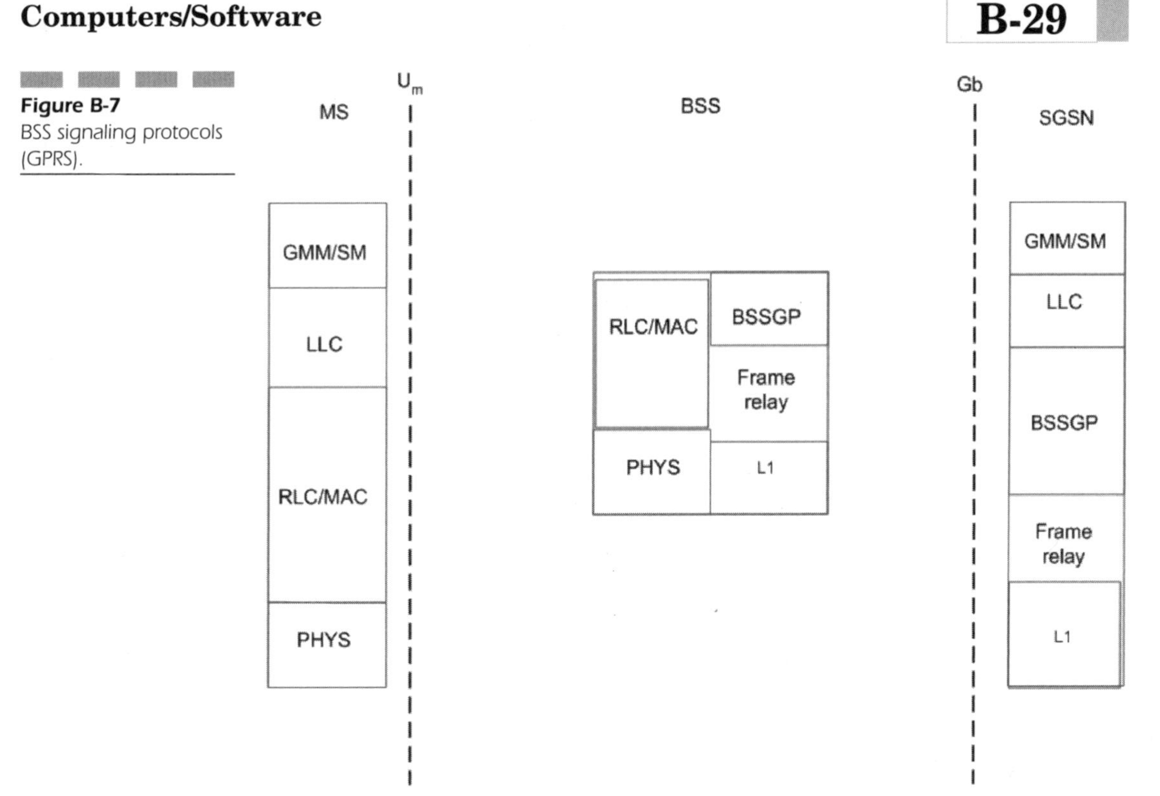

Figure B-7
BSS signaling protocols (GPRS).

Backtracking

A process of retracing a user's path of interaction by, for example, using the Back button in the application or browser window.

Bandwidth

1. A data transfer rate supported by physical or wireless media.
2. A range of frequencies supported by media.

Batch File Compression

A technique by which files may be compressed for distribution, archiving, or backup purposes. It is appropriate for modem-based file transfer and DSM-based distribution. It can be used to compress any binary file. In the context of video distribution, its main disadvantage is the fact that end users have to decompress or unpack the files before they can be played.

Decompression can be carried out using an installation program, a program such as PkUnzip, or in the case of a self-extracting compressed file, the user simply types the name of the compressed file. Standard batch file compression programs include WinZip.

Bend

A bend in an optical fiber results in increased attenuation. Such losses can be used to determine the degree to which a fiber is bent. This forms the basis of operation for many gloves and suits used in VR, where the fibers run along the lengths of fingers or limbs.

BER (Bit Error Rate)

A measurement of how error-free the storage or transmission of data is. Typically expressed as the average number of bits in which one bit-error will occur. CD-ROM has appropriate error detection and correction codes.

In Mode 1 CD-ROM data blocks, 4bytes are reserved for error detection and 276bytes are reserved for error correction.

The three layers of error detection and correction integrated in the CD-ROM format include CIRC, EDC, and ECC. Typically the bit error rate of CD-ROM equates to 10^{18}, which amounts to one error for every 1,000,000,000,000,000,000 bits.

Berners-Lee, Tim

The architect of the World Wide Web, and inventor of the HTTP server at CERN in Geneva. That is why the original server is referred to as the CERN server.

The conceptual birth of the Web might be accredited to the visionary Vannevar Bush through his momentous article, "As We May Think." Theodore Nelson is also significant (*but much more contemporary*) through his work *Literary Machines*, and the project Xanadu. If Vannevar Bush and Ted Nelson were responsible for putting forward the concept of the Web, then Tim Berners-Lee may be considered its architect.

Beta Copy

A test copy of a software product that has yet to be commercially released.

Bi-directional

1. A highly compressed frame in an MPEG-1 data stream
2. A link that offers upstream and downstream data transmission
3. A link in an information structure that can be following in either direction

Big Blue

A nickname for IBM. It originates from the fact that early IBM mainframes were painted blue.

BIN (Bank Identification Number)

A three- or five-digit code that adheres to the ISO 8663 recommendation and dictates account numbers that are owned by card companies.

Binary

A counting system comprising only two states: either "1" or "0." All electronic files are stored in binary form. Binary files generally contain executable programs and program data, and have the .EXE or .COM extension.

Bit

A single, indivisible item of binary data that might be "1" or "0."

Black Box

A conceptual view of software or hardware where the internal architecture and operation are ignored. All that is considered are input and output values.

BLOB (Binary Large Object)

An item of binary data that is of no specific type, but is identified simply as containing some sort of digital data. It may be a graphic, a video file, a midi file, a wave audio file, a program file, or any type of digital data.

Block

1. A segment of code that is enclosed within opening "{" and closing "}" braces
2. A block of 2352bytes on a CD/CD-ROM track

BMP

A graphics file format that usually creates larger files than more compressed variants such as Compuserve's GIF and JPEG.

BNC

A connector consisting of a round socket and plug, which are locked together with a twist. It was invented by the engineers Neill and Concelman at Bell Laboratories.

Bolt-on Application

An application that is coupled with another, for example, a SET implementation that is added to a Merchant Commerce Server.

Bookmark

A marker inserted at a specific point in a document so that it may be revisited easily.

Boolean

A variable type that has one of two states: either yes or no, although an indeterminate state that is neither "yes" nor "no" is possible. Named after Irish mathematician George Boole, who pioneered logic-based *Boolean algebra*, Boolean variable types feature search engines and databases in programming languages.

AND, OR, NOR, NAND, and NOT are Boolean operators. They are also logic gates used in electronics and in the architecture of digital components, and their behavior can be described using a truth table. A truth table is a simple table that shows the output obtained for each and every combination of inputs.

Bottom-up Analysis

A design approach, where the process begins with the design of low-level components, and progresses to the design of higher-level components; top-down analysis represents the opposite.

Broadband

A term used to describe access technologies and networks that typically offer bandwidths of 2Mbits/s and more, though narrow bandwidth networks and access technologies may also be described as broadband. Broadband offers high-speed data transfer and is useful for multimedia networks.

Broadcast Quality

1. A video recording whose quality approximates that of broadcast television
2. A camcorder that can produce broadcast-quality video recordings

Browse (Browsing)

A process of following the intricate paths through a hypertext-based information structure such as a Web site. The user passes to and from nodes or objects. In the context of the Internet, it is known as *surfing*.

Browser

An application that permits the user to browse the World Wide Web. The most popular Web browsers are Netscape Navigator and Microsoft Internet Explorer. Earlier implementations include Cello and Mosaic. A modern browser allows users to:

- Add Web sites and/or pages to an address book or a folder called Favorites
- Navigate backward and forward through visited Web pages
- Open a URL that is entered by using the keyboard
- Play back streaming audio and video using a plug-in or ActiveX control

- Play back streaming multimedia using an appropriate plug-in or ActiveX control
- Send e-mail messages, although a browser is not an e-mail client/application
- Specify various preferences including the appearance of displayed Web pages
- Open HTML files that might be local or remote
- Chat in real time, using an appropriately enabled browser
- View Web pages that contain Java applets (using a Java-enabled browser)
- Download files
- Make telephone calls over the Internet using an appropriate Web phone
- Take part in videoconferences using appropriate hardware and software

Miscellaneous

Bankcard

A card that is linked to one or more bank accounts. It may have an assigned PIN, in which case it may be used with cash dispensers, and it may perform transactions using international standards that include Switch, EFTPOS, and Cirrus.

Bastion Host

A host that is critical to a network's security and firewall architecture. This is the focus of network management, security monitoring, and is a network's principal defense against illegal usage. A dual-homed host may play the role of a bastion host.

Batch Settlement

An accrued number of card transactions that may be processed.

Baud Rate

A rate at which data is transferred from one point to the next, which broadly equates to bps. More precisely, it is a measure of logic and/or bit changes per second over media that may be physical or wireless.

It is the namesake of French telegraphic communications pioneer J. M. E. Baudot (1845–1903). It is a rarely used term nowadays, and is replaced by bits/second (bps), which is not quite the same thing.

(*See* Modem under letter M.)

Bel (B) A ratio between the power transmitted and the power received:

$$B = \log_{10} P_0 / P_i$$

$$P_0 = \text{output power}$$

$$P_i = \text{input power}$$

Logarithms to the base 10 are used because:

1. Human hearing uses a logarithmic scale; so when a sound doubles in volume and/or amplitude, the transmitted power increases by a factor of 10.
2. Changes to a signal gains or losses are additive; for example, a 15-B signal that loses 10-B and is amplified by 30-B results in a signal strength of (15 – 10 + 30) 35.

Bell, Alexander Graham

The inventor of the telephone and the photophone, as well as the founding father of telecommunications. The photophone was described by Bell as one of his most significant inventions, and he and his assistant Sumnar Tainter successfully implemented the photophone that used line of sight, freespace, and lightwave communications.

BIN (Bank Identification Number)

A three- or five-digit code that adheres to the ISO 8663 recommendation and dictates account numbers that are owned by card companies.

B-ISDN (Broadband ISDN)

An access technology that offers a wider bandwidth than conventional (narrow bandwidth) N-ISDN, which offers data transfer rates of 64kbits/s per single connection. Low-end B-ISDN implementations include multiple 64kbits/s channels. For instance, videoconferencing architectures that feature FMFSV might include 6 × 64kbits/s channels, yielding a collective bandwidth of 384kbits/s.

Black List

A list of barred equipment held in an EIR (Equipment Identification Register) of a mobile network. The EIR has a gray list of faulty equipment or devices that are registered for no services, and a white list for registered users.

bps (Bits per Second)

A measurement of data transfer rate. Modems are frequently specified in terms of their data transmission and reception speeds. Baud rate differs from bits per second, because it is a measure of bit changes per second.

(*See* Baud Rate above.)

Brand Certificate Authority

A party authorized by a credit card company like Visa or MasterCard to carry out digital certificate management.

Branded Payment Card

A credit card or charge card that has the brand of a sufficiently large company like American Express, Visa, or MasterCard.

Acronyms

BCCH	Broadcast control channel
BCH	Broadcast channel
BER	Bit error rate
BLER	Block error rate

BMC	Broadcast multicast control
BPSK	Binary phase shift keying
BRAN	Broadcast radio access network
BS	Base station
BSC	Base station controller
BSS	Base station system
BTS	Base transceiver station

C

C-2

cdmaOne

cdmaOne was designed initially to increase the capacity of 1G AMPS networks operating in the 800MHz band in the United States. Using cdmaOne, AMPS operators were able to increase capacity by replacing 30kHz carriers with 1.25MHz cdmaOne carriers, allowing dual-mode cdmaOne/AMPS mobile stations to switch between either carrier, depending on which one was in use in a particular geographic area. The technology was developed by Qualcomm under the name CDMA and was referred to as IS-95; it was renamed cdmaOne in 1997. cdmaOne evolved to become CDMA-PCS operating in the 1900MHz band so that it could be used with the Personal Communication System (PCS) in the United States.

cdmaOne Radio Interface

CDMA or IS-95 operates in the U.S. cellular frequency band that is divided into five blocks and distributed between two operators, allowing for two different cellular systems in the same vicinity. CDMA uses frequency division duplex (FDD), meaning that the forward and reverse link transmissions occur in different frequency bands. The duplex separation is 45MHz and the carrier spacing 1.25MHz. CDMA PCS is designed to operate in the 1.9GHz PCS band with a duplex spacing of 80MHz, and it is divided into three 2 × 15Mhz blocks, including 15MHz for reverse and forward links, and three 2 × 5Mhz blocks. See Table C-1.

TABLE C-1

U.S. Cellular Bands, PCS Spectrum Allocations, PCS Channel Numbers

U.S. Cellular Bands		
	Frequency (MHz)	
System	**Reverse Link**	**Forward Link**
A″	824.040–825.000	869.040–870.000
A	825.030–834.990	870.030–879.990
B	835.020–844.980	880.020–889.980
A′	845.010–846.480	890.010–891.480
B′	846.510–848.970	891.510–893.970
PCS Spectrum Allocations		
	Frequency (MHz)	
Block	**Reverse Link**	**Forward Link**
A	1850–1865	1930–1945
D	1865–1870	1945–1950
B	1870–1885	1950–1965
E	1885–1890	1965–1970
F	1890–1895	1970–1975
C	1895–1910	1975–1990
PCS Channel Numbers		
Band	**Frequency (MHz)**	**Channel Numbers**
Reverse link	1850.000 + 0.050N	$0 \leq N \leq 1200$
Forward link	1930.000 + 0.050N	$0 \leq N \leq 1200$

cdmaOne Forward Link The *forward link* is the collective name given to the base station (BS) transmitter, the MS receiver, and the radio channel, including the pilot channel, synchronization channel, paging channel, and traffic channel.

- The pilot channel is transmitted constantly by CDMA carriers and allows the MS to identify the BS, while also allowing MSs in other cells to determine the suitability of the cell for handover.
- The synchronization channel permits the MS to time synchronize with the BS, and at the same time it carries information such as system timing and BS internal register contents that are used for coding, spreading, and encryption.
- The paging channels alert MSs when there are incoming calls, and also carries network information.
- The traffic channels are assigned to users and operate at 9.6kbps or 14.4kbps for cdma-PCS.

CDMA carriers are assigned 64bit Walsh codes. See Fig. C-1.

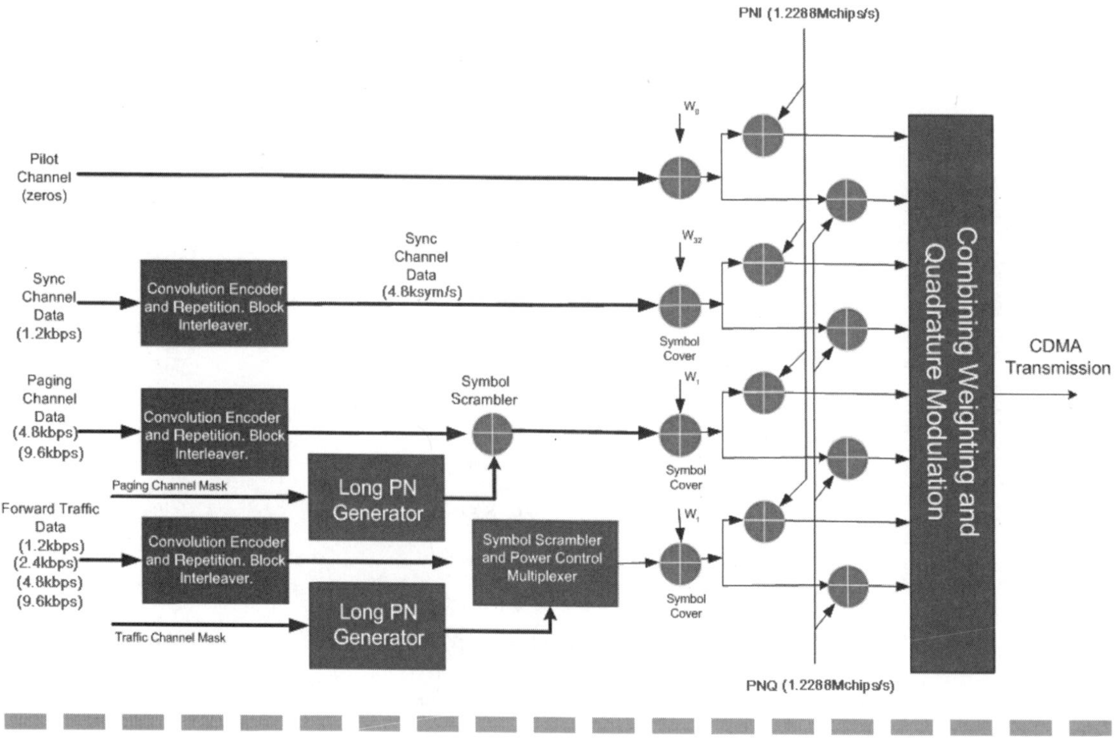

Figure C-1 cdmaOne BS transmitter.

Pilot Channel The pilot channel's simplicity is derived from the fact that it merely carries an all-zero bit stream that is EXORed with Walsh code that has an index of zero—that is, a series of logical 0s. This culminates in an all-zero bit stream that is divided into two parts that are EXORed with one of two PNI pseudo-random noise sequences that are 2^{15}bits long and include:

- PNI in-phase component
- PNQ quadrature component

The sequences are generated by a 15bit feedback register and a rate of 1.2288Mchips/s, so the period equates to:

$$2^{15}/122{,}880 = 32{,}768/122{,}880 = 26.666\text{ms}$$

This results in 75 PN sequence repetitions every 2s.

CDMA Signals

Signal Strength Signal strength is the power received by a copolarized dipole antenna into a 50′Ω load. ERP (Effective Radiated Power) and path loss drives signal strength, where the loss is a culmination of propagation characteristics, distance, and obstructions. The sensitivity of the mobile user is the signal strength required in an open area under normal atmospheric and environmental conditions such as a rural area where less than 10 to 12dB attenuation allowance is adequate. Densely populated metropolitan areas, on the other hand, may require 15 to 20dB.

Cell Configurations The cell configuration depends upon the classification of environment, and in rural areas 300ft towers with high-gain antennas may be appropriate. In suburban areas 100 to 200ft towers may be used with low-gain antennas. Dense metropolitan districts include tall buildings with 10 or more stories with a population density of 20,000 subscribers per square mile. Urban areas include residential and office areas with buildings about 5 to 10 stories high, and with a population density of 7500 to 20,000 subscribers per square mile. Suburban areas have housing with one to two stories that are 50m apart, shops and offices that are two to five stories, and population densities between 500 and 7500 subscribers per square mile. Rural areas include farming communities with a population density as low as 500 subscribers per square mile.

Cell Overhead Channels CDMA cells transmit pilot, synchronization, and paging signals (*see* cdmaOne Radio Interface earlier in this chapter.) Power allocations for the forward link as a percentage of total transmitted power are:

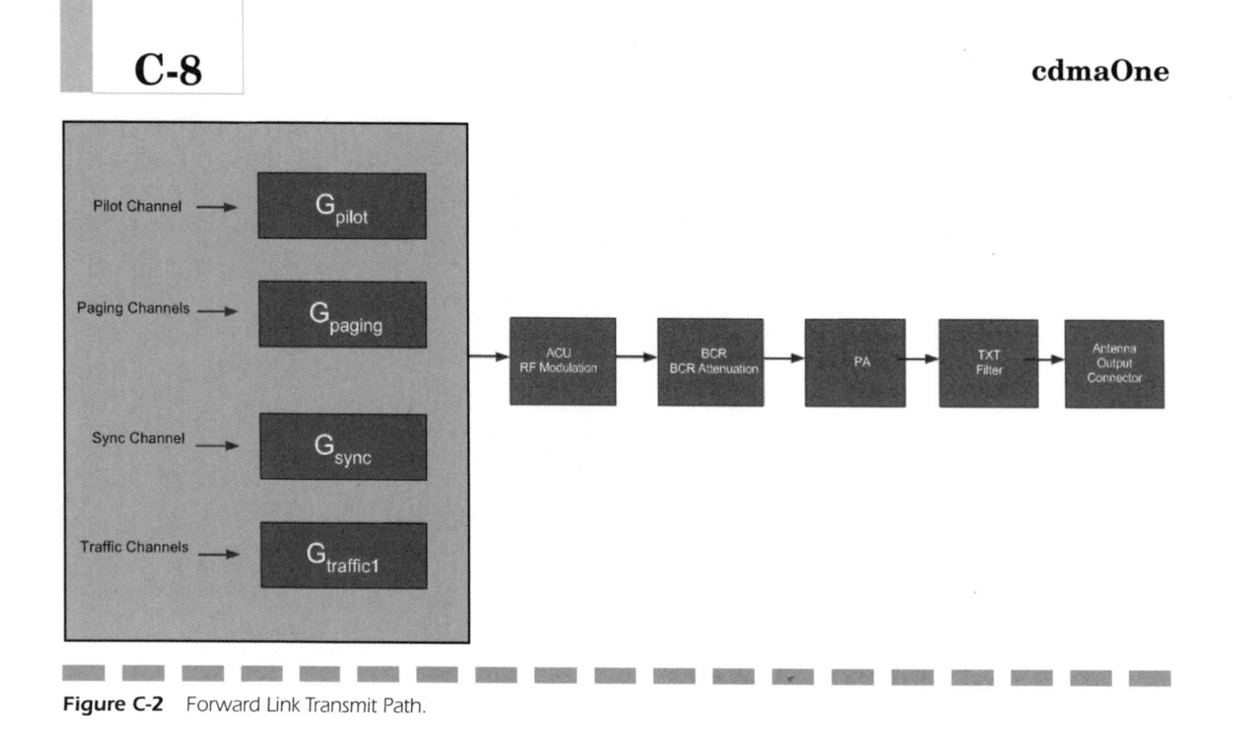

Figure C-2 Forward Link Transmit Path.

- Paging channels: 6–7%
- Pilot channel: 15–20%
- Synchronization channel: 2%
- Traffic channels: 71–77%

Power Allocation Base station output power calibration ensures that when the BCR and digital gains (dg) are established, the antenna receives the required output power. When the transmit path is calibrated, the power is adjusted electronically, so when the BCR attenuation falls, more power is given to the traffic channel. And when attenuation increases, traffic channels are allocated less power. The transmit amplifier is not overdriven because maximum output power controls forward link overload control. See Fig. C-2.

Power allocations for the pilot, paging, sync, and traffic channels may be calculated using the formula:

$$\text{Power} = \text{scale} \times 10^{-(\text{bcr})/10} \times \left[v_{pl} G_{pilot}^2 + v_{pg} G_{page}^2 + v_{sy} G_{sync}^2 + v_{tr} G_{traffic,m}^2 \right]$$

G_{pilot} and v_{pl}	Pilot channel gain and channel activity
G_{sync} and v_{sy}	Sync channel gain and channel activity
G_{page} and v_{pg}	Paging channel gain and channel activity
$G_{traffic,m}$ and v_{tr}	Traffic channel gain and channel activity

Cdma2000

Cdma2000 uses a wideband, spread-spectrum CDMA radio interface that accommodates 3G requirements and those specified by the ITU and by the IMT-2000 (International Mobile Telephony), and it is also compatible with cdmaOne standards. Cdma2000 may be deployed in indoor/outdoor environments, wireless local loops (WLLs), vehicles, and hybrid vehicle/indoor/outdoor scenarios. Outdoor megacells are less than 35km in radius, and outdoor macrocells may be 1km to 35km in radius. Indoor/outdoor microcells can be up to 1km in radius, and indoor/outdoor picocells are less than 50m in radius.

Cdma2000 may be summarized as supporting:

- Data rates from a TIA/EIA-95B compatible rate of 9.6kbps to greater than 2Mbps
- Circuit-switched and packet-switched
- Channel sizes of 1, 3, 6, 9, and 12 × 1.25MHz
- Modern antennas
- Large cell sizes to reduce cell numbers per network
- Higher data rates for all channels
- High-speed circuit data, B-ISDN, or H.224/223 teleservices
- Mobility for speeds up to 300mph

Cdma2000 reuses the following standards:

- IS-127 enhanced variable rate codec
- IS-634A
- IS-637 short message service (SMS)
- IS-638 over-the-air activation and parameter administration
- IS-707 data services
- IS-733 13kbps speech coder
- IS-97 and IS-98
- TIA/EIA-41D
- TIA/EIA-95B channel structure
- TIA/EIA-95B extensions
- TIA/EIA-95B mobile station and radio interface specifications

Cdma2000 Upper Layers

The upper layers shown in Fig. C-3 address voice, signaling, and end-user data-bearing services. Voice services include voiced telephony, IP telephony,

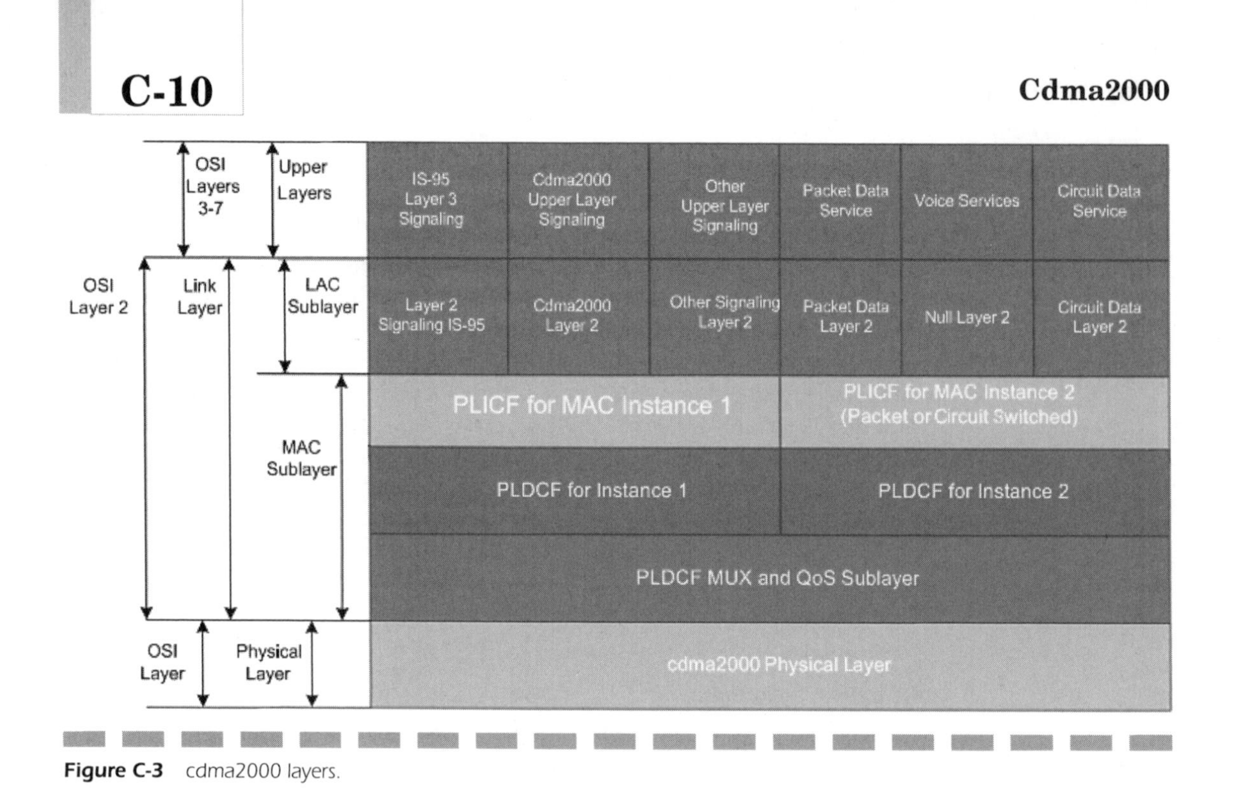

Figure C-3 cdma2000 layers.

PSTN access, and mobile-to-mobile. End-user data-bearing services deliver packet data like Internet protocol/TCP services, circuit data services like B-ISDN emulation services, and short message services. Packet data services may use IP, TCP, UDP, and ISO/OSI connectionless internetworking protocol (CLIP). Circuit data services include asynchronous dial-up access, fax, V.120 rate adapted ISDN, and B-ISDN. Signaling services control the operation of the mobile.

Lower Layers (LAC and MAC)

The link layer provides features that drive the QoS for upper-layer services, providing functions used to map the data transport to the upper layers. The Link layer includes the Link Access Control (LAC) and Media Access Control (MAC) sublayers. The LAC sublayer addresses point-to-point communication for upper-layer entities, and has a framework supporting end-to-end link layer protocols.

The MAC sublayer holds multiple instances of an advanced-state machine, and delivers 3G media-rich services with QoS management per service, and provides three basic functions:

- Media access control state: Controls access of services to the physical layer, and also handles contention between users and services

TABLE C-2

Channel Acronyms

First Letter	Second Letter	Third Letter
F: forward (BS to MS)	D: dedicated	T: traffic
R: reverse (MS to BS)	C: common	M: media access control
		S: signaling

- Best-effort delivery: Provides delivery guarantee over the radio link with protocol (RLP)
- Multiplexing and QoS control: Enforces QoS by prioritizing access requests, and by mediating conflicting requests from services

Cdma2000 Channels

A logical channel is designated a three- or four-letter acronym followed by "ch." (See Table C-2.)

The naming convention for physical channels is the use of uppercase letters, with the first letter indicating direction, except for paging and access channels where direction is implied. See Tables C-3 and C-4.

Physical Layer

The physical layer is a source of coding and modulation services for logical channels that are used by the PLDCF MUX and QoS layer. Physical channels may be F/R-DPHCH (Forward/reverse dedicated physical channels) where they carry information in a point-to-point model between the base station and a mobile user. They can also be F/R-CPHCH (Forward/reverse common physical channels), or a number of physical channels carrying information in a point-to-multipoint model between the base station and multiple mobile stations.

Charge Mechanisms

There are many revenue streams obtained in mobile applications or services with the most apparent being the actual billing of the customer that may be time charged, packet data charged, fixed charged, and/or subscription invoiced. A user's bill could contain monthly subscriptions for services that include pay-per-view movies, online games, and call charges that may have been used for voice, multimedia, or for connection to remote enterprise networks over VPN links. The billing system creates this information in the form

TABLE C-3

Physical Channel Names

Name	Physical Channel
F/R-FCH	Forward/Reverse Fundamental Channel
F/R-SCCH	Forward/Reverse Supplemental Coded Channel
F/R-SCH	Forward/Reverse Supplemental Channel
F/R-DCCH	Forward/Reverse Dedicated Control Channel
F-PCH	Forward Paging Channel
R-ACH	Reverse Access Channel
R-EACH	Reverse Enhanced Access Channel
F/R-CCCH	Forward/Reverse Common Control Channel
F-DAPICH	Forward Dedicated Auxiliary Pilot Channel
F-APICH	Forward Auxiliary Pilot Channel
F/R-PICH	Forward/Reverse Pilot Channel
F-SYNC	Forward Synchronization Channel
F-TDPICH	Forward Transmit Diversity Pilot Channel
F-ATDPICH	Forward Auxiliary Transmit Diversity Pilot Channel
F-BCH	Forward Broadcast Channel
F-QPCH	Forward Quick Paging Channel
F-CPCCH	Forward Common Power Control Channel
F-CACH	Forward Common Assignment Channel

of CDRs from the network, calculates the cost of services, and dispatches statements to customers via the operator or service provider.

GPRS Billing

A GPRS billing system relies on the SGSN and GGSN to register usage and provide appropriate information using CDR (Charging Data Records) that are routed to the billing gateway. GPRS packet-based charging may be based on:

- Volume or bytes transferred
- Duration of a PDP context session
- Time, including day and date, so as to provide lower tariffs at off-peak periods
- Destination address of the network or proxy server holding the service that the subscriber has been accessing

TABLE C-4 Logical Channels Used by PLICF

Channel	Description
Dedicated traffic channel f/r-dtch	This carries user data and is a point-to-point channel and is leased for the time period of an active data service.
Common traffic channel f/r-ctch	This carries short data bursts of a data service in a dormant/burst substate.
Dedicated media access control channel f/r-dmch_control	This carries media access control messages and is a point-to-point channel. It is leased for the time period of an active data service.
Reverse common media access control channel r-cmch_control	This carries media access control messages, and is used by the mobile while the data service is dormant or idle.
Forward common media access control channel f-cmch_control	This is used by the Base station when the data service is dormant or idle.
Dedicated signaling channel dsch	This carries upper-layer signaling data for a single PLICF instance.
Common signaling channel csch	This carries upper-layer signaling data shared access.

- Quality of service assigned to the subscriber, giving predefined priority and bandwidth rights
- SMS usage where the SGSN produces specific CDRs
- Served IMSI/subscriber classes that have different tariffs for business users, private users, and frequent users
- Reverse charging where the sender is billed
- "Free of charge" where the data or received service is not billed to the user
- Flat rate where a fixed fee (perhaps monthly) is billed to the user, and is appropriate for mass market telecommunications
- Bearer service used where the operator is offering multiple network types that may include GSM900, GSM1800, and perhaps wireless LAN for areas where the operation of other mobile networks are considerably less viable

Current GPRS networks may not be equipped to bill users based on applications used, but the network may be upgraded to do so through the addition of a payment server on the service network. (*See* Applications under letter A.)

Prepaid

Unlike postpaid billing, prepaid billing requires users to buy airtime for a SIM card, and does not require that the user have a subscription with an operator.

Comparative advantages of prepaid services include:

- Credit checks are redundant because payment is made in advance.
- The phone and SIM can be purchased together and activated immediately.
- The user is unknown to the operator.
- The user has no contract with the operator, and may switch to another compatible operator merely by purchasing an appropriate SIM.
- There is no monthly subscription charge.
- If the phone is unused, the user is not charged.
- Prepaid lends itself to international travel by changing SIMs so as to use overseas (that are perhaps European) mobile networks.
- Prepaid is ideal for younger users.
- The operator requires no billing system.
- The user may purchase prepaid SIMs to receive calls only.

Disadvantages of prepaid billing include:

- Operators may not allow international roaming because they do not have real-time credit checks on prepaid-card credit status.
- New airtime vouchers must be physically purchased by the user.
- Anonymity of the customer (from the operator's perspective) creates security problems as it is difficult to trace fraudulent usage.
- Airtime vouchers can be counterfeited, stolen, and duplicated.

Prepaid mobile phones are popular throughout the world; 80 percent of users in Italy and Portugal have operators on prepaid billing. Finland, however, with its high level of mobile phone usage has an unusually small percentage of its mobile users on prepaid billing schemes, because handset subsidies in Finland are illegal, and buyers must pay the market price. As a result, network usage is relatively inexpensive because the operator does not subsidize the cost of handsets, and it follows, therefore, that prepaid billing would make little or no difference to usage costs.

Service Providers

Service providers are essentially third-party companies and occasionally WASPs that host a given mobile service, or wireless application service, that is secure and offers an agreed QoS, and an agreed mobile device standard such as WAP, GPPRS, EDGE, or 3G. The service provider can, therefore, charge subscribers for use of its services on a monthly basis, or on a packet data or volume basis, or on a time charged basis. The service provider may provide various types of free and conventional marketed and sold services including:

- Lifestyle
- Traffic
- Travel
- Finance
- Weather
- City
- News items—news flashes and updates
- Text to speech
- Speech to text

The physical installation of a service provider may include:

- LAN
- Firewall
- Router
- Data link such as ISDN
- Real-time data feed from the information provider—if this is not generated on-site
- Non-real-time data feed—that may be free of charge
- Storage—usually RAID
- Secure backup servers
- Application server
- Software server to interface with the mobile network
- Core software solution that may be a software client/server architecture designed to process information that may be traditional SMS messages
- WAP gateway

In countries where service providers represent a large market share, they typically partner the established telcos or operators where they collaborate on marketing and selling mobile application services. And, to a lesser extent, there is collaborative research and development involving both the operator and the service provider.

The service provider has a number of options for billing or for mining revenues that it obviously requires to exist. These include:

1. The service provider may charge the operator directly a flat rate—or a flat payment usually settled in agreed phases—for a service that is then available to the user of the operator's network. In this instance the service provider has no real-time billing system applied to the service. And the operator may not itself bill for the use of the service, but merely use the service as a value add, as a promotion used to grow its user base. The service may be a full-fledged wireless application

used by the operator's entire user base, or it may be a pilot scheme where the service is being tested for technical and commercial viability.

2. The service provider may have its own billing system, or charging gateway, that monitors usage and generates billing information, or it may simply sell flat-rate subscriptions, perhaps over the Internet, that eliminates the need for time or packet volume billing systems.

Classification of Environment

A mobile network is typically deployed over numerous environments including dense metropolitan, urban, suburban, and rural. Dense metropolitan districts include tall buildings with 10 or more stories with a population density of 20,000 subscribers per square mile. Urban areas include residential and office areas with buildings about 5 to 10 stories high, and with a population density of 7500 to 20,000 subscribers per square mile. Suburban areas have housing with one to two stories that are 50m apart, shops and offices that are two to five stories, and population densities between 500 and 7500 subscribers per square mile. Rural areas include farming communities with a population density as low as 500 subscribers per square mile.

Classification of GPRS Handsets

A GPRS terminal phone is classified as:

- Class A: Supports GPRS, GSM, and SMS, and may make or receive calls simultaneously over two services. In the case of CS services, GPRS virtual circuits are held or placed on busy as opposed to closed down.
- Class B: Supports either GSM or GPRS services sequentially, but monitors both simultaneously. Like Class A, the GPRS virtual circuits will not be closed down with CS traffic, but switched to busy or held mode.
- Class C: Make or receive calls using the manually selected service, with SMS an optional feature.

CGI (Common Gateway Interface)

A protocol that provides bidirectional information flow within the active or dynamic Web model, and may be perceived as permitting users to interact with remote applications such as E-commerce implementations. It is a protocol that provides bidirectional information flow between an HTTP server and an

HTTP client. The resulting interactivity on the client side permits data entry and the editing of HTML documents. The Common Gateway Interface (CGI) may connect the HTTP server and its applications and databases. CGI scripts are created using a scripting language or programming tool.

CGI may be used to:

- Query databases and post the output to HTML documents
- Generate HTML forms for data entry
- Interact with the indexes of on-line documents to produce searching and retrieval features
- Interact with email

CGI programming is possible using Unix, Windows, and Macintosh servers. CGI scripts may be created using:

- Perl
- Apple Script

CGI programs may be created using almost any high-level programming language including:

- C++
- Visual Basic

The protocol that is CGI bases itself on standard environment variables that are sometimes extended by the Web server used.

CGI Environment Variables

A set of variables that defines the CGI (Common Gateway Interface) is normally set when a CGI script or program is called by using the:

- GET method where the URL defines the CGI program (such as credit.cgi for example) and the accompanying data used by the server that follows the question mark:

  `www.FrancisBotto.com/cgi-bin/credit.cgi?subject=transaction`

- POST method in which the program is specified as part of the URL, passing data using the requester path: which is a uni-directional link from the client to the server.

  `www.FrancisBotto.com/cgi-bin/credit.cgi`

HTTP_ACCEPT Holds the "Accept:" headers from the client.

HTTP_COOKIE Holds the contents of "Cookie:" headers from the client.

HTTP_FROM Holds the contents of the "From:" header from the client that may be the client's:

- Correct e-mail address if not withheld
- Incorrect e-mail address that is simply false, or entered in error.

HTTP_REFERER Holds the contents of the "Referer:" header from the client, containing a URL.

HTTP_USER_AGENT Holds the contents of the "User-Agent:" header from the client, containing the Browser's name.

PATH_INFO Holds the URL's suffix or that data that follows the script's name.

QUERY_STRING Holds the "query" part of an HTTP GET request that is the URL's suffix portion following "?".

REMOTE_ADDRESS Holds the client's or proxy's IP address from where the request is being made.

REMOTE_HOST Holds the hostname of the client or proxy making the request, or its IP address only when NO_DNS_HOSTNAMES is defined in the config.h file.

SCRIPT_NAME Holds the name and path of the CGI script being executed.

SERVER_SOFTWARE Holds the name and perhaps version of the server software.

SERVER_NAME Contains name of host on which server is running.

SERVER_PROTOCOL Contains "CGI 1.1".

SERVER_PORT Holds the port on which server is running.

Computers/Software

C++

An object-oriented version of the C programming language. Like modern programming languages such as Java, it provides the programmer with OO methodologies. Bjarne Stroustrup evolved C++ from C that has links with BCPL (Basic Combined Programming Language). C++ includes the following OO concepts:

- Inheritance
- Polymorphism
- Encapsulation
- Data hiding

ANSI C++ is an internationally agreed standard for the C++ programming language.

#include <file> When compiled, the #include statement is implemented by the preprocessor that reads the contents of a named file.

main () A C++ program must have a main () function that begins and ends with open { and close } braces. This is the first function called when the program is run, and may be used to define variable types.

Comments Single-line comments in a C++ program must begin with //, and multiple line comments begin and end with /* and */.

Syntax (Basic) All statements have a semicolon (;) as their suffix. White space may be included that is ignored by the compiler.

Compound statements such as those of a function or a subroutine begin with a single open brace "{", and end with a closing brace "}".

C++ variables C++ variable types may be defined as follows:

```
#include iostream()
main()
{
  char find;
  float prime;
  double prime_large;
  short int xx;
  long int xxxx;
  unsigned short int yy;
  yy=35;                    //assign the unsigned
  long int yyyy;
}
/* the character variable find, may store 256
  character values */
```

```
// the variable prime, may store signed 4Byte
  values
/* the variable prime_large, may store signed 8Byte
   values */
// the variable xx, may store signed 2Byte values
// the variable xxxx, may store signed 4Byte values
// the variable yy, may store unsigned 2Byte values
// the variable yyyy, may store unsigned 4Byte
   values
```

Defined variables may be equated to values using the statement:

```
yy=35;
```

Variables may be defined, and assigned values using the statement:

```
unsigned short int xx=45;
```

Multiple variables of the same type are defined using a comma as a separator:

```
unsigned long int yyyy, yflow
```

typedef A method of defining types and variables. Using typedef, mnemonics may be assigned to the statements used to define variables and their types. The following statement assigns the word xxxx to the unsigned short int statement:

```
#include <filename>
typedef unsigned short int xxxx;
int main ()
{
xxxx coordinate;
// define coordinate as an unsigned short integer
  variable
}
```

C++ Literal Constants A variable may be assigned a value that is considered a literal constant:

```
int yearsAfter=25;
```

A literal constant may also be used when performing arithmetic operations on variables. In the following statement, where the time variable is assigned to the product of the variable present and 10, 10 is considered a literal constant:

```
time=present*10
```

C++ Symbolic Constant A symbolic constant has a name and is assigned an unchanging value. It may be used just like an integer constant. Symbolic constants improve program maintenance and updating; a single change made to a symbolic constant is echoed at every point it may occur.

A symbolic constant multiplier may be assigned the value 10 using the statement:

```
#define multiplier 10
```

or,

```
const unsigned short int multiplier = 10
```

C++ Enumerated Constants Enumerated constants take the form of a type, and are a useful shorthand for defining a number of what might be related constants. The following code defines the constants back, forward, left, and right, where Move is the enumeration.

```
enum Move { back=4, forward, left=6, right=3 };
```

The forward constant is assigned the value 5, an increment (of one) relative to the previously defined constant back.

C++ Precedence In C++, arithmetic operators have a precedence value. These indicate the order in which such operators are implemented and is significant with expressions such as:

```
dev = xx + yy * zz + yy;
```

Control over such arithmetic operations is obtained using parentheses, for example:

```
dev = (xx + yy) * zz;
```

Parentheses may be nested.

C++ If Statement

```
The If statement determines whether or not the ensuing statement
is executed, based on a single condition:
```

```
{
  if (xxx = yyy)
    transform = Scale;
}
```

C++ If... else Statement The If... else statement is used to implement either one of two statements:

```
{
  if (xxx = yyy)
      transform = Scale;
  else
      transform = Scale * adjust;
}
```

C++ Logical Operators Logical AND, OR, and NOT are implemented using the syntax '&&', '||', and '!'

Cabbing

A method of compressing objects such as ActiveX controls and Java objects into a single CAB file. This optimizes their rate of transfer across networks.

Cache

1. A segment of SRAM (Static Random Access Memory) that drives processor performance gains. Its rationale is to expedite the rate at which data may be read from, and written to, memory. It may be an integral part of the processor (internal), or external in the form of dedicated SRAM chips on the PC motherboard. The fast speed of SRAM overcomes the slower speed of DRAM (Dynamic Random Access Memory) making up the system memory.
   They significantly improve system performance. External memory cache sizes are relatively small, ranging from just 128kbytes to 1Mbyte in size. An algorithm is used to estimate what portions of system memory should reside in the memory cache. The Pentium Pro has an internal cache accommodated on a single die or chip.
2. An area of memory or hard disk used as a temporary storage for downloaded HTML files and data, including URLs. The size of the cache may be specified.
3. A hard disk controller that expedites hard disk performance. A hard disk cache controller typically comprises Mbytes of RAM, and is usually expandable. It speeds up read/write operations by using its on-board RAM as an intermediate data store between disk and system memory.
   Based upon which data is most often requested, a caching algorithm estimates which portions of hard disk should reside in on-board RAM,

thus making it more readily available. The ingenuity of this technique simply takes advantage of the inescapable fact that a small percentage of disk data is rewritten and accessed most frequently.

The decision-making process that is insulated from the system processor fuels the view that it is an intelligent controller. Cache controllers are the most expensive of all variants, and in terms of random access and data transfer rate they may be assumed to outperform all others.

4. A RAID often features a cache for improved performance.

Capture Reversal

An event that sees the reversal of a capture response when goods are returned after the completion of a sale.

Card Association

A bank that supports a franchise for shown card brands.

Card Issuer

A card company or bank that has powers to grant credit or bank cards.

Card-not-present

A card transaction where there is no physical evidence of the card; this typically exists in a MOTO (Mail Order/Telephone Order) scenario.

Cardholder

An authorized owner and user of a credit card or bank card.

Cardshield

A service provided by Shielded Technologies, Inc., that may be applied in the development of a Web commerce solution that includes credit card transactions.

Casting

A process where one data type is converted into another.

CD-R (CD-Recordable)

A drive capable of writing to blank CD-R discs, usually in a variety of different formats including Video CD, Photo CD, CD-ROM XA, CD-I, and CD-ROM. The mid-1990s saw the launch of more affordable CD-R drives, bringing low-volume CD-ROM publishing to the desktop. Important factors to consider when acquiring CD-R drives include:

- The maximum data capacities supported.
- The read rate of the drive that may be single-speed, double-speed, triple-speed, quad-speed, or faster.
- The disc recording speed that may be single-speed, double-speed, triple-speed, quad-speed, or even faster. High recording speeds yield savings in terms of person hours used.
- The disc formats supported that might include audio CD, CD-ROM, CD-ROM XA, CD-I, Photo CD, and Video CD.
- The interface type. Most operate over the SCSI bus variants.
- What type of interface software is provided? It is important that this be user friendly.

CD-ROM (Compact Disc—Read Only Memory)

A universal distribution medium based on the compact disc. It was the first viable multimedia distribution medium.

Announced in 1983 it is typically a 12cm-diameter optical disc offering data capacity in the hundreds of Mbytes range.

The standard 12cm-diameter CD-ROM supports up to about 660Mbytes (692,060,000 bytes) of data capacity. A single disc is equivalent to approximately 400 1.44Mbyte floppy disks or 1500 360kbyte floppy disks. 8cm-diameter CD-ROMs are also available.

A 12cm CD-ROM can store up to 250,000 A4 pages of text or approximately 100,000,000 words. *Note:* These methods of quoting data capacity are rather vague and not likely to satisfy many people.

Like audio CD, a CD-ROM disc physically consists of a metallic disc bonded to a polycarbonate base. This is coated with a transparent, protective lacquer. A track spiraling from its center measures some 5km long, and is arranged at a density of 16,000 tracks per inch.

The CD-ROM physical format includes:

- Mode 1 data blocks that are used to store code and data where accuracy is critical
- Mode 2 data blocks that are used to store data that might be impervious to minor errors

Data blocks supported by all fully specified CD-ROM drives. Mode 1 disc yields 527Mbytes data capacity and Mode 2 gives 602Mbytes data capacity.

A Mode 1 data block will yield just 2048bytes (2kbytes) of user data, while Mode 2 holds 2.28kbytes of user data.

CD-ROM Data Block

A CD-ROM data block has 2352bytes. User data yielded by each block is a function of the mode of operation.

CD-ROM Drive

A device for reading CD-ROM discs. It may be portable, external, or integral to the computer/multimedia system. Modern drives are able to read Mode 1 and Mode 2 discs, as well as audio CDs.

Principal factors that govern the performance of a CD-ROM drive include access time and data transfer rate. In general, a CD-ROM drive may be specified in terms of the following information:

- Access time: Highly specified drives may offer access times a little longer than 100ms
- Average data transfer rate may be generally specified in terms of how fast the disc is rotated; a single-speed drive will give a data transfer rate of around 150kbytes/s. This data rate is broadly doubled, tripled, and quadrupled using double-, triple-, and quad-speed drives.
- The physical interface type may be proprietary: IDE, SCSI, SCSI-2, or use may be made of a PCMCIA card or parallel port. Highly specified CD-ROM drives tend to be SCSI-2 based.
- Compatibility in terms of disc formats that may be read is generally specified in terms of 8cm-diameter CD-ROM, CD-ROM XA, linear CD-I, Video CD, and Photo CD.
- Physical characteristics include whether the drive is internal, external, or portable.
- The maximum number of drives that may be daisychained.

Chat

1. A real-time, text-based communications medium, carried out over a network, or over the Internet.
2. A Windows NT-based server that is part of the MCIS.

The Chat server provides real-time text-based communications. The communications may be private (one-to-one), one-to-many, or conferences. It has its own proprietary protocol and supports the IRC protocol.

A Chat SDK and ActiveX control permit the integration of Chat functionality, where a single server may support up to 48,000 users.

Checksum Validation

A method of validating credit card numbers by using the *mod 10* check digit algorithm and is implemented by:

1. Doubling the value of alternative digits of the credit card number by beginning with the second digit:

1	3	6	5	8	9	7	6	2	4	2	7	6	0	8	7
	6		10		18		12		8		14		0		14

2. Adding the product values to the alternate digits beginning with the first:

$$7 + 16 + 26 + 17 + 10 + 16 + 6 + 22 = 120$$

In this instance the credit card number has passed the validity check because the result is evenly divisible by 10.

Ciphertext

An input into a decryption algorithm that sees it returned to plaintext.

Class

A formal description of objects in terms of the methods and data they may use.

Class Diagram

A pictorial representation of the class hierarchy, including links of inheritance, revealing subclasses and their superclasses. It illustrates how interfaces and methods are inherited within the class hierarchy of an architecture.

Client

1. A collective portable or desktop system that provides the human machine interface to a client/server architecture, including:
   - E-mail client such as Microsoft Outlook for receiving and sending e-mail messages
   - Client software such as Web browsers that may be Netscape, Explorer, or HotJava
   - Client operating system, which is typically Windows 95/98/2000/NT

   Between clients and servers there may be a number of hardware and software entities, including:

   - Access technologies such as ISDN or wireless media such as GSM
   - Modems or NIC (Network Interface Cards)
   - Protocols such as TCP/IP at the transport layer, HTTP, and UDP
   - Middleware such as those based on the IDLs of DCOM or CORBA NS that provide a means of exchanging messages
   - ORB (Object Request Broker)
2. In the context of middleware based on the OMG Notification Services, such client applications are termed consumers, while the server applications become suppliers or publishers. In this context clients may operate according to the push-and-pull models within the client/server architecture.
3. A device or appliance that is driven by remote server applications and data. It may be a portable device such as a PDA or palmtop appliance manufactured by companies such as Psion, Casio, and 3Com.

Client/Server

A distributed system architecture where client systems are connected to server systems. The client provides an interface to applications and data that is stored on the server. The interface may be provided through a browser such as:

- Microsoft Explorer
- Netscape Navigator
- Sunsoft HotJava

Client activity and processing is said to be on the client-side, while server activity and processing is on the server-side. The network that provides connection between clients and servers might be:

- LAN
- WAN
- Internet
- Intranet

Industry client/server standards for database manipulation include:

- ODBC (Open Database Connectivity), which is the most common
- IDAPI (Integrated Database Application Programming Interface)

Client/server network protocols include:

- IP/TCP
- IPX (Internet Packet eXchange)

Using the three-element representation of an application, the client/server model (as observed by the Web) may be explained.

The five topologies are:

- Distributed presentation that distributes a portion of the user interface (UI), and may be equated to the inactive Web model, where the browser is used only to view documents.
- Remote presentation that distributes the entire UI to the client system.
- Distributed function that divides application logic between the server and the client. In Web context this processing distribution may be achieved using appropriate plug-ins and ActiveX controls with Netscape Navigator or Microsoft Internet Explorer.
- Remote data access that is a model that sees the so-called fat client. This means that the client system is substantial (or "fat") in terms of application logic.
- Distributed database that distributes the data management functions between the client- and server-side. This configuration is used in Webcasting, where users are served information that matches their predefined criteria.

The distribution of the three key application elements (namely, Presentation, Logic, and Data or Data Management) may be used to explain the many client/server models. See Fig. C-4.

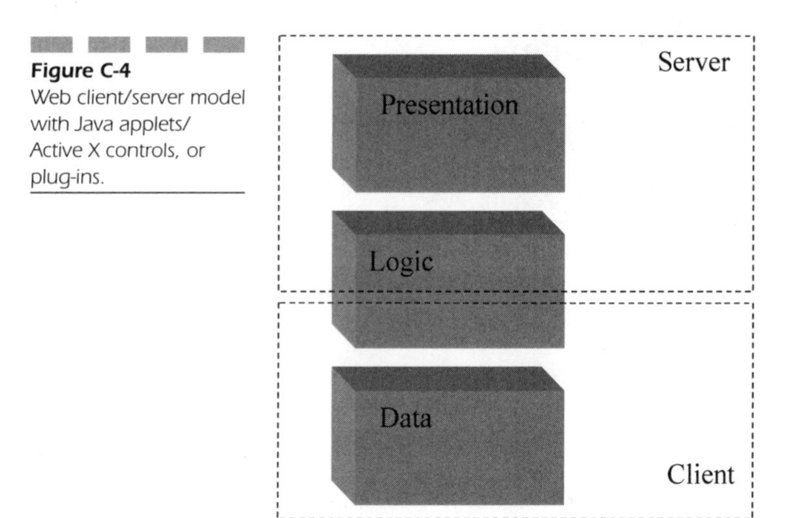

Figure C-4
Web client/server model with Java applets/ Active X controls, or plug-ins.

Client/Server Architecture

A hardware infrastructure used to platform client/server applications. It may be 2-tier, 3-tier or n-tier.

Client Side

A term that indicates the presence of software or data on client systems. Synonym: local.

Client Software

An application that resides on the client side within a client/server architecture.

Client System

A collective system with which users interact directly, and is physically located on users' desktops or in similarly close proximity.

Codec (Coder/Decoder)

1. A feature that digitizes a voice signal, and decodes a data stream to create an analog voice signal.
2. A codec is normally considered to be a hardware device or software driver able to compress and decompress audio, video, or both audio and video. It may operate using standard compressed file formats or proprietary formats such as those generated by the Cinepack codec. Codecs that operate in the Microsoft Windows environment exist as drivers.

Compiled

A process by which the source code of a high-level language is translated into machine-executable form or machine code. Generally, compiled languages offer better run-time performance than interpreted languages.

A compiler is able to convert source code into machine-executable code. Unlike an interpreter, which attaches precise code to high-level statements each time a program is run, a compiler produces machine-executable object code once. Inexpensive to develop, BASIC interpreters were widely used in the microcomputer industry in the 1980s.

Compound Document

A document that may integrate different document and media types that emanate from different sources. The various documents may be OLE objects provided by an appropriate OLE 2.0 server.

Alternatively, they may be objects of a similar architecture such as the more modern Microsoft ActiveX component architecture. Equally, they may be objects that comply with OpenDoc or JavaBeans component architectures.

Compressed Image

An image following compression through hardware and/or software means.

Compression Parameters

A video compressor setting may be used to optimize a video sequence for playback using a target system of a given bandwidth. MPEG compression parameters include the placement of I frames.

More general compression parameters might include interleave ratio, frame playback speed, and compression ratio requirements. Generally, the quality of video diminishes as the compression ratio increases.

Compression Ratio

A ratio that relates the size of a data file before and after compression. The video compression ratio using fully specified compressors may be altered.

Compressor

A hardware and/or software solution used to compress motion video or still computer graphics. Using video compressors, it is possible to specify a number of compression settings.

Computer

A system or appliance able to process and store digital information. Its many components and subsystems may include:

- Processor or CPU (Central Processing Unit)
- Modem
- DVD-ROM drive
- Graphics engine or card
- Hard disk
- Sound card
- Electronic memory devices, including RAM, SRAM, VRAM, ROM, and NVRAM
- Color display
- Video playback device such as MPEG-1 or MPEG-2 decoder
- QWERTY Keyboard
- Mouse
- Microphone
- Television tuner
- Radio tuner
- Scanner

Computer Graphics

A means of displaying images using a computer. The advancement of computer graphics has unleashed numerous computer applications, ranging from Computer Aided Design (CAD) to color desktop publishing (DTP), VR, multimedia, and 3-D graphics. Through the coupling of high-resolution color monitors and high specification graphics controllers, truly photographic quality images are now possible.

Built up of digitally defined pixel elements, computer images are invariably complex. For example, a 1024-by-768 pixel image yields 786,432 individual pixels. Digitizing such a black-and-white image requires a corresponding number of bits, or 98,304 bytes (786,432/8), or 96kbytes (98,304/1024). Progressing to a gray-scale arrangement using bits per pixel to give 256 (2^8) gray shades, the same 1024-by-768 pixel image requires 768kbytes—eight times the storage capacity of its black-and-white equivalent.

Such is the complexity of photographic quality images, a minimum of 24bit color graphics are required. Red, green, and blue are each represented by eight bits, thus facilitating the selection of 256 tones of each. By combining each color component, over 16.7 million ($256 \times 256 \times 256$) colors are made available.

Even higher-quality results are achieved using 32bit and 36bit graphics. Such 24bit graphics on a 1024-by-768 pixel resolution monitor means that a single frame consumes around 3072kbytes.

Large image files of this sort are costly to process, transmit, and store. They are also slow to transfer from computer to screen, as well as to and from a hard disk. A solution to these problems lies in image compression. Many popular image file formats such as JPEG feature image compression.

Computer Name

A name of computer/system connected to a network. All Windows 98/NT systems have names when connected to a network. Additionally, their users are given passwords that may be used to log on and retrieve their specified or default Windows configuration.

Concurrent Computing

An environment in which processes, or program elements, execute simultaneously.

Concurrent Programming

A programming model where processes are implemented in parallel.

Concurrent Programming Language

A programming language that may be used to implement processes in parallel.

Confidence Factor

A measure of the percentage probability of an event or circumstance being correct. In KBSs (Knowledge Based Systems) and rule-based systems, it may be applied so as to weigh facts and conclusions that exist in a knowledge or heuristics base.

Constant

An unchanging entity.

Content Authoring Tool

A development tool that permits the creation of Web and multimedia content.

Controller

A generic name for a hardware component that controls a peripheral device, such as a disk drive, CD-ROM drive, or monitor.

Cookie

A minor transaction that allows server-side components such as CGI scripts and programs to store and retrieve data from the client system.

It gives Web applications the ability to write data to the client that reflects usage habits. For example, the data may relieve the user from repetitive tasks, such as the re-entry of ID numbers or data each time a Web site is visited. Instead the server-side components may identify the user through cookies on the client system, extract them, and perform the necessary processes.

CORBA (Common Object Request Broker Architecture)

An object architecture featuring an IDL (Interface Definition Language) and is managed by the Object Management Group (OMG).

CORBA IDL (Common Object Request Broker Architecture Interface Definition Language)

A language that is based on C++, and may be compiled into Java and C++ using appropriate compilers such as IDL2JAVA and IDL2CPP.

CosNotification

An IDL module that defines the operations (or methods) used by the CORBA Notification Services, which supports push-and-pull models on networks. The collective IDL modules might be referred to as the API of the implementation.

```
module CosNotification {

// The following two are the same, but serve
   different purposes.

typedef CosTrading::PropertySeq
   OptionalHeaderFields;

typedef CosTrading::PropertySeq
   FilterableEventBody;

typedef CosTrading::PropertySeq QoSProperties;

typedef CosTrading::PropertySeq AdminProperties;

struct EventType {
   string domain_name;
   string type_name;
};

typedef sequence<EventType> EventTypeSeq;

struct PropertyRange {
   CosTrading::PropertyName name;
   CosTrading::PropertyValue low_val;
   CosTrading::PropertyValue high_val;
};
```

```
typedef sequence<PropertyRange> PropertyRangeSeq;
enum QoSError_code {
    UNSUPPORTED_PROPERTY,
    UNAVAILABLE_PROPERTY,
    UNSUPPORTED_VALUE,
    UNAVAILABLE_VALUE,
    BAD_PROPERTY,
    BAD_TYPE,
    BAD_VALUE
};
struct PropertyError {
QoSError_code code;
PropertyRange available_range;
};
typedef sequence<PropertyError> PropertyErrorSeq;
exception UnsupportedQoS { PropertyErrorSeq qos_err; };
exception UnsupportedAdmin { PropertyErrorSeq
    admin_err; };
// Define the Structured Event structure
struct FixedEventHeader {
EventType event_type;
string event_name;
};
struct EventHeader {
FixedEventHeader fixed_header;
OptionalHeaderFields variable_header;
};
struct StructuredEvent {
EventHeader header;
FilterableEventBody filterable_data;
any remainder_of_body;
}; // StructuredEvent
typedef sequence<StructuredEvent> EventBatch;
```

```
// The following constant declarations define the
    standard
// QoS property names and the associated values each
    property
// can take on. The name/value pairs for each standard
    property
// are grouped, beginning with a string constant
    defined for the
// property name, followed by the values the property
    can take on.
const string EventReliability = "EventReliability";
const short BestEffort = 0;
const short Persistent = 1;
const string ConnectionReliability =
    "ConnectionReliability";
// Can take on the same values as EventReliability
const string Priority = "Priority";
const short LowestPriority = -32767;
const short HighestPriority = 32767;
const short DefaultPriority = 0;
const string StartTime = "StartTime";
// StartTime takes a value of type TimeBase::UtcT when
    placed
// in an event header. StartTime can also be set to
    either
// TRUE or FALSE at the Proxy level, indicating whether
    or not the
// Proxy supports the setting of per-message stop
    times.
const string StopTime = "StopTime";
// StopTime takes a value of type TimeBase::UtcT when
    placed
// in an event header. StopTime can also be set to
    either
// TRUE or FALSE at the Proxy level, indicating whether
    or not the
// Proxy supports the setting of per-message stop
    times.
```

```
const string Timeout = "Timeout";
// Timeout takes on a value of type TimeBase::TimeT
const string OrderPolicy = "OrderPolicy";
const short AnyOrder = 0;
const short FifoOrder = 1;
const short PriorityOrder = 2;
const short DeadlineOrder = 3;
const string DiscardPolicy = "DiscardPolicy";
// DiscardPolicy takes on the same values as
    OrderPolicy, plus
const short LifoOrder = 4;
const string MaximumBatchSize = "MaximumBatchSize";
// MaximumBatchSize takes on a value of type long
const string PacingInterval = "PacingInterval";
// PacingInterval takes on a value of type
    TimeBase::TimeT
interface QoSAdmin {
QoSProperties get_qos();
void set_qos ( in QoSProperties qos)
raises ( UnsupportedQoS );
void validate_qos (
    in QoSProperites required_qos,
    out PropertyRangeSeq available_qos )
raises ( UnsupportedQoS );
}; // QosAdmin
// Admin properties are defined in similar manner as
    QoS
// properties. The only difference is that these
    properties
// are related to channel administration policies, as
    opposed
// message quality of service
const string MaxQueueLength = "MaxQueueLength";
// MaxQueueLength takes on a value of type long
```

```
const string MaxConsumers = "MaxConsumers";
// MaxConsumers takes on a value of type long
const string MaxSuppliers = "MaxSuppliers";
// MaxSuppliers takes on a value of type long
```

Counter Program

A program that records the number of occasions (or hits) a Web page or URL is opened. Such program variants may count Web pages that are opened and served to the client, and not merely count URLs. The program may be embedded in an HTML script.

Coupling

A term used to describe efficiency of communication between network hardware and software components. Tight coupling between two network components indicates comparatively high-speed communication capabilities. While loose coupling indicates the exact opposite.

Cray, Seymour

A computer scientist made famous by his work in the field of MPP.

Creative Labs

A Singapore-based company specializing in sound cards and video capture cards. Its SoundBlaster card became an industry standard. Its video capture cards include the VideoBlaster range, which extends to video conferencing. It also marketed and sold the VideoSpigot video capture card, although it did not develop it.

Critical Error

An error resulting from a hardware or software bug. Using DOS, the user will be prompted by R(etry), I(gnore), F(ail), or A(bort).

Cropping

A process of trimming an image or frame. In terms of video or picture editing, image or video data is cropped just as you would snip a photograph using a pair of scissors. Most editing programs provide an Undo Crop command (on the Edit menu) in order to cancel a previous cropping operation.

Cross Platform

A software program, module, or object that may be run on more than one platform. Java applications are cross-platform.

Such applications may be described as platform or hardware independent. For instance, a platform independent program might run on Windows, OS/2, and 386 Unix.

CRT (Cathode Ray Tube)

A display device used in desktop color monitors, consisting of a screen area covered with phosphor deposits (or pixels), each consisting of red, blue, and green phosphors. The CRT was the first optronic device.

The distance between the phosphors is termed the dot-pitch. Most monitors feature a dot-pitch of 0.26, while more highly specified versions offer a smaller dot-pitch.

An electron beam is projected from the back of the CRT onto the inner screen, using an electron gun. To help focus the electron beam, a fine mask is included behind the screen phosphors.

This fine gauze separates the three-color phosphors allowing the electron beam to shine more accurately upon them while improving picture definition in the process.

The electron beam scans each of the phosphor lines horizontally. The rate at which the electron gun scans a single line is termed the horizontal frequency, or the line frequency.

There are two methods of scanning the lines:

- Interlaced
- Non-interlaced

In a non-interlaced arrangement, all the lines are scanned one after another. The rate at which all lines are scanned is termed the refresh rate or the vertical frequency.

Using an interlaced configuration, the lines making up the screen are scanned in two separate fields. One field is used to scan even-numbered lines and the other to scan odd-numbered lines.

This interlaced technique was introduced in television broadcasting specifically to reduce screen flicker. Today, however, a monitor that operates at high resolutions in an interlaced mode will flicker.

Non-interlaced monitors with sufficiently high screen refresh rates are preferred. These provide flicker-free images, with improved stability, and are least likely to cause eye strain. The minimum acceptable refresh rate, or vertical frequency, for a non-interlaced monitor is around 70Hz.

Cryptoanalysis

A subject and/or science that addresses attacks on cryptosystems.

Cryptography

A process that ensures data or information is read or used only by its intended readers or users. This is achieved through:

- Encryption that disguises inputted information or data, so it may not be read or used. Resulting encrypted information or data may only be read or used following decryption.
- Decryption that returns the decrypted data or information to its original usable and readable form.

Implementations of cryptography are called cryptosystems, and take the form of algorithms. Cryptosystems may be categorized in two main groups:

1. Secret-key, where the processes of encryption and decryption each require the use of the same single key. The key is a number, and preferably a large one, hence the phrase 56bit key, etc. Unless the recipient of the encrypted data already knows the key, it may be left to the sender to transmit its details unencrypted. This is a notable flaw of secret-key encryption, because it exposes the key to unintended users such as eavesdroppers. A remedy is found in public-key encryption that is described below.
2. Public-key, where the sender need only know the recipient's public key. This may be obtained in unencrypted form, because it may not be used to decrypt data; rather, all it may do is encrypt data. In order to decrypt data, the recipient uses a private key that is the mathematical inverse of the public key. It may be considered impossible to determine the private key from the public key insofar as most security requirements are concerned.

The mathematics that underline public-key encryption have a simple goal: namely, to make difficult the derivation of the private key from the public key.

This is achieved through a one-way function that describes the difficulty of determining input values when given a result.

RSA is among the best-known cryptosystems or algorithms. This was developed by MIT professors Ronald L. Rivest, Adi Shamir, and Leonard M. Adleman.

Cryptology

A subject and/or science that addresses cryptography and cryptoanalysis.

Cryptosystem

A means of securing data so that it is read only by its intended users.

Crystal Reports

A reporting engine.

Cut

A technique where video footage is switched from one sequence to another.

Cut and Paste

A process by which a section of a screen image or video sequence is removed (cut) and implanted (pasted) elsewhere.

Cyberspace

A term used to describe the Internet (or Net).

Miscellaneous

Cable Modem

A modem that may operate over cable TV networks. The speed of operation is many times greater than the fastest analog modems. Typically a cable modem's data transfer rate is considerably greater downstream than it is upstream. For example, the Motorola CyberSurf cable modem offers an upstream rate of 768kbps and a downstream rate of 10Mbps. Competing cable modems have downstream rates approaching 30Mbps and faster. Cable modems offer high-speed access to the Internet, and are offered as extras by such ISPs as Telstra Big Pond (Australia).

Calling-Line Identification Presentation

A facility that reveals the caller's number.

Card

A WML page that is one of a collective deck.

Carrier

A carrier signal is used to transport a signal over media that may be physical or wireless. The carrier might be encoded using frequency modulation (FM), amplitude modulation (AM), or another modulation technique.

CCIR 601

A standard for uncompressed digital video, also known as D1. Using CCIR 601 in order to digitize a 525-line NTSC signal running at 30frames/s, its chrominance elements U and V, and its luminance Y elements are digitized individually. The Y element is digitized using 858 samples per line, and the U and V elements each are digitized using 429 samples per line. Each pixel is generated using 10 bits per sample. The digital video is coded at 270Mbits/s, which is derived as follows:

$$
\begin{aligned}
Y &: 858 \times 525 \times 30 \times 10 = 135\text{Mbits/s} \\
U &: 429 \times 525 \times 30 \times 10 = 67.5\text{Mbits/s} \\
V &: 429 \times 525 \times 30 \times 10 = 67.5\text{Mbits/s} \\
&\overline{\qquad\qquad\qquad\qquad\qquad 270\text{Mbits/s}}
\end{aligned}
$$

CCITT (Committee Consultatif International Telephonique et Telegraphic)

An international standards organization that issues recommendations and standards for communications.

CDMA (Code Division Multiple Access)

A wireless solution where signals are spread across a wide frequency band known as broadband or spread spectrum.

cdmaOne

A brand name for an application of CDMA.

Cell

A geographic area that is served by the base station.

Cell Handoff

A process where down- and up-link frequencies used by a subscriber are altered as the mobile user moves from one cell to the next.

CELP (Code Excited Linear Prediction)

An 8.55kbps voice coder that transports in a 9.6kbps data stream over CDMA.

Compression

1. A method by which data of any sort (often image and video data) is scaled down in size, eventually consuming less storage space and requiring a narrower bandwidth.
2. Video compression optimizes both the bandwidth and data storage capacity of media. Popular video compression schemes include Intel Indeo, MPEG-1, MPEG-2, and M-JPEG.
3. Audio compression serves to reduce the data storage requirements of wave audio files, and optimize the bandwidth of distribution media.

4. Disk compression increases the data storage capacity of hard disks. Commercial disk compression programs include Stacker (Stac Electronics), which is also available in hardware form that gives improved performance over software-only solutions. Stac Electronics made international news when it won a $100 million lawsuit, resulting from Microsoft infringing on its patents for compression algorithms.
5. Batch file compression is useful for archiving files and compressing them for distribution purposes. Compressed program files have to be unpacked or uncompressed before they may be run. Popular batch file compression programs include Pkzip and Lharc.
6. Data compression to reduce the size of data parcels transmitted and received using a modem. Standard data compression in this context include V.42bis.

Content Provider

A company or individual that provides usually copyrighted material for inclusion in a multimedia production. Content providers typically include publishers, recording companies, photo libraries, and so on.

Credit Card Merchant Account

A POS feature that enables an E-commerce Web site to process credit card transactions. Numerous companies offer such facilities over the World Wide Web where users are required to complete on-line forms, and produce relevant evidence of their on-line business.

Acronyms

CA-ICH Channel assignment—indicator channel
CAMEL Customized applications for mobile networks enhanced logic
CB Cell broadcast
CBR Constant bit rate
CBS Cell broadcast service
CC Call control
CCBS Completion of calls to busy subscriber
CCCH Common control channel
CCH Control channel

CCPCH	Common control physical channel
C	Codes
CAC	Channel access control
CAF	Channel activity factor
CAP	CAMEL application part
CAS	Call-associated signaling
CASE	Common application service element
CATU	Central access and transcoding unit
CC	Connection control
CC	Call control
CC2	Channel coding 2
CCCH	Common control channel
CCIR	Comite Consultatif International des Radio Communications
CCITT	Comite Consultatif International Telegraphique et Telephonique
CCK	Complementary code keying
CCPCH	Common control physical channel
CCR	Commitment concurrency and recovery
CC-Tr-CH	Coded composite transport channel
CDF	Cumulative distribution function
CDG	CDMA development group
CDG-13	CDMA Development Group 13
CDL	Coded digital control channel locator
CD-P	Collision-detection preamble
CDPD	Cellular digital packet data
CELP	Code-excited linear prediction
CN	Core Network
CS	Circuit switched
CSD	Circuit-switched data
CMIP	Common management information protocol
CORBA	Common Object Request Broker Architecture
CTG	Cell Trunk Group

D

D-AMPS

In 1977 Illinois Bell introduced AMPS (Advanced Mobile Telephone System) that was developed by AT&T's Bell Laboratories and operated between the 800MHz and 900MHz bands. It was the most used service in the United States until the early 1990s. The year 1982 marked the beginning of the development of an international standard called GSM (Global System for Mobile Communications), and a rapid rise in compatible networks emerged globally by 1993. This marked the beginning of wireless telecommunications for the mass market and was the impetus for the many growing cell phone services we see today.

The American AMPS standard used the 800MHz frequency band and was also used in South America, the Far East, and in the Asia Pacific region, including Australia and New Zealand. In the Asia Pacific country of Japan was NTT's MCS system that was the first commercial delivery of a mobile 1G Japanese network. While most first-world countries are closing 1G networks, many Less Developed Countries (LDCs) are actively investing in and upgrading them.

With 2G cellular networks we saw the emergence of Digital-AMPS (D-AMPS), Code Division Multiple Access (CDMA—IS95), and Personal Digital Cellular (PDC). GSM is the most popular globally, and in August 2000 approximately 372 GSM networks were in operation with a collective mobile user base of 361.7 million.

Specifications of D-AMPS are as follows:

- Multiple access parameters and physical layer
- Multiple access: TDMA
- Duplexing: FDD
- Channel spacing: 30kHz
- Carrier chip/bit rate: 48.6kbps
- Time slot structure: 6 slots/frame
- Spreading: N/A
- Frame length: 6.667ms
- Multirate concept: Multislot
- FEC code: Convolutional
- Data modulation: $\pi/4$-DQPSK
- Spreading modulation: N/A
- Pulse shaping: Root raised cosine, Roll-off = 0.35
- Detection: Coherent
- Other diversity means: N/A
- Power control: N/A

Data

Data transmission and reception may take place over wireless and physical data links/networks that may be packet-switched–based (PS), which are displacing the traditional circuit-switched (CS) networks. PS networks require protocols like IP and defined packets or data units that are sometimes referred to as frames or cells. IP is an example protocol used to provide WAP mobiles users with access to the Internet through an appropriate WAP gateway. Maximum sustainable data rates and assigned QoS levels are the key differences between 1G, 2G, 2.5G, and 3G networks. See Table D-1.

Packet-Switched Network

A packet-switched network uses data streams that are divided into packets coded with origination and destination information. The packets may be interleaved with different data transmissions. For instance, the packets that provide a two-way voice communication link in IP telephony might be interleaved with other streams such as videoconferencing data. Packets may follow dissimilar routes over a network, and are directed over what are perceived as the quickest and least congested paths.

If available routes or logical channels are congested, then packets are buffered before transmission. The buffer is a FIFO (first-in first-out) storage, where the first packet placed in the buffer is the first to be retrieved and transmitted when the appropriate virtual channel is available. The X.25 protocol standard dictates that a packet may contain between 3 and 4100 octets or bytes. (*See* X.25 later in this chapter.)

Up to 4095 logical channels might be accommodated on a single physical link (1997). The logical channel followed by a packet is determined by its

TABLE D-1

Data Rates

Network	Data Unit	Maximum User Data Rate	Technology	Resources Used
SMS	Single 140 octet packet	9bps	Simplex circuit	SDCCH or SACCH
CS	30 octet frames	9600bps	Duplex circuits	TCH
HSCSD	192 octet frames	115kbps	Duplex circuits	1-8 TCH
GPRS	1600 octet frames	171kbps	Virtual circuit–packet-based	PDCH(1-8 TCH)
EDGE		384kbps	Virtual circuit–packet-based	1-8 TCH

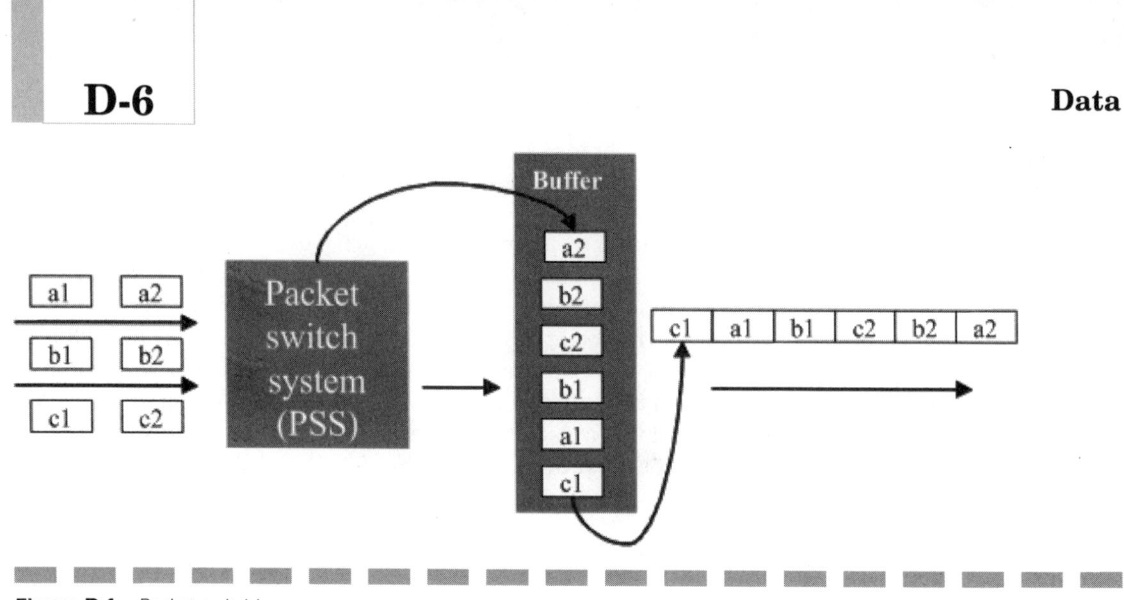

Figure D-1 Packet switching.

header information. There is also error correction, where the receiver might request that a particular packet be retransmitted.

The original packet switching standard for public data networks is CCITT X.25. (*See* X.25 later in this chapter.) This is a multitiered recommendation embodying everything from physical connectors to data formatting and code conversion.

Packet switching is rather like the logistics involved in shipping an automobile part by part. The disassembled parts are sent and assembled at the factory of destination. Equally, if a part is damaged, the factory will request that it be sent again. See Fig. D-1.

The packets may have one of two identities:

- *Multicast* packets (or items of transmitted information) that may be delivered to more than one destination
- *Unicast* packets that have one destination only

Packet-switched networks (that use IP) are currently displacing switched networks in the telecommunications industry, and drives the growing use of IP telephony or Internet telephony.

Comparative advantages of IP telephony include reduced costs, and reduced cost of ownership for telcos, and for corporations running IP-compatible networks such as intranets. The reduced costs are largely brought about by the fact that IP and Internet traffic is unregulated.

Protocol

A protocol is a format used to transmit and to receive data. Examples of industry standard protocols include IP, Ethernet, SMTP, HTTP, etc. Each protocol is optimized for the information it is intended to carry, and for the intended network. A protocol often consists of:

- An information field for data
- Destination address
- Error detection and correction codes
- Originating address

All of this information is held together in a single unit that might be a packet, cell, or frame. In IP networks, such as the Internet and intranets, they're called packets. The packets are assembled at the point of transmission, and sent over various different paths to their destination. When received, they are checked for errors, and then appropriately assembled. Network protocols are analogous to the mail service; the packets are comparable to envelopes, and they have destination and origination addresses, etc.

Coding

Data rates attainable over a transmission path may be linked to Claude Shannon's theorem that dates back to the 1940s. The theorem is the result of studying the maximum practical data rate of media that may be noisy, and provides proof that high data rates may not be linked proportionately to high error rates. The maximum channel capacity may be derived using:

$$\mathrm{C} = W \log_2(1 + S/N)$$

where W = channel bandwidth, S = signal strength, and N = power of noise.

The theorem shows that increasing either the bandwidth or the signal strength results in an increase in the channel capacity. Equally an increase in noise level requires an appropriate increase in signal strength in order to maintain data rates.

Channels may be coded using:

- Block code
- Convolutional code
- Turbo codes

Block Codes A block code encoder adds error detection data to blocks prior to transmission, so the output is therefore larger than the input. See Fig. D-2. The code rate is the ratio between the input block information bits (k) and the output block of channel-coded bits (n):

$$\mathrm{Rc} = \mathrm{k}/\mathrm{n}$$

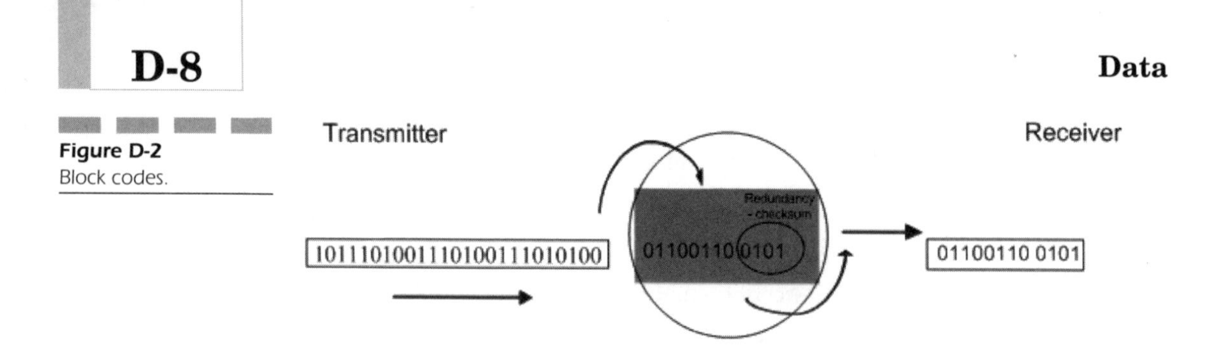

Figure D-2
Block codes.

The addition of redundant data determines whether the resulting code is systematic or nonsystematic. A systematic code has redundant bits appended to the end of the code word, and with a nonsystematic code they are mixed with the information bits.

Hamming distance between code words a and b is represented by d(a,b), and equates to the number of bit differences between the two words. The smallest Hamming distance is referred to as the minimum distance d_{min}. This indicates the level of error detection, where a minimum distance of i indicates that the channel decoder may detect i-1bit errors.

CRC (cyclic redundancy check) may be used with block coding, and also with WCDMA, where the CRC field added to a transport block may be 0, 8, 12, 16, or 24bits.

Convolutional Codes Convolutional codes operate continuously on streams of data, and the coder has memory that is used to influence its output by considering several preceding input bits. Convolutional codes are described using (n, k, m):

n: output bits per data word

k: number of input bits

m: length of coder memory

And the code rate is:

$$Rc = k/n$$

The convolutional encoder shown in Fig. D-3 is a (3, 1, 9) variant combining shift registers (D) and EXOR gates.

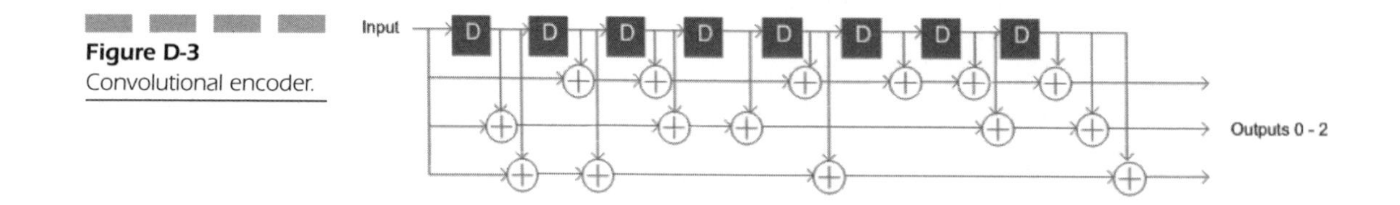

Figure D-3
Convolutional encoder.

A convolutional decoder may take the form of an MLSE (Maximum Likelihood Sequence Estimator) where the receiver:

- Generates all possible sequences that could be sent
- Compares the received bit sequence with possible sequences, and generates the Hamming distance
- Pinpoints the most probable transmitted sequences using the minimum Hamming distance

Convolutional decoders are effective against random errors, but not with bursts of errors that are prevalent in mobile radio systems. A solution to this is provided by interleaving, which distributes bit errors over a longer period.

There is a saturation point using the MLSE as it cannot handle large transmitted sequences. This problem is solved by the Viterbi algorithm that estimates the MLSE algorithm. Based on the way the decoder receives bit information, the convolutional decoder may be hard or soft decision variants. Hard decision implementations see the demodulator output as either 1 or 0, while soft decision implementations see the addition of an estimation of the reliability parameter. This weight is then used by the decoder algorithm.

Turbo Codes Turbo codes operate close to the threshold given by Shannon's Law. The UTRAN turbo encoder illustrated in Fig. D-4 is a PCCC (Parallel Concatenated Convolutional Code) with two parallel convolutional encoders separated by an interleaver. The encoders could also be arranged in serial (SCCC) with the interleaver separating them. The interleaver randomizes data and performs inter-row and intra-row permutations for input bits. The output bit sequence is streamlined by omitting the bits that were not part of the inputted bits.

The shown turbo decoder has two SISO (soft input soft output) decoders connected by deinterleavers and by an interleaver. The second decoder's output in terms of extrinsic information is relayed back to the input of the first decoder. This iterative process sees a gradual improvement of the estimate of extrinsic information and decoded data. The iteration occurs on a block basis with block sizes ranging from 40 to 5114bits.

UTRAN Channel Coding If the UTRAN uses FEC schemes, they may be either convolutional or turbo. Convolutional is used for low data rates, and turbo coding for higher data rates, with the practical threshold division between these schemes assumed to be around 300bits per TTI. Turbo coding is unsuitable for low data rates because it is inefficient with short data blocks.

The UTRAN may be without error correction; in that case FEC coding may be omitted because it gives latencies and increases the number of bits sent and so increases interference. The channel coding schemes used by UTRAN are shown in Table D-2.

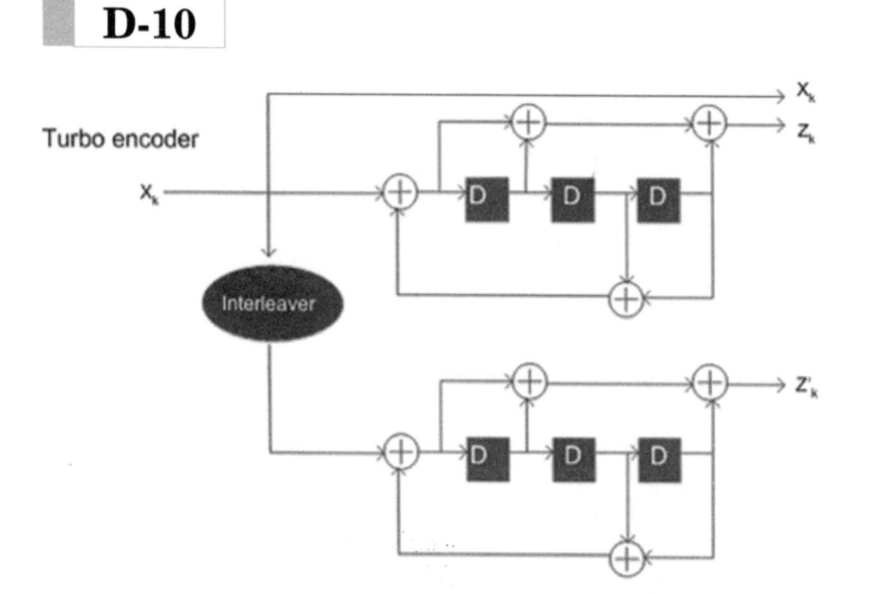

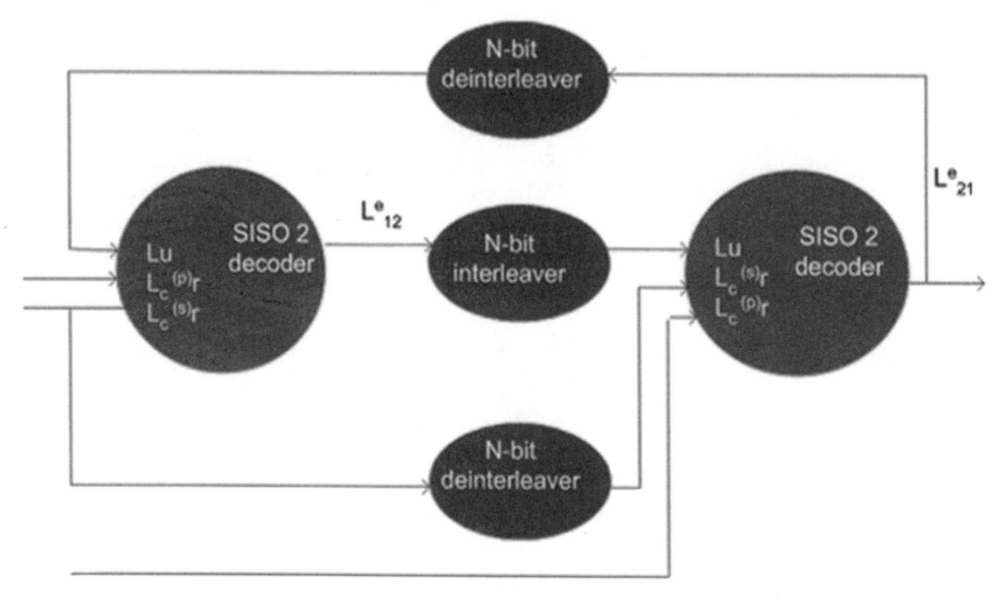

Lu - priori values for information bits u
L^e_{12} - 1st to 2nd decoder extrinsic information
L^e_{21} - 2nd to 1st decoder extrinsic information
$L_c^{(p)}r$ - parity information
$L_c^{(s)}r$ - systematic information

Figure D-4 Turbo encoder/decoder.

TABLE D-2

Channel Coding Schemes Used by UTRAN's Air Interface

TRCH Type	Coding Scheme	Code Rate
BCH PCH RACH	Convolutional coding	1/2 1/2 or 1/3
CPCH, DCH, DSCH, FACH	Turbo coding No coding	1/3

PCM (Pulse Code Modulation)

PCM is a method of encoding data in digital form, for transmission over a network, or for storage on DSM. Used in the Integrated Services Digital Network (ISDN) standard, multiplexing involves creating a data stream consisting of 8bit PCM blocks. The blocks are created every 125μs. By interleaving the blocks with those from other encoders, the result is time division multiplexing (TDM).

In North America, ISDN typically interleaves data from 24 64kbits/s sources or channels. This results in connections that provide 1.536Mbits/s. However, because each channel's frame has a marker bit "F," adding 8kbytes/s, the connection actually has a bandwidth of 1.544Mbits/s. Europe sees ISDN that typically interleave 30 64kbits/s channels, giving 2.048Mbits/s. This and the 1.544Mbits/s connection are known as primary rate multiplexes.

Further interleaving of primary rate multiplexes sees:

- 6, 45, 274Mbits/s in North America
- 8, 34, 139, and 560Mbits/s in Europe

PCM was conceived in 1937 by Alec Reeves but was not applied widely for many years.

Sampling Using ISDN, a 3.4kHz analog signal is sampled at 8kHz. The sampling rate is less than twice the bandwidth of the analog signal, in accordance with Nyquist's sampling theorem, and prevents aliasing.

A sampling frequency in a multiple of 4kHz was used because the existing networks used 4kHz carriers, and would cause audible interference in the form of whistles.

Coding The amplitude of each sample is measured, and encoded using 12bit values that give $\pm$2048 possible values.

Compression The 12bit samples are reduced to eight bits using logarithmic compression that may be:

- "mu-law" in North America
- A-law in Europe

These compression standards permit the system to be embedded anywhere in an analog network.

X.25

A standard set of protocols for packet-switched networks was introduced by the CCITT, but now comes under ITU-T. It covers the protocols between DTE (data terminal equipment) and DCE (data circuit terminating equipment).

X.25 was developed in the 1970s, when data transfer rate requirements were slow in comparison to today's. High-speed data transmission using the X.25 protocol is possible, but increasingly modern communications networks integrate frame relay.

The X.25 error-correction is accommodated using a scalable acknowledgment window that may typically include seven packets. This means that the sending device must wait for an acknowledgment for each group of seven packets.

The maximum packet size is defined as 256Bytes, so the transmitting device may send n × 256bytes of data before receiving an acknowledgment that verifies data reception.

The error correction that is integrated into X.25 is robust because earlier networks were unreliable. Today's digital networks are much more reliable; thus there is an opportunity to develop more efficient protocols.

These need not include the intensive error detection and correction of previous packet-switched protocols. Frame relay is one such relatively contemporary protocol designed for modern communications networks.

Frame Relay

Frame relay is designed for modern communications networks, and may typically operate at speeds between 9600bits/s and 2Mbits/s, although higher speeds are possible. Compared to X.25 it makes better use of network bandwidth as it does not integrate the same level of intense error detection and correction.

That is not to say that frame relay is unreliable; it is simply optimized for modern networks, which do not impose the same level of error on transmitted data—which is the case with older network technologies for which X.25 was designed. The frame relay protocol may be applied in WAN and backbone implementations, and integrated into solutions that require high data transfer speeds.

Each frame consists of a(n):

- Flag that separates contiguous frames
- Address field that stores the data link connection identifier (DLCI) and other information

- Control field that contains the frame size, and receiver ready (RR) and receiver not ready (RNR) information
- Information field that contains up to 65,536bytes
- Frame check sequence that is a CRC for error correction

DCT

DCT (Discrete Cosine Transform) is a widely used mathematical technique for image, audio, and video compression, and is key to 3G multimedia applications. It provides the basis for lossy image compression where redundant image data is omitted. It is part of the JPEG algorithm, and is also used in videotext. The DCT process operates by converting image data from the *spatial* or temporal domain to the *transform* domain. The complex underlying mathematics are transcribed to matrix manipulations. Image energy in the *spatial* domain is defined as the square of the pixel values. This energy is spread evenly over pixel blocks and resulting coefficients. Following the transformation, the energy is confined to fewer coefficients.

The resulting intensive arithmetic operations are best implemented using dedicated image processors, or general-purpose processors that have multimedia capabilities. With modifications, the hardware/software used for forward transforms may be used for inverse DCT transforms. Equally a DCT chip may be used to implement different size transforms including 4×4, 4×8, 8×4, 8×8, 8×16, 16×8, and 16×16.

DCT applications are as follows:

- Broadcast TV
- Ceptstral analysis
- Decimation (subsampling)
- Digital filter banks
- Digital Storage Media (DSM)
- HDTV
- Hierarchal image retrieval
- Image coding
- Image enhancement
- Image filtering
- Image-compression boards
- Infrared image coding
- LMS filters
- Low bit rate codecs, videoconferencing

- Packet video
- Pattern classification
- Pattern recognition
- Printed color image coding
- Progressive image transmission
- SAR image coding
- Speech coding
- Speech encryption
- Speech recognition
- Subband decomposition (filter banks)
- Texture analysis
- Videophone
- Videotext
- VQ codebook design

DCT features include:

- Picture/signal energy distributed across few coefficients
- Fast forward and inverse implementation
- Multiple size transforms possible
- Adoption by numerous internationally agreed standards, including ITU-T, ISO, IEC, JPEG, MPEG, and HDTV

Digital Multimedia

Multimedia Messaging Service

Multimedia messaging service (MMS) is a significant application of UMTS and provides services with voice, text, audio, video, images, and sound synthesis. An MSS message may have multiple components (or message elements) that are combined at the user interface to produce a multimedia presentation. This introduces a non-real-time facet where network latencies may cause no significant QoS degradation. This means that dropped packets, packet retransmission, long interleaving do little to lower the quality of service. 3GPP MMS specifications cover the technology required to distribute multimedia over mobile networks. Applications of MMS include e-cards, e-business cards, multimedia news, traffic information, music on demand, POI, e-business, and many others.

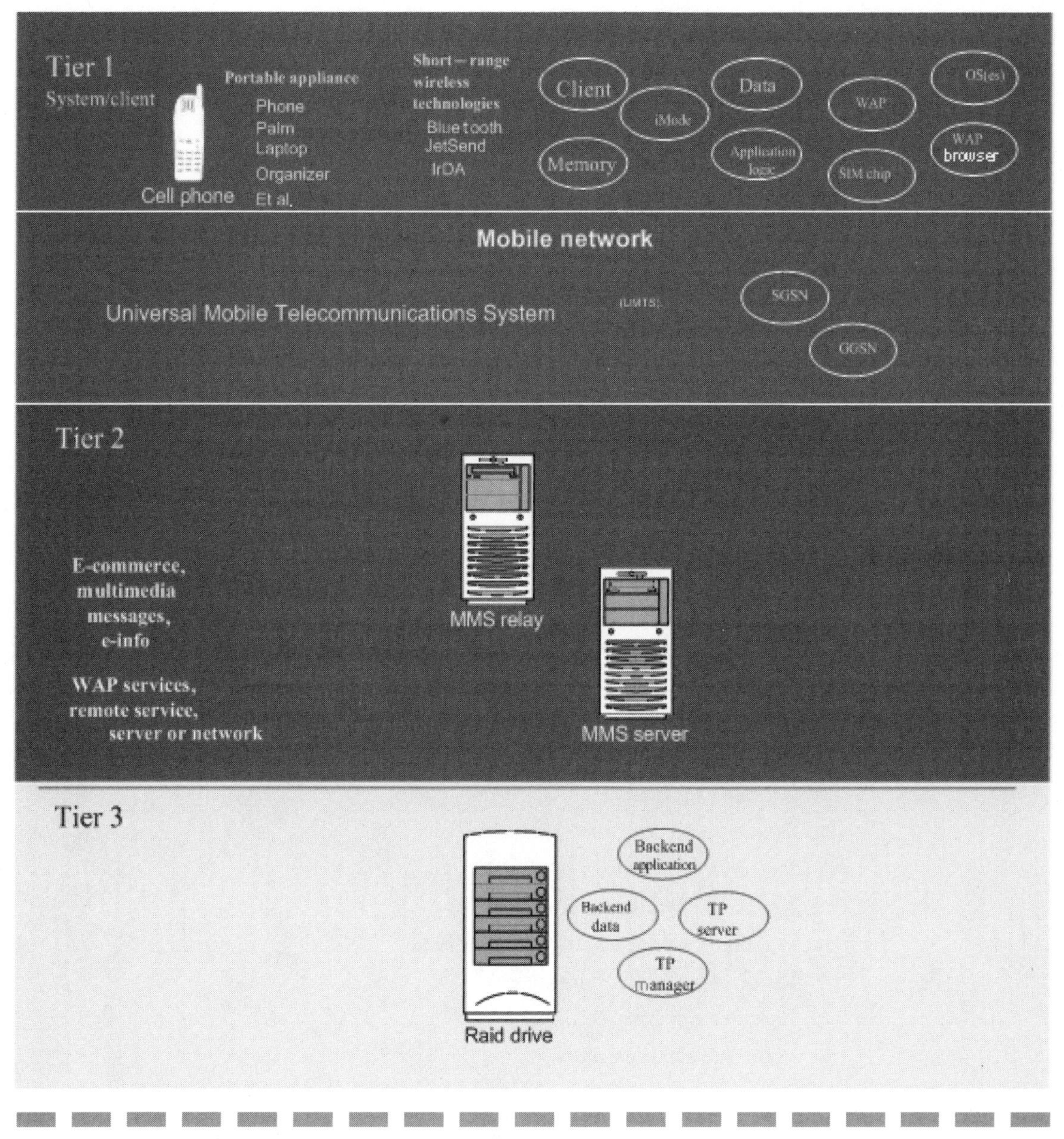

Figure D-5 MMS network.

MMS architecture Figure D-5 shows some of the components used in a multimedia messaging service environment (MMSE) and could conceivably be implemented using 2G and 2.5G networks like GPRS, EGPRS, and EDGE. The MMS user agent is part of the UE, and functions in the application layer

TABLE D-3

Multimedia Elements Supported by the MSS User Agent

Character Sets That Have a Subset of the Logical Unicode Characters	
Audio	AMR/EFR speech MP3 MIDI WAV
Image	JPEG GIF89a
Video	MPEG4 ITU-t H.263 Quicktime

providing MMS message composition, presentation, retrieval, and notifications. See Table D-3.

The MMS server stores and processes messages and may use a separate database for message storage, and may be dedicated to a message type. The MMS relay is an intermediate and control mechanism that is between the user agent and the MMS server and services.

3GPP MMS relay specified functions are as follows:

- Check terminal availability
- Convert media format
- Convert media types
- Enable/disable MMS functions
- Erase MMS messages based on user profile
- Forward messages
- Generate CDRs
- Message notification
- Negotiate terminal capabilities
- Personalize MMS using user profiles
- Receive/send MMS messages
- Retrieve message content
- Screen MMS messages
- Translate addresses

The MMS user database stores:

- MMS subscriptions
- Access control to MMS
- Data to control the extent of services
- Rules that determine handling of incoming messages and their delivery
- Data about user terminal capabilities

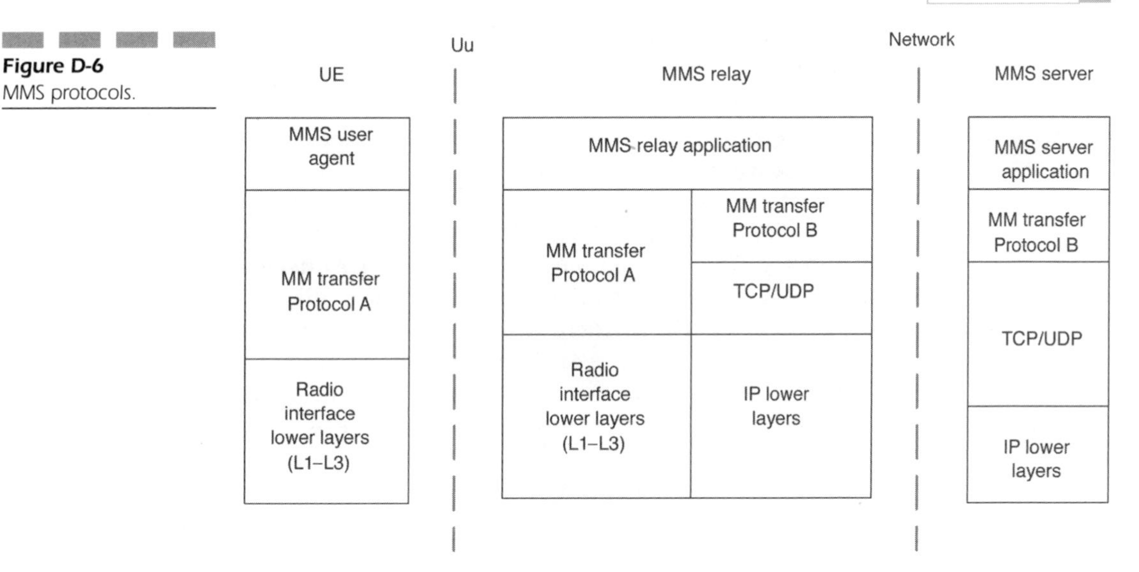

Figure D-6
MMS protocols.

MMS Protocols Figure D-6 illustrates the MMS protocol framework where the MMS transfer protocol A may include WAP and TCP/IP. WAP is now an established protocol for deploying wireless applications, and is still being used by 3G users where fast access is provided to Internet sites. The faster data transfer rates provided by UMTS will evolve WAP, and drive the inclusion of additional features that are not possible using GSM and other 2G networks. See www.wap

The other technology that fits into the MMS protocol framework is Java, and the use of Java virtual machines (JVMs) on the user's device. This gives true device independence, and renders a service universally available. The mobile device may be extremely light in terms of processing and memory because the JVM or software processor does have excessive hardware requirements. This technology model is a most challenging one, as it defies the notion of having a single browser on a mobile device.

For example, PalmPC, WAP, Windows CE, are fixed user interfaces. The JVM concept means that a UI object may be downloaded to the mobile device, so each service could have a suitable user interface. The Java framework that makes this concept a reality is called Jini, which is discussed in Discovery below. The MMS message transfer protocol layer may include SMTP, POP3, IMAP4, HTTP, or SMPP.

Digital Audio

Digital audio is an audio signal in digital form. The most common standard digital audio is that defined in the CD-DA Specification, and the most popular online digital audio is provided by the MP3 standard that uses layer 3 of the MPEG-1 audio/video compression standard. Digital audio is used widely in modern multimedia through wave audio files.

Audio may be digitized using either video capture boards or sound cards. Audio sources can take the form of a microphone, CD player, audiotape, audiocassette, and even electronic musical instruments. Audio cards may be regarded as Analog to Digital Convertors (ADCs) where the accuracy of digitization and subsequent quality achieved largely depends on the sample rate and the number of bits used per sample. The audio quality required can also be preset from within many authoring programs.

The memory capacity consumed by a sequence is a function of quality. If it is necessary to calculate the exact memory and/or data capacity consumed, then the following simple formula may be applied:

Memory capacity required (bits)

$$= \text{sequence duration (secs)} \times \text{sampling rate (Hz)} \times \text{bits per sample}$$

For example, if an 8bit sound digitizer with a sample rate of 11kHz were used to digitize a 15-s sequence, then:

$$\begin{aligned}\text{Data capacity required (bits)} &= 15 \times 11{,}000 \times 8 \\ &= 1{,}320{,}000\text{bits} \\ &= 165{,}000\text{bytes} \\ &= 161.13\text{kbytes}\end{aligned}$$

Memory or disk data capacity required naturally increases linearly with increased sample rates.

Digital Video A video sequence that is stored and played in digital form. Digital motion video is the most animating feature of modern multimedia. Using videodisc players it has been possible to incorporate color full-motion, full-screen video (FMFSV) in a computer environment for some time.

Because multimedia is a blend of concurrent processes, its storage on a single optical disc like CD or DVD requires various elements to be interleaved on the same track. Before this concept could be addressed, the inability of conventional (serial) desktop computers to play motion video stored on CD-ROM represented a significant hurdle.

Reasons as to why this is not possible lie in the inadequate rate at which data is transferred from CD-ROM to computer, and in inadequate data storage capacity. A blanket solution to both problems lies in image compression. For example, if frames of video are compressed significantly, then the need for large data storage capacity and, more importantly, high rates of data transfer is reduced.

Intel refined a technology that it acquired from General Electric in October 1988. Called Digital Video Interactive (DVI), its home was the Intel Princeton Operation that is part of the Microcomputer Components Group. It originally began in the David Sarnoff Research Center in New Jersey—once the RCA laboratories.

Using DVI up to 72 minutes of FMFSV (at 30fps) may be stored on a single 12cm-diameter CD-ROM disc.

DVI (Digital Video Interactive)

A largely obsolete, but nonetheless pioneering, video compression and decompression technology, DVI was aimed at early PC designs, including the PC ATs and IBM PS/2 systems. Intel Indeo superseded DVI. It is specified as being able to generate full-color full-screen, full-motion video (FSFMV). The original specification embodied 8bit digital video.

The MPEG was presented with the DVI compression algorithm but it was rejected. However compression techniques used in DVI were influential in the development of the MPEG compression schemes. Digital Video Interactive (DVI) was demonstrated at the second Microsoft CD-ROM conference of 1987. An image-compression technology, DVI permits full-screen, full-motion video in the PC environment. DVI offered full-color FMFSV at 10 to 30fps and a frame size of 512-by-480 pixels resolution.

Digital Multimedia

Multimedia services consisting of mixed media such as voice, audio, and video fit naturally into the 3G services/applications area given the higher user data rates offered by the networks. Video delivery over mobile networks requires compression using MPEG-1/2/3/4 or other standard algorithms that the targeted mobile devices can decode and play using an appropriate codec. Typically the data rate required to deliver streaming video to a mobile devices is high:

$$\text{Data\_rate} = \text{pixels} \times \text{bits\_per\_pixel} \times \text{playback\_rate}$$

The MPEG solution to this is basically the use of a *lossy* algorithm that omits redundant picture information from video when compressing or encoding it using an MPEG encoding solution. This encoding may occur in real-time as video is captured from analog or digital sources such as D1. The MPEG video stream consists of intermittent full frames (or I-frames) that are followed by partial screen frames called P-frames that hold only changes to the previous frame.

Videoconferencing Videoconferencing allows users in remote locations to communicate in real time both visually and verbally.

Systems may be divided into the following categories:

- Mobile videoconferencing
- Desktop videoconferencing using conventional desktop or notebook computers
- Conference room videoconferencing that typically includes appropriately large displays

Videoconferencing systems include a camera, microphone, video compression/decompression hardware/software, and an interface device that connects the system to an access technology.

The interface device may be:

- A wireless 3G interface
- A conventional modem used to connect with an ISP or intranet server, and thereafter use an Internet-based videoconferencing solution such as CU See-Me
- A cable modem that might provide high-speed internet access via cable
- An ISDN interface that provides connection to the Internet or appropriate IP network
- A Network Interface Card (NIC) that connects to a LAN
- A wireless interface that provides connection over GSM or other mobile communications network

Access technologies for videoconferencing include 3G, PSTN, ISDN, ADSL, cable, GSM, ATM, T1 frame relay, and proprietary wireless technologies. Point-to-point videoconferencing involves communication between two sites, while multi-point videoconferencing involves interaction between more than two sites. The latter might require a chairperson to conduct proceedings. Also the collective system might be voice activated, switching sites into a broadcasting state when the respective participant begins speaking.

MPEG-1 MPEG-1 is an established and internationally agreed digital video compression standard used widely in fixed and mobile networks. It is used for local playback and for streaming multimedia over the Internet, and other IP and multimedia networks. The early days of digital video were plagued by the problem of just how digital video data should be compressed, thus illuminating the need for international standards for the digital storage and retrieval of video data.

Sponsored by the then ISO (International Standards Organization) and CCITT (Committee Consultitif International Telegraphique et Telephonique), the Motion Picture Experts Group (MPEG) was given the task of developing a standard coding technique for moving pictures and associated audio.

The group was separated into six specialist subgroups, including Video Group, Audio Group, Systems Group, VLSI Group, Subjective Tests Group, and DSM (Digital Storage Media) Group.

The first phase of MPEG work (MPEG-1) covered DSMs with up to 1.5Mbits/s transfer rates, for storage and retrieval, advanced Videotex and Teletext, and telecommunications. The second phase (MPEG2) of work addressed DSMs with up to 10Mbits/s transfer rates for digital television broadcasting and telecommunications networks. This phase would cling to the existing CCIR 601 digital video resolution, with audio transfer rates up to 128kbits/s. MPEG1 was finally agreed on, developed, and announced back in December 1991.

MPEG participants included leaders in computer manufacture (Apple Computer, DEC, IBM, Sun, and Commodore); consumer electronics; audio visual equipment manufacture; professional equipment manufacture; telecom equipment manufacture; broadcasting; telecommunications; and VLSI manufacture. University and research establishments also play an important role.

It provided a basis for the development of Video CD, which was specified publicly by Philips in late 1993. This is an interchangeable format that may be played using both PCs fitted with appropriate MPEG video cards and compatible CD-ROM drives, as well as Philips CD-I players fitted with Digital Video cartridges. Its development is constant so as to accommodate the increasing data transfer rates of both DSMs and other video distribution transports.

MPEG-1 compression is optimized for DSMs with data transfer rates of up to 1.5Mbits/s. MPEG-2 accommodates DSMs and video distribution transports capable of supporting higher data transfer rates of up 10Mbits/s. MPEG-4 video compression is designed to transmit video over standard telephone lines.

An MPEG video stream generally consists of three frame types:

- Intra
- Predicted
- Bidirectional

Central to MPEG encoding is the use of reference or intra (I) frames that are complete frames and exist intermittently in an MPEG video sequence.

The video information sandwiched between I frames consists of that which does not exist in the intra frames. Information that is found to exist in the intra frames is discarded or "lossed." Intra frames can act as key frames when editing or playing MPEG video because they consist of a complete frame.

Generally, compressed MPEG video is difficult owing to the paucity of authentic access points. However, editable MPEG files do exist, one of which is backed by Microsoft. Additionally, an MPEG video stream composed entirely of I frames lends itself to nonlinear editing.

The quality of MPEG video depends on a number of factors ranging from the source video recording quality to the use of important MPEG parameters that affect the overall compression ratio achieved.

Contrary to popular belief, the logical operations that provide a basis for obtaining high-quality MPEG video are by no means the preserve of expensive video production bureaus.

Equipped with a reasonably specified PC and a basic understanding of MPEG video, there is nothing to stop you from producing good-quality White Book–compatible video on your desktop.

Probably the most obvious elements that influence MPEG video quality include the analog or digital source recording, the video source recording format, and the video source device specification.

It may be assumed that the higher resolution S-VHS format will provide slightly better results than VHS, but there will not be a dramatic improvement in resolution because the MPEG SIF is standardized at 352-by-288 pixels for PAL.

If you are digitizing the sound track of the source video recording also, then you will probably obtain the best results with camcorders and VCRs that offer hi-fi–quality stereo sound.

When capturing a video file so that it may eventually be compressed, it is important to choose an appropriate capture frame rate, capture frame size, and image depth.

The capture frame rate should be set for 25 frames/s for PAL and 30 frames/s for NTSC. Frame rates that differ from these will cause the MPEG video sequence to run at the wrong speed, and it will not be White Book compliant.

The capture frame size should correspond with the MPEG-1 SIF, which is 352-by-288 pixels for PAL and 352-by-240 pixels for NTSC. Authentic MPEG requires a truecolor image depth of 24bits per pixel, giving a total of over 16.7 million colors, which are generated by combining 256 shades of red, green, and blue.

The quality of captured audio that is used as an input audio stream obviously depends on the sample size, recording frequency, and whether mono or stereo is chosen.

You can assume that your wave audio recorder or video capture program will provide sampling rates of 11kHz, 22kHz, and 44.1kHz, and samples sizes of 8bit and 16bit. While higher sampling rates and larger sample sizes yield improved audio quality, the resultant audio stream can consume an unacceptably large portion of the available MPEG-1 bandwidth.

With regard to careful adjustment of the MPEG compression parameters, there is not much you can do if the MPEG encoding software provides no control over them. If it does, then it may be assumed that a greater number of I frames can improve the quality slightly, though this will introduce an overhead in terms of lowering the compression ratio.

A SIF frame has an MPEG-1 frame resolution of 360-by-240 pixels for NTSC, and 360-by-288 for PAL. This resolution equates to the MPEG Source Input File (SIF) that is achieved by omitting odd or even lines from a standard interlaced PAL (Phase Alternating Line) signal.

This is an exceptionally "lossy" procedure, omitting a great deal of picture information and losing video quality. It is this single operation that limits the quality of video that can be achieved using MPEG-1, though it has to be implemented in order to confine the video stream to the narrow bandwidth of about 1.5Mb/s. The MPEG claim that this is VHS-quality is an area of debate.

MPEG-1 video production has also become increasingly popular using comparatively inexpensive video capture hardware and encoding software. An alternative to such video production is to use the services of a fully equipped bureau. The decision as to whether or not the services of a bureau should be used is driven by a number of obvious key factors that include the amount of encoding you require, for which you will be charged on a per-minute basis.

The production of MPEG-1 video using encoding software begins with the capture of a video sequence from a source recording that might be in the VHS or S-VHS formats. Film studios and production companies might rely on professional and broadcast-quality formats such as Digital Betacam or D1 for their source recordings.

The capture process can be carried out using an appropriate video capture card that provides adequate control over capture parameters, allowing you to a set a capture frame rate of 25 frames/s, a truecolor image depth of 24bits, providing 16.7 million colors per captured pixel, and an acceptable frame resolution of 352-by-288 pixels.

This resolution equates to the MPEG Source Input Format (SIF), which is achieved by omitting odd or even lines from a standard interlaced PAL (Phase Alternating Line) signal. This is an exceptionally "lossy" procedure, omitting a great deal of picture information and losing video quality.

It is this single operation that limits the quality of video that can be achieved using MPEG-1, though it has to be implemented in order to confine the video stream to the narrow bandwidth of about 1.5Mb/s. If you are unable to capture video at 25 frames/s, then you can increase the frame rate following video capture using Video for Windows VidEdit, or an equivalent digital video editing program. The increased frame rate is achieved merely by duplicating frames, but it does mean that the finally encoded MPEG video stream will at least be an authentic one.

A video editing program can be used to synchronize audio and video streams, usually by introducing a time offset for the audio track. By then separating the file into video and wave audio files, once again using the video editing program, their play times should become equal.

Some MPEG encoders will automatically alter the length of the input audio file so that it matches the length of the input video file. However, this does not guarantee that the audio and video material will be synchronized correctly when multiplexed. It should be added that the synchronization of audio and video information can also be carried out at the decoding stage.

Video Capture The video source recording might be analog or digital when capturing video for transmission over wireless networks. The latter requires that the video capture card incorporates an appropriate input. The three general types of video capture include:

- The real-time video capture technique involves digitizing the incoming video source signal on the fly, and the video source device is not stopped or paused at any moment during capture.
- Automatic step-frame capture requires that the source device is stopped, paused, and even rewound so as to digitize a greater amount of the source recording. It offers certain advantages, namely, it is possible to achieve a greater number of colors (or greater image depth), higher capture frame rates, and larger capture frame resolutions than would normally be possible using the same video capture hardware and software configuration to record video in real time.
- Manual step-frame capture usually depends upon the operator clicking a button on screen in order to capture selected video frames.

Available color depths using fully specified video capture card and capture program partnerships include 8bit, 16bit, and 24bit. The 8bit format gives a maximum of 256 colors stored in the form of a color palette that can be edited using programs such as PalEdit. 16bit and 24bit formats are described as true-color, giving a maximum of 65K (2^{16}) and 16,777,216 (2^{24}) colors, respectively, and when using appropriately specified video capture hardware and software they can produce impressive results.

Using many video capture systems, the data throughput required to capture 16bit and particularly 24bit video in real time limits both the capture frame rate and frame size. The frame dimensions chosen hinge largely on the specification of the capture card, though the image depth chosen is also influential as is the capture frame rate.

Though the video frame dimensions can be scaled using video editing programs and even multimedia authoring tools, enlargement can result in a blocking effect as the individual pixels are enlarged. However, certain graphics cards, particularly those that enlarge Video for Windows video sequences, will apply a smoothing algorithm during playback in an attempt to minimize the blocking effect.

Video editing techniques also can be used to increase the playback frame rate (through frame duplication). This, other digital video editing techniques, and hardware/software features of the playback system can help improve the quality of video playback.

However, capturing and compressing optimum quality digital video relevant to the intended playback platform remains the most important process. There are limitations in what can be achieved through digital video editing, and through playback hardware that enhances digital video playback.

The original video sequence may be enhanced, even enlarged through duplication, but it cannot be used to play video information present in the source recording that it simply does not contain. Even though numerous algorithms can enhance digital video, and numerous other will emerge, it is reasonable to assume that if the video file does not contain a particular frame then that frame cannot be played.

The quality levels available using wave audio recorders together with mainstream sound cards, also can be achieved through fully specified Windows video capture programs. 8bit or 16bit sample sizes are available, recorded at frequencies of 11.025kHz, 22.05kHz, and 44.1kHz in mono or in stereo. The size of the sound track, which increases in relation to the recording quality chosen, can be monitored using many video editing programs.

MPEG Frames An MPEG video sequence consists of partial frames in the form of Predicted (P) frames and Bi-directional (B) frames, and full frames or Intra (I) frames.

I frames are compressed in a similar way to JPEG (Joint Photographic Experts Group) images and do not rely on image data from other frames. They exist intermittently, perhaps between 9 and 30 frames, and provide nonlinear entry points.

Increasing the frequency of I frames provides a greater number of valid entry points, but the compression ratio of the overall file diminishes proportionately. Realistically, the compression ratios achieved using MPEG may be assumed to be around 50:1. Higher compression ratios lead to an unacceptable loss of quality, and it is wise to forget the 200:1 ratio that MPEG is supposedly capable of producing. Normally this is achieved through a pretreatment process that dramatically reduces the number of frame pixels.

I frames and the following P and B frames are termed Groups of Pictures (GOPs), and the occurrence of each frame might be predefined through the careful adjustment of MPEG parameters prior to encoding. However this fine level of control over compression parameters may not be provided by low-cost MPEG encoding programs.

MPEG-2 An improved version of MPEG-1 video compression, MPEG-2 is supported by DVD technology. It was developed for media and networks able to deliver 10Mbits/s data transfer rates.

MPEG-1 was developed for narrow-bandwidth media, such as the original single-speed CD drive variants that offered average data transfer rates of approximately 150kbytes/s or 1.2Mbyts/s.

MPEG-2 video may contain considerably more audio and video information than MPEG-1. The most noticeable improvement is the higher playback screen resolutions that are possible, making possible D1 or CCIR 601 quality. DCT is key to MPEG-2, as it is to MPEG-1 and JPEG (or even M-JPEG). As is the case with MPEG-1, MPEG-2 requires decoding solutions that may be hardware based such as set-top boxes (STBs), or equivalent hardware implementations integrated in computers. Applications of MPEG-2 video include Video-on-demand, multimedia, videoconferencing, etc. It may also be stored and delivered using DVD variants.

Discovery (Jini)

Services or entities that join a *djinn* use the *discovery* protocols to find discoverable lookup services or to find specified lookup services only, and register with them, or obtain references to them. These stages and their substages are precursors to the many others defined by the *join* protocol that render the service an operational entity in the context of those that make the *djinn*. The *djinns* that are not local, or used on a regular basis by a service, are best joined using the *unicast discovery protocol* that permits communication with a specific lookup service. In contrast those entities that require the discovery of local lookup services use the *multicast request protocol*–which is another discovery protocol. The lookup services themselves, if started or restarted following a network or communications failure, may make use of another discovery protocol called the *multicast announcement protocol* that is used to broadcast their presence to potential clients. These three discovery protocols are architecture neutral in the Jini perspective, and information is exchanged using the Java platform's object serialization.

Hosts

To become part of *djinn,* a host must have a JVM implementation and have access to all the appropriate Jini or Jini-related packages, and a protocol stack whose configuration will depend on the protocols used. In the typical scenario where we use the IP protocol the host must:

1. Support unicast TCP for RMI, and multicast UDP for discovery
2. Have an IP address that is assigned statically, or dynamically using DHCP
3. Include a running entity like an HTTP server that may download RMI stubs and other code to remote entities, if it is not platformed elsewhere on the network

Discovery

Multicast Request Protocol The Multicast Request Protocol may be used over physical or wireless Local Area Networks (LANs) that are typically the fabric of Jini services. The network's interwoven technologies are transparent to the Discovery and Join protocols, and may include traditional 10 or 100Mbps or more modern *Gigabit Ethernet* NICs in the home or in the office and may use the multicast request protocol so that djinns may be discovered by an entity that:

1. Creates a multicast request client that sends packets to the multicast request service at the network's multicast endpoint.
2. Creates a TCP multicast response server socket that listens to the connection used by the unicast discovery protocol.
3. Creates net.jini.core.lookup.Service ID objects that hold service IDs for lookup services that have responded and been heard.
4. Periodically sends multicast requests that contain the recent service IDs for lookup services heard, and data to connect with the multicast response server. The interval between requests is not specified but 5 s is recommended.
5. Adds the service ID for lookup services received from the multicast response service.
6. Sends multicast requests until a time when it unexports its multicast response server.
7. Stops the discovery process if it has received adequate references to lookup services, else it begins using the multicast announcement protocol.

These multiple processes and subprocesses are initiated by the discovering entity, and may be perceived as existing in a client/server architecture that has a peer-to-peer personality, and is naturally an OO system, and also a distributed event-driven one. The discovery process is nothing more than a conversation between the discovering entity and its JVM and the lookup service and JVM as illustrated in Fig. D-7.

Multicast Discovery Request Packet Format A packet is merely a standard way of sending information from one point to the next over a physical or wireless Jini connection, and is the basis of a protocol like those of Jini's *Discovery* and *Join*. The packet itself has a payload of 512bytes and is transported using the UDP protocol that has become popular for streaming media applications and for those solutions where dropped packets are not mission critical. Its other key features focus on architecture-neutral Jini concepts, and on the ease of encoding and decoding. Transmitted data is,

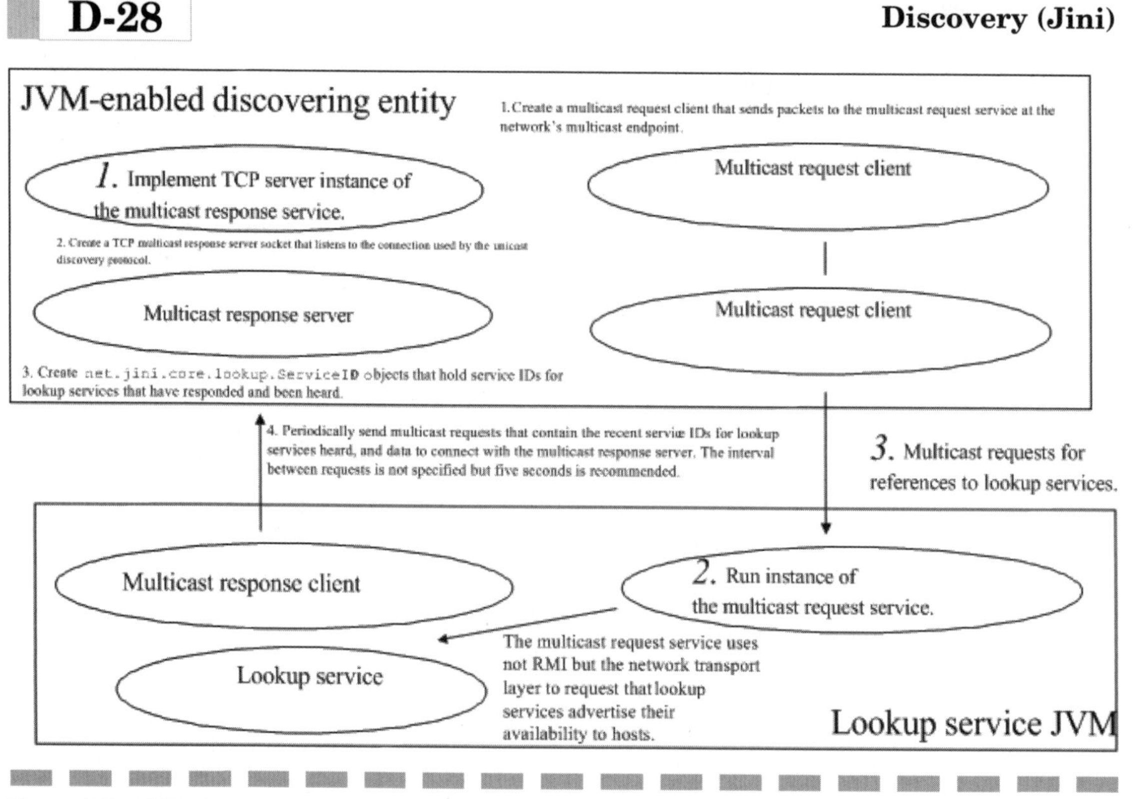

Figure D-7 Multicast request protocol.

of course, generated or encoded, transmitted, and decoded as a serial sequence of bytes, and may be encoded into a byte array, and sent using a java.io.ByteArrayOutputStream object.

A packet is merely a standard way of sending information from one point to the next over a physical or wireless Jini connection, and is the basis of a protocol like those of Jini's *Discovery* and *Join*. The packet itself has a payload of 512bytes and is transported using the UDP protocol that has become popular for streaming media applications and for those solutions where dropped packets are not mission critical. Its other key features focus on architecture-neutral Jini concepts, and on the ease of encoding and decoding. Transmitted data is, of course, generated or encoded, transmitted, and decoded as a serial sequence of bytes, and may be encoded into a byte array, and sent using a java.io.ByteArrayOutputStream object. See Table D-4.

Multicast Request Server

When an instance of the *multicast request service* is invoked, the hosting server's first process ensures that incoming multicast requests may be received, by establishing a datagram socket that is bound to the multicast

TABLE D-4

Multicast Request Packet Contents

Count	Type	Description
1	Int	The protocol version number must currently be 1, and the inclusion of this `protoVersion` variable permits a simple test for compatibility between Jini devices and networks.
1	Int	The Port field holds the TCP port that responses must address.
1	Int	A variable that holds the number of lookups heard.
variable	Net.jini.core.lookup.ServiceID	The *heard* variable has an array of `Net.jini.core.lookup.ServiceID` objects, and provides a record of lookup services that have already responded, and therefore do not need to respond to the current request.
1	Int	A variable that holds the number of groups.
variable	Java.lang.String	The *groups* variable has a string array that names the groups that are to be discovered. This is useful for discovering specific services. If the array is empty, then all lookup services on the network will be searched.

endpoint (where the multicast request service resides). A received multicast request has a service ID set that may be used to determine whether or not to respond to a sending entity. It must respond if the Multicast Request Server's own service ID is not in the set, or if it is a member of any requested groups, or if the set of requested groups is empty. When a response is required, the multicast request server connects with multicast response server, and sends a lookup service registrar object. If more than one *djinn* sends calls to an entity's discovery response service, an appropriate interface may allow the user to join one or more *djinns* by choosing them from a list.

Multicast Announcement Protocol and Service

Jini lookup services use the multicast announcement protocol to advertise themselves on a network. The multicast announcement service communicates from a single multicast announcement client (that is on the same system as the lookup service) to (one or more) multicast announcement servers that are run by listening entities. The longevity of an instance of the multicast announcement client approximates that of the lookup service it serves, because without it the lookup service may not operate.

The entity holding the lookup service performs key processes that encapsulate a number of others, and begin by making a datagram socket object to transmit to the endpoint where the multicast announcement resides. This process is followed by the creation of the server-side of the unicast discovery service, and finally multicasts of announcement packets are transmitted at a (recommended) frequency of 120s.

Meanwhile entities may listen to multicast announcements by creating service IDs of lookup services that it has already heard from. They then create a datagram socket that is bound to the multicast endpoint where the multicast announcement service is operating. They then become listening entities and when they receive announcements they determine whether or not they have heard from the lookup service by checking service Ids. Those that have been heard already or are irrelevant are discarded, while those that have not been heard before or are of interest may be discovered and their service ID added to the entity's set. This process requires unicast discovery using the host and port information held in the announcement that is used to get a reference for the announced lookup service. See Table D-5.

Security and Multicast Discovery The `net.jini.discovery.DiscoveryPermission` class is the basis for security files that give permissions to creators of `LookupDiscovery` objects. A security file grants the creator with permissions to attempt the discovery of a group in a set or all groups.

net.jini.discovery.DiscoveryPermission ""	A permission to discover the public group.
net.jini.discovery.DiscoveryPermission "*"	A permission to discover all groups.
net.jini.discovery.DiscoveryPermission "Hifi"	A permission to discover the group Hifi.
net.jini.discovery.DiscoveryPermission "*.slam.com"	A permission to discover those groups ending in "slam.com."

Unicast Discovery Protocol

The unicast discovery process is where an entity receives a multicast announcement packet, or a series of packets, and if the entity could talk, it would say, "I'm interested in that service," or "My owner or user is going to be interested in that service, and I have not heard from it before so I want it right now!" All unicast discovery does is to obtain a reference to a lookup service of what is possibly a remote djinn. A key technical difference between it and the multicast announcement hinges on the use of reliable unicast TCP instead of the less reliable UDP.

A lookup service's response to a multicast request is to make a TCP connection with the address and port held in the request; it then receives a unicast

TABLE D-5

Jini Multicast Announcement Protocol

Count	Type	Description
1	Int	The protocol version number must currently be 1, and the inclusion of this protoVersion variable permits a simple test for compatibility between appliances. It is written in a manner similar to the `java.io.DataOutput.writeInt` method.
1	Java.lang.String	This field contains the host name used by recipients that will perform unicast discovery. It is written similar to the `java.io.DataOutput.writeUTF` method.
1	Int	The Port field holds the TCP port of the host, and is written in a manner that echoes the `java.io.DataOutput.WriteInt` method.
1	Net.jini.core.lookup.ServiceID	The Service ID field tracks the services that have sent announcements thereby avoiding unnecessary unicast discovery. The field is written in a manner that echoes the `java.jini.core.lookup.ServiceID.writeBytes` method.
1	Int	This field indicates the number of groups a lookup service is a member of, and is written similar to the `java.io.DataOutput.writeInt` method.
variable	Java.lang.String	This is a sequence of strings that denotes the groups that the lookup service is a member of. Instances of this field may be written in manner that echoes the `java.io.DataOutput.writeUTF` method.

discovery request, and the lookup service replies by sending a proxy object. A discovering entity may have a djinn's lookup locator object that specifies the host and port that it must connect to using TCP and send a request in order to receive the appropriate response. See Fig. D-8.

Computers/Software

Daemons

A daemon is a program or process dedicated to perform what is usually a singular given task, such as sending mail. TCP/IP daemons include those added by third parties that include SCO.

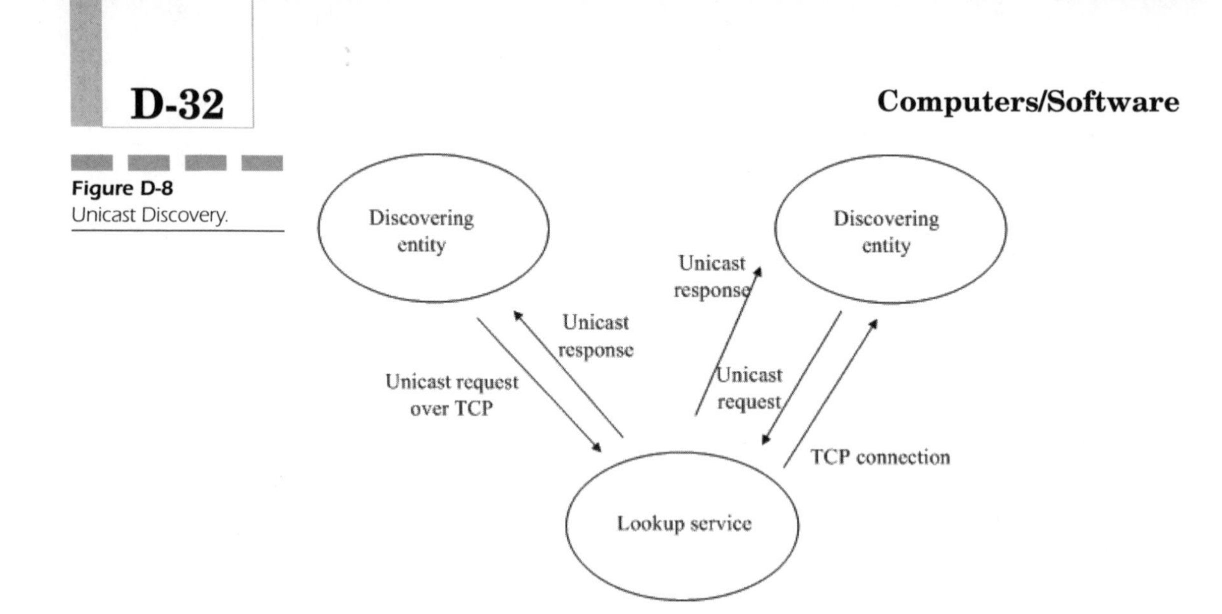

Figure D-8
Unicast Discovery.

Data Cube

A data cube is an information storage model. In the context of a data warehouse, data cube area evolved as a result of extractions from operational data. They may be assumed to be static entities that do not change, and may not be altered or even built from query data. A cube cache is used to store them in memory. If grown beyond three dimensions, the cube becomes a hyper-cube.

Data Dictionary

A data dictionary is a type of metadata that defines stored data along with its relationships. Typically the database dictionary is dynamic, updating its contents as data structural changes occur.

Data Extraction

A data extraction process abstracts data from one or more sources, in order to build a static database of unchanging data.

Data Mart

A datamart is a single-subject (and generally small-scale) data warehouse that provides DSS for a limited number of users.

Data Partitioning

Data partitioning is a method of segregating data, so that it is distributed across different systems. It may serve to store selected records in more secure (and often expensive) mass storage (such as SRAM or an appropriate level of RAID), while storing less important data in conventional storage media, namely, a hard disk.

Data Schema

A data schema describes a database structure, such as the entity relationship (E-R) diagram of an RDBMS. The E-R diagram shows the links that unite the database tables.

Data Sonification

Data sonification is a general term used to describe the process of enhancing data through the addition of audio.

Data Transfer Rate

A rate that data is transferred from a mass storage device, such as a hard disk, or from removable media, or over a physical or wireless medium.

Data Type

A data type is a classification for data. Modern relational databases commonly store the following different data types: currency, numeric, date, alphanumeric, Boolean, graphical, and BLOB (Binary Large Object).

Data Warehouse

"An integrated, subject-oriented, time-variant, nonvolatile database that provides support for decision making" (Bill Inmon).

A data warehouse is a unified data repository extracted from multiple data storage structures that may emanate from various data sources. It provides a single interface with relational and/or multidimensional data. It is the rebirth of what IBM termed the Information Warehouse in the 1980s.

Data warehouses form the information storage methodology in modern decision support systems (DSS). Collectively, these systems provide a means of querying data that emanates from disparate information storage models.

On-Line Analytical Processing (OLAP) is a crucial facet of the data warehouse architecture, providing a means of abstracting and analyzing data in a manner that makes transparent the multiple data sources and data storage models used.

The data mining system (DMS) is also a key DSS component. Data mining is an attempt to embed intelligence into the interrogation of stored data, and may automate the querying of data, and provide user access to new data structures whose information is in close proximity in terms of related subject matter, and may assist in solving defined problems.

The underlying storage metaphors of a data warehouse may be:

- Two dimensional, where values are stored using the table metaphor, adhering to the established formal RDBMS model for information storage
- Multidimensional, where data is perceived as a three-dimensional cube or a data cube, where values have x, y, and z coordinates

Data cubes evolved as a result of extractions from operational data. They may be assumed to be static entities that do not change, and may not be altered or even built from query data. A cube cache is used to store them in memory. If grown beyond three dimensions, the cube becomes a hypercube. See Fig. D-9.

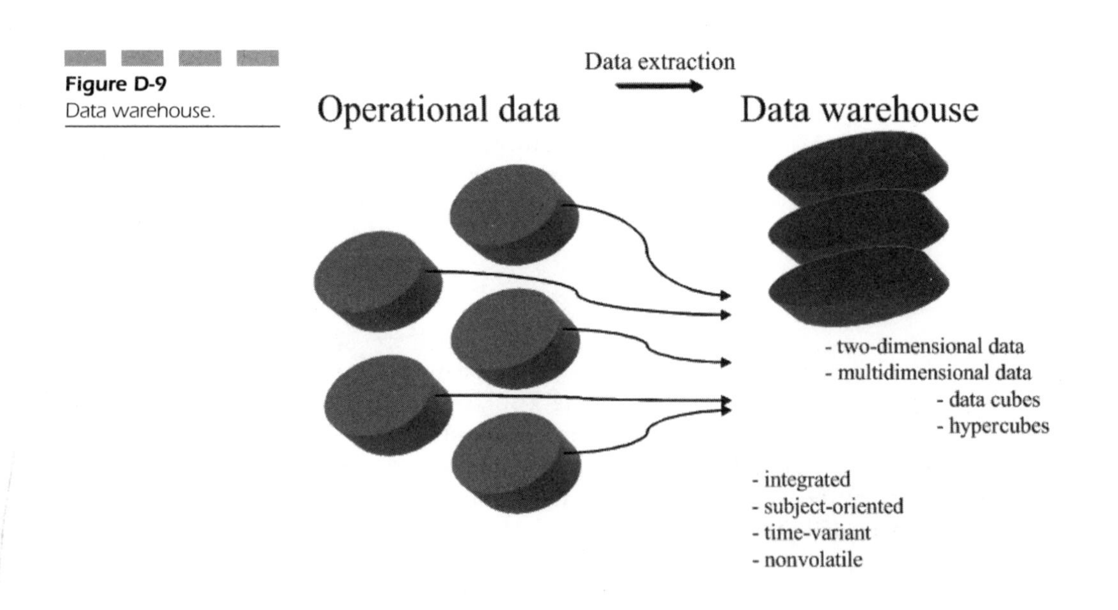

Figure D-9
Data warehouse.

According to Inmon's definition, a data warehouse is characterized as:

- Integrated, providing a unified interface to multiple data sources that may use disparate information storage models
- Subject-oriented, revealing data which is in close proximity in terms of subject matter, providing related information that may be dedicated to specific analysis
- A time-variant, permitting data retrieval and analysis using the dimension of time
- Nonvolatile, making the collective data entities static in definition, except during the periodic instances where updates are driven through the data by the integrated operational systems. On-line updates are impermissible, and the data warehouse may be considered as being read-only

Database

A database is an electronic information storage system offering data storage and retrieval.

A generic term that describes the storage of information on a record-by-record basis. Records are divided into fields of different types including text, numeric, date, graphic, and even BLOB (Binary Large Object).

The records are stored in tables or files. Database types include flat-file and relational. The flat file database model embodies no links between different files or tables.

A relational database is quite different in that records from one file may be linked to records stored in a separate file or table. Codd's standard text about relational databases published in the 1960s specified different types of relational links.

Types of links include one-to-one, one-to-many, and many-to-many. There are many commercial examples of relational databases that base their design on the original writings of Codd.

Relational databases are formally referred to as RDBMSs (Relational Database Management Systems); flat-file databases are termed simply DBMSes (Database Management Systems). Commercial examples of software products that permit the development of RDBMSs include Paradox for Windows, dBase, Microsoft Access, Oracle, and Ingress.

Relational databases are used to store tabular information in the form of records, and useful versions are able to generate graphs. Popular PC relational databases include Microsoft Access, Borland Paradox, dBase, Q & A, and DataEase. Because they are relational, an invoice can extract information from a number of different tables or files.

Flat-file databases are used to store isolated records, and cannot be used to link files or tables. They are used for simple applications such as card files.

Text databases are used to store documents such as articles and even complete books. Documents may be indexed where the user interface simply allows users to search for documents that contain target words, phrases, or sentences.

DBMS (Database Management System)

A DBMS provides the operations necessary to manage stored data that may be two-dimensional or multidimensional. A DBMS:

- Requires a data dictionary that defines stored data along with relationships. The database dictionary is dynamic, updating its contents as data structural changes occur.
- Ensures that entered data undergoes predefined validity checks.
- Transforms entered data so it may be stored by the underlying data structure.
- Provides storage for data, its relationships, forms, reports, queries, and miscellaneous files.
- Includes security features, such as the password protection of files, allocated user access rights, prohibits certain users from accessing certain files and from making data changes.
- May maintain data integrity in a multi-user environment.
- May provide a database communications interface that might permit users to submit forms-based queries through Web browsers, and to publish reports and data using various media that include the Web, Email, and Lotus Notes.
- Provides features pertaining to backup and recovery.
- Provides access to data using a query language (such as SQL or a variant thereof), or a querying mechanism which might involve the completion of tables using defined query statements [such as the Borland Query By Example (QBE) technique].

DCOM (Distributed Component Object Model)

A Microsoft technology or protocol that permits interaction between entities over a network. An open standard, DCOM is operable with standard Internet protocols that include HTTP and TCP/IP.

DDE (Dynamic Data Exchange)

DDE is a standard technique for data exchange between running Windows applications. For example, a database tool might have a DDE interaction with a spreadsheet in order to draw graphs based on spreadsheet data. A DDE interaction is occasionally called a conversation. Nowadays most Windows users harness OLE (Object Linking and Embedding) rather than DDE. OLE 2.0–compatible applications may be assumed to be considerably less difficult to link.

Debugger

A debugger program or feature that permits program code to be corrected or debugged. It assists the process through appropriate prompts and indications as to where the bugs exist in the source code listing.

Decode

Verb: A process by which encoded data that may be compressed is interpreted and delivered to the receiving system or device. For example, the process may involve the decoding of MPEG video.

Decryption

A decryption process unlocks encrypted data so that it becomes readable.

Defragmentation Program

A defragmentation program is used to defragment a hard disk. It ensures that used data blocks are arranged in a contiguous stream.

Density

Density is a measure of how densely packed data bits are on a storage medium.

DES (Data Encryption Standard)

An encryption technique; a symmetric cryptosystem. Both senders and receivers use a common 56bit key to encrypt and decrypt messages and data. The U.S. government backed DES in 1977, and has since recertified every five years. Variants of DES include DESX.

DHTML (Dynamic Hypertext Markup Language)

DHTML is an object-based version of HTML and was released jointly by Microsoft and the World Wide Web Consortium (W3C).

Dialup Password File

A dialup password file stores passwords, and authenticates access to networks via dialup links.

Dictionary Attack

A dictionary attack on a cryptosystem using the iterative technique of comparing the key with a dictionary of possibilities, usually beginning with those that are most likely to match the key.

Digital

A device such as a computer that processes and stores data in the form of ones and zeros. In a positive logic representation, "one" might be "+5" volts and zero might be "0" volts. This lowest of levels at which computers operate is known as machine code. Binary arithmetic and Boolean algebra (named after Irish mathematician George Boole) permit mathematical representation. Boolean algebra and Karnaugh maps are used widely for minimization of logic algebraic expressions. Though digital signals exist at two levels (one or zero), an indeterminate state is possible.

Digital Certificate

A digital certificate links an entity's identity with a public key and is carried out by a trusted party.

Digital Signature

A digital signature may be applied to an encrypted message. A message digest is ciphered using the sender's private key and then appended to the message, resulting in a digital signature.

Direct Connection

A direct connection is a modem connection without error connection, compression, and overflow control. It may be assumed that in such a situation the modem rate equates precisely to the connection rate.

Distributed Computing

A distributed computing environment spreads processing, applications, and resources across different platforms. It is typically a client/server environment, and may be heterogeneous where disparate operating systems are integrated.

Distributed Debugging

A methodology for debugging client/servers, where the collective distributed system is perceived as a single system.

Distributed Glue

A distributed glue is the name given to the collective entities that bind together (dynamically) running components that are on the client and on the server. As is the case with local glues, standard OO component architectures use different distributed glues.

Djinn (pronounced "gin")

A djinn is a community of users, devices, and resources that are held together with Jini software infrastructure, and have agreed policies of trust and administration.

DLL (Dynamic Link Library)

A DLL file contains a number of functions that may be called by different applications. The Windows architecture is itself based on DLLs. DLLs may be:

- Dynamic, where programs interact at runtime
- Static, where the DLL is embedded into the application when compiled.

Static libraries tend to make applications fat, requiring more memory than their dynamic counterparts.

A DLL has a:

- File that contains its source code and entry and exit functions
- Module definition file
- Resource definition file

Typically DLLs offer:

- Leverage program investment through improved reusability
- Better code compatibility
- Easier migration paths
- Cost-effective system renovations
- Better program performance
- Improved memory management

DNS (Domain Name Service)

A server that converts domain names (such as www.digital.com) into IP addresses.

DNS Negotiation

A process by which the DNS address is determined by the PPP server and passed to the PPP client.

Domain Category

A collection of servers on the Internet that share the same suffix in their URLs. For example, http://www.cia.com.au is in the domain com.au (which is a mnemonic for a commercial site in Australia). Other domains include .edu (educational), .gov (government), .mil (military), and .net (network).

Domain Name

A name of a domain. For example, in the URL www.microsoft.com, Microsoft is a domain name.

Dot Pitch

A measurement of the distance between addressable pixels on a monitor screen, indicating the clarity of picture and maximum resolution supported.

Double Double

An item of data that consists of 64 contiguous bits. It is twice as long as a double word.

Double Word

An item of data that contains 32 contiguous bits. It is twice as long as a 16bit word.

Downloading

A process of copying files from a remote server to a local computer. The reverse process is called uploading.

Dropped

1. A packet that does not reach its destination
2. A frame in a video source recording that does not appear in a captured digital video file

DSM (Digital Storage Medium/Media)

A medium used to store digital data. Commercial examples include audio CD, CD-ROM, CD-ROM XA, CD-I, Digital Versatile Disc (DVD), floppy disk, Sony Mini disc, Philips DCC (Digital Compact Cassette), and DAT (Digital Audio Tape).

DSN (Data Source Name)

A means of identifying, and connecting to, a database. A DSN is required for many Web applications that interact with and query databases that are typically ODBC compliant.

Dumb Terminal

A client device that is restricted to the presentation element of the application. It has no more application logic than that which is required to send requests and receive visual information. Physically, it consists of a keyboard, display, and a network interface.

DUN (Dial-up Networking)

A connection to a remote computer or network.

DVD Video

An alternative term of MPEG-2 video stored on DVD disc.

Dynamic

A language that may accept a class while running; Java has this capability.

Miscellaneous

D Channel

A 16kbps signaling channel that supports two 64kbps data channels, according to the ISDN standard.

DBS (Direct Broadcast Satellite) A communication and broadcasting technology, where information is transmitted (from a geostationary satellite) and received by a satellite dish that is typically 18in to 3ft in diameter.

It can also be applied as an access technology that offers downstream bandwidths of perhaps 400kbps. Hughes Network Systems (US) offer such service and implementation.

Up to 200 television channels may be chosen using many DBS or Direct To Home (DTH) services. MPEG-2 encoding is used for many DBS services.

DECK

A deck of cards metaphor is used in WML applications.

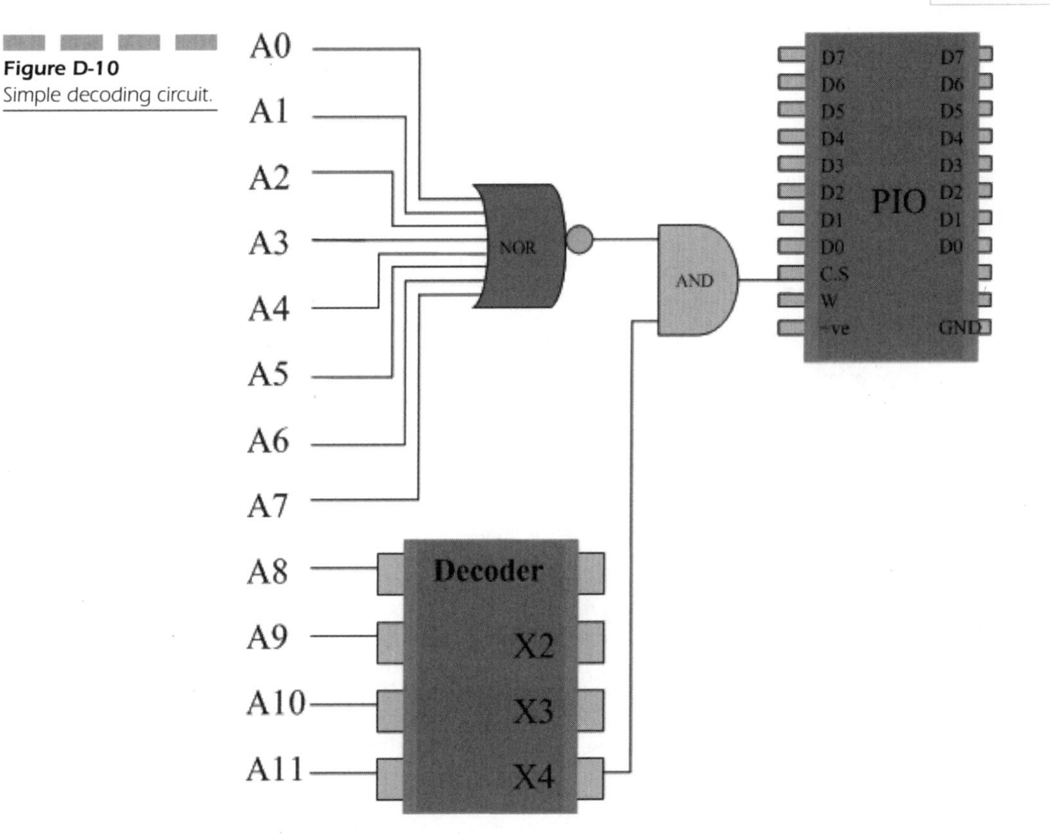

Figure D-10
Simple decoding circuit.

Decoder

A decoder is able to interpret an encoded signal. An MPEG decoder is able to uncompress digital video, as in an MPEG-2 STB. A decoder may also take the form of an electronic device able to decode digital addresses. A simple two-input device may set up to four digital outputs. Such devices may form part of the address decoding between the processor and connected electronic devices. See Fig. D-10.

DVD

DVD is an optical disc technology that provides a sufficiently wide bandwidth so as to play MPEG-2 video. DVD was once an acronym for Digital Video Disc and Digital Versatile Disc. It offers maximum data capacities of 4.7Gbytes, 8.5Gbytes, and 17Gbytes, and exists in three forms:

- DVD-ROM that provides the same functionality as CD-ROM, but with a wider bandwidth and considerably more data capacity
- DVD that is aimed at the consumer market as a replacement for VideoCD and VHS video
- DVD-RAM that is a rewritable format able to support data capacity of 2.6Gbytes
- DVD+RW that is a rewritable format offering a data capacity of 3.0Gbytes

The general DVD specification includes the following:

- 1.2mm thick, 120mm diameter disc
- 4.7Gbytes for a single-layered, single-side
- Track pitch of 0.74μm
- 650/635nm laser
- RS-PC (Reed Solomon Product Code) error correction scheme
- Variable data transfer rate yielding an average of 4.69Mbits/s

DVD-ROM drives offer backward compatibility with CD-ROM, and the important factors that apply to the performance of a dedicated CD-ROM drive are applicable. Features that drive the DVD-ROM specification include:

- Supported data capacities, i.e., 4.7Gbytes, 8.5Gbytes, and 17Gbytes
- Interface type
- Ability to record CD-R discs
- Burst transfer rate
- MTBF (Mean Time between Failures)
- DVD disc average access time
- CD-ROM disc average access time
- Average CD-ROM data transfer rate, i.e., 16-speed, 24-speed, 32-speed, etc.
- CD-ROM burst transfer rate
- Disc spin modes that may be either CAV and CLV
- Meets MPC3 requirements
- Vertical or horizontal installation

Acronyms

D-AMPS Digital AMPS

DCCH Dedicated control channel

DCE Distributed computing environment

DCF	Distributed coordination function
DCH	Dedicated channel
DCHFP	DCH frame protocol
DCN	Dual communications network
DCOM	Distributed component object model
DCR	Dedicated common router
DCT	Discrete Cosine Transform
DCU	Digital connection unit
DECT	Digital enhanced cordless telecommunications
dg	Digital gains
DHCP	Dynamic host configuration protocol
DII	Dynamic Invocation Interface
DLCI	Data link control identifier
dmch	Dedicated media access control channel
DNS	Domain name server
DPCCH	Dedicated physical control channel
DPCH	Dedicated physical channel
DPDCH	Dedicated physical data channel
DPE	Distributed processing environment
DQPSK	Differential quadrature phase-shift keying
DR	Direct sequence
DRNC	Drift RNC
DS	Direct spread
dsch	Dedicated signaling channel
DSCH	Downlink shared channel
DSI	Dynamic skeleton interface
DSP	Digital signal processor
DSSS	Direct sequence spread spectrum
dtch	Dedicated traffic channel
DTCH	Dedicated transport channel
DTMF	Dual tone multifrequency
DTX	Discontinuous transmission

E

E-2

E-business

Electronic Payment

Mobile e-commerce or m-commerce is driving the deployment of secure transaction technologies using protocols like WAP's WTLS, for example. Radicchio is an electronic payment service launched by Sonera SmartTrust,

Gemplus, and EDs. This used a SIM card that also acts as a credit card, and the credit card reader is the mobile handset itself. The mobile device may have two SIM cards: one to act as a normal SIM card, and another to act as bank or credit card that would be given to the user by an issuer like a bank or credit card company. Ericsson, Motorola, and Nokia have developed MeT (Mobile Electronic Transactions), which relies on the mobile user possessing a SIM-like card to prevent tampering.

E-commerce Site Development Life Cycle

The e-business site development life cycle is a collection of processes and sub-processes. A Web server facility may be:

- An acquired, leased, or rented in-house Web server solution featuring an ISDN or T1 connection, necessitating personnel to maintain and run the server. It may be chosen for security reasons, or when it is important to evolve the Web server in house.
- Platformed on a Web host, or a company dedicated to providing turn-key Web server solutions. The Web sever is hosted on a remote site.
- Platformed on a public server such as those offered to its subscribers by AOL, Compuserve, Prodigy, and many others. Other public servers include Geocities and Angelfire, offering access on the World Wide Web to businesses with a low-cost migration path to architecting a low-cost e-commerce presence.
- Colocated on a server farm, where ISDN or T1 connection technologies, maintenance, and day-to-day running take place off-site.

Dell, Hewlett-Packard, IBM, Olivetti, Compaq, and the many other computer manufacturers produce server implementations, many of which are turn-key solutions. Such servers are specified in terms of processor types, number of processors, mass storage capacity, bundled operating systems, and server software, as well as many other such common features. Generally it may be assumed that much of the low-level technical descriptions of servers may be ignored when purchasing from the major producers such as those mentioned earlier; it may be assumed that they will always bring the latest technologies to market—at a time when perhaps budget computer makers are not.

Web servers run either the Windows NT or UNIX operating systems (OSes), and there are many differentiating features that separate them.

A Web server hosted off-site should allow you to include a domain name of your choice, and for security purposes, to restrict user access to directories or files. It must also allow you to conduct transactions in a secure mode using mainstream encryption techniques such as SHTTP. This may be verified by adding the "S" prefix to the collective Web address, and then opening the site using a Web Browser You may also require a site certificate, confirming your ownership rights. It may also be necessary to permit you to run your own programs and scripts, including CGI variants, as well as accommodate any additional requirements imposed by the Web site authoring software you may have used such as the FrontPage extensions.

More complex e-commerce sites require programming in languages such as Perl, C++, Java Visual Basic, et al. Such languages may be used to create feedback forms, etc. An e-commerce site typically comprises many components like CGI scripts, counters, and applets. Many of these may be gathered from public domain resources on the World Wide Web or from shareware.

An e-commerce site may be secured using many different technologies including SSL, RSA, and the many products that offer everything from password protection to firewalling.

Certificate Authority

A Certificate Authority is a third party that issues digital certificates for creating digital signatures and public/private keys. The CA seeks to guarantee that individuals granted certificates are identified. The CA may have an agreement with a financial institution that provides confirmation of an individual's identity.

Checksum Validation

A method of validating credit card numbers by using the *mod 10* check digit algorithm and is implemented by:

1. Doubling the value of alternative digits of the credit card number by beginning with the second digit:

1	3	6	5	8	9	7	6	2	4	2	7	6	0	8	7
	6		10		18		12		8		14		0		14

2. Adding the product values to the alternate digits beginning with the first:

$$7 + 16 + 26 + 17 + 10 + 16 + 6 + 22 = 120$$

In this instance the credit card number has passed the validity check because the result is evenly divisible by 10.

Cookie

A minor transaction that allows server-side components such as CGI scripts and programs to store and retrieve data from the client system.

It gives Web applications the ability to write data to the client that reflects usage habits. For example, the data may relieve the user from repetitive tasks, such as the re-entry of ID numbers or data each time a Web site is visited. Instead the server-side components may identify the user through cookies on the client system, extract them, and perform the necessary processes.

Credit Card Merchant Account

A POS feature that enables an e-commerce Web site to process credit card transactions. Numerous companies offer such facilities over the World Wide Web where users are required to complete on-line forms, and produce relevant evidence of their on-line business.

Cryptography

A process that ensures data or information is read or used only by its intended readers or users. This is achieved through:

- Encryption that disguises inputted information or data, so it may not be read or used. Resulting encrypted information or data may only be read or used following decryption.
- Decryption that returns the decrypted data or information to its original usable and readable form.

Implementations of cryptography are called cryptosystems, and take the form of algorithms. Cryptosystems may be categorized in two main groups:

1. *Secret-key*, where the processes of encryption and decryption each require the use of the same single key. The key is a number, and preferably a large one, hence the phrase 56bit key, etc. Unless the recipient of the encrypted data already knows the key, it may be left to the sender to transmit its details unencrypted. This is a notable flaw of secret-key encryption, because it exposes the key to unintended users such as eavesdroppers. A remedy is found in public-key encryption that is described below.
2. *Public-key*, where the sender need only know the recipient's public key. This may be obtained in unencrypted form, because it may not be used to decrypt data; rather, all it may do is encrypt data. In order to

decrypt data, the recipient uses a private key that is the mathematical inverse of the public key. It may be considered impossible to determine the private key from the public key insofar as most security requirements are concerned.

The mathematics that underline public-key encryption have a simple goal: namely, to make difficult the derivation of the private key from the public key. This is achieved through a one-way function that describes the difficulty of determining input values when given a result.

RSA is among the best-known cryptosystems or algorithms. This was developed by MIT professors Ronald L. Rivest, Adi Shamir, and Leonard M. Adleman.

(*See* RSA below; see also www.rsa.com.)

EDI (Electronic Data Interchange)

EDI is a standard set of formats and protocols for exchanging business information over networks and systems. Translation programs may play the role of converting extracted database information into the EDI format so that it may be transmitted to appropriate entities such as banks that have appropriate EDI-capable IT implementations.

- *EDI Trading Partner:* An EDI entity/establishment able to receive or transmit EDI data
- *EDI Transaction Set:* A message or block of EDI information that relates to a business transaction
- *EDI Transaction Set Standards:* A formal standard that dedicates syntax, data elements, and transaction sets or messages
- *EDI Translation:* An EDI conversion to and from the X12 format
- *EDI Translator:* An entity that converts between the flat file to EDI formats
- *EDIFACT* (Electronic Data Interchange for Administration, Commerce, and Transportation): A standard for electronic data interchange that is approved by the UN
- *Financial EDI:* An exchange of payment information in standard formats between business partners

POI (Point of Information)

POI provides a means of exhibiting products electronically through streaming video to 3G appliances. Traditionally, consumer education has consisted of publishing product brochures, advertising, allowing the potential customer to peruse in a shop or showroom, and product demonstrations.

These generally accepted ways in which the consumer chooses an appropriate product may be aided or replaced using multimedia in a point-of-information (POI) guise. POI presents the customer with the ability to browse through product ranges, or experience just those items that fit a user-defined profile.

POI may be a powerful marketing and advertising tool providing the means to display products. The benefits are clear: It provides an opportunity to promote products in a medium that cannot be rivaled by (current) television advertising. It also allows small and medium-size companies to promote products on terms that only large companies and corporations could previously afford.

Furthermore, if products may be demonstrated adequately through multimedia, the need to exhibit them physically becomes unnecessary and so floor space may be saved. Research also indicates that users of POI terminals spend more money than those shoppers using conventional means, and the possibility of fewer sales staff is raised.

Museums of various kinds throughout the world could provide POI to mobile users. These give visitors the opportunity to follow user-defined guided tours. This approach also allows visitors to experience interesting items that might otherwise be cataloged and hidden away from public view.

Many other areas such as career advice, geographic information systems, and surrogate travel (brochures) are also possible through POI.

RSA

RSA is a public-key or asymmetric cryptosystem or algorithm developed by MIT professors Ronald L. Rivest, Adi Shamir, and Leonard M. Adleman in 1977. It is used by numerous e-commerce site developers and e-commerce product vendors.

Its aim is to make difficult the derivation of the private key from the public key using a one-way function. For example, if the public key is a known function of x, f(x), it may be made theoretically difficult to determine the unknown x that is the private key. The same cannot be said of the reverse, where x is known and f(x) is unknown, particularly in the case of factorizing.

This was illustrated in 1977 when RSA-129, a 129-digit integer, was published by Martin Gardner in *Scientific American*. He laid down the gauntlet, challenging readers to factorize it, for which they would receive a small cash prize. Not until March 1994 was it factorized by Atkins et al., using the resources of some 1600 computers and the quadratic sieve factoring method that has been superseded by the more economical general number field sieve.

(*See* RSA One-Way Function below.)

RSA One-Way Function

The so-called one-way function is appropriately effective when attempting to factorize the product of two large primes that, using RSA, is implemented as follows:

1. Two prime numbers are selected: p and q.
2. Calculate their product n, or the public modulus.
3. Another chosen number $e < n$.
4. e is relatively prime with $(p - 1)(q - 1)$.
5. e and $(p - 1)(q - 1)$, therefore, have only 1 as their common factor.
6. Calculate $d = e^{-1} \bmod [(p - 1)(q - 1)]$.
7. e is the public exponent.
8. d is the private exponent.
9. The public key is the pair (n,e).
10. The private key is the pair (n,d).

The chosen prime numbers p and q may be kept with the private key, or destroyed. Using PGP, p and q are retained in encrypted form and help expedite operations through the Chinese Remainder Theorem.

The reverse process is difficult, thus obtaining the private key (n,d) from the public key (n,e) is deemed secure. Factorizing n would result in p and q, leading to the private key (n,d).

The encryption process involves:

- Dividing the target message into blocks smaller than n
- Modular exponentiation: $c = m^e \bmod n$
- Decryption or the inverse is driven by: $m = c^d \bmod n$

(*See* Cryptography earlier in this chapter, RSA earlier in this chapter.)

SET (Secure Electronic Transactions)

SET is a standard means of securing payment transactions made to online merchants. Integrating cryptosystem techniques, it is perceived as a credit card security system, and was initiated by Visa and Microsoft. SET implementations are an amalgam of cryptosystems, protocols, secure protocols, and techniques.

- SET application: An application that uses the SET internationally agreed technologies and methodologies.
- SET ASN.1 (Abstract Syntax Notation One): A standard that defines the encoding, transmission, and encoding of data and objects that are architecture neutral.

- SET Baggage: A method of appending ciphered data to a SET message.
- SET CDMF (Commercial Data Masking Facility): A ciphering technique based on DES that is used to transfer messages between the Acquirer Payment Gateway and the Cardholder in SET implementations.
- SET Certificate Authority: A trusted party that manages the distribution of SET digital certificates, where layers of the Tree of Trust has the representation of a digital certificate.
- SET certificate chain: A group of digital certificates used to validate a certificate in a chain.
- SET certificate practice statement: A group of rules that determine the suitability of certificates to particular applications and communities.
- SET certificate renewal: An event that sees the renewal of a certificate for continued transacting purposes.
- SET Consortium (Secure Electronic Transactions): An international organization that was formed when Mastercard and Visa announced SET, whose initial objective was to create an agreeable standard, and to consider STT and SEPP.
- SET Digital certificate: A means of linking an entity's identity with a public key and carried out by a trusted party.
- SET Digital signature: A digital signature may be applied to an encrypted message. A message digest is ciphered using the sender's private key and then appended to the message, resulting in a digital signature.
- SET E-wallet: An element of a cardholder that creates the protocol and assists in the acquiry and management of cardholder digital signatures.
- SET Hash: An element that reduces the number of possible values using a hashing function such as the Secure Hashing Algorithm (SHA-1).
- SET Idempotency: An attribute of a message that sees repetition yield a constant result.
- SET Message authentication: A process or usually subprocess that verifies that a message is received from the appropriate or legal sender.
- SET Message pair: Messages that implement the POS and certificate management in a SET implementation.
- SET message wrapper: A top-level data structure that conveys information to message recipients.
- SET Order inquiry: A pair of set messages used to check the status of orders.
- SET Out-of-Band: An activity that is not within the bounds of the SET recommendations, guidelines, and standards.

SET PKCS Public Key Cryptography Standards

SET PKCS is a set of public-key cryptography standards used by SET that includes:

- RSA
- Diffie-Hellman key agrements
- Password-based encryption
- Extended certificate syntax
- Cryptographic message syntax
- Private-key information syntax
- Certification request syntax

SSL (Secure Sockets Layer) Protocol

SSL is a secure channel or protocol that is supported by the TCP transport protocol, and includes the higher level protocol SSL Handshake Protocol that authenticates the client and server devices, and allows them to decide upon an encryption algorithm and keys before data reception or transmission commences. Its ability to allow high-level protocols to sit on top of it is perceived as an advantage. Used alone, SSL is not perceived as a complete security solution, though it does present one significant security perimeter in the eyes of many security analysts. It ensures a secure connection by authentication of peer's connection and uses integrity checks and hashing functions, to secure the channel between applications.

It was designed to prevent:

- Information forgery
- Eavesdropping or sniffing
- Data changes

Private or symmetric key is the basis of the SSL's cryptography, and authenticates a peer's identity using asymmetric cryptography.

SLL's flaws are documented widely and include:

- When a browser connects with an SSL server, it receives a copy of its public key wrapped in a certificate that the Browser sanctions by checking the signature. The flaw here is that the Browser hasn't the means to authenticate the signature, because no verification is performed up the hierarchy because many certificates used by SSL are root certificates.
- It consumes client and server processing resources.
- It has operational difficulties with proxies and filters.

- It has operational difficulties with existing cryptography tokens.
- Its key management tends to be expensive.
- Expertise to build, maintain, or operate secure systems is in short supply (as of 1999).
- It creates network traffic when handshaking.
- The migration path from nonpublic key infrastructures is arduous.
- It requires Certificate Authority with appropriate policies.
- Its communication data does not compress and therefore steals network bandwidth.
- It is subject to certain international import restrictions.

Symmetric Cryptosystem Operation

A series of processes and subprocesses that:

- Converts plaintext into ciphertext using a cipher or encryption algorithm
- Returns ciphertext to plaintext using a decryption algorithm

Using a symmetric key and the transposition technique, the processes include the following:

Encryption Send the key, such as UNLOCK, for example, to the recipient using a secure channel. Arrange the key in a columnar fashion with numerals indicating their alphabetical sequence:

U 6
N 4
L 3
O 5
C 1
K 2

Arrange the plaintext, such as ATTACK VESSEL, in columns as shown below:

U 6 A V
N 4 T E
L 3 T S
O 5 A S
C 1 C E
K 2 K L

Create the ciphertext by writing the row values in sequence dictated by the numerical value, i.e.:

CE KL TS TE AS AV

Send the ciphertext to the recipient where a secure channel is optional.

Decryption Again the key characters are assigned numerals indicating their alphabetical sequence:

U 6
N 4
L 3
O 5
C 1
K 2

The decryption algorithm takes the ciphertext and, based on the numerical value, it creates the plaintext:

U 6 A V
N 4 T E
L 3 T S
O 5 A S
C 1 C E
K 2 K L

Transaction Processing ACID Properties

Atomicity, Consistency, Isolation, and Durability (ACID) properties define the real-world requirements for transaction processing (TP) that are supported by Jini:

- Atomicity ensures that each transaction is a single workload unit. If any subaction fails, the entire transaction is halted, and rolled back.
- Consistency ensures that the system is left in a stable state. If this is not possible, the system is rolled back to the pretransaction state.
- Isolation ensures that system state changes invoked by one running transaction do not influence another running transaction. The changes must only affect other transactions, when they result from completed transactions.
- Durability guarantees the system state changes of a transaction are involatile, and impervious to total or partial systems failures.

Transaction Server A transaction server is allocated the task of transaction processing (TP), and often invokes the application logic necessary to perform database interactions and manipulations. The process(es) invoked directly or indirectly by the client are collectively referred to as the transaction.

Transaction servers may include UI logic, driving the client UI, relegating the client device as little more than a dumb terminal. Typically mainframe-based transaction systems might adhere to this model. Alternatively, the UI logic, or presentation, may be distributed to the client.

The server consists of a TP monitor that performs transaction management and resource management. Transaction management ensures the so-called ACID properties of transactions. These include Atomicity, Consistency, Isolation, and Durability. ACID property compliance is achieved through the two-phase commit protocol.

Trusted Site

A trusted Web site that is secure and safe and includes a secure connection.

Netscape and Microsoft browsers have lists of Certificate Authorities.

To view the lists using Netscape:

1. Click the *Security* button on the toolbar.
2. Click the *Signers* link under *Certificates*.

Using Microsoft Explorer:

1. Click the *View* menu.
2. Click *Internet Options*.
3. Select the *Security* tab.
4. Click the *Zone* menu.
5. Select *Trusted Site Zone*.
6. Click the *Add Sites* button to add or remove sites.

EDGE (Enhanced Data Rate for GSM Revolution)

EDGE upgrades 2G networks like GSM to provide a maximum data rate of 384Kbps able to deliver many 3G service types. A key advantage of EDGE is the reuse of existing 2G network infrastructures and technologies including GSM, GSM carrier bandwidth and time slots, IS-136 and PDC, and at the same time it provides a smooth migration path to 3G networks. The result of integrating the EDGE overlay is a 2.5G network that widens user bandwidth

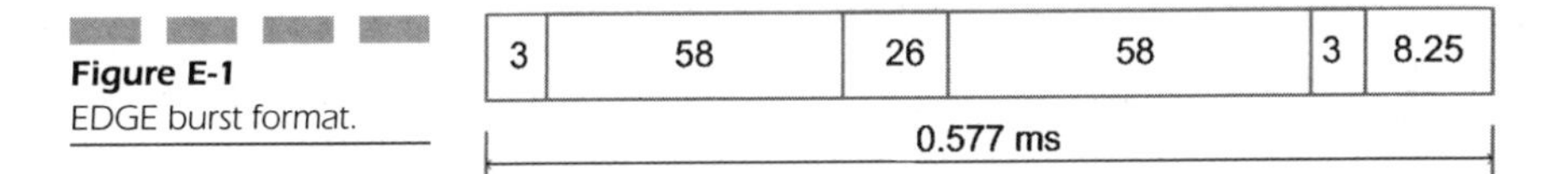

Figure E-1
EDGE burst format.

GPRS bandwidth by using higher-level modulation. EDGE may also provide a generic air interface for other TDMA systems, and it was adopted by UWCC (Universal Wireless Communications Consortium) as 136 HS's outdoor component.

EDGE is essentially a radio interface upgrade and allows GSM and IS-136 to offer new services. The modulation scheme is based on 8-PSK2.5G EDGE. EDGE is an overlay solution for existing ANSI-136/TDMA networks, and may use the existing ANSI-136 30kHz air-interface. EDGE is on the migration path to UMTS, and may even coexist with it so as to provide services for wide area coverage. EDGE standards support mobile services in ANSI-136/TDMA systems with data rates of up to 473kbps.

A significant change in the ANSI-136/TDMA standards to support higher data rates is the use of modulation schemes including 8-PSK (Phase Shift Keying) and GMSK (Gaussian Minimum Shift Keying). GMSK provides for wide area coverage, while 8-PSK provides higher data rates but with reduced coverage. EDGE provides high data rates over a 200kHz carrier, giving up to 60 kbps/timeslot that may equate to 473kbps. EDGE is adaptive to radio conditions, giving the highest data rates where there is sufficient propagation.

8-Level Phase Shift Keying is an enhanced modulation method that is used in the EDGE radio interface. In this context 8-PSK has three bits per symbol and gives a gross bit rate per slot of 69.2kbps (including overhead, and given that the symbol rate is 271ksymbols/s).

All rates per time slot include 22.8, 34.3, 41.25, 51.6, 57.35, and 69.2kbps for code rates of 0.33, 0.50, 0.60, 0.75, 0.83, and 1. See Fig. E-1.

See Tables E-1 and E-2.

EDGE Services

- PS services: GPRS provides connection to IP and X.25 networks, so the mobile user may connect with external IP networks. Each service has an accompanying QoS profile whose parameters include priority, reliability, delay, maximum bit rate, and mean bit rate.
- CS services: EDGE does not change the service definitions, but the channel coding is naturally different; a 57.6kbps nontransparent service may have a ECSD TCS-1 coding and require just two time slots. GSM defines eight transparent services offering constant bit rates that may range from 9.6 to 64kbps. Nontransparent services may use radio

TABLE E-1

Channel Coding Scheme—EDGE (PS Transmission)

Coding Scheme	Gross Bit Rate (kbps)	Code Rate	Modulation	Radio Interface Rate per Time Slot (kbps)	Radio Interface Rate per Eight Slots (kbps)
CS-1	22.8	0.49	GMSK	11.2	89.6
CS-2	22.8	0.63	GMSK	14.5	116.0
CS-3	22.8	0.73	GMSK	16.7	133.6
CS-4	22.8	1.0	GMSK	22.8	182.4
PCS-1	69.2	0.329	8-PSK	22.8	182.4
PCS-2	69.2	0.496	8-PSK	34.3	274.4
PCS-3	69.2	0.596	8-PSK	41.25	330.0
PCS-4	69.2	0.746	8-PSK	51.60	412.8
PCS-5	69.2	0.829	8-PSK	57.35	458.8
PCS-6	69.2	1.000	8-PSK	69.20	553.6

link protocol (RLP) for error free delivery, and there are eight such services with bit rates ranging from 4.8 to 57.6kbps.

- Asymmetric services due to terminal implementation: ETSI has established the following mobile classes:
  1. GMSK for the UL, and 8-PSK for the DL providing EDGE bit rates for the downlink.
  2. 8-PSK for the UL and DL.

TABLE E-2

EDGE (CS Transmission) Channel Coding

Channel Coding	Code Rate	Modulation	Radio Interface Rate per Time Slot (kbps)
TCH/F2.4	0.16	GMSK	3.6
TCH/F4.8	0.26	GMSK	6.0
TCH/F9.6	0.53	GMSK	12.0
TCH/F14.4	0.64	GMSK	14.5
ECSD TCS-1 (NT + T)	0.42	8-PSK	29.0
ECSD TCS-2 (T)	0.46	8-PSK	32.0
ECSD TCS-3 (NT)	0.56	8-PSK	38.8

Migration to EDGE

EDGE is a layer that may be added to existing GSM networks and requires few network medications since BTSs, along with other infrastructures, are reused. GPRS PS nodes such as the 2.5G GGSN (Gateway GPRS Server Node) and the 2.5G SSGN (Serving GPRS Server Node) may be upgraded using software.

The wideband data adoption of EDGE provides a smooth migration to IMT-2000 as there can be a gradual deployment of the 3G air interface over the GSM core network. This leverages the initial investment in the GSM while guaranteeing the customer base. More important, GSM operators with 2GHz licenses can deploy IMT-2000 wideband coverage in hotspots where there is likely to be highest demand. This will be supported by dual-mode EDGE/IMT-2000 mobile terminals that can hand off and map services from both network types. It is reasonable to assume that building 3G networks on existing GSM core network infrastructures offer improved ROI and increased profitability.

EGPRS (Enhanced General Packet Radio Service)

The dominant data networking protocol, on which most data network applications are running, is TCP/IP, the Internet Protocol. All Web applications are run on some form of TCP/IP, which is by nature a protocol family for packet-switched networks. This means that EGPRS is an ideal bearer for any packet-switched application such as an Internet connection. From the end user's point of view, the EGPRS network is an Internet subnetwork that has wireless access. Internet addressing is used and Internet services can be accessed. A physical (Internet) IP address consists of 32bits that identifies hosts, servers, networks, and connected computers. The syntax for such addresses consists of 4bytes each written in decimal form, and separated by a full stop: 118.234.165.124. This physical address is obtained by a DNS server when given a Web address.

The three types of IP address are as follows:

- IP Address Class A: A networks may have between 2^{16} (65,536) and 2^{24} (16.7 million) hosts.
- IP Address Class B: B networks may have between 2^{8} (256) and 2^{16} (65,536) hosts.
- IP Address Class C: C networks may have up to 253 hosts, and not 255, because two values are reserved.

The addresses consists of a network address (netID) and a host address (hostID). The leftmost digits represent the netID address. This is set to zero when addressing hosts within the network.

IPv6 is an advancement of the IP protocol that introduces 128bit addressing, and other improvements. The scaling of IPv4 32bit addressing to 128bit is intended to accommodate future growth of the Internet, in terms of the growing number of network addresses.

IPv6 is also called IPng (IP Next Generation), and is specified by the Internet Engineering Task Force (IETF).

IPv6 supports addressing of the types:

- Multicast that connects a host to multiple addressed hosts
- Unicast that connects a host to a single other addressed host
- Anycast that connects a host to the nearest of multiple hosts

Ericsson

Ericsson is a leading company in the wireless telecommunications sector communications and produces telecom and datacom solutions. It has over 100,000 employees in 140 countries (as of November 2001).

Ericsson R380

The Ericsson R380 is a popular WAP phone in the mass market telecommunications sector. It has a grayscale touchscreen with a resolution of 360 × 120 pixels and a 0.23 pitch, and an active screen size of almost 83 × 28mm. Its browser includes a Browser Area, Card Title bar, and Toolbar. The Browser area presents the card content and measures 310 × 100 pixels. When a card is too large for the screen, only the beginning of the card is shown at first, and a vertical scroll bar appears. Graphical components, text, and images are displayed in the top left corner, and in the same order as in the WML code.

Card Title Bar The Title bar shows which card is displayed and also browsed cards. The navigation history is reset when a loaded Card's newcontext attribute is *true*.

Toolbar The Toolbar has buttons required by the browser, and text input is done with a screenpad or character recognition screen. Three different keyboard layouts include: alpha, numeric, and national characters.

To navigate to a WAP service:

- Activate a bookmarked site using the Bookmark.
- Use the History list.
- Enter the URL in the Open Location dialog.

Design Components

Font The font used in R380 is a proportional font called Swiss A and it is used in small, normal, and big sizes.

Font Size/Height/Number of Lines

- Small/9 pixels/7.5
- Normal/10 pixels/7
- Big/12 pixels/5.5

WML The browser supports the emphasis elements big and small to change font sizes, and the elements em, strong, i, b and u.

Text Formatting

```
<wml>
<card id="first" title="Fast Burgers" newcontext="true">
<p align="center">
<br/>
<b><big><a href="#second">Welcome</a></big></b><br/>
to<br/>
Welcome to Fast Burgers.<br/><br/>
<a href="#fifth">[Contact Us]</a>
</p>
</wml>
```

Line Spacing and Line Breaks Single line spacing is used with one pixel before and after lines. Long lines are word-wrapped onto multiple lines. Text lines may include images, select lists, buttons, input fields, and hyperlinks.

WML New lines start with the br element that affects all contents in the browser, and the current alignment is used to position the line's entities.

A Line Break Example

```
<wml>
<card id="init" title="BR tag">
<p>
```

```
This sentence continues until the end of the line and
then wraps to the next line.<br/>
This phrase begins on a new line.<br/><br/>
This phrase follows two br elements.
</p>
</card>
</wml>
```

Paragraphs Paragraphs start on a new line, follow a 3-pixel gap, and have a default left alignment, but may be aligned right or centered.

WML Paragraphs are defined and aligned using the p element.

Attribute Description The align attribute may have the values: left, right, and center. The mode attribute specifies the line-wrap and makes the values *wrap* or *nowrap*.

A Paragraph Example

```
<wml>
<card id="init" title="P tag">
<p>
<b>LEFT</b><br/>
This text is left aligned, and will continue until
the end of the line and then wrap to a new line.<br/>
</p>
<p align="center">
<b>CENTER</b><br/>
This text is centered, and will continue until the
end of the line and then wrap to a new line.<br/>
</p>
<p align="right">
<b>RIGHT</b><br/>
This text is right aligned, and will continue until
the end of the line and then wrap to a new line.
</p>
</card>
</wml>
```

Indented Paragraphs Text and components may be grouped, and groups may be nested. Groups have a 20-pixel indentation and follow a 3-pixel gap.

WML A group's beginning is defined using the fieldset element.

Attribute Description The title attributes value is used as leading text, and the text following the title appears on a new line.

Fieldset

```
<card id="burgerinfo" title="What's On">
<p>
<fieldset title="Big">
Lettuce, tomato sauce, mustard, pickle, tomato and cucumber
</fieldset>
<fieldset title="Double">
tomato sauce, mustard, pickle, tomato and cucumber
</fieldset>
<fieldset title="Double Big">
Onions, tomato sauce, mustard, pickle, tomato
and cucumber
</fieldset>
.
.
.
.
</p>
</card>
</wml>
```

Card Title The title is defined using the title attribute in the card element.

Single Choice Lists A single choice list is shown as a drop-down listbox, showing the currently selected value in angled brackets. It is 15 pixels high, has 5 pixels of white space on either side, and adapts to the length of the text in brackets.

WML A single choice list is specified using the select element with the multiple attribute set to *No*, and each list item is specified by an option element.

A Single Choice List

```
<p>
<b>Select Burger</b>
<select>
<option>Big</option>
<option>Double</option>
<option>Double Big</option>
</select>
<a href="#burgerinfo">What's On</a>
</p>
<p>
<b>Select Meal</b>
<select>
<option>Snack</option>
<option>Medium</option>
<option>Big</option>
</select>
</p>
```

Multiple Choice Lists A multiple choice list has check boxes on separate lines that are followed by 2 pixels of white space. They may form a hierarchy of groups that have a 20-pixel indention.

WML Multiple choice lists are created using the select element with the multiple attribute set to *Yes*, and each check box item is included using the option element.

A Multiple Choice List

```
<p>
<b>Extras</b>
<select multiple="true">
<option>Fast special sauce</option>
<option>Bread crumb fries</option>
<option>Pineapple</option>
</select>
</p>
```

Buttons Buttons are displayed using the Normal Bold font, and are defined using the do element.

A Do Example

```
<p align="center">
<do type="accept" label="Continue">
<go href="#third"/>
</do><br/>
</p>
```

Three Do Elements of the Same Type

```
<p align="center">
<do type="accept" label="Continue" name="cont">
<go href="#third"/>
</do><br/>
<do type="accept" label="Contact Us" name="contact">
<go href="#fifth"/>
</do>
<do type="accept" label="Go to Start" name="start">
<go href="#first"/>
</do>
</p>.
```

Input Fields Input fields are defined using the `input` element that holds the shown attributes:

Attribute Description

- Type: This may have the values text or password, and if set to password, entered characters are shown as asterisks.
- Value: The value of the `value` attribute is used as default text if no preload value is defined for the input object.
- Format: Specifies an input mask for user input entries. The string has mask control characters and static text.

An Input Example

```
<p>
Name<input type="text" title="Enter name:" name="name"/><br/>
Address<input type="text" title="Enter address:"
name="address"/>
</p>
```

Images The browser supports images in WAP bitmap (WBMP) and also in the GIF format that have the colors: white, 0 percent black; light-gray, 25 percent black; mid-gray, 50 percent black; and black, 100 percent black. The WML img element indicates that an image is included in the text flow.

The R380 supports the following attributes:

- Alt: A text name for the image used in the placeholder.
- Src: The image's source (URI).
- Vspace: Specifies the amount of white space inserted above and below the image.
- Hspace: Specifies the amount of white space on either side of the image.
- Height: Specifies the vertical size of an image.
- Width: Specifies the horizontal size of an image.
- Align: The *align* attribute may have the values *top*, *middle*, and *bottom*.

An Image Example

```
<p align="center">
<img alt="baker"src="baker.gif" vspace="5" width="40"
height="30"/><br/>
<b>The burger is sizzling and being expertly prepared, and
will be delivered soon.</b><br/>
</p>
```

EVRC (Enhanced Variable-Rate Codec)

Maximizing concurrent calls, and optimizing the mobile network requires speech compression between the mobile and base station. The EVRC, in the context of the CDMA Development Group's (CDG) implementation, provides 13kbps voice quality at 8kbps data rate. Other industry implementations include the 16kbps LD-CELP, 13kbps CELP, and ADPCM.

The EVRC speech compression algorithm bases itself on relaxed code-excited linear predictive (RCELP) coding that is based on CELP. CELP processes 20ms speech frames for coding and decoding, and for each 20ms time interval the encoder processes 160 speech samples. The volume, pitch, and rate of the voice waveform are used by the coder to represent speech at 1, 4, or 8kbps. For each 20ms speech frame, the CELP generate 10 linear-prediction coding filter coefficients, and with EVRC these are represented by vectors.

CELP speech coders use long-term pitch analysis to generate a 3bit pitch gain and a 7bit pitch period. The pitch analysis occurs on either four 5ms subframes, two 10ms subframes, or four 5ms subframes. This results in a number of bits per frame and gives pitch information. The EVRC algorithm categorizes speech full-rate (8.55kbps), half-rate (4kbps), and eighth-rate (0.8kbps), which occur every 20ms. See Fig. E-2 and Table E-3.

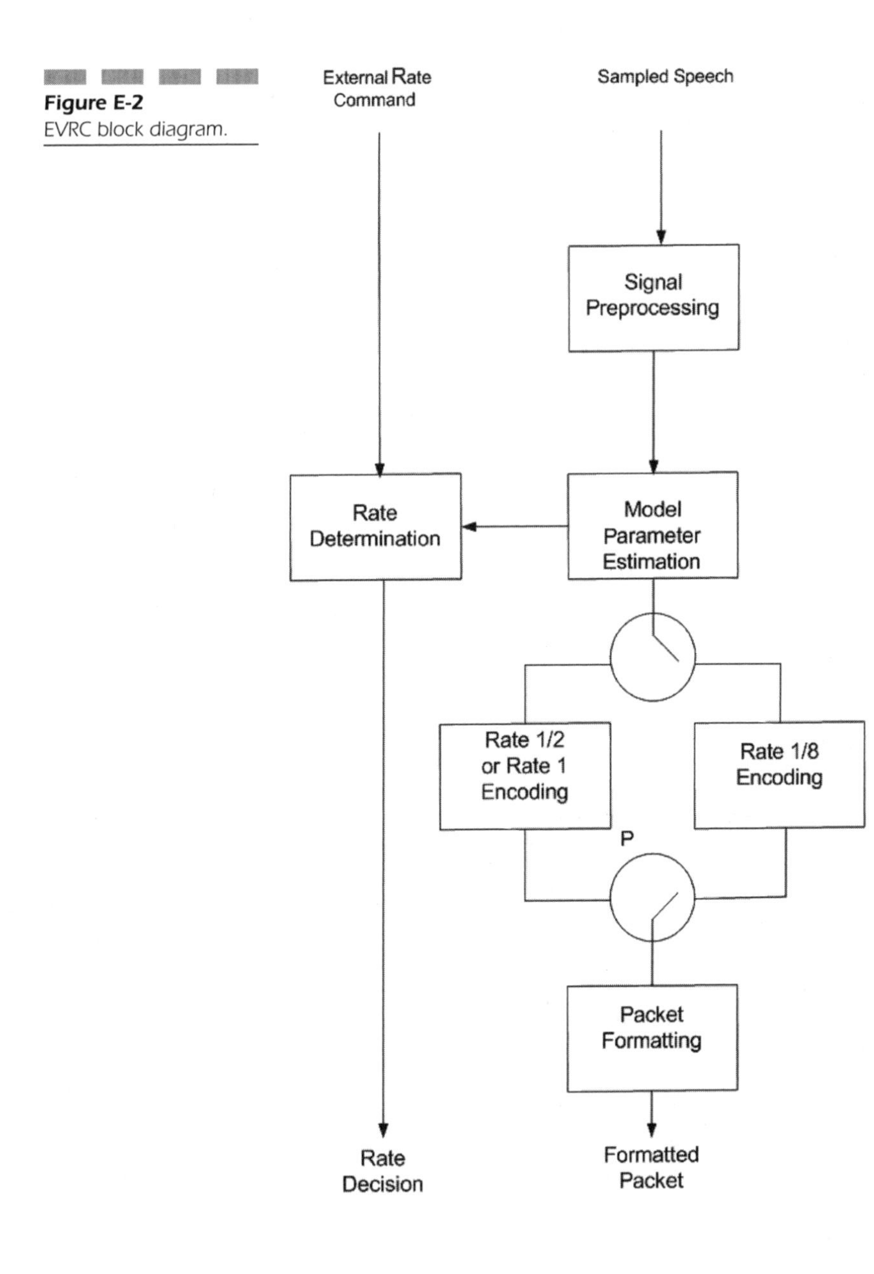

Figure E-2
EVRC block diagram.

TABLE E-3

Bit Allocations according to Packet Type

EVRC Bit Allocations	Packet Type			
	Rate 1	Rate 1/2	Rate 1/8	Blank
Spectral transition indicator	1			
Line spectral pair (LSP)	28	22	8	
Pitch delay	7	7		
Delta delay	5			
Adaptive codebook gain (ACB)	9	9		
Fixed codebook shape (FCS)	105	30		
FCB gain	15	12		
Frame energy			8	
Unused	1			
Total encoded bits	171	80	16	
Mixed mode bit	1			
Frame quality indicator	12	8		
Encoder tail bits	8	8	8	8
Total bits	192	96	24	8
Rate kbps	9.6	4.8	1.2	0.4

Computers/Software

E-business

E-business is a generic term used to describe business processes implemented in electronic or virtual environments like the World Wide Web.

E-mail

A method of communicating documents and digital files electronically; a computer-based equivalent of a letter.

E-mail addresses generally conform to

name@domain.domain_category.country

name might be a login name. *domain* might be a company name such as

Microsoft. *Domain_category* is the type of domain (*see* Domain Category under letter D). *country* is the geographic location of the server that might be uk (United Kingdom), nz (New Zealand), au (Australia) and so on.

For instance, subscribers to Compuserve have e-mail addresses that have the syntax: 123456.7654@compuserve.com. Other ISPs (Internet Service Providers) allow users to use their name as an ID. Examples include F_Botto@compulink.co.uk, or fbotto@cia.com.au.

E-mail messages may be sent using browsers (such as Netscape Navigator and Microsoft Explorer), though these are not e-mail applications or clients such as Microsoft Outlook or Endora Mail. The latter offers folders such as inbox, outbox, and sent messages, and are dedicated applications that support such e-mail protocols as SMTP and POP3. Compuserve (owned by AOL) offers e-mail functions and features, as well as options dedicated to its own services. The MIME (Multipurpose Internet Mail Extensions) are applicable to such transmission, permitting the integration of program and video files within e-mail documents and communications.

Typically, a computer fitted with a modem (Modulator Demodulator) is used for transmission and reception of e-mail, although within organizations NICs are more common. E-mail messages may be sent over LANs, intranets, and the Internet. Users generally read their e-mail messages by downloading them from a server, and there is often an option within the e-mail program that allows them to choose whether or not to leave a copy of the e-mail message on the mail server.

E-mail Autoreply

A reply to an e-mail message that is created automatically, using an e-mail autoresponder such as MReply. In a e-commerce context, such responses are useful for conveying the receipt of orders, advertising related products and promotions, publicizing trading hours, and so on.

E-mail Responder

A program that replies automatically to received e-mail messages.

Encapsulation

A term which describes hiding the internal workings of an object. Resulting objects encapsulate code and data that are hidden from the user and the remaining collective OO system. Essentially it becomes a black box, and all that matters are its responses to stimuli, such as defined events that are intercepted and processed by the object's public interface.

Enterprise Computing

A general term used to describe the application of computers and Information Technology (IT) in medium-size to large businesses. Only larger small businesses are considered to be enterprises.

Entity Relationship Diagram

A diagram that illustrates the design structure of a relational database, together with all its data tables and links. Programs that may be used to draft such diagrams include EasyCase. Entity relationship diagrams rarely included reports and query information, though some relevant notes might be included.

Entry

A group of object references in a class package such as `net.jini.core.entry.Entry` interface.

Error Log

A log of errors experienced by a server.

Ethernet

A Local Area Network (LAN) standard. Ethernet adapters included on computers may comprise thin-Ethernet or more expensive thick-Ethernet connectors and cables. Ethernet is considered to first have been discussed in 1974 by Robert Metcalfe through his Harvard Ph.D. thesis.

Event

A change in state that may invoke responses or a series of processes and subprocesses that may be implemented by objects. The event may be a simple message sent from one object to another, and its origins may be anything from another message from an agent to a physical mouse click or key press. Applications and operating system environments that respond to such events are termed event-driven.

Event Generator

An object whose state changes are relevant to another object, and that may send notification messages to compliant objects when events are generated.

Event Listener

An object that responds to events or, more specifically, responds to one or more event types.

Event-Driven

An environment or program that responds to external events such as mouse clicks. Modern event-driven applications may be assumed to be object-oriented. Objects such as buttons respond to events, triggering a method or item of code that is attached to them. Windows is an event-driven environment.

Expansion Bus

A bus used to provide a means of expanding a PC to include various peripheral devices that might range from graphics cards to MPEG players. Standard expansion buses include 16bit ISA (Industry Standard Architecture), IBM MCA (Micro Channel Architecture), and EISA (Enhanced Industry Standard Architecture).

Explorer

A program that is part of Windows 95 and Windows NT, and is used to peruse files, open files, launch programs, and perform file management functions.

It shows file details such as their size in Bytes, the date and time they were last modified, and their attributes, including whether they have read, write, or read/write status.

It is commonly used to move, rename, copy, and delete files and even complete directories. The move, copy, and delete commands work with multiple selected files, so you can copy and move batches of files without having to go through the monotony of dealing with one file at a time.

Windows applications may be run from Explorer by double-clicking them, or by double-clicking files that were created with them.

Explorer may be used to:

- Connect to shared directories on other network users' drives
- Declare directories as shared
- Give shared directories password protection
- Monitor who on the network is using shared directories
- Stop sharing shared directories.

Miscellaneous

E Interface

The E interface exists between different MSCs. In the context of a GSM network, an operator may use one or a number of such MSCs as gateways (GMSCs) that provide the link between the PLMN and external networks. When a call is received from another network, the GMSC communicates with the appropriate network database so as to direct the call to the appropriate MS.

E-purse

A value that may be stored on a SmartCard and represents an amount that may be used to make small purchases.

EIGRP (Extended Interior Gateway Routing Protocol)

A protocol developed by Cisco for routers.

Encode

A process of converting data, or an analog signal, into another form in terms of data representation. For example, Video-on-Demand services often use MPEG-2 video that is encoded using uncompressed source recordings that may be analog or digital. Equally, streaming video/multimedia sites store video encoded according to the MPEG-1 specification.

Encryption

A process of ciphering messages or data so that it may be deciphered and read only by intended recipient(s).

Encryption techniques include:

- DES
- TripleDES
- DES X
- RSA
- DSS

EVRC (Enhanced Variable-Rate Codec)

A codec that may be used with CDMA, operating at 13.25kbps in a 14.4kbps data stream.

Acronyms

EACH	Enhanced Access Channel
ECC	Error control coding
EDGE	Enhanced Data Rates for GSM Revolution
EDI	Electronic Data Interchange
EGPRS	Enhanced GPRS
EIB	Erasure Indicator Bit
EMI	Electromagnetic Interference
EML	Element Management Layer
ERIP	Equivalent Isotropic Radiated Power
ERP	Effective Radiated Power
ESN	Electronic Serial Number
ESS	Extended Service Set
ESSID	Extended Service Set Identification
ETSI	European Telecommunications Standard Institute
EVRC	Enhanced Variable-Rate Codec

F

FDD (Frequency Division Duplexing)

FDD is a form of multiplexing where transmission and reception occur simultaneously over the same channel using different frequencies. The weak received signal is protected from the effects of the strong transmitted signal using a sufficiently large f_{dup}, ensuring transmitted energy is low at the received carrier frequency. GSM uses FDD as well as FDMA carriers that each have eight channels as a TDMA time frame. The frames have time slots, each carrying a single channel with traffic or signaling information. See Fig. F-1.

When an MS is assigned a time slot such as Slot-2 for example, it transmits for the duration of Slot-2. The traffic that is acquired over a complete frame is speeded up for transmission, resulting in a *traffic burst*. See Fig. F-2.

Fiber Channel

Fiber channel is a high-performance communications pathway, which was introduced by the Fibre Channel Association (FCA). An open standard, it is a protocol that supports data transfer rates from 133Mbits/s up to 100Mbytes/s.

Fiber channel can be used to connect sites up to 10km apart using a 9μm single-mode optic fiber. The fiber-channel protocol may also propagate along traditional copper-based transmission media.

Typical data transfer rates (and maximum transmission distances for a 9μm single-mode optic fiber) are:

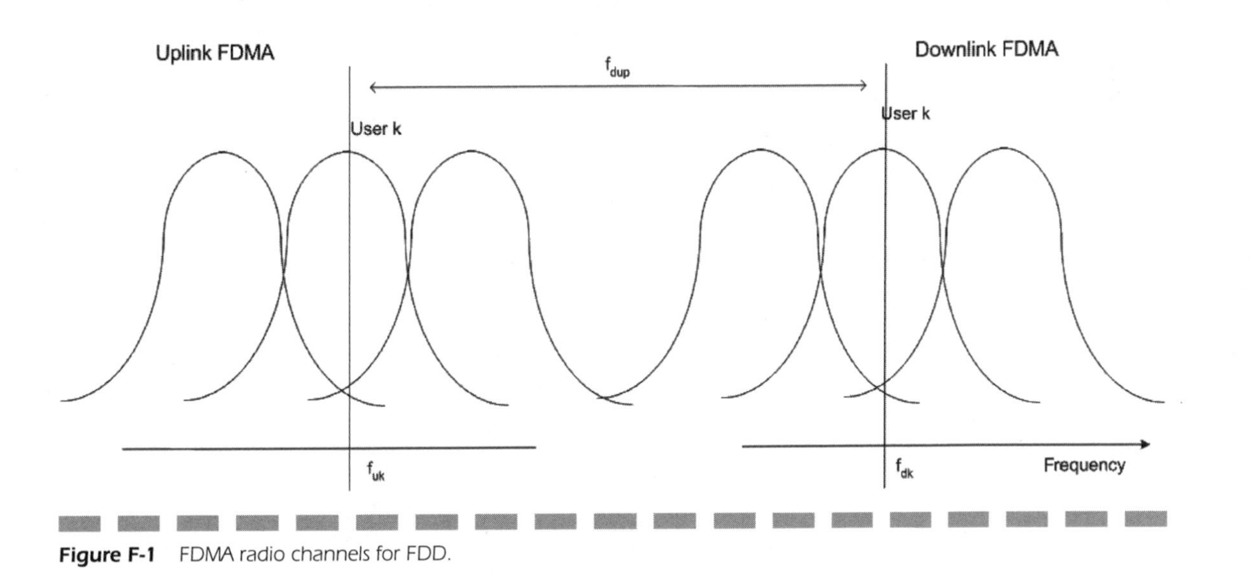

Figure F-1 FDMA radio channels for FDD.

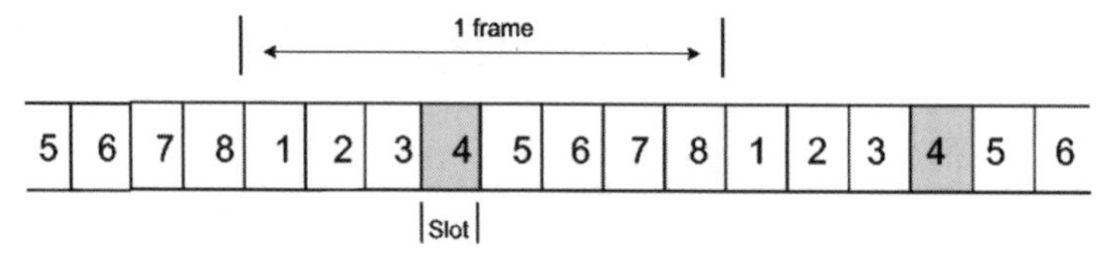

Figure F-2
Transmitter TDMA frame.

- 100Mbytes/s (10km)
- 50Mbytes/s (10km)
- 25Mbytes/s (10km)

For a 50μm multimode optic fiber:

- 100Mbytes/s (0.5km)
- 50Mbytes/s (1km)
- 25Mbytes/s (2km)

For a 62.5μm multimode optic fiber:

- 25Mbytes/s (500m)
- 12.5Mbytes/s (1km)

For video coax:

- 100Mbytes/s (10km)
- 50Mbytes/s (10km)
- 25Mbytes/s (10km)

Applications of fiber channel include mass storage interface and control, and high-speed networks. Network topologies may be point-to-point, ring, or a Fiber Channel—Arbitrated Loop (FC-AL), which requires neither switches nor hubs. See Figs. F-3 and F-4.

Frames are used to send and receive data, each having the fields:

- Start-of-frame delimiter
- Frame header
- Optional header
- Payload, which is the user data, and may be between 0 to 2112 bytes
- 32bit CRC error detection
- End-of-frame delimiter

Figure F-3
FCC frame.

4 bytes	24 bytes	0 to 2112 bytes (payload)		4 bytes	4 bytes
Start of frame	Frame header	64 bytes Optional header	2048 bytes (maximum payload with *optional header*)	32bit CRC	End of frame

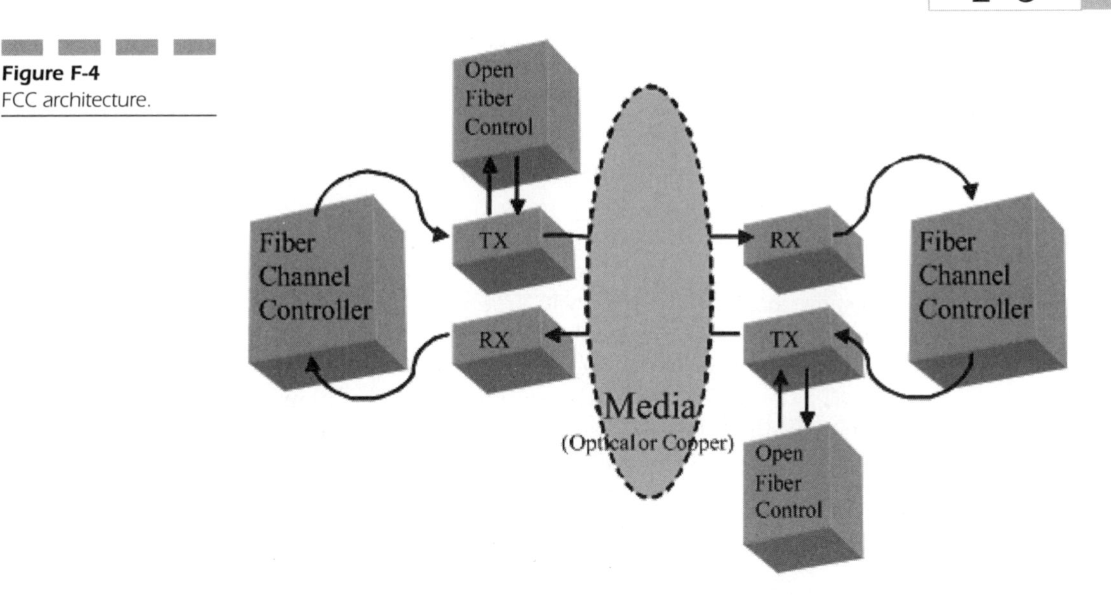

Figure F-4
FCC architecture.

The Fiber Channel—Physical (FC-PH) standard consists of the following levels:

- FC-0, which covers physical media, cable connectors, LEDs, short- and long-wave lasers, transmitters and receivers
- FC-1, which covers the encoding and decoding protocol, to cater for the adopted serial transmission techniques
- FC-2, which covers the signaling protocol, and defines the shown frame format (or *framing protocol*) for data transfer

Upper FC layers include FC-3, whose common services include:

- *Multicast,* for transmissions to multiple destinations.
- *Striping,* for transmitting to multiple N_ports concurrently, and supports multiple links.
- *Hunt groups,* which is a collection of N_ports, which is assigned an alias. Frames containing the alias are routed to any non-busy N_port within the defined group.

Upper layer protocols (ULPs) are defined by FC-4, covering industry network standards, which may be transported using Fiber Channel. These include:

- Internet Protocol
- ATM Adaption layer
- IEEE 802.2

Channel protocols supported by FC-4 include:

- SCSI (Small Computer Systems Interface)
- High Performance Parallel Interface (HIPPI) framing protocol
- Intelligent Peripheral Interface (IPI)
- Single Byte Command Code Set Mapping (SBCCM)

First-Generation (1G) Networks The 1G category of public mobile network is now consigned to history in first-world countries. 1G networks were analog and offered mainly telephony services, including AMPS (Advanced Mobile Phone System) that operated in the 800MHz cellular band.

By the late 1970s early cellular networks began to emerge, and formed the basis for the wireless communications used today. In 1977 Illinois Bell introduced a cellular network in Chicago. Called AMPS (it was developed by AT&T's Bell Laboratories and operated between the 800MHz and 900MHz bands, and was the most used service in the United States until the early 1990s. In 1982 the development of an international standard called GSM (Global System for Mobile Communications) was begun, and by 1993 a rapid rise in compatible networks had emerged globally. This marked the beginning of wireless telecommunications for the mass market and drove the many growing cell phone services available today.

The capacity of 1G networks was naturally small, because authentic 1G networks were not cellular. Later cellular networks provided greater geographic coverage through cells using the same frequencies. An early internationally agreed standard analog network did not exist, and so countries and continents had disparate network systems. These included Nordic Mobile Telephone (NMT–Scandinavia), Total Access Communications System (TACS), C-Netz (West Germany), Radiocomm 2000 (France), and, of course, AMPS.

NMT was also used in central and southern Europe, and was introduced in Eastern Europe during the late 1990s. NMT had two variants, namely, NMT-450 and a later NMT-900 system using the 900Mhz frequency band. Like some of the later networks, NMT offered the option of international roaming—although this was not as seamless as it is today.

The United Kingdom adopted TACS, which was based on AMPS, but used the 900MHz frequency band; TACS became successful in the Middle East and in Southern Europe. The American AMPS standard used the 800MHz frequency band and was also used in South America, Far East, and in the Asia Pacific region, including Australia and New Zealand. In the Asia Pacific country of Japan was NTT's MCS system that was the first commercial delivery of a mobile 1G Japanese network. While most first-world countries are closing 1G networks, many Less Developed Countries (LDCs) are actively investing in and also upgrading them.

Telecommunications networks remain a foundation for the World Wide Web, and connectivity between tier 1 and tier 3/4 devices and servers and hosts, but it was the advent of the TCP/IP (Transmission Control

Protocol/Internet Protocol), which emerged from the early DARPA and ARPANET networks, that introduced a common transport protocol for communications. In 1983 all ARPANET networks were running TCP/IP, and later 32bit IP addressing meant that networks could include millions of addressable hosts. Some 18 years later, the growing global network is now driving the development of IPv6, that is, a 128bit addressing scheme specified by the IETF.

Firewall

Firewalls are software/hardware implementations that partition networks or systems that restrict access to selected users and, appropriately, isolate networks and network segments. It may perform the simple functions of checking client connections and requests, securing server-side applications and data. It may intercept inbound data packets, and perform a number of security checks. These may revolve around the origins of the packet, checking such packet information as its:

- Source IP address
- Source IP port that identifies the originating application

Firewalls are key to many organization's security strategy. Other adopted security facets include:

- Passwords for logging on to networks
- Client-side password checks for connecting to Web sitesClient-side password checks for connecting to e-mail applications and services
- Password-protected compressed hard disks, made possible using Stac Electronics disk compression programs

Firewalls may also include the ability to virus check and to screen incoming documents and executables such as ActiveX controls, Plug-ins, Java applets, and any other code that is downloaded and intended to be processed. Cookies may also be filtered.

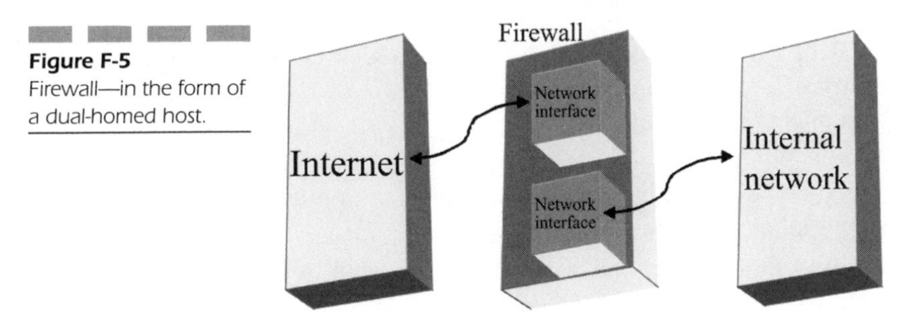

Figure F-5 Firewall—in the form of a dual-homed host.

Firewalls may be at the network level that harness packet filtering techniques using routers. The routers are intelligent in that they may be programmed to behave as a selective barrier to unwanted network traffic. See Fig. F-5.

Dual-Homed Host Firewall

A dual-homed host has two network interfaces that connect with disparate networks, while a multi-homed host typically interfaces with two or more networks. The term *gateway* was used to describe the routing functions of such dual-homed hosts. Nowadays the term gateway is replaced by router. A dual-homed host may be used to isolate a network, because it acts as a barrier to the flow of TCP/IP traffic. The implementation of a Unix dual-homed firewall requires (among other things) that

- IP forwarding is disabled, thus yielding a protective barrier. Unrequired network services are removed.
- Programming tools are uninstalled.

Bastion Host

A bastion host is critical to a network's security. This is the focus of network management, security monitoring, and is a network's principal defense against illegal usage. A dual-homed host may play the role of a bastion host.

Screened Subnets

A screened subnet restricts TCP/IP traffic from entering a secured network. The screening function may be implemented by screening routers.

Commercial firewall products include

1. FireWall-1 that is a commercial gateway product, from the Internet Security Corporation, and uses:
   - Application gateway
   - Packet filtering
2. ANS InterLock that is a commercial gateway product from Advanced Network Services.
3. Gauntlet that is a firewall product from Trusted Information Systems.

FMA (FRAMES Multiple Access)

FRAMES (Future Radio Wideband Multiple Access System) was set up by ACTS (Advanced Communication Technologies and Services) to define a UMTS radio interface that eventually became FMA that had two modes:

- FMA1 for wideband TDMA
- FMA2 for direct sequence wideband CDMA

FMA1

FMA1 user bit rates up to 2Mbps are achieved by allocating time slots and/or spreading codes to the user. FMA1 users are separated orthogonally into time slots that are assigned on a whole frame basis to either the uplink or downlink when in FDD mode. Time slots are divided dynamically between uplink and downlink in TDD mode. The TDD frame duration is 4.615ms and may have 1/64, 1/16, and 1/8 slots, and UL and DL parts have a minimum duration of 1/8 of frame (577μs). FMA1 nonspread bursts are assigned 1/64 and 1/16 slots, and spread bursts are assigned 1/8 slots. 1/8 and 1/64 slots are used for all services ranging from speech to broadband data services, and the 1/16 slot is used for medium- to high-rate data services.

Synchronization and handover measurements use either a slotted discontinuous wideband broadcast control channel (BCCH) or a continuous narrowband (200kHz) BCCH.

Slotted wideband BCCH bursts include correction, synchronization, and access:

- The frequency correction burst is of 72μs with 171 fixed symbols.
- The synchronization bursts may be either 72μs or 288μs duration.
- The access bursts may be either 72μs or 288μs duration.

The guard periods of the access bursts allow reception of initial random access messages for a cell radius of 5km to 36km.

FMA2 (W-CDMA)

W-CDMA features include the following:

- Dual-mode UMTS/GSM terminals
- Efficient packet access
- Interfrequency handovers
- Support multiple parallel variable-rate services on each connection

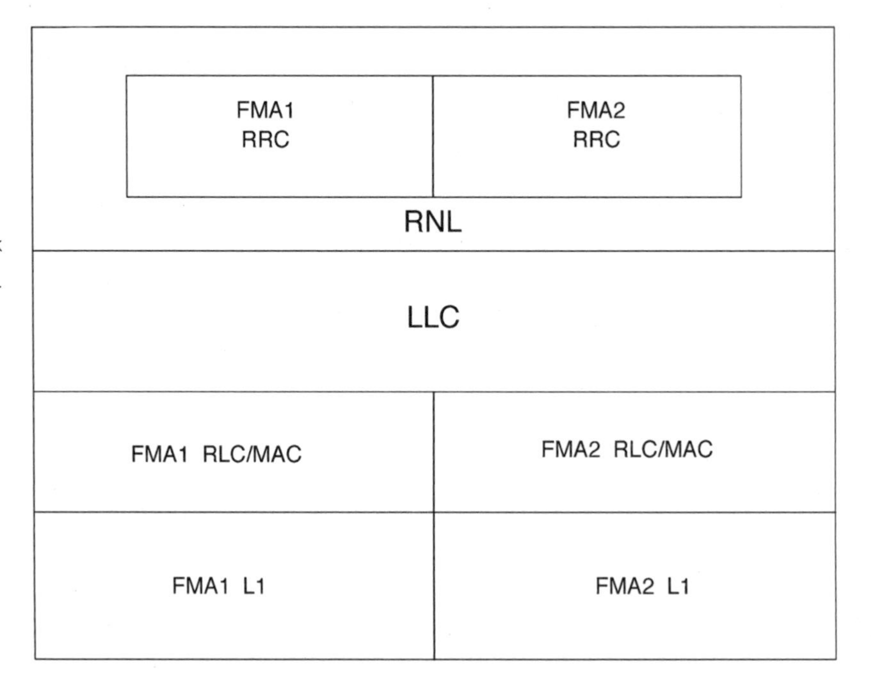

Figure F-6
FMA1 and FMA2 layers. RLC = Radio Link Control, MAC = Media Access Control, RRC = Radio Resource Control, RNL = Radio Network Layer, LLC = Logical Link Control.

- User data rates of >384kbps for wide areas and up to 2Mbps for indoor/local coverage

FMA1 and FMA2 protocol stacks have obvious similarities in an effort to reuse protocol designs. However, differences exist in the radio link control (RLC) and media access control (MAC) layers between FMA1 and FMA2, and the radio network layer (RNL) is also different. The FMA2 radio resource control (RRC) is also different when used in handovers. See Fig. F-6.

UMT5 Terrestrial Radio Access Network (UTRAN) layers consist of the:

- *Physical layer:* The Physical layer provides an information transfer service over the W-CDMA radio medium.
- *Media access control sublayer:* The MAC provides data transfer services for RLC, and provides information to the higher layers regarding traffic volume and quality.
- *Radio link control sublayer:* The RRC provides general broadcast and notification services to all mobiles.

The W-CDMA data flow is similar to that of GPRS. W-CDMA data flow involves segmenting network layer packet data units (N-PDUs) into smaller packets, transforming them into link access control (LAC) PDUs. An additional LAC data (of 3 octets) has a service access point identifier, and a sequence number for higher-level ARQ. Segmenting LAC PDUs results in small packets, RLC-PDUs, like physical layer transport blocks. RLC-PDUs have a sequence number for the low-level ARQ, and error detection data in the form of cyclic redundancy check (CRC) that is appended by the physical layer.

Fraud

As growing networks bring a greater number of subscriber services to mobile users, larceny and fraud present themselves as significant problems, costing operators valuable revenues, and at the same artificially inflating service costs for mobile users. Antifraud measures therefore are becoming widespread with many of the mobile operators investing in automated solutions that are based on intelligent decision-making software that operates without human intervention. More practical solutions have emerged and these include prepaid phone schemes where the user may purchase air time, and relieve the operator from architecting and managing billing services. (*See* AI under letter A, Anti–Money Laundering under letter A.)

The oldest form of antifraud and theft prevention is the SIM chip; mobile devices have SIM chips holding user identification and configuration data. SIM chips permit an authorization procedure to be implemented between MSCs and EIRs (Equipment Identification Registers). The EIR has a blacklist of barred equipment, a gray list of faulty equipment or for devices that are registered for no services, and a white list for registered users and their service subscriptions. However, the SIM chips themselves can be replicated illegally and used, and there is also a reliance on the user to report the theft of the phone immediately.

Fraud measures begin when users apply for mobile contracts and subscription services, when credit checks may be carried out and a weighting given to the user in terms of the possibility that a fraud may occur. Once a user is connected, the operator may be equipped with antifraud systems that monitor usage, and alert personnel about irregularities. For example, if the user pays bills late, then the possibility of a user fraud may arise. Equally if the phone usage suddenly spirals, then it is probable that the phone is stolen, and the user may be contacted. Antifraud systems that are able to provide such alerting are being deployed by many of the larger operators that use intelligent decision making based on Neural networks, rule-bases, and even genetic algorithms.

The fraud scenario is actually much broader and wider ranging than the mere illegal use of operator networks and services. For example, if the operators provide m-commerce where the user is equipped with an electronic credit card given by an issuer, then the threat of mobile fraud or larceny increases. The credit card company would also require antifraud solutions. More complex is the scenario where the operator is itself a credit card issuer, providing the mobile user with the options of m-commerce—where the user may purchase certain products and services.

This could be achieved with the operator working in partnership with large vendors that may be high street chains, airlines, travel agencies, etc. This m-commerce loop would bypass the credit card company, and would therefore place greater demands on the operator in terms of antifraud measures that are typically employed by banks and credit card issuers.

Frequency

See Frequencies in Numerals chapter.

Frequency Hopping

Using frequency-hopping spread spectrum, the carrier signal for transmission changes periodically. This system uses a bandwidth part that changes according to the spreading code. CDMA systems are divided into slow- and fast-hopping systems, where the former transmits several symbols at the same frequency. Fast hopping requires the frequency to change several times for the transmission of one symbol. GSM is a slow-hopping system where frequency changes with the time slot rate of 217hops/s.

Bluetooth Frequency Hopping

Bluetooth uses the Industrial Scientific Medical (ISM) frequency band (2402Hz–2480Hz), and operates at 2.4GHz, using 79 channels that are 1MHz wide. The 2.4GHz band is unlicensed so other devices may operate in the same vicinity. Spread spectrum technologies are used to remedy interference between radio technologies, and frequency hopping means that a Bluetooth device changes frequency in pseudo-random at 1600 changes per second. If interference does cause errors, the packet-based solution allows erroneous packets to be detected and corrected by retransmitting them. See Fig. F-7.

A Bluetooth channel has a master and one or more slaves, and the master usually initiates the connection and selects a hopping scheme that is consistent with its internal clock. The slaves calculate the difference between the master and slave clock, and then a hopping frequency is determined. As a result a master and slaves will hop in unison to the same frequencies. The uplink and downlink channels are separated using TDM (Time Division Multiplexing), and both channels use the same frequency hopping scheme.

Link Types The delivery of different traffic like voice and bursty traffic is accommodated by two link types:

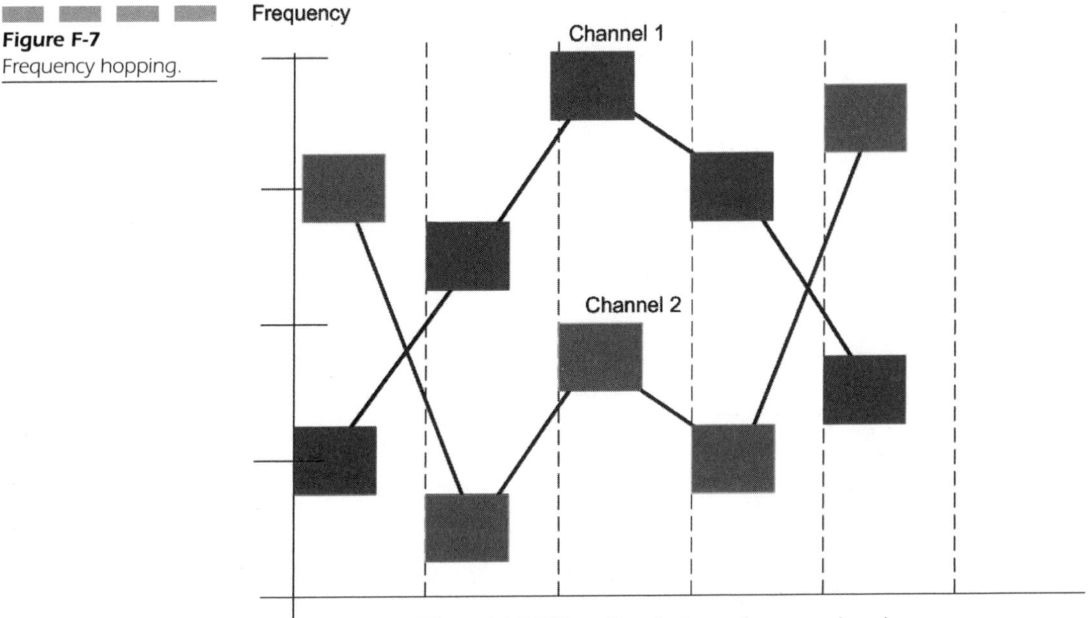

Figure F-7 Frequency hopping.

- *SCO (Synchronous Connection-Oriented)* is applicable using CS services with short latencies and high levels of QoS. This offers symmetric channels that are synchronous, and is used mainly for voice, though data is possible over the 64kbps channel.
- *ACL (Asynchronous Connectionless)* is used for data transfer and other asynchronous services, and offers PS channels. Transmission slots are not reserved but are granted using a polling access scheme.

A connection may have several links of ACL and/or SCO, and when using a piconet a master may have SCO and ACL links with several slaves, but there is a three-voice call limit.

Bluetooth's analog voice may be digitally encoded using PCM (Pulse Code Modulation) or CSVD (Continuously Variable Slope Delta). PCM uses eight-bit samples at 8kHz, and so requires 64kbps in order to transmit the digitally encoded signal. CSVD uses 1bit samples that indicate whether the slope of the voice signal is increasing or decreasing. Bluetooth voice calls are ISDN quality using 64kbps, and this is better than many 2G handsets that offer 8kbps.

Bluetooth packets can be sent one slot at a time, or using multislot, where a large packet is sent over several slots and the frequency remains the same. Multislot bit rates may reach 721kbps but are without protective coding.

TABLE F-1

Bluetooth Bit Rates

Type	Symmetric (kbps)	Asymmetric Downlink (kbps)	Asymmetric Uplink (kbps)
DM1	108.8	108.8	108.8
DH1	172.8	172.8	172.8
DM3	258.1	387.2	54.4
DH3	390.4	585.6	86.4
DM5	286.7	477.8	36.3
DH5	433.9	721	57.6

Computers/Software

Failover

Failover is a contingency measure that provides an alternative service provider, should a failure occur.

Fat Client

A fat client system within a client/server architecture (such as that of the Web) features:

- Presentation, which is typically in the form of a Web browser
- Complete application(s)
- A data cache, which is used to store information from a server-side database, or back-end database.

Many systems connected to the Web may be described as thin clients. Fat clients depend heavily on client-side processing and resources, while thin clients do not. This higher demand for hardware results in higher client system costs.

Generally, fat clients may be integrated by:

- Improved intelligence, because the user's interaction can be personalized through the local customization of the application. Additionally, intelligence features such as those associated with kilobytes per second are more feasible.
- Additional local applications, such as industry standard products.
- Data verification, prior to sending messages to the client side, thus improving system responsiveness, while reducing network traffic.
- Security on the client-side, through password checks, and restricted access to documents, data, and applications.

FAT32 (File Allocation Table)

FAT32 is a filing system introduced in the Windows 98 operating system. It is an advancement of the FAT16 implementation, and is able to address hard disks with up to a 2Gbyte formatted data capacity. It is more efficient than FAT16, because it uses smaller clusters of 4kbytes that are used to store files. Clusters are used to store data from a single file. The larger 32kbyte clusters of FAT16 are comparatively inefficient. For example, when a 34kbyte file is written to the hard disk, two 32kbyte clusters are used. The second cluster has some 30kbyte of unused payload. So even though the file is just 34kbytes, it consumes 64kbytes of hard disk that equates to two of its 32kbyte clusters. Clearly, FAT32's dependence on 4kbyte clusters helps eradicate the unused data capacity of clusters. This yields considerable storage capacity gains.

Feedback Form

A feedback form may be used to gather information about visitors to e-commerce sites and may be created simply by using HTML.

FHSS

Frequency Hopping Spread Spectrum

Field

A column in a database table or a container for data entry in a form. Entries within fields are termed *field values*.

Field Value

A data item in a database.

FIFO (First In–First Out)

A queue whose operation hinges on regurgitating items in the order in which they were deposited. An analogy is that of a vending machine that sells chocolate bars that are stored in a vertical dispensing tube.

Fifth-Generation Language

A fifth-generation language is nonprocedural. They are declarative in that actions are not implemented through fixed procedures. They are also known as AI languages and include PROLOG (Programming Logic).

Find-and-Replace

A phrase used to describe the automated process of replacing a specified word or phrase with another. The phrases find-and-replace and search-and-replace are interchangeable.

Firewire

A high-performance interface that permits the connection of peripheral devices such as mass storage devices, modems, and printers. It is otherwise known as IEEE1394, and as such it is an internationally agreed standard.

Firmware

A program or data stored using a ROM variant Firmware is thus involatile and permanent.

Form

1. A metaphor for a paper form that is used by client Browsers in order to interact with programs and data that may be on the client- or server-side. Typically, forms permit users to enter:
   - Signup details with Web sites
   - Contact details
   - Password details
   - Credit or Switch card details for purchase from e-commerce sites
2. A metaphor for a paper form, used for data entry and viewing data in a database. RDBMS development tools, such as Excel, DataEase for Windows, or Paradox for Windows, may be used to create table-based applications.
3. A data sector type on a CD-I disc. Similar to CD-ROM blocks, CDI sectors are 2352bytes long—including headers, sync information, error detection, and correction data. Similar to Mode 1 block, Form 1 sector yields 2048bytes user data. Unlike Mode 2 block, however, Form 2 sector yields 2324bytes user data.

Form Method

A method of gaining customer information and for taking orders. Forms may be created using HTML and by using scripting languages:

```
<FORM> NAME="Customer"
       ACTION="http://botto.com/cgibin/form/cgi
       METHOD=get>
</FORM>
```

The <FORM> tag may have the attributes:

- NAME, which is the form's name
- ACTION, which indicates the URL where the form is sent to
- METHOD, which indicates the submission method that may be POST or GET
- TARGET, which indicates the windows or frame where the output from the CGI program is shown

Forward Channel

A channel that a base station uses to communicate with users.

fps (Frames per Second)

A measure of the speed at which frames making up a video sequence are played or captured.

Frame

1. A tiled area of a Browser's window. A frame provides an efficient method of presenting information without using a separate Web page. For example, a frame might be used to play a video sequence or animation. A frame-enabled Web application reduces the complexity of designing multiple pages at design time, and is toured more easily by users. Frames are supported by many Web page designs and Web application development tools such as Microsoft FrontPage.
2. A single image making up a video sequence. Digital video sequences may consist predominantly of partial frames called interframes, or full frames called intraframes.

3. A single item of transmitted data using the Frame Relay protocol that is designed for modern digital networks, and does not integrate the demanding error detection and correction schemes prevalent in older protocols.

Front End

A name given to the client application or system that may be served by a server-side or back-end application. Between the back- and front-end applications is middleware, or glues, that exist at a number of levels. These may bind together and coordinate application logic, data, and presentation distributed across the back and front ends.

FTP (File Transfer Protocol)

A protocol used to transfer files between FTP servers and client systems. It is a standard method for distributing files across IP/TCP networks. Using an FTP client program, users are able to link with FTP sites, and browse the remote directories and files as if they were on a local hard disk. Users can then download files from the ftp server.

Full Duplex

A simultaneous bidirectional transmission of different data streams.

Full Frame Updates

A video sequence that is composed of full frames. Any such frame can provide a valid entry point for nonlinear playback or editing. Such video sequences are also referred to as intraframe.

Miscellaneous

FDDI (Fiber Distributed Data Interface)

FDDI is a computer-to-computer fiber link technology, and an internationally agreed ANSI standard. The topology comprises a dual multi-mode optic fiber, Led (or laser), and Token ring network. Data rates of up to 100Mbps are possible. Without repeaters, transmission distances up to 2 km are attainable, at a data transfer rate of 40Mbps.

FMFSV (Full Motion Full Screen Video)

The term *FMFSV* is used to describe video that may be assumed to fill the entire screen, or a greater part of it, and provides the illusion of a frame rate of not less than 25 frames per second (fps) without the use of duplicated frames. MPEG-2 or DVD video is FMFSV. 25fps is the frame rate delivered by PAL and SECAM broadcast standards. The American NTSC broadcast standard provides 30fps. Ideally the frame rate should be greater than 25 to 30fps. The frames that make up an FMFSV may be full frames as in the case of an M-JPEG video stream or a combination of full frames and partial frames as is the case with MPEG video. The full frames or reference frames occur at regular intervals, and dictate the number of authentic random access points provided by an encoded MPEG video sequence. The frame resolution of what may be described as FMFSV varies, but it should not fall below around 720-by-360 pixels. Larger standard frame resolutions may broadly equate to 640-by-480 pixels, 800-by-600 pixels, 1024-by-768 pixels, 1240-by-1024 pixels, and 1600-by-1240 pixels.

Frame

1. A single item of transmitted data using the Frame Relay protocol that is designed for modern digital networks, and does not integrate the demanding error detection and correction schemes prevalent in older protocols.
2. A single image making up a video sequence. Digital video sequences may consist predominantly of partial frames called interframes, or full frames called intraframes.

Frame Relay

The frame relay protocol was designed for modern communications networks. Typically it may be operated at speeds between 9600bits/s and 2Mbits/s, though higher speeds are possible. Compared to X.25 it makes better use of network bandwidth as it does not integrate the same level of intense error detection and correction. That is not to say that frame relay is unreliable; it is simply optimized for modern networks which do not impose the same level of error on transmitted data—which is the case with older network technologies for which X.25 was designed. The frame relay protocol may be applied in WAN and backbone implementations, and integrated into solutions that require high data transfer speeds.

Each frame consists of a:

- Flag that separates contiguous frames
- Address field that stores the data link connection identifier (DLCI) and other information
- Control field that contains the frame size, and receiver ready (RR) and receiver not ready (RNR) informationInformation field that contains up to 65,536 bytes
- Frame check sequence that is a CRC for error correction

Acronyms

F-APICH forward auxiliary pilot channel
F-ATDPICH forward auxiliary transmit diversity pilot channel
F-CPCCH forward common power control channel
F-DAPICH forward dedicated auxiliary pilot channel
F-DCCH forward dedicated control channel
F-PICH forward pilot channel
F-SCCH forward supplemental code channel
F-SCH forward supplemental channel
F-SYNC forward sync channel
F-SYNCH forward sync channel
F-TDPICH forward transmit diversity pilot channel
F/R-DCCH forward/reverse dedicated control channel
F/R-FCH forward/reverse fundamental channel
F/R-PICH forward/reverse pilot channel
F/R-SCCH forward/reverse supplemental coded channel
F/R-SCH forward/reverse supplementary channel
FACCH fast associated control channel
FACH forward access channel
FACHFP FACH frame protocol
FAUSCH fast uplink signaling channel forward broadcast channel feedback indicator
FCH frame controller head

FDD	frequency division duplex
FDMA	Frequency division multiple access
flr-csch	forward and reverse common signaling channel
flr-dmch	forward and reverse dedicated media access control channel
flr-dsch	forward and reverse dedicated signaling channel
FMA	FRAMES multiple access
FPLMTS	Future Public Land Mobile Telephony
FR	frame relay
FRAMES	Future Radio Wideband Multiple Access System
FSK	frequency shift keying
FTP	file transfer protocol

G

GPRS

The GPRS (General Packet Radio Service) overlay increases the GSM and TDMA user data rate to a maximum of about 171kbps, and includes the following physical network overlay core solutions:

- IP-based GPRS backbone
- GGSN (Gateway GPRS Serving Node)
- SGSN (Server GPRS Serving Node)

Principal features of GPRS include:

- Packet-based transmission
- Packet-based charging as opposed to time-based charging
- Always-on—so a VPN connection, for example, does not require constant logging in to the remote/enterprise network/intranet
- Eight time slots
- A maximum user data rate of about 171kbps when all time slots are used

GPRS architecture (see Fig. G-1):

- Alleviates capacity impacts by sharing radio resources among all mobile stations in cells.
- Lends itself to bursty traffic.

GPRS applications include:

- E-mail, fax, messaging, intranet/Internet browsing
- Value-added services (VAS), information services, games
- m-commerce: retail, banking, financial trading, advertising
- Geographic: GPS, navigation, traffic routing, airline/rail schedules
- Vertical applications: freight delivery, fleet management, sales-force automation

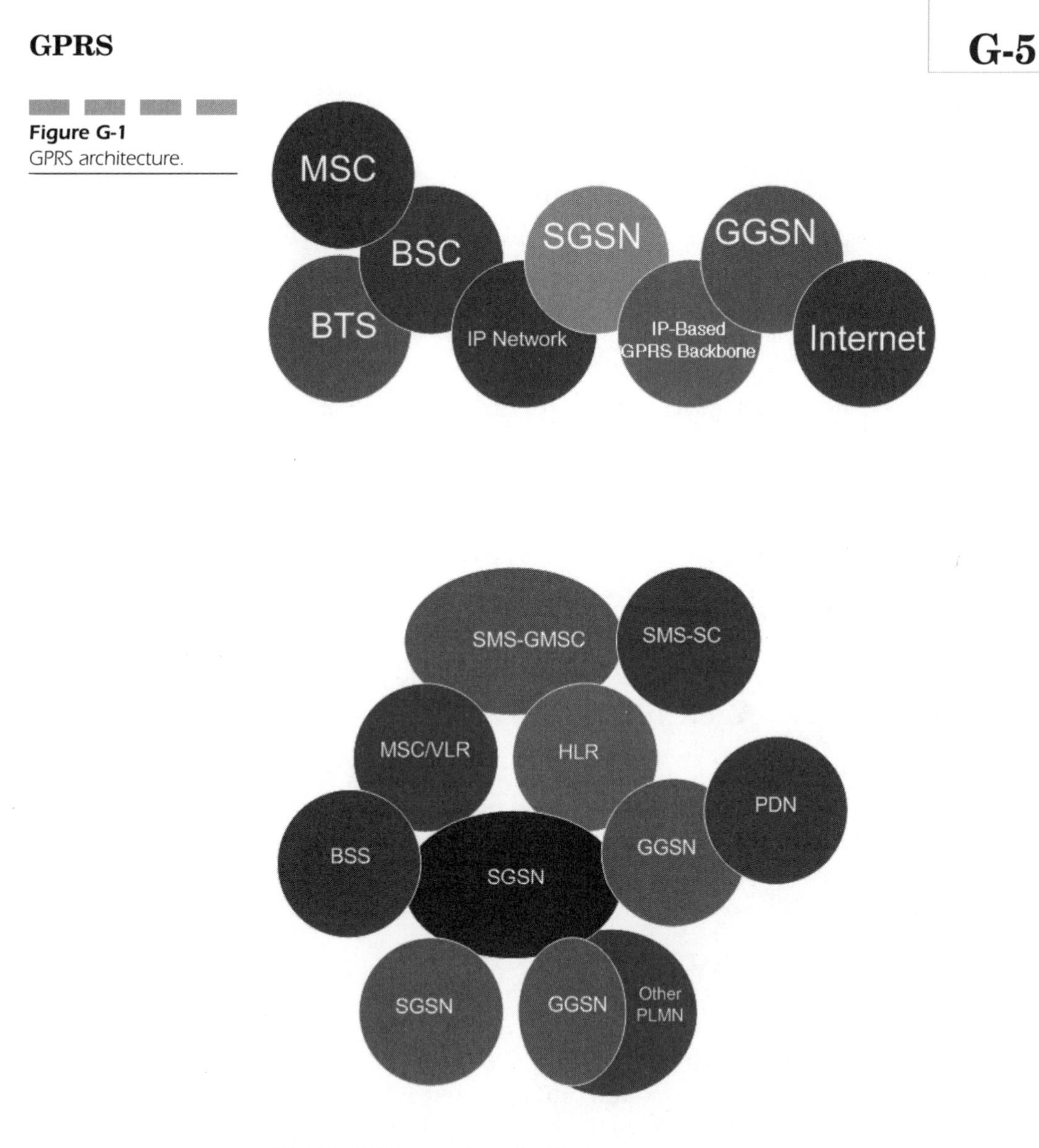

Figure G-1
GPRS architecture.

GPRS Terminal Classes

A GPRS terminal phone may be:

- Class A: Supports GPRS, GSM, and SMS, and may make or receive calls simultaneously over two services. In the case of CS services, GPRS virtual circuits are held or placed on busy as opposed to closed down.
- Class B: Supports either GSM or GPRS services sequentially, but monitors both simultaneously. Like Class A, the GPRS virtual circuits will not be closed down with CS traffic, but switched to busy or held mode.

TABLE G-1

GSM to GPRS Migration Elements

GSM Entity	GPRS Overlay
BSC	BSCs require software upgrades and packet control unit (PCU) hardware that directs data to the GPRS network.
Core Network	Nodes and gateways that are essentially packet-switched MSCs are required, including:
	Serving GPRS Support Node (SGSN)
	Gateway GPRS Support Node (GGSN)
Databases (HLR and VLR)	Databases require software upgrades to handle GPRS functions and call models.
Subscriber Terminal	New subscriber terminals are required to access GPRS services.

- Class C: Make or receive calls using the manually selected service, with SMS an optional feature.

GPRS Architecture

The GPRS architecture is an overlay network for 2G GSM networks, and provides packet data for user data rates between 9.6 and 171kbps. GPRS's air-interfaces are shared among multiple users.

See Table G-1.

GPRS BSS

To become a GPRS BSC, each GSM BSC requires the installation of one or more PCUs that give interfaces for packet data out of the BSS. Like GSM when either voice or data traffic originates at the subscriber terminal, it goes over the air interface to the BTS, from where it goes to the BSC. Unlike GSM, however, the traffic is separated, voice is sent to the MSC (like GSM), and data is sent to the SGSN, through the PCU using a Frame Relay interface.

GPRS Network

2G GSM MSCs are circuit-switched (CS) and cannot therefore handle packet data, so nodes are introduced:

- Serving GPRS Support Node (SGSN)
- Gateway GPRS Support Node (GGSN)

The SGSN acts like a packet router or packet-switched MSC that directs packets to MSs.

SGSN

A 2.5G GPRS SGSN performs the following functions:

- Queries home location registers (HLRs) to get subscriber profiles
- Detects new GPRS MSs
- Processes registration of new subscribers
- Keeps records of subscribers' locations
- Performs mobility management functions such as mobile subscriber attach/detach and location management
- Connects to the base-station subsystem using a Frame Relay connection to the BSC's PCU.

GGSN

A 2.5G GPRS GGSN performs the following functions:

- Interfaces with IP networks like the Internet and enterprise intranets, and with other mobile service providers' GPRS provisions
- Maintains routing information used to tunnel protocol data units (PDUs) to SGSNs that serve MSs
- Networks and handles subscriber screening
- Addresses mapping
- Supports multiple SGSNs

GPRS Mobility Management

1. GPRS mobility management functions track the location of an MS as it moves within a given area.
2. Visitor location registers (VLRs) store the MS profiles that are accessed by SGSNs via the local GSM MSC.
3. The MS and the SGSN are connected via a logical link in each mobile network.
4. When transmission ends, or when an MS moves out of an SGSN's area, the logical link is released.

GPRS Applications

Applications are hosted services that ideally should be available to mobile users irrespective of the network or handset that they use. Historically this has not been the case because of the lack of international standards regarding the networks and accompanying mobile devices. 3G wireless systems remedy this problem using a horizontally layered and logically separated architecture including Applications/Services, Control, and Transport planes.

- The Application plane holds services such as e-mail, Voice mail, and real-time information.
- The Control plane sets up calls, tracks mobiles, and manages billing information.
- The Transport plane transports calls set up in the Control plane, including routing, coding, and switching.

Applications or services can be hosted on a LAN sometimes referred to as a DC (data center) that may hold various services, including 3G multimedia services such as VOD. However, the data center may also hold old-fashioned SMS services that may be hosted using a core solution that provides backend data and messaging. This type of data center would typically have a real-time data feed from perhaps a weather bureau, traffic information center, news bureau, or a city firm providing shared data. This data can then be relayed from the data center to subscribers. The data feed may consist of an ISDN connection, and may not be real time. See Figs. G-2 and G-3.

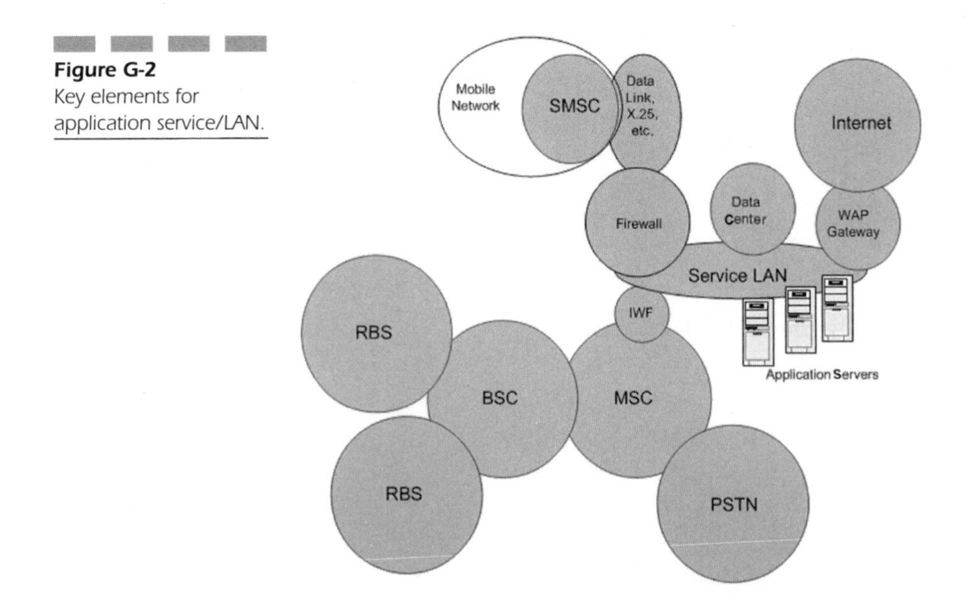

Figure G-2 Key elements for application service/LAN.

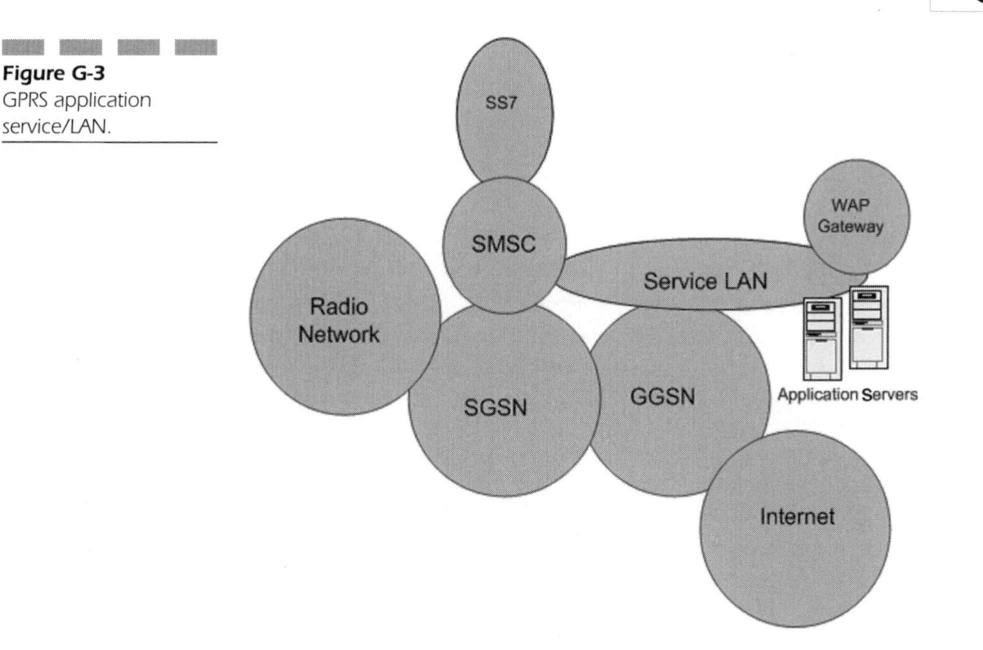

Figure G-3
GPRS application service/LAN.

The LAN should be tightly coupled using perhaps the fastest Ethernet technologies, and the hosted core application should be easily scaled to grow with usage. One solution to this is to use distributed computing applications/services running on the LAN, where it is possible to include an infinite number of servers–in theory, of course. For example, an SMS solution written in Objective-C using software servers and software clients can deliver such a scalable solution. Services such as these may be architected so as to provide interfaces with different types of fixed and mobile networks. And multiple data centers may be linked, perhaps using IP data links.

Billing GPRS A GPRS billing system relies on the SGSN and GGSN to register usage and provide appropriate information using CDR (Charging Data Records) that are routed to the billing gateway. GPRS packet-based charging may be based on:

- Volume or bytes transferred
- Duration of a PDP context session
- Time, including day and date, so as to provide lower tariffs at off-peak periods
- Destination address of the network or proxy server holding the service that the subscriber has been accessing
- Quality of service assigned to the subscriber, giving predefined priority and bandwidth rights
- SMS usage where the SGSN produces specific CDRs

- Served IMSI/subscriber classes that have different tariffs for business users, private users, and frequent users
- Reverse charging where the sender is billed
- "Free of charge" where the data or received service is not billed to the user
- Flat rate where a fixed fee (perhaps monthly) is billed to the user, and is appropriate for mass market telecommunications
- Bearer service used where the operator is offering multiple network types that may include GSM900, GSM1800, and perhaps wireless LAN for areas where the operation of other mobile networks is considerably less viable

Current GPRS networks may not be equipped to bill users based on applications used, but the network may be upgraded to do so through the addition of a payment server on the service network.

GPS (Global Positioning System)

GPS Introduction

To determine the coordinates of a device/user, GPS uses the propagation delays from satellite transmissions. The Interagency GPS Executive Board (IGEB) (http://www.igeb.gov) governs GPS and was set up up by President Clinton in 1996. GPS application by the U.S. military led to the use of encryption, though this reduced the accuracy of the GPS receiver to 100m. In 2000 President Clinton stopped this jamming/jittering or selective interference (SA). The inaccuracy caused by SA was remedied by differential GPS (DGPS), but is now largely redundant since 2000—although DGPS beacons are located in coastal areas for maritime use.

DGPS operates as follows:

- GPS receivers are planted in known positions.
- They will receive the GPS signals from satellites and make location calculations.
- The receivers calculate the needed corrections to the received signals.
- These correction values are broadcast repeatedly (or several times per minute) to other GPS receivers in the area.

Other global navigation satellite systems (GNSS) include the Russian GLONASS that is available for civilian use. And the European Union Galileo system aims to reduce the European dependence on U.S./Russian systems.

GPS—Network Assisted Network-assisted GPS may use the:

- UE-based method when the UE has a full GPS receiver
- UE-assisted method when the UE has a lite GPS receiver.

The UTRAN gives the UE an estimate of the GPS signal's time of arrival, and it gives it GPS parameters including the visible satellite list. This information may also be obtained from the GPS although this is slower.

GPS Techniques

- The angle of arrival (AOA) method is used by the BS to estimate the direction of the UE. The RNC uses this information to estimate the location of the UE.
- The observed time of arrival (OTOA) method bases itself on propagation delay measurements made on the time of arrival of signals.
- Using the reference-node–based positioning (OTDOA-RNBP), the network operator can plant equipment in a cell that is difficult for LCS (location services). The network uses signals from the equipment as a reference.
- Using OTDOA positioning elements (OTDOA-PE), elements are placed in known network locations, and they transmit secondary synchronization codes (SSC). The UE then measures the time difference between the arrivals of these signals to estimate the UE's location.

LCS Types LCSs may be commercial or "value added," internal, emergency, or lawful-intercept. Commercial services are used by service providers that provide applications based on UE location such as Navigation.

Internal LCS enhances the UMTS network perhaps for location-assisted HO, for example.

Emergency LCS locates user whose locations is relayed by the operator to the emergency service. The accuracy according to the FCC is 125m.

Graphics

Macromedia Flash Player 4 for Pocket PC

Macromedia Flash Player 4 for Pocket PC targets mobile devices and supports the Compaq iPaq, Casio Cassiopeia, and the HP Jornada. Macromedia Flash is a cost-effective and standard method for creating multimedia-based wireless applications. The Macromedia Generator—server-side component—is used to manage content, create charts, and deliver data-driven content from a wide variety of standard data sources, application servers, and middleware like ASP, JSP, PHP, ODBC/JDBC, and XML.

Flash Development The component-based nature of mobile applications requires attention to reusability where Flash files and Generator templates may be used for multiple platforms.

Personalization There are many methods of personalizing applications, like date-time stamping content exchanges and updates, for example. This requires a custom Macromedia Generator Object (MGO) from the Macromedia Exchange to date-time stamp the applications using server data. The MGO may also be used for selective advertising using the users' preferences.

User Interface Data input may be achieved using the input devices: stylus, screen-based keyboard, hardware keys that may perform specific functions, and external keyboards. The application may allow the user to choose between clicking screen icons or pressing a key. The stylus is most ergonomic, as are navigational controls and buttons. Because Pocket PCs do not use mice, a pen tap acts as a mouse click.

Screen Size The Internet Explorer for Pocket PC measures 240 × 320 pixels and gives away 52 pixels for the:

- Menu bar at the bottom
- Caption bar at the top

This limits content to 240 × 268 pixels, and exceeding this causes the appearance of scroll bars that require 11 pixels. Frames should be avoided for obvious reasons, or they should be at least limited to not more than two per screen.

Movie Size The Pixels option sets WIDTH and HEIGHT values in the HTML Publish Settings. Note that the Internet Explorer for Pocket PC ignores the "%" using the width and height in the HTML embed tag.

Text The Macromedia Flash Player 4 for Pocket PC has:

- Tahoma (variable width fonts)
- Courier (default fixed width font)
- Bookdings
- Frutiger Linotype

Colors Pocket PCs display 4096 to 65,535 colors, or from 4 to 16 gray-scale levels.

Vectors Flash's vector graphics rely on the mobile device's processing to render graphics and animations. Because of this, bitmaps may be chosen for increased speed.

Bitmaps Anti-aliasing smooths an image's edges, while compression reduces the file size. These properties may be altered using the Bitmap Properties dialog box. Compression settings may also be established in Publish Settings.

Media

Audio MP3 files may not be included with Macromedia Flash Player 4 for Pocket PC. APDCM 11kHz 4-bit audio may be used for voice or low-quality sound, and RAW 11kHz or 22kHz audio may be used for music and higher-quality speech.

Animation Animation may be included using tweens, key frame animations, and ActionScript versions.

Frames per Second Depending on frame size, 8 to 15 frames per second are possible, and the ActionScript permits the refinement of movies for best results on the targeted platform.

```
n = 0;
a = getTimer();
while (Number(n)<1000) {
n = Number(n) +1;
}
b = getTimer();
CPUlag = b-a;
if (Number(CPUlag)>100) {
response = "device";
} else {
response = "desktop";
}
```

The above script may determine device speed, where desktop systems return 10–80ms.

- Pentium II/G3, 250MHz+ (average PC/Mac system): 20–60ms
- Intel StrongARM, 206MHz: 225–250ms
- SH3 7709, 133MHz: 520–550ms
- NEC VR4121 MIPS, 133MHz: 1100–1200ms

Quality Settings HTML quality may be set to low, high, autolow, autohigh, or best. Quality Setting in the Publish area control anti-aliasing used in movies. Anti-aliasing needs intensive processing to smooth frames.

- Low favors playback speed and does not use anti-aliasing.
- Autolow sets Low.
- Medium uses anti-aliasing and does not smooth bitmaps, and is used most often.
- High, Auto-High, and Best map to Medium.

Generator Macromedia Flash templates may be created using Generator that may deploy Flash content. Sniffing for the browser and embedding a Macromedia Flash template, allows the use of the same Macromedia Flash template to produce content for devices supporting JPG, GIF, PNG, and Macromedia Flash.

Internet Explorer for Pocket PC and HTML Internet Explorer for Pocket PC supports:

- HTML 3.2
- Secure Sockets Layer (SSL) versions 2.0 and 3.0
- Microsoft JScript
- Cookies for visitor tracking, etc.
- ActiveX controls that reside on the Pocket PC
- Client-side JavaScript 1.1 (ECMA-262)
- Internet Explorer 3.02 Document Object Model (DOM)

Macromedia Flash movies are embedded in HTML pages. The <Object> tag requires an ID attribute whose value should be the filename of the SWF movie. ID causes Internet Explorer to instantiate a Macromedia Flash player ActiveX Control. Between <Object> and </Object>, there is a tag <Param name="movie" value="moviename.swf"> where "moviename.swf" should be the SWF movie filename.

The Pixels option sets WIDTH and HEIGHT in the HTML Publish Settings. Internet Explorer for Pocket PC will not display SWF files if width and height are inconsistent with the SWF file's aspect ratio. The following is a typical example of a Macromedia Flash movie embedded in an HTML page to be used for the Pocket PC.

```
<<HTML>
  <HEAD>
  <TITLE> c e </TITLE>
  </HEAD>
  <BODY bgcolor="#CCCCCC">
  <OBJECT classid="clsid:D27CDB6E-AE6D-11cf-96B8-444553540000"
  codebase="http://active.macromedia.com/flash2
  /cabs/swflash.cab#version=4,0,0,0"
  ID=rapier4 WIDTH=230 HEIGHT=255>
```

```
<PARAM NAME=movie VALUE="mymovie.swf">
<PARAM NAME=menu VALUE=false>
<PARAM NAME=quality VALUE=best>
<PARAM NAME=bgcolor VALUE=#CCCCCC>
<EMBED src="mymovie.swf" menu=false quality=best
bgcolor=#CCCCCC WIDTH=230 HEIGHT=255 TYPE="application/x-
shockwave-flash"
PLUGINSPAGE="http://www.macromedia.com/shockwave/download/
index.cgi?P1_Prod_Version=ShockwaveFlash"></EMBED>
</OBJECT>
</BODY>
</HTML>
```

Sniffing for Internet Explorer for Pocket PC on the Server A file named BROWSCAP.INI in the directory \WINNT\system32\inetsrv is included with Microsoft Internet Information Services 4.0 or later, and has descriptions of all browsers. Apache supports similar files. When Internet Explorer for Pocket PC sends a request to a server, it uses the HTTP request header:

```
UA-pixels: {i.e. 240 × 320}
        UA-color: {mono2 | mono4 | color8 | color16 | color24 | color32}
        UA-OS: {Windows CE (POCKET PC) - Version 3.0}
        UA-CPU = {i.e. MIPS R4111}
```

Middleware Optimized pages may be created using middleware language like Cold Fusion, ASP, JSP, and PHP.

Internet Explorer for Pocket PC Security Internet Explorer for Pocket PC supports all common security schemes:

- SSL 2.0, SSL 3.0, and Server Gate Cryptography (SGC).
- By default, Internet Explorer for Pocket PC supports 40-bit security encryption; a 128-bit enhancement pack is available for download.
- NTLM authentication as well as clear text authentication.

WAP-Based Multimedia

The assumption that WAP will be consigned to history given the impact of 3G services is erroneous, as it is in the case of i-Mode. 3G usage habits show that consumers use many of the same services available to them over 2G networks; it is also possible to provide multimedia services (although crude in appearance compared to 2G and 2.5G WAP devices using bitmaps, animations, effects, and text.

The Ericsson R380 is a popular WAP phone in the mass market telecommunications sector. It has a gray-scale touchscreen with a resolution of 360 × 120 pixels and a 0.23 pitch, and an active screen size of almost 83 × 28 mm. Its browser includes a Browser Area, Card Title bar, and Toolbar. The Browser area presents the card content and measures 310 × 100 pixels. When a card is too large for the screen, only the beginning of the card is shown at first, and a vertical scroll bar appears. Graphical components, text, and images are displayed in the top left corner, and in the same order as in the WML code.

Card Title Bar The Title bar shows which card is displayed and also browsed cards. The navigation history is reset when a loaded Card's newcontext attribute is *true*.

Toolbar The Toolbar has buttons required by the browser, and text input is done with a screenpad or character recognition screen. Three different keyboard layouts include: alpha, numeric, and national characters.

To navigate to a WAP service:

- Activate a bookmarked site using the Bookmark.
- Use the History list.
- Enter the URL in the Open Location dialog box.

Design Components

Font The font used in R380 is a proportional font called Swiss A and it is used in: small, normal, and big sizes.

Font Size/Height/Number of Lines

- Small/9 pixels/7.5
- Normal/10 pixels/7
- Big/12 pixels/5.5

WML The browser supports the emphasis elements big and small to change font sizes, and the elements em, strong, i, b, and u.

Text Formatting

```
<wml>
<card id="first" title="Fast Movies" newcontext="true">
<p align="center">
<br/>
<b><big><a href="#second">Welcome</a></big></b><br/>
to<br/>
Welcome to Fast Movies.<br/><br/>
```

```
<a href="#fifth">[Contact Us]</a>
</p>
</wml>
```

Line Spacing and Line Breaks Single line spacing is used with one pixel before and after lines. Long lines are word-wrapped onto multiple lines. Text lines may include images, select lists, buttons, input fields, and hyperlinks.

WML New lines start with the br element that affects all contents in the browser, and the current alignment is used to position the line's entities.

A Line Break Example

```
<wml>
<card id="init" title="BR tag">
<p>
This sentence continues until the end of the line and
then wraps to the next line.<br/>
This phrase begins on a new line.<br/><br/>
This phrase follows two br elements.
</p>
</card>
</wml>
```

Paragraphs Paragraphs start on a new line and follow a 3-pixel gap, and have a default left alignment, but may be aligned right or centered.

WML Paragraphs are defined and aligned using the p element.

Attribute Description The align attribute may have the values *left*, *right*, and *center*. The mode attribute specifies the line-wrap and makes the values *wrap* or *nowrap*.

A Paragraph Example

```
<wml>
<card id="init" title="P tag">
<p>
<b>LEFT</b><br/>
This text is left aligned, and will continue until
the end of the line and then wrap to a new line.<br/>
</p>
<p align="center">
```

```
<b>CENTER</b><br/>
This text is centered, and will continue until the
end of the line and then wrap to a new line.<br/>
</p>
<p align="right">
<b>RIGHT</b><br/>
This text is right aligned, and will continue until
the end of the line and then wrap to a new line.
</p>
</card>
</wml>
```

Indented Paragraphs Text and components may be grouped, and groups may be nested. Groups have a 20-pixel indentation and follow a 3-pixel gap.

WML A group's beginning is defined using the fieldset element.

Attribute Description The title attributes value is used as leading text, and the text following the title appears on a new line.

Fieldset

```
<card id="movieinfo" title="What's On">
<p>
<fieldset title="Number 1">
The Hollywood Brat Packers.
</fieldset>
<fieldset title="Number 2">
We are Not Amused.
</fieldset>
<fieldset title="Number 3">
Hurley Burley Crosses the Road
</fieldset>
.
.
.
.
</p>
</card>
</wml>
```

Card Title The title is defined using the title attribute in the card element.

Single Choice Lists A single-choice list is shown as a drop-down listbox showing the currently selected value in angle brackets. It is 15 pixels high, has 5 pixels of white space on either side, and adapts to the length of the text in brackets.

WML A single choice list is specified using the select element with the multiple attribute set to *No*, and each list item is specified by an option element.

A Single Choice List

```
<p>
<b>Select Movie</b>
<select>
<option>Number 1</option>
<option>Number 2</option>
<option>Number 3</option>
</select>
<a href="#movieinfo">What's On</a>
</p>
<p>
<b>Select Movie</b>
<select>
<option>Number 1</option>
<option>Number 2</option>
<option>Number 3</option>
</select>
</p>
```

Multiple Choice Lists A multiple choice list has check boxes on separate lines and are followed by 2 pixels of white space. They may form a hierarchy of groups that have a 20-pixel indention.

WML Multiple choice lists are created using the select element with the multiple attribute set to *Yes*, and each check box item is included using the option element.

A Multiple Choice List

```
<p>
<b>Video choice</b>
<select multiple="true">
<option>DVD</option>
```

```
<option>CD/MPEG</option>
<option>AVI</option>
</select>
</p>
```

Buttons Buttons are displayed using the Normal Bold font, and are defined using the do element.

A Do Example

```
<p align="center">
<do type="accept" label="Continue">
<go href="#third"/>
</do><br/>
</p>
```

Three Do Elements of the Same Type

```
<p align="center">
<do type="accept" label="Continue" name="cont">
<go href="#third"/>
</do><br/>
<do type="accept" label="Contact Us" name="contact">
<go href="#fifth"/>
</do>
<do type="accept" label="Go to Start" name="start">
<go href="#first"/>
</do>
</p>.
```

Input Fields Input fields are defined using the `input` element that holds the shown attributes:

Attribute Description

- Type: This may have the values: `text` or *password*, and if set to *password*, entered characters are shown as asterisks.
- Value: The value of the `value` attribute is used as default text if no preload value is defined for the input object.
- Format: Specifies an input mask for user input entries. The string has mask control characters and static text.

An Input Example

```
<p>
Name<input type="text" title="Enter name:"
  name="name"/><br/>
Address<input type="text" title="Enter address:"
name="address"/>
</p>
```

Images The browser supports images in WAP bitmap (WBMP) and also in the GIF format that have the colors: white, 0% black; light-gray, 25% black; mid-gray, 50% black; and black, 100% black. The WML img element indicates that an image is included in the text flow.

The R380 supports the following attributes:

- Alt: A text name for the image used in the placeholder.
- Src: The image's source (URI).
- Vspace: Specifies the amount of white space inserted above and below the image.
- Hspace: Specifies the amount of white space on either side of the image.
- Height: Specifies the vertical size of an image.
- Width: Specifies the horizontal size of an image.
- Align: The align attribute may have the values *top*, *middle*, and *bottom*.

An Image Example

```
<p align="center">
<img alt="Raxi"src="raxi.gif" vspace="5" width="40"
height="30"/><br/>
<b>The movie is on its way to you via Media Cable.</b><br/>
</p>
```

4:3 A standard aspect ratio adopted in broadcast television, video, and graphics display technology. The IBM VGA graphics standard, and the MPEG-1/2/3/4 video standards offer resolutions that have 3:4 aspect ratio.

8bit Image Depth An 8bit image depth gives a maximum of 256 colors for digital video and computer-generated animations and images. The color information for each pixel (or dot) is stored using eight bits giving a maximum of 256 (2^8) colors.

The 8bit color information can be edited using a palette editor such as Microsoft PalEdit so as to:

- Alter the order of color cells in a palette
- Reduce the number of colors in a palette by deleting unwanted color cells
- Alter brightness
- Alter color contrast
- Fade and tint colors
- Copy color cells from one palette to another
- Merge two or more palettes into one
- Develop common color palettes that can be used with a number of different 8bit video sequences so as to reduce any flicker that may occur as a result of palette switching, which occurs when one image, animation, or video sequence is exchanged for another. This operation may also be implemented using a palette optimizer.

Palettes can be pasted into 8bit video sequences using a video-editing program such as Adobe Premier, Asymetrix Digital Video Producer, and Microsoft VidEdit (which is part of the full implementation of Microsoft Video for Windows). Palettes can be applied to a complete video sequence or a preselected portion of a video sequence, or even to a single frame. They can be pasted in still 8bit images using an editing program such as Microsoft BitEdit, which is supplied with Microsoft Video for Windows.

24bit Image Depth A 24bit digital video, computer-generated image or animation is generated and stored using 24bits of color information for each pixel (or dot). This results in a maximum of over 16.7 million (2^{24}) colors. 24bit digital videos, animations, and images are described as truecolor. Red, green, and blue are each represented by eight bits, giving 256 tones of each, which in turn leads to over 16.7 million ($256 \times 256 \times 256$) colors. 24bit graphics make possible near-photographic-quality images.

25 The playback frame rate of a PAL or SECAM broadcast television/video signal, and prevails in most countries except the United States and Japan.

30 The playback frame rate of an NTSC broadcast television/video signal. It is used in the United States and in Japan.

30bit Image Depth A 30bit digital video or computer-generated image or animation that is generated and stored using 30 bits of color information for each pixel (or dot). This results in a maximum of about 1 billion (or 2^{30}) colors.

32bit An extension of the 24bit image depth, an additional Byte (or Alpha channel) provides control over the transparency of pixels. Red, green, and blue are each represented by eight bits, giving 256 tones of each, which in turn leads to over 16.7 million (256 × 256 × 256) colors. The additional eight bits (the Alpha Channel in Apple parlance) are used to control transparency. 32bit graphics make possible photographic-quality images. The Apple Macintosh is remembered as the first platform upon which the 32bit graphics capability became commercially available.

36bit An image depth.

352-by-240 Pixels A frame resolution that is described as the SPA (Significant Pel Area) for an MPEG-1 video sequence encoded using an NTSC broadcast television/video source. The playback frame rate is standardized at 30 frames/s.

352-by-288 Pixels A frame resolution that is described as the SPA (Significant Pel Area) for an MPEG-1 video sequence encoded using a PAL broadcast television/video source.

360-by-240 Pixels A frame resolution that may be used as an SIF (Source Input Format) for an MPEG-1 video sequence encoded using an NTSC broadcast television/video source. The playback frame rate is standardized at 30 frames/s.

360-by-288 Pixels A frame resolution that is described as the SIF (Source Input Format) for an MPEG-1 video sequence encoded using a PAL broadcast television/video source.

640-by-480 Pixels The standard resolution of SVGA.

GSM

The Global System for Mobile (GSM) public mobile network offered 2G wireless telecommunications. In 1982 CEPT (Conference Europeene des Postes et Telecommunications), the main governing body of European telecommunications operators, formed the GSM (Groupe Special Mobile) that was given the task to specify a pan-European cellular radio system operating in the 900MHz band. See Fig. G-4.

As many as 20 to 25 MSCs (Mobile Switching Centers) or control elements are included in a national network, and each will cover a given geographic area covered by BTSs (Base Transmitter Stations). These naturally determine the coverage of the network and as you drive, the receiving device seamlessly switches between BTSs. It operates in the uplink band between 890MHz and 915MHz, and provides a data transfer rate of 9.6kbps that is very slow

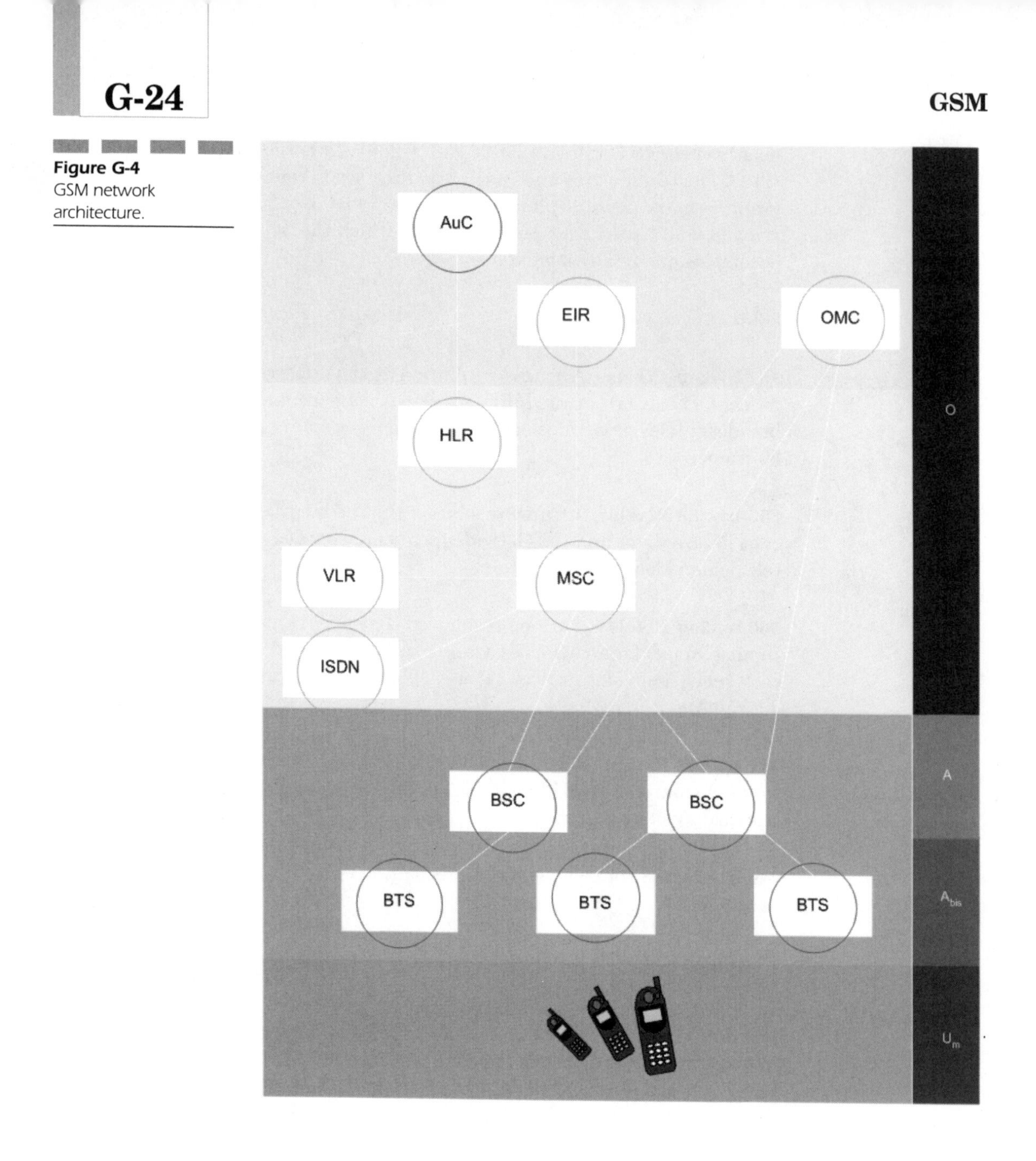

Figure G-4 GSM network architecture.

compared to even analog land services. The speed restriction on the mobile station (or phone or GSM modem-based device) is 250 km/h.

Personal Communications Network (PCN) or DCS-1800 network is comparable with GSM. It operates in the uplink radio band between 1710MHz and 1785MHz, and like GSM provides a data transfer rate of 9.6kbps. Also like

GSM, the speed of the mobile station (or phone or GSM modem-based device) is 250 km/h.

Network Operation

When a mobile phone is switched on, it registers its presence with the nearest MSC that is then informed of the location of the mobile user.

If the user is outside the geographic area of the home MSC, the nearest MSC will implement a registration procedure. This procedure uses the home MSC to acquire information about the mobile device. This information is held by the home MSC in a database called the home location register (HLR) that holds mapping information necessary so that calls can be made to the user from the PSTN. The local MSC duplicates part of this information in the VLR (Visitor Location Register) for as long as the caller is in the MSC area.

Normally one HLR and one VLR are associated with each MSC that provides switching and a gateway to other mobile and fixed networks.

Mobile devices have SIM chips holding user identification and configuration data. SIM chips permit an authorization procedure to be implemented between MSCs and EIRs (Equipment Identification Register). The EIR has a blacklist of barred equipment, a grey gray list of faulty equipment or devices that are registered for no services, and a white list for registered users and their service subscriptions.

In 1982 CEPT (Conference Europeene des Postes et Telecommunications) assembled the Groupe Special Mobile (GSM) committee so as to specify a pan-European cellular radio system that would increase the capacity of the analog systems like the Nordic Mobile Telephone system (NMT). A pan-European bandwidth of 890 to 915MHz and 935 to 960MHz was agreed on.

Eight system proposals included:

- Bosch proposed the S900-D system that used four-level frequency shift keying (FSK) modulation.
- ELAB's (Norway) proposed system employed adaptive digital phase modulation (AD PM) and a Viterbi equalizer to combat the effects of intersymbol interference (ISI).
- Ericsson proposed the DMS90 system that used frequency hopping, GMSK modulation, and an adaptive decision feedback equalizer (DFE).
- Televerket proposed the Mobira system and the MAX II system that were similar to the DMS90 system.
- SEL proposed the CD900 wideband TDMA system that is also used in conjunction with spectral spreading.
- TEKADE proposed the MATS-D system that incorporated three different multiple access schemes, namely, code division multiple

access (CDMA), frequency division multiple access (FDMA), and time division multiple access (TDMA).

- LCT proposed the SFH900 system that used frequency hopping in combination with Gaussian minimum shift keying (GMSK) modulation, Viterbi equalization, and Reed-Solomon channel coding.

The proposed systems were piloted in Paris in 1986, when ELAB's offering was chosen. By June 1987 a narrowband TDMA system based on ELAB's was agreed, and would support eight (and eventually 16) channels per carrier.

Six different codecs at 16kbps were considered, where a residual excited linear prediction (RELP) codec and a multipulse excitation linear prediction codec (MPE-LPC) proved best. Eventually these were merged to produce a regular pulse excitation LPC (RPE-LPC) with a net bit rate of 13kbps.

GMSK was chosen because of its improved spectral efficiency, and the initial drafts of the GSM specifications were published in mid-1988 when it became clear that it would not be possible to fully specify every feature before the 1991 launch. So the system specification was given two phases: The most common services (including call forwarding and call barring) were in Phase 1; and remaining services (including supplementary services and facsimile) were in Phase 2.

At the request of the United Kingdom, a version of GSM operating in the 1800 MHz dual band was included in the specification for Personal Communications Networks (PCN) that became Digital LPC CellularSystem at 1800MHz (DCS 1800 or GSM 1800). In Phase 2 of the specifications (June 1993), GSM900 and DCS 1800 were merged in the same documents.

Because of the complexity of processes required for a third (Phase 3) revision, it was decided that an incremental advancement strategy would be adopted as and when new features surfaced; this is known as Phase 2+. Significant Phase 2+ proposals include an increase in the maximum mobile speed, and the half-rate speech coder.

1988 saw GSM become a Technical Committee within the European Telecommunications Standards Institute (ETSI). In 1991 the GSM Technical Committee was renamed Special Mobile Group (SMG) and given the task of specifying a successor to GSM. The group SMG5 was assigned the task of specifying the Universal Mobile Telecommunication System (UMTS), but has since been discontinued and the task of specifying UMTS has been given to other committees.

GSM1900

In 1991 ETSI defined GSM1800 for 1800MHz frequencies, and TAG-5 modified it for the U.S. market as GSM1900. The proposed standard was completed in January 1995 for adoption by ANSI as J-STD-007, and recognized as GSM based on Phase 2 work. It supports call forwarding, SMS, emergency calls, video text, fax, data services up to 9.6kbps, digital bearer services with a net rate of 12kbps, and T1 backhaul facility.

The data rate over the radio channel is 270kbps, and GMSK modulation is used with a bandwidth (B) multiplied by bit period (1) equal to 0.3 and channel spacing of 200kHz. Frequency hopping is optional and occurs at the TDMA frame rate of 217hops/s. Frequency hopping reduces multipath fading that has already occurred through channel coding, interleaving, and antenna diversity.

The security functions for authentication of subscriber-related information are held in the mobile station's subscriber identity module (SIM) or in a smart card. Cells range up to 35km in rural areas, and up to 1km in urban areas, and may be extended to 120km. Mobility speed is about 125km/h, and the data rate per traffic channel is 22.8kbps, and 9.6kbps for CS data. The power output is 1W for a handheld and 2W for a vehicle, and the BS power output is within the FCC limits of 1640W EIRP.

In 1996 the GSM MoU GSMI900 was officially named GSM North America, and by 1997 U.S. GSM digital service subscriptions rose to 100,000. GSM North America supports a commercially available multirate codec to avoid proliferation of voice codecs.

Computers/Software

Garbage Collection

A memory management feature that reclaims the space occupied by an object at such times when there are no references to it.

Gcc

A compiler program.

Geocities

A resource on the World Wide Web that offers among other things hosting services for Web sites.

GET Method

A means of running a CGI script or program where the URL defines the CGI program (such as credit.cgi, for example) and the accompanying data used by the server that follows the question mark:

```
www.FrancisBotto.com/cgi-bin/credit.cgi?subject=transaction
```

GIF (Graphics Interchange Format)

A standard graphics file format that produces relatively compact files.

GIOP (General Inter-ORB Protocol)

A protocol or set of message formats and data structures for communications between ORBs.

Global Roaming

A term used to describe the process of reading e-mail messages other than by using your local ISP's point-of-presence. The ability to access email for subscribers for international ISPs such as Compuserve is unimportant, due to the availability of worldwide points of presence. Web-based, global roaming e-mail services are available that simply provide users with a PIN. The term *global roaming* is also applicable to mobile telephony that major digital carriers offer in specified countries that may be assumed to include all first-world countries.

Glue

A term given to the entities that provide communications between distributed and local application components. In a client/server context, glue is an alternative name for middleware. The underlying client/server system architecture may be that of the Web.

Object-oriented glues include all the collective entities that provide the communications between distributed components.

Glues in the Web architectural model include the protocols:

- TCP/IP
- HTTP
- SMTP
- Miscellaneous low-level protocols including UDP

Glues in LANs might include Ethernet and even proprietary protocols.

Protocols are the lowest level glues in both traditional and modern OO systems.

The next level is the programming models that are of concern to systems programmers, systems architects, and programmers. This dictates the method of communications between components that include:

- Remote procedure call (RPC)

- Message queuing, where messages are exchanged between components normally using queues, buffers, or even pipes, which interface with more loosely coupled components, perhaps via a WAN
- Peer-to-peer, where either component may be the server (sending a message) or the client (receiving the message)

Local glue is a collection of entities that unite client components, so that they may operate collectively. OLE, OpenDoc, ActiveX, JavaBeans components require local glues so that their running operations may be coordinated.

These common OO component architectures use different local glues, where:

- OLE uses ODL (Object Definition Language)
- ActiveX uses COM
- OpenDoc uses CORBA IDL (Interface Definition Language)
- JavaBeans uses a subset of the Java programming language

Distributed glues bind together (dynamically) running components that, are on the client and on the server. As is the case with local glues, standard OO component architectures use different distributed glues.

Gold Code

A final build of a program that is released for end users. It is the final stage of development, and will have been alpha and beta tested. Programs that are sold conventionally, such as those from Microsoft, and those that are shareware or freeware are termed gold code.

Graphics Card

An electronic assembly used to generate graphics and text. Occasionally it is referred to as a graphics engine or graphics controller. A VGA card is a graphics engine, but is more commonly referred to as a graphics adapter or card.

Standard IBM graphics cards include Monochrome Display Adapter (MDA), Color Graphics Adapter (CGA), Enhanced Graphics Adapter (EGA), Video Graphics Array (VGA), and Multi Color Graphics Array (MCGA—used on PS/2 30) and 8514/A. The fastest graphics controllers are of the local bus variety. These connect more directly to the processor's data bus.

The graphics card specification of a PC is influential in determining the quality digital video playback attainable. A video card comprising dedicated hardware for decoding and playing MPEG, VideoCD, or Intel Indeo generally will yield improved video playback.

Graphics Engine

An alternative name for a graphics card or for the chipsets responsible for generating graphics.

Graphics Format

An image file may be produced and stored according to a number of different graphics file formats that include CompuServe GIF, PCX, Windows BMP, PIC, TIFF, IMG, EPS, and others. The efficiency of various image file formats in terms of the data capacity they consume tends to vary significantly.

Groupware

A name given to a sofware implementation that provides collaboration and communication across an enterprise's (business's) network solution, or even over the Web.

Orfali, Harkey and Edwards define groupware as: "Software that supports the creation, flow, and tracking of nonstructured information in direct support of collaborative group activity."

Conventional modern groupware integrates:

- E-mail
- Conferencing such as whiteboards
- Telephony, including voice mail
- Scheduling
- Workflow
- Shared document databases
- Internet access

The best known groupware product is Lotus Notes.

GUI (Graphical User Interface; pronounced "gooey")

A user interface consisting of icons, usually facilitating interaction via a mouse, resulting in minimal keyboard use. Sometimes referred to as graphical front-end. The most widespread commercial examples include those of the Microsoft Windows continuum, although others exist in the form of Apple System, OS/2 Warp, and X Windows. Originally when the Windows concept was originated at Xerox PARC (Palo Alto Research Center), the UI was called a WIMP (Windows, Icons, Mouse, and Pointer) environment.

Miscellaneous

Gaussian Noise

An unwanted signal that appears during the transmission over networks and media.

Gigabit Ethernet

Gigabit Ethernet is an upscaled version of the Fast Ethernet network standard. It may deliver up to 1000Mbps access speeds, and is backwardly compatible with 10Base-T and 100Base-T Ethernet standards.

It may be used over the media:

- Multimode fiber optic cables over a maximum distance of 500m
- Single- or mono-mode fiber optic cables over a maximum distance of 2km
- Coaxial cable over a maximum distance of 25m

Gigahertz

A unit of frequency that equals 1 billion cycles per second.

Guard Time

Guard time is the latency that separates adjacent channels or time slots in order to minimize interference between them.

Acronyms

GBS GPRS backbone system
GDMO Guideline for Definition of Managed Objects
GGSN Gateway GPRS Support Node
GMSC Gateway MSC
GMSK Gaussian minimum shift keying
GPS Global Positioning System
GSM Global System for Mobile Telecommunications
GSN GPRS Support Node

H

Handovers

Cellular mobile networks require frequent and theoretically seamless handovers between cells as the user makes the transition from one to the next. This requires the new cell's BS to communicate with the mobile. This "macrodiversity" or soft handover or soft handoff (SHO) eventually results in the old cell's BS ceasing to communicate with the mobile. When the mobile travels sufficiently far into the new cell, the old BS ceases to communicate with the mobile. Softer handovers (SSHOs) are when soft handovers occur in the boundary zones of adjacent sectors of the same cell site. This section addresses the various types of handovers, and includes some GSM to UTRAN examples.

CDMA Handover

Network operation and design become complex when the operator is using many allocated carriers, and when the user is roaming between different mobile networks. For example, an adjacent BS may not support carrier f_1 and an SHO may not be performed. This scenario requires a hard handover (HHO) to a different CDMA carrier when the strength of the f_1 carrier is insufficient. This leaves the mobile transmitting and receiving at higher power levels, causing interference and decreasing system capacity.

Intersystem Handovers When a UE moves from one mobile network standard to another, an intersystem or an inter-RAT HO occurs.

Intersystem HO: GSM to UTRAN An intersystem GSM to UTRAN HO first requires UTRAN neighbor cell monitoring during idle GSM slots. The GSM-BSS receives the results of this monitoring and may then perform an HO. See Fig. H-1.

HO procedure:

- The UE requires the UTRAN's frame timing to communicate with the new network.
- The SRNC-RRC configures the appropriate Node B and the lower layers of the RNC.
- The GSM-BSS sends a *handover to UTRAN* command to the UE. The message is decoded and sent to the handset.
- The RRC makes a DCCH link.
- The *handover to UTRAN complete* message is sent to the UTRAN using the created link.

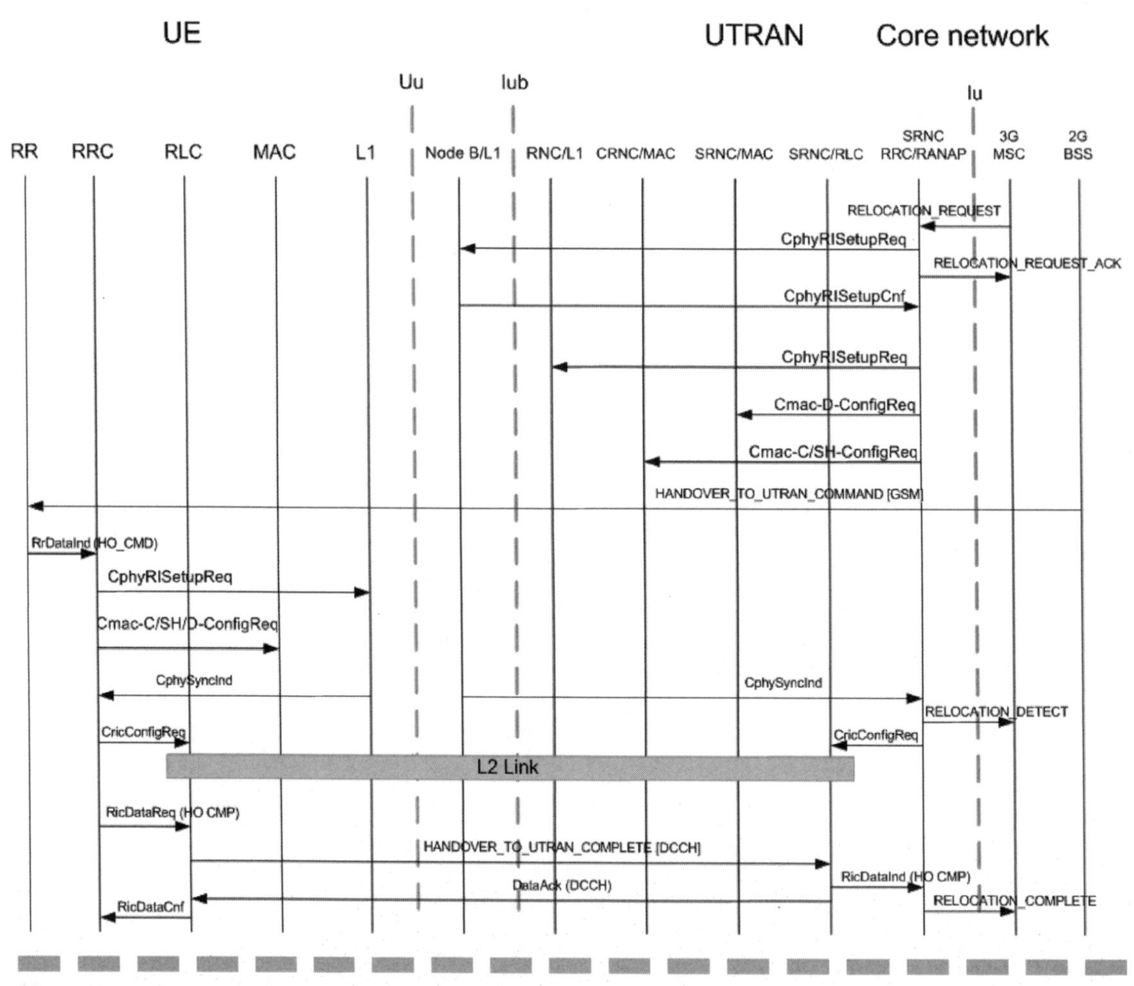

Figure H-1 Intersystem HO: GSM to UTRAN.

Intersystem HO: UTRAN to GSM CDMA mobile stations are constantly in reception mode, so compressed mode is used to measure other systems: The UTRAN BS includes gaps in its transmissions, during which the UE can monitor other frequencies and other systems. Equally the UE may use a dual receiver that accommodates the FDD resource and the GSM resource. See Fig. H-2.

An HO from the UTRAN TDD mode to GSM may occur without two receivers, because of gaps in the downlink transmission. Measurements may be carried out using idle slots or by using assigned periods, and then reported back to the network that may then perform an intersystem HO. The *handover from UTRAN* message is sent to the RRC, and then directed to the UE that performs a normal GSM HO to the specified GSM cell.

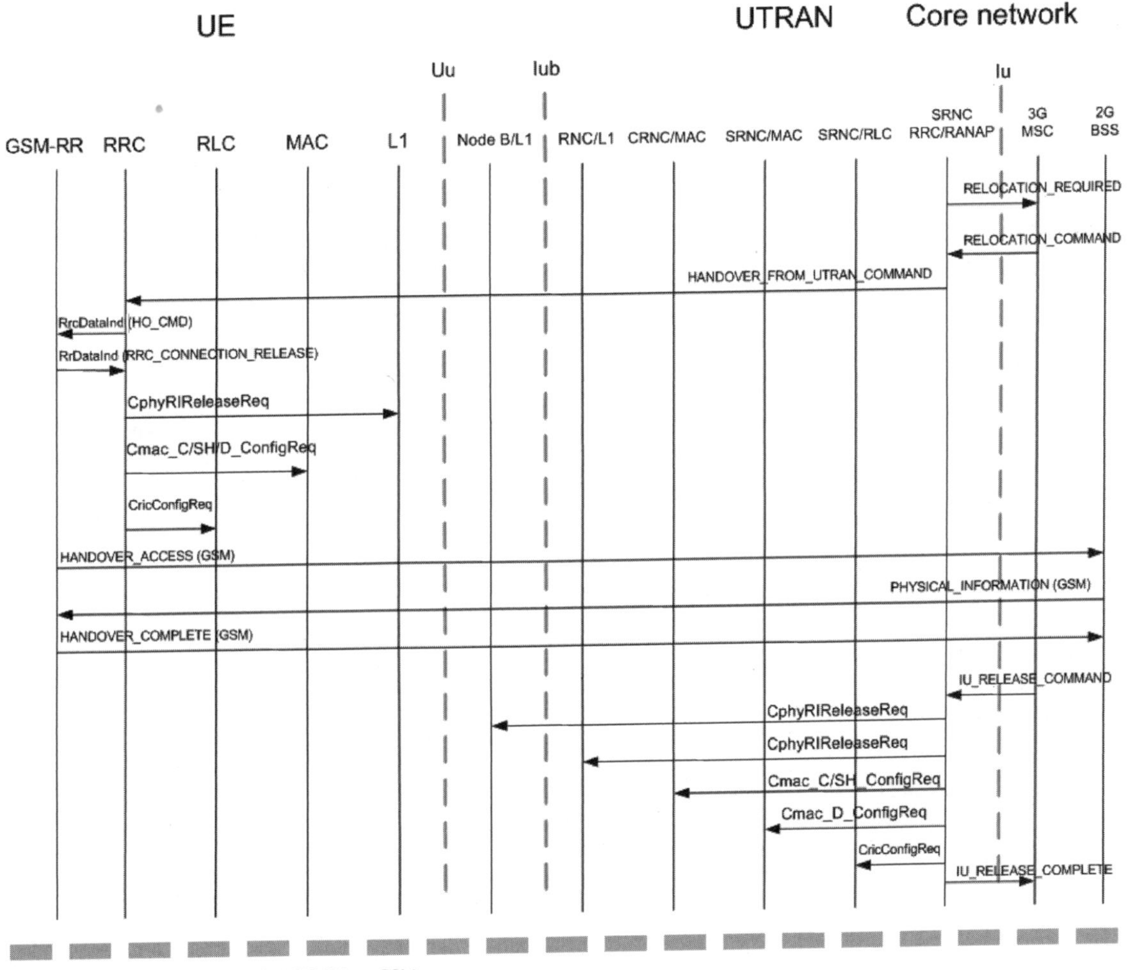

Figure H-2 Intersystem HO: UTRAN to GSM.

Hard Handover A hard handover (HHO) or an inter-frequency HO requires measurements on the new channel. But because a CDMA handset transmits and receives continuously, there are no slots available for measurements. The HHO procedure begins with the network sending one of the following:

A hard handover (HHO) or an inter-frequency HO requires measurements on the new channel. But because a CDMA handset transmits and receives continuously, there are no slots available for measurements. The HHO procedure begins with the network sending one of the following:

- HO command via the old cell
- Physical channel reconfiguration message
- Radio bearer reconfiguration message

- Radio bearer establishment message
- Radio bearer release message
- Transport channel reconfiguration message

A *nonseamless* HO occurs if the UE stops transmitting over the old channel, and recommences over the new channel. The user will hear a clicking sound if the compressed mode slots are not overlapped during the HO. The HO is also more seamless if the UE has dual radio parts. See Fig. H-3.

Figure H-3 Hard handover.

HTTP HTTP is a standard protocol that allows Web browsers to communicate with Web servers. The transport protocol is provided by TCP. HTTP is regarded as part of the TCP/IP (Transmission Control Protocol/Internet Protocol) network protocols that may transmit or route data from one IP address to the next. Protocols that use TCP/IP include:

- HTTP (HyperText Transfer Protocol)
- HTTPS (HyperText Transfer Protocol Secure) which integrates cryptography and forms the basis of a secure connection and secure site.
- SMTP (Simple Mail Transfer Protocol)
- FTP (File Transfer Protocol)

Unix servers provide numerous commands and daemons that relate to TCP/IP, a standard set of protocols used in packet-switched networks. It consists also of standard and nonstandard files, utilities, and daemons. It interprets a standard set of commands. TCP/IP originated from DARPA and ARPANET, and is one of the most established internationally agreed standard protocols. Occasionally, however, it includes proprietary files and programs through specific implementations that include that of Santa Cruz Operation (SCO).

TCP/IP Daemons A daemon is a program or process dedicated to performing what is usually a singular given function, such as sending mail. TCP/IP daemons include those added by third parties, including SCO. The daemons include:

- DNS (Domain Name Server) that is used to provide IP address for given host names.
- SYSLOG (System Logger) that stores messages pertaining to various operational events, including status, detected errors, and debugging.
- SNMP that is an implementation of the Simple Network Management Protocol, and is capable of receiving information from such compatible agents.
- INETD (Super Server) that monitors TCP/IP ports for incoming messages.
- BOOTP that implements an Internet Boot Protocol server.
- ROUTE that manages Internet routing tables, and is invoked when booted. The netstat command is used to print the routing tables. Among other details, the resulting listing shows gateways to networks.
- RARP (Reverse Address Resolution Protocol) that is able to provide a 32bit IP address in response to a 48bit Ethernet address.
- LINE PRINTER that accepts incoming print jobs, and queues them for remote printing.

- SLINK that links STREAMS modules and is included within Unix implementations that use STREAMS-TCP/IP.
- LDSOCKET that initializes the System V STREAMS TCP/IP Berkeley interface.

Configuration Interfaced devices are configured in terms of IP address, netmask, and operational status using the command ifconfig.

Configuration files include:

- /ETC/HOSTS that provides a lookup table for finding IP addresses for host names.
- /ETC/ETHERS that provides a means of converting IP addresses into Ethernet hardware addresses. An alternative conversion method is provided by ARP (Address Resolution Protocol).
- /ETC/NETWORKS that provides a lookup table for IP addresses and their respective network names.
- /ETC/PROTOCOLS that provides a list of DARPA Internet protocols.
- /ETC/SERVICES that lists services that are currently available to the host.
- /ETC/INETD.CONF that monitors a specified port, and invokes daemons when required.

Network access files include:

- /ETC/HOSTS.EQUIV that contains a list of trusted hosts, and is significant to system security. Each entry is trusted, in that users access their accounts without a password.
- RHOSTS that lists system and user names, and users are permitted to log in using any name in the file /ETC/PASSWORD.

HTTP CGI variablesVariables

HTTP_ACCEPT A CGI variable that holds the "Accept:" headers from the client.

HTTP_COOKIE A CGI variable that holds the contents of "Cookie:" headers from the client.

HTTP_FROM A CGI variable that holds the contents of the "From:" header from the client that may be the client's correct e-mail address if not withheld or incorrect e-mail address that is simply false, or entered in error.

HTTP_REFERER A CGI variable that holds the contents of the "Referer:" header from the client, containing a URL.

HTTP_USER_AGENT A CGI variable that holds the contents of the "User-Agent:" header from the client, containing the browser's name.

HSCSD

2.5G systems utilize high-speed circuit-switched data (HSCSD) as a cost-effective means of increasing data rates where MSs may use up to four time-slots, each carrying 9.6kbps or 14.4kbps—and an HSCSD connection is limited to 64kbps over the A interface. It is possible for operators to upgrade to HSCSD using software upgrades and HSCSD handsets. Key disadvantages, however, include demands on radio resources because timeslots are allocated constantly even when nothing is transmitted. HSCSD is rare today, but in some countries it is found particularly useful for real-time services whereas the GPRS alternative is not, and it was/is also a more expensive overlay for GSM.

HSCSD Technology

The HSCSD service improves CS data services supported by early GSMs, and requires no changes to the physical layer interfaces between the different network elements required for HSCSD. The MS and the network multiplex and demultiplex data for transmission over both the Abis interface and the radio interface. HSCSD services may support symmetric transmissions or asymmetric transmissions, where the latter refers to a different number of timeslots allocated for one direction. HSCSD connections may only have downlink-biased asymmetry, and the uplink timeslot numbers must be a subset of those for the downlink timeslot. The provision of the asymmetric air interface permits mobile equipment (type 1) to receive data at higher rates than symmetric variants.

HSCSD Operation When a call is set up, the MS informs the network of the HSCSD connection type. The MS's multislot class indicates the maximum number of accessible timeslots, and the interval between them for neighbor cell measurements. This information defines the MS's specification for HSCSD and GPRS services. Table H-1 shows the multislot classes that can be type 1 or type 2 that may transmit and receive simultaneously. The table also shows the maximum number of receive and transmit timeslots that the MS may occupy per TDMA frame, and the sum or total number of transmit and receive timeslots the MS may access per TDMA frame.

TABLE H-1

Multislot Classes

Multislot Class	Maximum Number of Slots			Minimum Number of Slots				Type
	Rx	Tx	Sum	T_{ta}	T_{tb}	T_{ra}	T_{rb}	
1	1	1	2	3	2	4	2	1
2	2	1	3	3	2	3	1	1
3	2	2	3	3	2	3	1	1
4	3	1	4	3	1	3	1	1
5	2	2	4	3	1	3	1	1
6	3	2	4	3	1	3	1	1
7	3	3	4	3	1	3	1	1
8	4	1	5	3	1	2	1	1
9	3	2	5	3	1	2	1	1
10	4	2	5	3	1	2	1	1
11	4	3	5	3	1	2	1	1
12	4	4	5	2	1	2	1	1
13	3	3	NA	NA	a	3	a	2
14	4	4	NA	NA	a	3	a	2
15	5	5	NA	NA	a	3	a	2
16	6	6	NA	NA	a	2	a	2
17	7	7	NA	NA	a	1	0	2
18	8	8	NA	NA	0	0	0	2
19	6	2	NA	3	b	2	c	1
20	6	3	NA	3	b	2	c	1
21	6	4	NA	3	b	2	c	1
22	6	4	NA	2	b	2	c	1
23	6	6	NA	2	b	2	c	1
24	8	2	NA	3	b	2	c	1
25	8	3	NA	3	b	2	c	1
26	8	4	NA	3	b	2	c	1
27	8	4	NA	2	b	2	c	1
28	8	6	NA	2	b	2	c	1
29	8	8	NA	2	b	2	c	1

A is set to 1 for frequency hopping.
B is set to 1 for frequency hopping or when there is a change from Rx to Tx.
C is set to 1 for frequency hopping or when there is a change from Tx to Rx.
The T_{ta} parameter is the time taken to make a neighbor cell measurement prior to an up-link transmission.
The T_{ta} parameter is the minimum number of timeslots between the previous down-link timeslot and the next up-link timeslot, or the time between two consecutive down-link timeslots on different frequencies. This occurs when the MS is not required to make measurements on neighboring cells.
The T_{ra} parameter is the number of timeslots required to make a neighbor cell measurement before the down-link burst reception.
The T_{rb} parameter is the number of timeslots required between the previous up-link transmission and the next down-link reception, or the time between two consecutive down-link receptions when the frequency is changed between receptions.

Computers/Software

Hacking

Hacking is an illegal intrusion into a system where its services, programs, and data are used without authority, and may involve:

- Eavesdropping or sniffing where the hacker taps into a connection
- Brute force factoring of a public key in a cryptosystem
- Dictionary attacks

Brute Force Attacks Using public key cryptography, the attacker usually has the public key, and attempts to gain the private key. Brute Force Factoring RSA is a means of deciphering RSA. Attackers may be armed with the public key (n,e), and then attempt to determine d in order to gain the private key (n,d). This process begins by factoring n in order to yield the two large primes; this is a common method of deciphering RSA. Other methods, such as calculating $(p - 1)(q - 1)$, and attempting to determine d through iterative techniques are deemed equally difficult.

Factoring may be carried using the algorithms of:

- Trial division that attempts to find all the prime numbers <= sqrt(n)
- Quadratic Sieve (QS) that is deemed fastest for numbers that are less than 110 digits
- Multiple Polynomial Quadratic Sieve (MPQS)
- Double Large Prime Variation of the MPQS
- Number Field Sieve (NFS) that is the fastest algorithm for numbers larger than 110 digits

Hard Disk

A hard disk is a magnetic mass storage device consisting of fixed disks. Removable versions are available but most are fixed. Storage capacities are increasing all the time. The usefulness of a stand-alone PC is greatly enhanced following the installation of a magnetic hard disk drive.

All hard disks must be paired with an appropriate controller, with which they must be 100 percent compatible. Popular commercial variants include IDE, E-IDE or ATA-2, SCSI, SCSI-2, Fast Wide SCSI, and Ultra SCSI. Controllers capable of accepting multiple devices will provide an economical path to vast data storage capacity in the future. An inexpensive array of drives may be built up, thus lowering the considerable cost of a single high-capacity drive bought at the outset. Where a number of drives in an array exhibit comparatively lengthy access times, it may be more practical to replace them with a single large disk, or several larger ones.

More expensive controllers are often expandable in terms of additional daughter boards. For example, SCSI daughter boards can increase the number of drives in standard multiples of seven. Such controllers can easily yield tens of gigabytes using inexpensive drives.

Some controllers are also capable of mirroring, i.e., writing the same data to two disk drives simultaneously, thus making the data more secure.

Controller technology and performance have advanced considerably in recent years, giving rise to an array of commercial devices ranging from scant MFM implementations to caching variants comprising on-board processors.

The main thrust of advancement is based on the need to expand data capacities, lower access times, and increase data transfer rates. In addition, the emergence of multiple device controllers reveals a secondary aim.

Cache controllers speed up read/write operations by using on-board RAM as an intermediate data store between disk and system memory. Based on which data is requested most often, a caching algorithm estimates which portions of hard disk should reside in on-board RAM.

The ingenuity of this technique simply takes advantage of the inescapable fact that a small percentage of disk data is rewritten and accessed most frequently.

The decision-making process regarding which data should reside in RAM may suggest that they are "intelligent controllers."

Cache controllers are the most expensive of all variants, and will outperform standard implementations. It is most often these types of controllers that are able to support increasing multiples of drives through the addition of daughter boards. High performance cache controllers can offer access times as low as a fraction of 1ms.

A RAID (Redundant Array of Inexpensive Disks) mass storage device has many individual disks. Identifying features of RAID may include:

- High levels of fault tolerance
- Scalability through the addition of hard disks
- Hot-swappable disks, meaning they may be removed and replaced without the need to power down the RAID
- Redundant power supplies for improved fault tolerance
- Shared mass storage, serving disparate computers/networks
- Heterogeneous characteristics, where they may be integrated into environments comprising multiple OSes
- High-speed interfaces such as Fiber Channel, Ultra SCSI, et al.

The original RAID specification originated from UC Berkeley in 1987, and was named Redundant Array of "Inexpensive" Disks. The aims of the Berkeley group were threefold:

- Improve fault tolerance of mass storage
- Reduce mass storage costs
- Improve mass storage performance

Realizing the inescapable fact that no single mass storage system could be optimized in all three of the aforementioned areas, the group defined what were to become a number of industry standard solutions.

Achieving its objectives to varying degrees, the Berkeley group defines a series of RAID levels employing several tried and tested data storage techniques. One of these was mirroring where data is written to, and read from, pairs of disks concurrently in order to deliver fault tolerance.

Modern RAID systems may be specified in terms of:

- Maximum data storage capacity that is typically in the GByte range for a single RAID unit, and is in the TByte range for multiple connected units
- Average access time measured in milliseconds (ms)
- Average and burst data transfer rates
- Cache size
- Interface type
- Multiplicity of host types that may be connected
- OS compatibility

RAID performance has obvious effects, and high performance echoes performance gains that are felt locally and remotely.

The five levels of RAID defined by the Berkeley group include:

- Level 0 that stripes data across multiple disks, but provides no error correction or redundancy.
- Level 1 that uses duplexing or mirroring, where data is written concurrently to pairs of independent disks, promoting a high degree of fault tolerance.
- Level 2 that stores and reads data by dividing it into bits, and storing them on different drives, otherwise known as striping. It also stores ECC codes on dedicated disks.
- Level 3 that divides data into blocks, storing them on different independent disks. One additional disk contains parity data.
- Level 4 stripes data blocks across multiple disks. One additional disk contains parity data.
- Level 5 stripes data blocks across multiple disks, while parity data is stored on multiple disks.

Other RAID configurations include Level 0 and Level 7, neither of which were devised by the Berkeley group. Level 7 offers improved fault tolerance, and is patented by Storage Computer Corporation.

HDML (Handheld Device Markup Language)

A scripting language, HDML was originally developed by Universal Planet (now named Phone.com) and is considered a forerunner to WAP/WML. It was based on HTML and HDML and is naturally designed for operation on small-screen, narrow-bandwidth handheld devices such as cellular phones and PDAs.

HDSL (High Bit Rate DSL)

A data transmission line that uses two pairs of copper wire as its medium. It offers T1 data speeds of up to 1.544Mbps.

Help System

An on-line information system that provides guidance on software usage through hypertext, hypermedia, or multimedia. Such systems are usually context sensitive where information regarding a current program operation may be produced immediately. Windows Help systems are essentially hypermedia applications. They may be authored using a word processor that is able to produce standard RTF (Rich Text Format) files together with a Help compiler such as that supplied with Borland programming tools. Numerous other Help compilers exist.

Hexadecimal

A base-16 counting system that is used widely in computing. Four binary digits represented by a single number or letter: 0, 1, 2, 3, 4, 5, 6, 7, 8, 9, A, B, C, D, E, F.

Hidden Node

A wireless LAN station, a hidden node fails to detect that media is busy because it cannot detect transmissions from stations.

Hit

1. An event when a Web site is visited by a user.
2. In terms of processor cache memory, hit rate is the percentage of memory requests that may be satisfied by the cache memory.

Home Page

A highest-level page in the hierarchy of Web pages at a Web site. It has a URL such as www.homepage.com. A home page may consist of a single page or a number of linked pages. It may include links to other sites, graphics, sound bites, video, an email address, and various forms for user feedback; it may also include a counter that records the number of hits or times it is visited.

Host-Based Processing

An architecture where a host computer is connected to dumb terminals. Typically the terminals do not have GUIs such as Windows, but are text-based. They are sometimes termed green screens, because many earlier terminals had green phosphor screens. The terminals are said to be dumb, because they lack processing capabilities. They merely accept user commands, pass requests to the host, and receive information from the host. Many host-based processing architectures are being renovated, or migrated to client/server architectures. A coexistence strategy is also being adopted, using mainframe and client/server-based architectures to form collective IT solutions.

Host Name

A name designated to a network device that permits it to be addressed without using its full IP address. The Internet Request for Comments (RFC) No. 1178 provides guidelines for naming hosts. Using host names there is a requirement to perform translations between host names and their respective IP addresses, using a lookup file containing host names and related IP addresses, or the Domain Network Service (DNS).

HotDog Pro

A Web site development tool.

HotJava

A Web Browser produced by Sun Microsystems. It does not enjoy the popularity of Netscape Navigator or Microsoft Internet Explorer, but is nonetheless equally sophisticated.

HP-UX

A Unix operating system variant.

HTML (Hypertext Markup Language)

A standard language consisting of formatting commands and statements that may be used to create Web pages.

HTML may be used to include hyperlinks leading to Web pages, frames or sites, and many other functions including visitor counters.

HTML has its roots in SGML, and is the standard language of the World Wide Web. When the Web was first introduced, almost all Web sites depended heavily on HTML. Today, however, HTML is almost a framework on which to hang other components such as:

- ActiveX controls
- Java applets
- JScript programs
- VBScript programs

The HTML syntax is similar to old word processor formatting languages such as LaTex and even that which was included in the Borland Sprint word processor.

The Web browser interprets the HTML first by reading the tags:

```
<HTML>
<HEAD>
<BODY>

</BODY>
</HEAD>
</HTML>
```

These basic tags form the basis of all HTML listings, and encapsulate such entities as VBScript code, JScript code, ActiveX Controls, and Java applets. Such components are enclosed between the <BODY> tags.

HTML Help

An on-line Help development tool from Microsoft.

HTML Template

A template file that a Web server uses to display information. The information may originate from a query submitted to a database.

HTML Validator

A testing program used to validate HTML documents at various levels, including 2.0, 3.2, and future versions of HTML as specified by the W3C.

HTTP (Hypertext Transfer Protocol)

A standard protocol that allows Web browsers to communicate with Web servers. The transport protocol is provided by TCP.

HTTP Server

A Web server that may connect with client systems and with back-end applications and data.

Hypermedia

An extension of the hypertext concept where text is combined with images. The terms hypermedia and multimedia are often regarded as interchangeable but they are *not*.

In French media circles the ludicrous and ridiculously extravagant term *hypermediatization* was coined in 1991. It was used to describe the immediacy with which news began to be transmitted, brought about by satellite broadcasting technology. With the sacrifice of time normally required by the reporter to prepare an informed report, the concept of the resulting, often confusing reports became known as hypermediatization.

Available to Macintosh users through HyperCard since 1987, hypermedia is a relatively mature area of multimedia. HyperCard for the Apple Macintosh may be considered as the earliest commercially successful hypermedia authoring tool that combined text, graphics, animated sequences, and sound.

It made the Macintosh an effective personal computer for multimedia. A plethora of hypermedia authoring tools has since emerged, including ToolBook for the Microsoft Windows environment on the IBM PC and compatible machines. Hypermedia applications developed using such tools may be thought of as interactive books that combine images, text, and sound.

Hypertext

"It seemed so clear to me right from the very beginning that writing should not be sequential . . . the problems we all have in writing sequential prose derives from the fact that we are trying to make it all lie down in one long string . . . if we could only break it up into different chunks that readers could choose. . . . " —Ted Nelson.

A term coined in the 1960s by Ted Nelson to describe the concept of linking textual information and presenting it in a nonlinear fashion so that it may be navigated and browsed. The Web is synonymous with hypertext. A hyperlink in a hypertext-based navigational scheme that permits the user to browse from one document to another, or from one Web site to the next.

Miscellaneous

HDR

High Data Rate (HDR) is Qualcomm's proprietary high-speed standard for IS-95 networks, providing a 2.4Mbps data rate.

HDSL (High Bit Rate DSL)

A data transmission line that uses two pairs of copper wire as its medium. It offers T1 data speeds of up to 1.544Mbps.

Hertz

A unit of frequency measurement that equals 1 cycle per second. Frequency describes the number of cycles, waves, or oscillations per second that an analog signal undergoes when transmitted. The length of each cycle or wave is called the *wavelength*, and the time taken for the signal to be transmitted over one wavelength is called the *period*.

The period (T) is a function of the frequency (f):

$$T = 1/f$$

Hold Mode

A Bluetooth operating mode where a device's clock is synchronized with that of the master's, but it does not participate with the network.

Acronyms

HCI	Host controller interface
HCM	Handoff completion message
HDLC	High-level data link control
HDM	Handoff direction message
HDML	Handheld Device Markup Language
HDR	High data rate
HDSL	High Bit Rate Digital Subscriber Line
HLL	High Level language
HLR	Home location register
HS	High speed
HSCSD	High-speed circuit-switched data
HSD	High-speed data packet
HSPD	High-speed packet data
HTML	Hyptertext Markup Language
HTTP	Hypertext transport protocol

I

I-Mode

I-Mode was introduced in February 1999 by Japan's NTT DoCoMo and is used primarily in 2G PDC networks; 3G handsets in Japan retain I-Mode functionality. I-Mode is an alternative to WAP or to be more precise an equivalent to WAP over GPRS, and features packet data, and packet volume charging. I-Mode is used to access Internet sites using a scripting language based on HTML, and is very successful in Japan. Like WAP, I-Mode permits limited Web browsing using I-Mode sites, and provides e-mail message services between I-Mode users. As is the case with many mobile services, the I-Mode user base is people between the ages of 24 and 35, and the highest usage levels are recorded among women in their late twenties.

I-Mode was introduced in Japan by NTT DoCoMo and is used primarily in 2G PDC networks, but 3G handsets in Japan retain I-Mode functionality. I-Mode is an alternative to WAP, or to be more precise, an equivalent to WAP over GPRS, and features packet data and packet volume charging. I-Mode is used to access Internet sites using a scripting language based on HTML and is very successful in Japan. I-Mode is essentially a precursor to 2.5G and 3G services, and offers many of the advantages of GPRS 2.5G, but at the same time it uses inexpensive and established robust 2G networks. It has been proven in the field, unlike many of the newer services such as GPRS/2.5G services that experienced rather less demand than was initially predicted.

When compared with WAP, we see I-Mode winning a technical battle in the perspective of 2G networks. However, WAP over GPRS is a rather different proposal, with WAP being a little more attractive; however, there is the overhead of GPRS handsets and GPRS contracts. When 3G is rolled out more completely, the momentum will pick up and should consign these technologies to history. Why? Because they are essentially 2G narrow-bandwidth services and do not have the display that is required to deploy and deliver 3G services, including multimedia.

Essentially both WAP and I-Mode are solutions for 2G networks, and they remedy bottlenecks that are redundant in 3G networks and services, and much of their infrastructure will no longer be required. I-Mode could still provide a cost-effective North American solution to deploying mobile Internet services in the mass market telecommunications sector. The notion that everyone has the means or the desire to buy and to own leading-edge products that are normally the preserve of executives and high-level decision makers is an erroneous one. Trailing edge proven technologies like I-Mode could conceivably win North

American interest particularly when considering the slowdown of the U.S. economy (2001) and with emerging evidence that American mobile users are beginning to decline the opportunity to purchase the very latest technology, such as 3G handsets, for example.

I-Mode services may be official and NTT DoCoMo approved, or unofficial. The former appear on the i-menu, whereas the latter are accessed by entering their URL or by sending bookmarks to phones using e-mail. Official sites are of many genres (see http://www.nttdocomo.co.jp/i/corp/teikyou.html).

I-Mode displays may be monochrome or 256-level grayscale, and may be animated GIFs measuring from 96×108 pixels to 120×130 pixels. I-Mode phones and microbrowsers are optimized for the Japanese language, so wrapping is rather difficult as lines may break in the middle of a word.

POP3 e-mail may be read using I-Mode, and e-mail may be forwarded to an I-Mode e-mail address using the phone number followed by @docomo.ne.jp. Messages may be up to 250 (double-byte) Japanese characters or 500 Latin characters, and received attachments are deleted.

Internet

The Internet or global network of computers and services is possibly the most overly documented subject, and includes the World Wide Web. A client/server architecture, HTTP servers are used to host Web services using scripting languages: ActiveX controls, Java applets, Flash movies, and streaming media. Telecommunications is the foundation for the World Wide Web that uses the HTTP (Hypertext Transfer Protocol). The TCP/IP (Transmission Control Protocol/Internet Protocol) emerged from the early DARPA and ARPANET networks that introduced a common transport protocol for communications. In 1983 all ARPANET networks were running TCP/IP, and later 32bit IP addressing meant that networks could include millions of addressable hosts. Some 18 years on the growing global network is now driving the development of IPv6.

The mobile wireless networks naturally converged on the Web through the introduction of the WAP (Wireless Application Protocol).

IP Address

A physical IP address consisting of 32bits, which identifies networks and its connected computers. The syntax for such addresses consists of four bytes each written in decimal form, and separated by a full stop: 118.234.98.87.

The three types of IP address include:

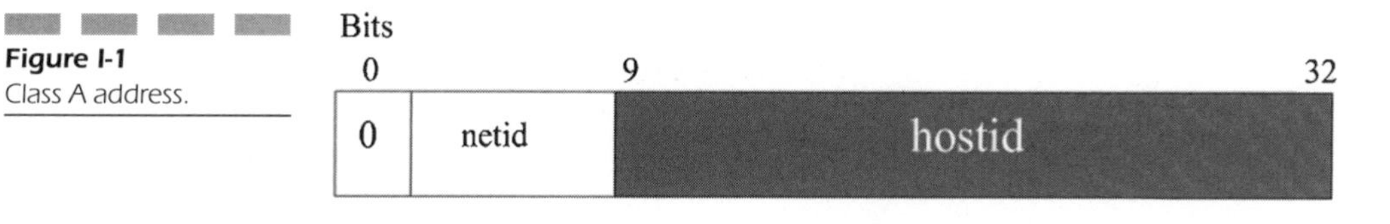

Figure I-1
Class A address.

- IP Address Class A: A network may have between 2^{16} (65,536) and 2^{24} (16.7 million) hosts. See Fig. I-1.
- IP Address Class B: B networks may have between 2^8 (256) and 2^{16} (65,536) hosts. See Fig. I-2.
- IP Address Class C: C networks may have up to 253 hosts, and not 255 because two values are reserved. See Fig. I-3.

The addresses consists of a network address (netID) and a host address (hostID). The leftmost digits represent the netID address. This is set to zero when addressing hosts within the network.

IP Multicast

An IP addressing system that is known as class D, and was developed at Xerox PARC. A multicast packet addresses a multicast group of nodes or hosts. The first multicast tunnel was implemented at Stanford University in 1988.

IPv6

An advancement of IP (Internet Protocol), IPv6 introduces 128bit addressing and numerous other improvements. The scaling of IPv4 32bit addressing to 128bit is intended to accommodate future growth of the Internet, in terms of the growing number of network addresses. IPv6 is also called IPng (IP Next Generation), and is specified by the Internet Engineering Task Force (IETF).

IPv6 supports addressing of the following types:

- Multicast that connects a host to multiple addressed hosts
- Unicast that connects a host to a single other addressed host
- Anycast that connects a host to the nearest of multiple hosts

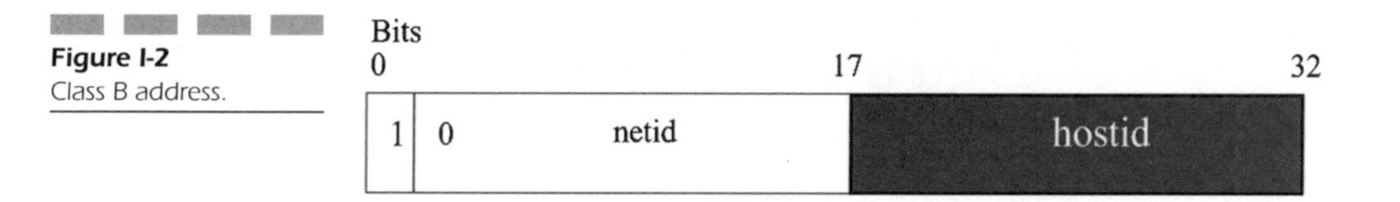

Figure I-2
Class B address.

Bits
0 17 25 32

1	1	0	netid	hostid

Figure I-3
Class C address.

IPv6 replaces IPv4 that is used by much of the Internet where IP addresses cannot keep up with growth. The IPv6 solution provides longer IP addresses and a considerably greater number. IETF working groups have developed IPv6 specifications and provided the following draft standards:

RFC 1886	DNS Extensions to Support IP Version 6
RFC 1887	An Architecture for IPv6 Unicast Address Allocation
RFC 1888	OSI NSAPs and IPv6
RFC 1981	Path MTU Discovery for IP Version 6
RFC 2080	RIPng for IPv6
RFC 2292	Advanced Sockets API for IPv6
RFC 2373	IP Version 6 Addressing Architecture
RFC 2374	An IPv6 Aggregatable Global Unicast Address Format
RFC 2375	IPv6 Multicast Address Assignments
RFC 2450	Proposed TLA and NLA Assignment Rules
RFC 2452	IP Version 6 Management Information Base for the Transmission Control Protocol
RFC 2454	IP Version 6 Management Information Base for the User Datagram Protocol
RFC 2460	Internet Protocol, Version 6 (IPv6) Specification
RFC 2461	Neighbor Discovery for IP Version 6 (IPv6)
RFC 2462	IPv6 Stateless Address Autoconfiguration
RFC 2463	Internet Control Message Protocol (ICMPv6) for the Internet Protocol Version 6 (IPv6) Specification
RFC 2464	Transmission of IPv6 Packets over Ethernet Networks
RFC 2465	Management Information Base for IP Version 6: Textual Conventions and General Group
RFC 2466	Management Information Base for IP Version 6: ICMPv6 Group
RFC 2467	Transmission of IPv6 Packets over FDDI Networks
RFC 2470	Transmission of IPv6 Packets over Token Ring Networks
RFC 2471	IPv6 Testing Address Allocation
RFC 2472	IP Version 6 over PPP
RFC 2473	Generic Packet Tunneling in IPv6 Specification
RFC 2491	IPv6 over Non-Broadcast Multiple Access (NBMA) Networks

RFC 2492	IPv6 over ATM Networks
RFC 2507	IP Header Compression
RFC 2526	Reserved IPv6 Subnet Anycast Addresses
RFC 2529	Transmission of IPv6 over IPv4 Domains without Explicit Tunnels
RFC 2545	Use of BGP-4 Multiprotocol Extensions for IPv6 Inter-Domain Routing
RFC 2553	Basic Socket Interface Extensions for IPv6
RFC 2590	Transmission of IPv6 Packets over Frame Relay
RFC 2675	IPv6 Jumbograms
RFC 2710	Multicast Listener Discovery (MLD) for IPv6
RFC 2711	IPv6 Router Alert Option
RFC 2740	OSPF for IPv6
RFC 29xx	Router Renumbering for IPv6

Addressing/Routing IPv6 Internet backbones for major enterprises and ISP networks use an address hierarchy, as does IPv4 when sorting traffic toward networks linked to the Internet backbone. This relieves backbone routers from having to store route table information about global networks. The hierarchy allows backbone routers to use IP address prefixes to determine how traffic should be routed through the backbone. IPv4 may use Classless Interdomain Routing (CIDR) that uses bit masks to allocate sections of 32bit IPv4 addresses to networks, subnets, and hosts. IPv6 has a hierarchical global routing architecture, and CIDR style prefixes allow route summarization, and control expansion of route tables in backbone routers. The great number of IPv6 addresses means that ISPs will have sufficient addresses for businesses and for certain dial-in users.

Address Autoconfiguration Autoconfiguration gives mobile devices forwarding addresses, irrespective of network connection location. IPv6 nodes create local addresses for themselves using stateless autoconfiguration that also allows nodes to configure globally routable addresses. The node may append a 48 or 64bit MAC address to a network prefix it learns from a neighboring router. This reduces administration costs as it is unnecessary to configure each workstation.

IPv6 Header Format The 128bit IPv6 header requires twice the overhead of that of IPv4. Optional header information occupies "extension headers" located between the header's end and transport-layer header. These extension headers are not processed by intermediate nodes. The option fields may carry routing information created by the source node, for mobility, authentication, and encryption.

Multicast Streaming of video, audio, and financial and news services transmit data to groups of endstations, which is achieved using multicast. Multicast routes packets to subscribers in multicast groups using optimum paths. Multiple networks with multicast group members require a packet distribution tree for multicast groups. The routers use the protocols:

- DVMRP (Distance Vector Multicast Routing Protocol)
- PIM (Protocol Independent Multicast)
- MOSPF (Multicast Open Shortest Path First)

These make the packet distribution tree that connects group members with a server.

Anycast Anycast services combine unicast and multicast. A group's anycast address receives a packet and delivers it to one group node that is often the one nearest to the group's interface. Group nodes are configured to recognize anycast addresses taken from the unicast address space.

Security IPv6 offers security and authentication header extension headers. IPv6 determines the authenticity of packets at the network layer, so as to gain product interoperability. Implementations require support for MD5 and SHA-1 algorithms for authentication. IPv6 security headers may be used between hosts or in conjunction with a security gateway.

Mobility IPv6 is optimized for mobile computing, and provides dial-up support, option processing for destination options, autoconfiguration, routing headers, encapsulation, security, and anycast addresses.

Quality of Service The IPv6 packet has a 20bit traffic-flow identification field for quality-of-service (QoS) network functions. QoS products that use this field may have functions for bandwidth reservation and delay bounds. A flow label is also included in the IPv6 header for optimum routing.

IPv6 DNS An IPv6 DNS allows dual-stack hosts to communicate with IPv6 nodes. A dual-stack host may query a DNS to obtain a 32bit address (IPv4) or a 128bit address (IPv6). The IPv6 DNS extensions help manage a site's addresses and simplify DNS updates. New DNS records permit the division of addresses into logical components that are stored separately in the DNS.

Routing There are IPv6 versions of Open Shortest Path First (OSPF) and Routing Information Protocol (RIP). Administrators may separate the IPv6 topology from the IPv4 network. IPv6 hosts may connect directly using IPv4 routers. Transition mechanisms are used for IPv6 hosts to communicate over IPv4 networks. This is achieved using IPv6 over IPv4 tunnel carrying IPv6 packets in IPv4 packets.

Internet Telephony

Internet Telephony or IP telephony is used to make telephone calls using packet-switched IP networks. In 1997, low-cost Internet telephony began to change the face of telcos, Internet Service Providers (ISPs), and corporations; the largely unregulated Internet became interwoven with telephony.

For the first time ISPs could mine revenue-rich long-distance and international calls businesses that were once the preserve of telcos like BT and Mercury. Corporations and government departments may also "toll bypass" the telcos by using Internet telephony over their own networks such as intranets, significantly reducing their operating costs in the process. Similarly since VocalTec launched the Internet Phone in 1995, growing numbers of ISP subscribers have been making long-distance and international calls for the cost of a local phone call. See Fig. I-4.

Internet telephony theoretically means that an ISP could become an international telco. The domestic long-distance calls business presents ISPs with one opportunity to compete with telcos, as does the international calls business. The technology is also being embedded into existing switched networks, where it will become transparent to the user. The successful proliferation of

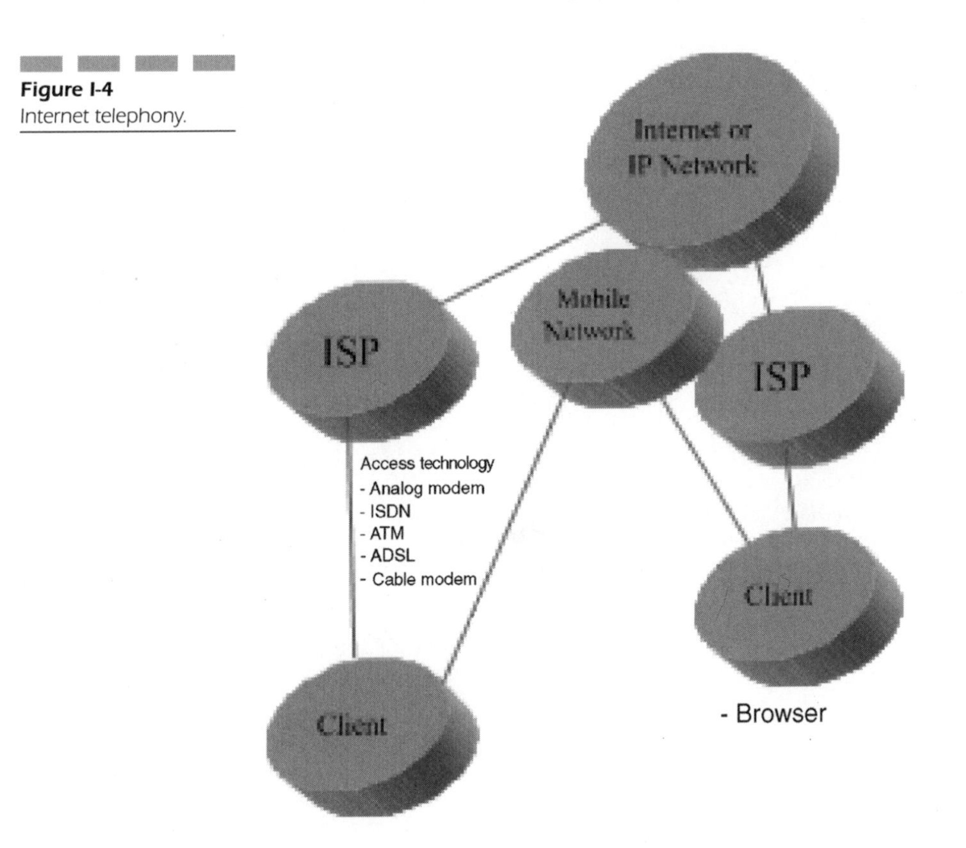

Figure I-4
Internet telephony.

Internet telephony also hinges on emerging internationally agreed standards, such as H.323, which collectively will unify ISP's services globally.

IP Mobile Networks

Internet access configuration using a mobile IP tunnel (see Fig. I-5) and Internet access configuration using voluntary L2TP tunnels and mobile IP present two solutions for Mobile IP network implementation. The mobile IP protocol gives mobility, and forward traffic is routed between home and foreign agents over a mobile IP tunnel, and reverse traffic is routed from foreign agents to remote servers.

Internet Access Network Mobile:

- An accounting server may interface with FA routers, collecting and store accounting records.
- Mobile IP-capable router gives the foreign agent (FA) function.
- Mobile IP-capable router gives the home agent (HA) function.

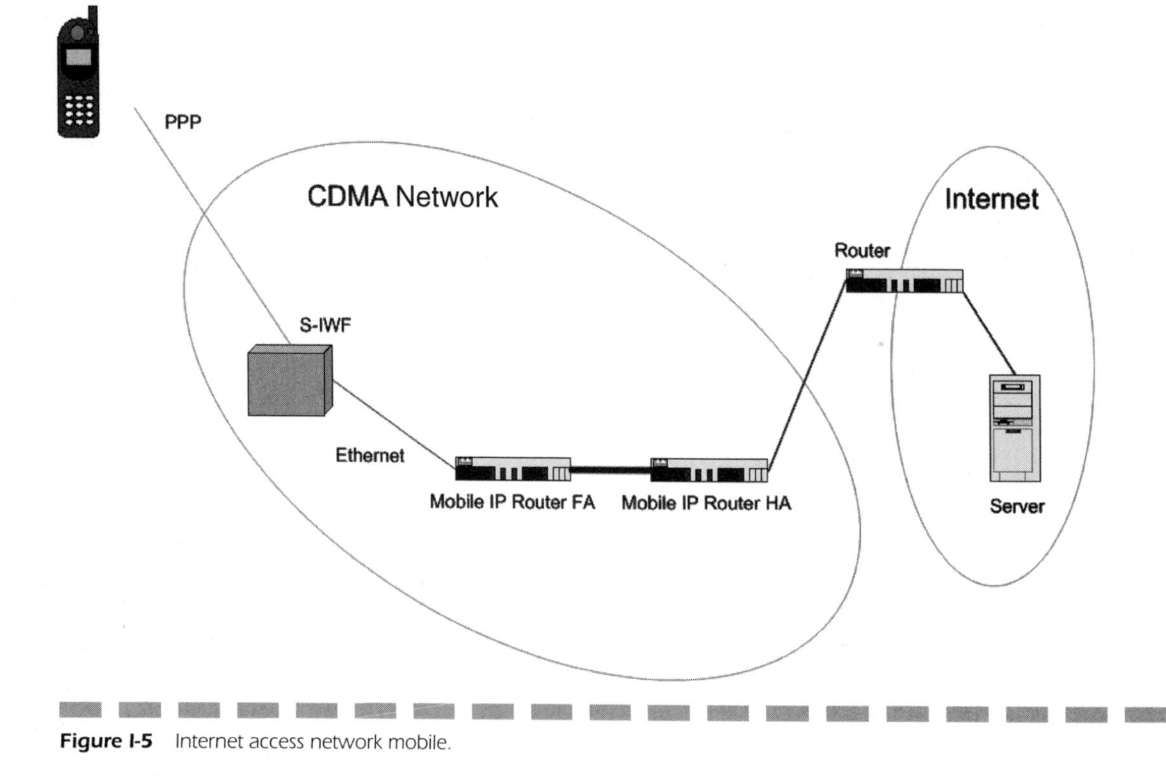

Figure I-5 Internet access network mobile.

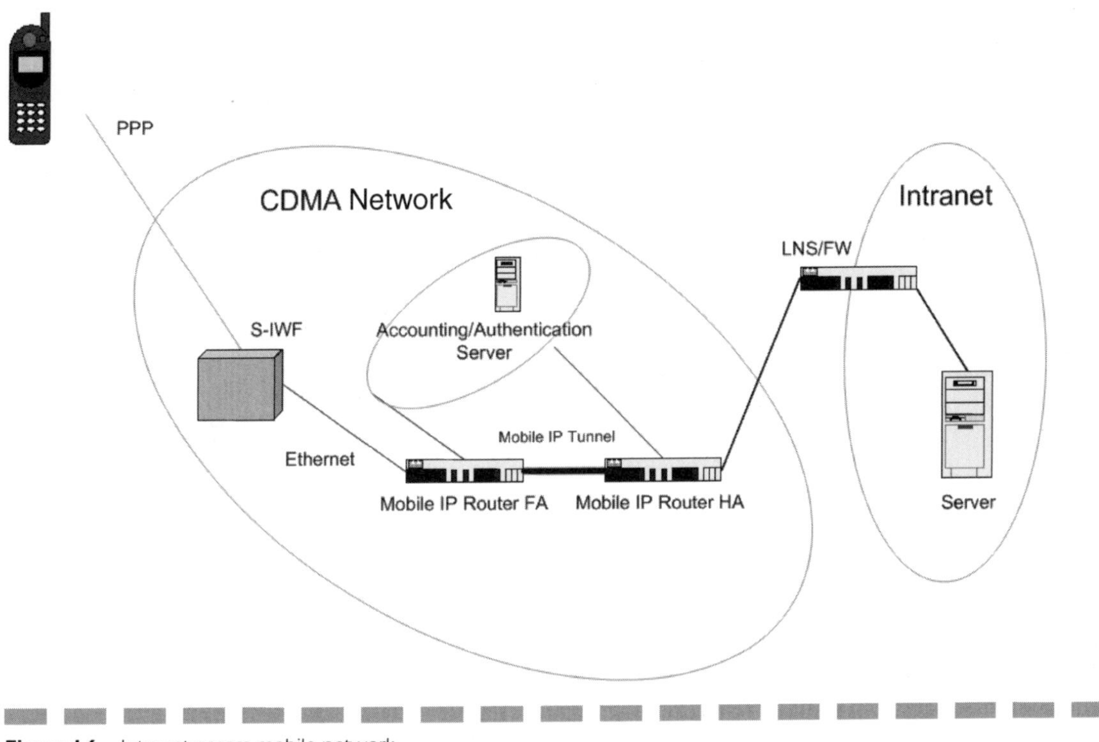

Figure I-6 Intranet access mobile network.

- The HA router's IP address is conveyed to the FA using the mobile IP registration message.
- The IWF in the serving system (S-IWF) and FA router are connected using Ethernet.
- The S-IWF terminates PPP protocol and relays received IP datagrams to the designated mobile-IP FA router on the local network.

Figure I-6 illustrates the intranet access mobile network. Figure I-7 shows the Internet access protocol stack.

Security

Data privacy over the air is optional and is implemented using RLP encryption (ORYX) between the mobile and packet-switching unit (PSU). Three levels of authentication and authorization validation are provided: IS-41 service authorization validation (mandatory), IS-41 authentication (optional), and mobile HA authentication for mobile IP registration/reply messages (mandatory).

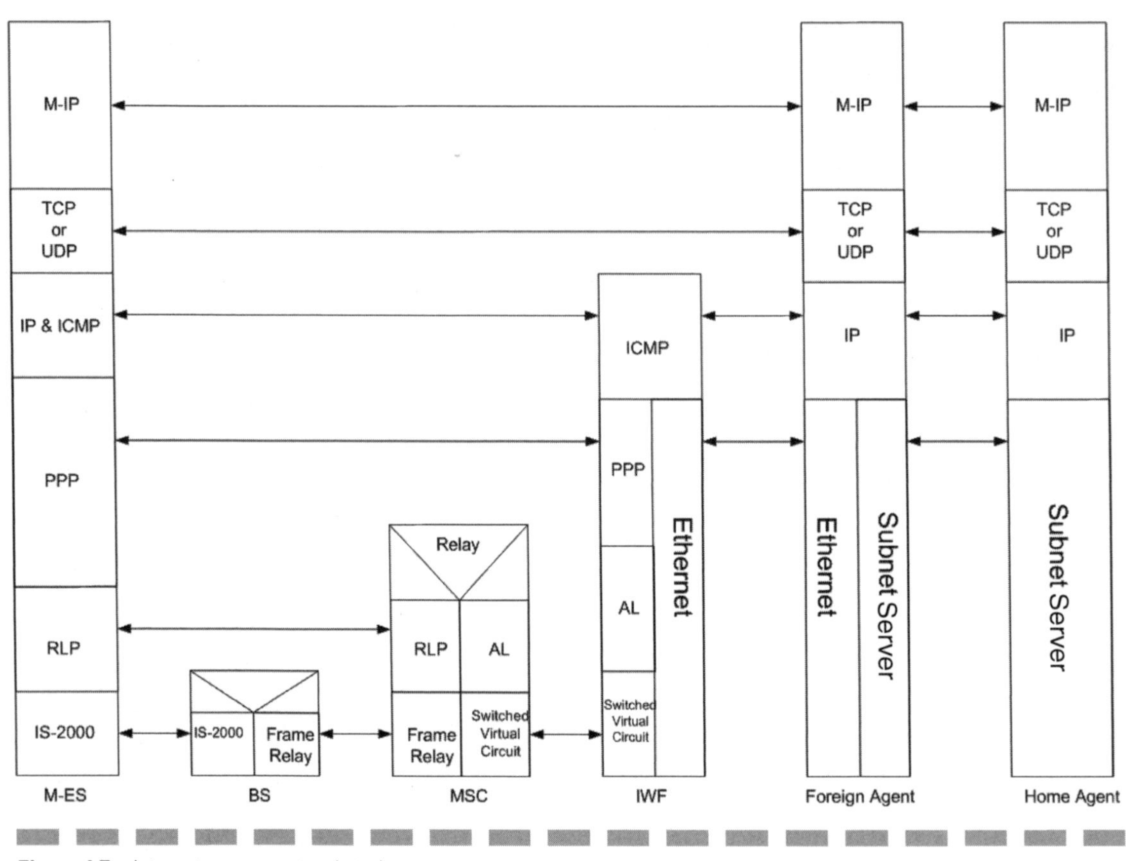

Figure I-7 Internet access protocol stack.

Security and Firewall

Authentication includes IS-41 and mobile HA during mobile IP registration. Identification in the mobile-IP registration gives antireply protection. LNS authenticates terminals using PPP authentication integrity like Challenge Handshake Authentication Protocol (CHAP). End-to-end authentication, integrity, and confidentiality can be provided if the terminal and LNS support IP.

ISDN

ISDN (Integrated Services Digital Network) is an access technology introduced by the CCITT, and is able to support reasonably sophisticated videoconferencing, and high-speed access to the Internet and other networks. A BRI

(Basic Rate Interface) ISDN line may have two 64kbits/s B-channels able to carry video, voice, or data. ISDN uses PCM for encoding data in digital form, for transmission.

Used in the ISDN standard, PCM involves creating a data stream consisting of 8bit PCM blocks. The blocks are created every 125μs. By interleaving the blocks with those from other encoders, the result is time division multiplexing (TDM). In North America ISDN typically interleaves data from 24 64kbits/s sources or channels. This results in connections that provide 1.536Mbits/s. Although in actual fact the connection has a bandwidth of 1.544Mbits/s, because each channel's frame has a marker bit "F", adding 8kbytes/s. Europe sees ISDN typically interleave 30 64kbits/s channels, giving 2048Mbits/s. This and the 1.544Mbits/s connection are known as primary rate multiplexes.

Further interleaving of primary rate multiplexes sees:

- 6, 45, 274Mbits/s in North America
- 8, 34, 139, and 560Mbits/s in Europe

While analog modem speeds increased steadily throughout the 1980s and early 1990s, the arrival of the Web placed greater demands on available bandwidth. This encouraged telcos to provide ISDN services that were defined by the CCITT in 1971, and published in 1984 in the Red Book. ISDN was based on PCM (Pulse Code Modulation) that was conceived by A. H. Reeves, and experimented with in the Second World War, and was used in American telecommunications in the 1960s so as to increase network capacity.

In 1986 a pre-ISDN service named Victoria was offered by Pacific Bell in Danville, California, offering RS232C ports that were configurable from 50bits/s to 9.6kbits/s. In the same year, official ISDN systems were introduced in Oak Brook, Illinios, and by 1988 some 40 similar pilot schemes were installed.

ISDN digital networks eventually developed into B-ISDN where multiple lines could be used to provide data rates in increments of 64kbits/s, and videoconferencing and high-speed Internet access were made possible. B-ISDN implementations could even be used to implement the lower data rates of 1.544Mbits/s offered by modern T1 digital links that arrived some time later.

Computers/Software

iBook

An Apple laptop computer that supports the Mac OS X.

Idempotency

An attribute of a message that sees repetition yield a constant result.

IDL (Interface Definition Language)

A language that may map to others and is neutral in terms of providing interfaces and operations to applications written in compliant languages. Examples include CORBA IDL that is used widely to create middleware implementations.

IDL to C++ Language Mapping

A mapping that equates IDL to the C++ equivalent.

IDL to Java Language Mapping

A mapping that equates IDL to the Java equivalent.

Idl2java

A compiler that converts IDL programs into Java.

IIOP (Internet Inter-ORB Protocol)

A standard specification for transmitting GIOP messages over TCP/IP networks such as the Internet.

Impact Statement

A document that outlines and describes the effects of introducing a new technology or solution.

Imported Style Sheet

A style sheet that can be imported to (combined with) another sheet, combining:

- Main sheet that apply to the whole site
- Partial sheets that apply to specific documents

Form:

```
<LINK REL=STYLESHEET HREF="main.css" TYPE="text/css">
<STYLE TYPE="text=css">
<!- -
@import url (http://www.botto.com/fast.css);
@import url (http://www.mcgraw.com/fast.css);
.... other statements
- ->
</STYLE>
```

Indeterminate

A Boolean floating condition that is neither yes nor no.

Information Property List

A configuration information file (Info.plist) for bundles, and is used in Cocoa and GNUStep development environments.

Inheritance

A feature of OOP languages that sees superclasses pass methods and data (or instance variables) to subclasses.

Inline Style

An attached style that affects one element and is specified in the start tag as a value of the STYLE attribute:

```
<P STYLE="text-indent: 14pt">Indented paragraph</P>
```

Inner Class

A class that is local to a given block of Java code.

Instance

An object created or instantiated at run time and adheres to the appropriate class definition.

Instance Variable

A data entity that may be inherited by objects using an OOP language such as Objective-C.

Instantiate

A process where an object is built using an OOP language.

Intelligent AML

An anti-money laundering solution that uses software intelligence.

Intelligent Business Solution

A business solution that may make intelligent decisions without human intervention. The decisions may lead to alerts informing human users, rather than leading to an automatic or preprogrammed action or response.

Internationalization

The implementation or modification of software so as to localize it.

Interoperability

A software or hardware attribute that indicates compatibility between products from different vendors, and is normally based on a standard specification.

Interpreted

A programming language where resulting programs are converted into an executable form each time they are run. Java, for example, is interpreted at run-time from byte-codes that are the result of compilation. Java then is both compiled and interpreted, but since it does not result in native machine code in the case of execution using JVMs, it is often considered an interpreted language.

Interpreter

A program that converts source code into runtime code each time a program is executed.

ISO/TC 154

A technical committee that addresses documents and data elements in administration, business, and industry.

ISO9736

A standard set of UN/EDIFACT syntax rules.

Miscellaneous

IETF

The Internet Engineering Task Force is an international standards organization dedicated to the maintenance, ratification, and publication of Internet-related technologies.

Interactive Television

A form of television broadcasting where the user is presented with a non-linear medium, and is able to select content using on-screen options to control what screened matter is shown. Depending on the implementation, the viewer may be able to select camera angles, and replay selected footage and scenes. Interactive television services may offer TV shopping services also, and Internet-related facilities that include e-mail.

Intranet

A TCP/IP network that is accessed by a user group that is usually from the same company, and is based on Web-based technologies.

IP Multicast

An IP multicast addressing system is known as class D, and was developed at Xerox PARC. A multicast packet addresses a multicast group of nodes or hosts. The first multicast tunnel was implemented at Stanford University.

IrDA (Infrared Data Association)

IrDA is a wireless interface technology that is able to drive compatible peripheral devices. It is supported by Windows, and is integrated in numerous laptop computer designs.

Acronyms

IC	Interference cancellation
ICMP	Internet Control Message Protocol
IDL	Interface definition language
IEEE	Institute of Electrical and Electronic Engineers
IETF	Internet Engineering Task Force
IHL	Internet Header Length
IIOP	Internet Inter-ORB Protocol
IMA	Inverse multiplexing
IMSI	International mobile subscriber identity
IMT-2000	International Mobile Telephony 2000 IN Intelligent network
IN/CAMEL	Intelligent network/customized applications for mobile enhanced logic
IP	Internet Protocol
IPR	Intellectual property rights
IR	Infrared
IrDA	Infrared data association
IrOBEX	Infrared object exchange protocol

IRP	Integration reference point
ISDN	Integrated Services Digital Network
ISP	Internet service provider
ITS	Intelligent telephone socket
ITU	International Telecommunications Union
ITU-R	International Telecommunications Union—Radio
ITU-T	International Telecommunications Union—Telecommunications

J

Java

J2ME

The Java 2 Micro Edition (J2ME) has the Java virtual machine specification and API specifications that define class libraries. The K virtual machine is a runtime JVM for thin devices like cell phones and PDAs that have a few hundred kilobytes. The J2ME Mobile Information Device Profile (J2ME MIDP) specification provides a standard platform for wireless devices that have 512K total memory (ROM + RAM), battery power, and a user interface.The J2ME Personal Digital Assistant (PDA) specification provides user interface and data storage APIs for PDAs and Palms with 512K for Java runtime and libraries, and with less than 16MB memory, and with UIs with a resolution of at least 20,000 pixels.

- The J2ME Connected Device Configuration (CDC) specification defines a full-featured runtime environment for sophisticated wireless/wired devices that have 512K ROM and 256K RAM.
- The J2ME Personal profile proposed specification is for devices with 2.5M ROM and 1M RAM and applet running capabilities.
- The J2ME Foundation Profile specification is for devices with the following characteristics: 1024K ROM and 512K RAM.
- The J2ME Remote Method Invocation (RMI) profile proposed specification is for inter-application RMI over TCP/IP.

The J2ME specification excludes the PersonalJava application environment and the JavaPhone API. The JavaPhone API extends the PersonalJava platform and includes:

- Direct telephony control
- Datagram messaging
- Address book and calendar information
- User profile access

J2ME provides a standard subset of the Java 2 Platform, Standard Edition (J2SE), that is a subset of Java technology for servers, the Java 2 Platform Enterprise Edition (J2EE). A J2ME configuration defines a minimum Java platform for a device type, where, for example, the CLDC specifies that it targets devices with 160K to 512K memory for Java technology. A *profile* is a collection of Java technology APIs that supplements a configuration to provide capabilities for a market or device type.

JAE (Java Application Environment)

A source code release of the JDK.

JAR

A Java Archive file format used for concatenating files; these files have the .jar extension.

JAR File Format A platform-independent file format that may be used to store applets and their components including images, sounds, resource files, and class files. JAR files may be used to download Java-related data more efficiently using HTTP, and offers support for compression and digital signatures.

Java

A general-purpose, high-level language (HLL) that is

- Not platform or operating-system sensitive, yielding "write-once-run-anywhere" code
- Object-oriented
- General purpose
- Multithreaded
- Class-based

The resulting compiled bytecode may be run on Windows 3.x, Windows NT, Windows 95/98, Macintosh environments, and UNIX et al. OS independence is a key characteristic of Java, making it suitable for deployment of applets, where client OSes are of a heterogeneous nature.

Web-based Java applets:

- Are interpreted by Java-enabled browsers
- May access code libraries on the client machine
- May download class libraries from the server

Development tools for Java include:

- The Java Development Kit (JDK)
- Microsoft J++ that is included in Microsoft Visual Studio
- VisualAge for Java

The Java language semantics, and high-level instructions are similar to C and C++. It is considered a static programming language, but is likely to be given a dynamic functionality through appropriate development environments and compilers.

Compilation of Java source code yields in a bytecoded instruction set and binary format. Java may also be used to develop client-side and server-side applications, perhaps using the CORBA-based Notification Services as its middleware or glue. Such Java code may be generated automatically from IDL files using an IDL2JAVA compiler.

Java Applet

A program created using the Java programming language, and typically deployed over the Web. It resides on the server side, and is downloaded to a Java-enabled Web Browser. It is then interpreted and run. The Browser must feature the Java Virtual Machine that is a software-based processor.

Java Array

A matrix of entities of the same type that may be simple or composite. The matrix or array may be multidimensional, and is declared using square brackets ([ ]).

```
int meters[];
char[] table;
long transform[][];
```

The size of an array is not specified at its declaration.

Java Blend

A database application development tool that is an environment for combining Java objects with enterprise databases. Applications may be developed by coding in Java, and resulting objects may be mapped to databases, and vice versa. It does not require knowledge of SQL. Java Blend was codeveloped by the Baan Company and Sun Microsystems.

Java Card

A smart card implementation that uses Java technology. The Java Card specifications can also be applied to devices that have:

- 16kbytes ROM
- 8kbytes EEPROM
- 256bytes RAM

Java Casting Types

A process of converting one data type into another. Casting often is necessary when a function returns a type different than the type you need to perform an operation. The int returned by the standard input stream (System.in) is cast to a char type using the statement:

```
char k = (char)System.in.read();
```

Java Comment and Whitespace

A textual comment, and whitespace consists of spaces, tabs and linefeeds.

```
/* multiple line comment */
// a single line comment
/** a multiple-line comment that may be used with the
      javadoc tool to create documentation**/
```

Java Data Type

A means of defining a storage method for information, such as the storage of variables in memory.

The following statement declares a variable, a variable type, and identifier:

```
Type Identifier [, Identifier];
```

The statement:

- Allocates memory to a variable type Type
- Names the Type "Identifier"
- Uses the bracketed identifier to indicate that multiple declarations of the same type may be made

Java data types may be:

- Simple, that include integer, floating-point, Boolean, and character
- Composite, which are based on simple types, and include strings, arrays, classes, and interfaces

Java Development Tool

A tool/environment that allows programmers to create Java applets, Java programs, JavaBeans, and possibly Java Servlets.

Java Electronic Commerce Framework

A point-of-sale (POS) application framework.

Java Floating-Point Data Types

A data type that may represent fractional numbers that may be the:

- Float type that is allocated a 32bit single-precision number
- Double type that is allocated a 64bit double-precision number

Such data types are implemented using the statements:

```
float altitude;
double angle, OpenRoad;
```

Java Floating-Point Literals A means of storing and processing fractional numbers that are expressed in decimal (i.e., 200.76) or in scientific notation (2.00.76e2).

Floating-point literals default to the double type that is a 64bit value. The "f" or "F" suffix harnesses the 32bit value.

Java Identifier

A Java token that stores names that are applied to variables, methods, and classes.

Java Integer Data Types

A means of representing signed integer numbers, and include:

- Byte (8bit)
- Short (16bit)
- Int (32bit)
- Long (64bit)

Integer variables are declared thus:

```
int x;
short ;scale
long lumin, light;
byte alpha, beta, gamma;
```

Java Integer Literals

A literal may be:

- Decimal (base 10)
- Hexadecimal (base 16) that have the "OX" prefix
- Octal that have the "0" prefix

By default, integer literals are stored in the int type that has a 32bit value. They may be stored in the double type that has a 64bit value, using the "l" or "L" suffix.

Java Keywords

A meaningful vocabulary of entities that perform specific functions, and include:

abstract	double	int	super
boolean	else	interface	switch
break	extend	long	synchronized
byte	false	native	this
byvalue	final	new	threadsafe
case	finally	null	throw
catch	float	package	transient
char	for	private	true
class	goto	protected	try
const	if	public	void
continue	implements	return	while
default	import	short	

Java Lexical Translation

A process by which Java source code is converted into Java tokens. It is implemented by the lexical analyzer facet of the compiler that:

- Translates Unicode escapes into Unicode characters, allowing the Java listing to be represented using ASCII characters
- Generates a stream of input characters and line terminators
- Generates Java input elements, or Java tokens, that are terminal symbols

Java Literal

An element that maintains a constant value, and may be:

- Numeric
  - Integer
  - Floating point
  - Boolean

- Characters
- Strings

Character literals refer to a single Unicode character. Multiple-character strings that are implemented as objects are also literals.

Java Management API (JMAPI)

A library of objects and methods used for the development of network and service management solutions targeted at heterogeneous networks.

JavaBeans

JavaBeans is a standard component architecture offering *beans* that may be used to build applets, servlets, and applications. The components are referred to as Beans, and compliant development tools provide access to the Beans using a toolbox. Visual programming plays an important role when architecting a Beans-based program; the developer simply selects Beans and modifies their appearance, behavior, and interactions with other Beans. JavaBeans-compatible development tools include:

- JavaSoft JavaBeans Development Kit (BDK)
- Lotus Development BeanMachine
- IBM VisualAge for Java
- SunSoft Java Workshop
- Borland JBuilder
- Asymetrix SuperCede
- Sybase PowerJ
- Symantec Visual Cafe

Java Package

A set of Java classes that address specific functions, where, for instance, `java.io` addresses input and output functions, and java.net addresses Internet and network operations.

Java Separator

A means of categorizing Java source code that directs the compiler appropriately, and includes:

```
{ } ; , :
```

Java String Literals A string, or number of characters, within a pair of double quotation marks. String literals invoke an instance of the String class that is assigned the character string.

Java Studio1.0 A development environment that does not require Java coding on a line-by-line basis. It harnesses the JavaBeans object architecture, and is typically used to build Web applications. It is a product of Sun Microsystems.
(*See* Java earlier in this chapter.)

Java Tokens A meaningful element of a Java program when compiled. The five categories of token include:

- Identifiers
- Keywords
- Literals
- Operators
- Separators

Tokens are compiled into Java bytecode that may be interpreted by a Java Virtual Machine.

Java Unicode A predominant character set with which Java source code is represented, and is:

- 16bit that gives up to 2^{16} or 65,536 possible characters
- Used exclusively by Windows NT at the system level
- A worldwide standard

In Java, three lexical translations[UNICODE1] convert a raw unicode character stream into a sequence of Java tokens.

Java.net

A package that has classes designed for networks, and includes the URL class that allows remote objects to be downloaded, and may be used to read and write streams to and from the object.

JavaHelp

A software product that allows the creation of on-line Help for Java applets, applications, OSes, and devices. It can also be used to deploy on-line Help over the Web and intranets.

JavaHelp is:

- Written using the Java language
- Implemented using JFC components
- Platform independent
- Browser independent
- Supported by browsers that comply with the Java Runtime Environment

JavaMail

An API used to build Java-based mail and messaging applications.

JavaOS A compact operating system dedicated to running Java programs/ applets. The JavaOS family includes:

- JavaOS for Network Computers (NCs) that is described as a standalone Java Application Platform for NCs
- JavaOS for Appliances that is intended for communications devices
- JavaOS for Consumers that is aimed at consumer electronics devices.

JavaPC

A software solution for migrating PCs to Java platforms. JDK 1.1-compliant Java applications may be stored locally or on a network, and may be run on DOS and Windows 3.x platforms.

JavaScript

An object-oriented scripting language optimized for the Web. Using JavaScript, Web pages and/or HTML documents may be given:

- Dynamic content such as animations
- Integrated Java applets and ActiveX controls
- Interactive content
- Data entry forms

Microsoft's implementation of JavaScript is JScript. The rationale behind JScript is echoed by VBScript: It is intended as a quick method of creating and tailoring applications. Unlike JScript, VBScript is not an OOP language. Like other objects, JavaScript objects have properties and methods, and include the:

- Window that is at the top of the HTML document's object hierarchy
- Frame that is a window
- Location that stores URL information
- Document that stores document characteristics such as its URL and title
- Form that stores form characteristics
- Text and text area that store text information
- Checkbox that is a standard Windows UI object
- Radio that refers to a single UI radio button
- Select that is an array of option objects
- Button that stores button information
- Password that is a text-entry box that disguises keyboard entries using asterisks
- Navigator that stores a visitor's version number of Netscape Navigator
- String that provides methods for string manipulation
- Date that is dedicated to calendar date information
- Math that facilitates common constants and calculations
- Image that indicates image information on the current page
- Array that is dedicated to arrays

JScript listings are integrated in HTML code by enclosing them between the following tags:

```
<SCRIPT LANGUAGE="JavaScript">
<SCRIPT>
```

Development environments and applications that support JScript are numerous, and include the Microsoft ActiveX Control Pad that also supports VBScript.

(*See* Visual Basic under letter V.)

JScript Commands

Comments Single and multiple line comments may be included using the syntax:

```
//    A single line comment
/*    Multiple lines comment
      require this syntax */
```

JScript Operators

++ increment

_ decrement

* multiplication

/ divide

% modulus

+ addition

– subtraction

≪ shift left

≫ shift right

> greater than

<= less than or equal to

>= greater than or equal to

== equal to

!= not equal to

&& logical AND

! logical NOT

|| logical OR

^ bitwise

| bitwise OR

& bitwise AND

for The for statement has three optional expressions:

```
for( initial.Expression; condition; update.Expression) {
   statement
   statement
   statement
}
```

- `Initial.Expression` initializes the counter variable that may be a new variable declared with var.
- `Condition` is evaluated on each pass through the loop. If the condition is true, the loop statements are executed.
- `Update.Expression` is used to increment the counter variable.

while A statement used to implement a conditional loop, based on a true or false validation:

```
while (condition) {
  statement
  statement
  statement
}
```

break A break statement stops for or while loops, and diverts program execution to the line following the loop statements.

for...in A for...in statement executes the statement block for each object property:

```
for (variable in object) {
  statement
  statement
}
```

function A statement which allows you to create a named JScript function together with parameters. The return statement may be used to return a value. Nested functions are not supported.

```
function name ([parameter] [...,parameter]) {
statements...
}
```

if...else A conditional statement that offers one of two conclusions.

```
if (condition) {
  statement
  statement
} [else {
  statement
  statement
} ]
```

return This is used to specify a returned value from a function.

var The var statement is used to declare a variable that may be local or global.

```
var variableName [=value] [..., variableName [=value]]
```

while Repeats a loop while an expression is true.

```
while (condition) {
statements...
}
```

with Declares a default object as the focus of a set of statements.

```
with (object) {
statement
statement
}
```

JavaSpaces

JavaSpaces is a SunSoft technology that provides a means of writing, reading, and transferring objects or entries between spaces.

The JavaSpace interface defines methods to read or copy entries from a space, and to take or copy entries from a space while removing them from spaces also. Locating entries in spaces is carried out using associative lookup, and a template is used to match entry contents. For example, you may use a template such as:

```
CarCreditcard anyCreditcardTemplate = new
  CarCreditcard();
AnyCreditcardTemplate.name = null;
AnyCreditcardTemplate.cost = null;
```

The null fields operate like wildcards and any *Creditcard* entry will be matched regardless of its name or cost. Null may also be used for the template so all entries in a space are matched. The match may be narrowed merely by adding values and name strings like:

```
AnyCreditcardTemplate.name = "Shell";
```

The JavaSpaces interface has two read methods:

```
Entry read(Entry tmpl, Transaction txn, long timeout)
```

And,

```
Entry readIfExists(Entry tmpl, Transaction txn, long
  timeout)
```

Both require three parameters: a template, transaction, and a timeout value that may be expressed in milliseconds.

When a matching entry is found, a copy of it is returned, or it is merely read, shall we say. Where multiple matches are found, the space returns a single arbitrary entry. This clearly must be borne in mind when developing JavaSpaces services. The Read method waits for the timeout if no matching entries are found in a space, or until a matching entry is found. If no matching entry is found, a `null` value is returned by the read method. The Long.MAX_VALUE may be used as a timeout value, causing the read operation to block until a matching entry is found. Alternatively, using the JavaSpaces.NO_WAIT value for the timeout causes the read operation to return immediately.

The `readIfExists` method is nonblocking, and returns immediately when no matching entry is found, irrespective of the timeout parameter value. The timeout parameter is relevant only when the read operation takes place under a transaction. So with a `null` transaction the timeout value is equivalent to NO_WAIT.

Both read operations may throw the:

- `InterruptedException`, when the thread implementing the read operation is interrupted
- `RemoteException`, when a failure occurs on the network or in the remote space
- `TransactionException`, when the supplied transaction is invalid
- `UnusableEntryException`, when an entry retrieved cannot be deserialized

The `take` method defined in the JavaSpaces interface is of the same form as the read operation:

```
Entry take(Entry tmpl, Transaction txn, long timeout)
```

The take method removes entries from spaces provided that no exception is thrown. The aforementioned exceptions for read and `readIfExists` may also be thrown by the `take` method. These include`InterruptedException`, `RemoteException`, `TransactionException`, and `UnusableEntryException`.

JavaSpaces Write Method

A JavaSpaces method that is used to write entries to spaces.

The following code segment uses the `write` method to copy a Creditcard requirement to a space:

```
public void writeCreditcard(CarCreditcard Creditcard) {
Try {
  space.write(Creditcard, null, Lease.FOREVER);
  } catch (Exception e) {
```

```
      e.printStackTrace();
    }
  }
```

The write method as defined in the JavaSpaces interface is of the form:

```
Lease write(Entry e, Transaction txn, long lease)
  throws RemoteException, TransactionException;
```

An entry is manipulated using the write method that may use the arguments transaction and lease time. If given a null value, the transaction is a singular operation and is detached from other transactions. The lease time argument simply specifies the entry's longevity in the new space. When the lease expires, the space removes it but if a Lease.FOREVER value is used, the entry exists in the space indefinitely, or until it is removed by a transaction process or operation. The lease time may also be expressed in milliseconds like the 3-minute lease shown below:

```
Long time = 1000 * 60 * 3 ; // three minute lease
Lease lease = space.write(entry, null, time);
```

Leases may also be renewed using the renew method:

```
void renew(long time)
, ,
```

The time value is added to the lease time remaining. The renew method may raise the exceptions: LeaseDeniedException, UnknownLeaseException, and RemoteException. The former is raised should the space be unable to renew the lease that may be caused by lack of storage resources. The UnknownLeaseException is thrown if the entry or object is unknown to the space, perhaps because its lease has already expired.

In real-world scenarios the use of Lease.FOREVER is considered an uneconomical use of resources. The intelligent solution to leasing is to use Lease.ANY that leaves it up to the space to decide when expiration occurs. But when building JavaSpaces services, the use of Lease.FOREVER may compress the development a life cycle, as it can reduce the probability of thrown exceptions.

The write method also throws two types of exceptions, including RemoteException and TransactionException. The former is raised when a communication breakdown occurs between the source process and the remote space, or when an exception occurs in the remote space while an entry is written to it. The RemoteException returns the detail field that holds the exception type. If the space is unable to grant a lease, a RemoteException is thrown. A TransactionException is thrown when the transaction is invalid or cannot be rolled forward or committed.

Jini

Jini is a SunSoft standard for distributing services over local and remote networks, and may be applied to implement self-configuring tier 0 devices sometimes (called smart devices) that include cell phones, DVD drives, palm PCs, and organizers. The connection between a Jini device and a Jini service takes place using the TCP/IP transport protocol and other Jini protocols that include Discovery that is based on UDP. Jini networks comprise lookup services that act as directories containing the network's registered services. A client may receive a proxy object from these services that is downloaded to a JVM, and may also use UIs specific to services. Applications of Jini are varied and exist at every tier, including tier 3 where transaction processing may be carried out, and may be applied to devices ranging from servers to simple household appliances as simple as a light switch with a single input and no JVM of its own. Docking stations are also porrible where Jini clients share a JVM. Mobile Jini applications may include smart devices able to connect with remote Jini services in the home or in the office. The Jini community at jini.org is dedicated to the development of Jini, its famework, its applications, its services, and marketplaces.

Jini Docking Bay

A docking bay that has a JVM that may be shared by multiple docked devices.

Jini Multicast Response Service

A transparent connection that occurs between Jini devices, services, and communities and that relies upon simply communicating messages and protocols. It is the intention of this section to provide an explanation of the protocols and communications that occur between new Jini entities and the network lookup services.

Jini has protocols to:

- Discover lookup services on LANs that may be in your office or in your home, or even installed in a hotel room.
- Broadcast the presence of lookup services.
- Forge communications with a lookup service using WANs.

A new entity uses the multicast request protocol to find lookup services, and then uses the multicast announcement protocol to receive multicast lookup announcements. This communication requires that the requesting entity has a multicast request client, and a multicast response server that listens to the lookup services. Both may be run on a single JVM implementation. The lookup service has a listening multicast request server and a multicast response client that responds to requesting entities by providing them with a proxy that is used to converse with the lookup service.

The multicast request service uses the transport layer of the network protocol so that lookup services may push their information to a requesting host. The actual protocol used is a version of UDP (User Datagram Protocol), and a multicast discovery packet has a maximum 512byte payload, may contain architecture neutral parameters, and should be simple to decode. The packets form a contiguous stream like that from a `java.io.DataOutputStream` to a `java.io.ByteArrayOutputStream` object.

Jini Nested Transactions A series of multiple transactions using Jini.

At times a transaction requires the implementation of subtransactions for its successful completion. These nested transactions provide a logical hierarchy where a parent transaction is dependent on a nested transaction. Nested transactions are implemented using `TransactionManagers` that use the `NestableTransactionManager` interface:

```
package net.jini.core.transaction.server;

import net.jini.core.transaction.*;

import net.jini.core.lease.LeaseDeniedException;

import java.rmi.RemoteException;

/**

* The interface used for managers of the two-phase
  commit protocol for

* nestable transactions. All nestable transactions must
  have a

* transaction manager that runs this protocol.

*

* @see NestableServerTransaction

* @see TransactionParticipant

*/

public interface NestableTransactionManager extends
    TransactionManager {

  /**

  * Begin a nested transaction, with the specified
    transaction as parent.

  *

  * @param parentMgr the manager of the parent
    transaction
```

```
* @param parentID the id of the parent transaction
* @param lease the requested lease time for the
  transaction
*/
TransactionManager.Created
  create(NestableTransactionManager parentMgr,
    long parentID, long lease)
  throws UnknownTransactionException,
  CannotJoinException,
    LeaseDeniedException, RemoteException;
/**
* Promote the listed participants into the specified
  transaction.
* This method is for use by the manager of a
  subtransaction when the
* subtransaction commits. At this point, all
  participants of the
* subtransaction must become participants in the
  parent transaction.
* Prior to this point, the subtransaction's manager
  was a participant
* of the parent transaction, but after a successful
  promotion it need
* no longer be one (if it was not itself a
  participant of the
* subtransaction), and so it may specify itself as a
  participant to
* drop from the transaction. Otherwise,
  participants should not be
* dropped out of transactions. For each promoted
  participant, the
* participant's crash count is stored in the
  corresponding element of
* the <code>crashCounts</code> array.
*
* @param id the id of the parent transaction
* @param parts the participants being promoted to
  the parent
```

```
* @param crashCounts the crash counts of the
  participants
* @param drop the manager to drop out, if any
*
* @throws CrashCountException the crash count of
  some (at least one)
* participant is different from the crash count the
  manager already
* knows about for that participant
*
* @see TransactionManager#join
*/
void promote(long id, TransactionParticipant[] parts,
    long[] crashCounts, TransactionParticipant drop)
  throws UnknownTransactionException,
  CannotJoinException,
    CrashCountException, RemoteException;
}
```

When a nested transaction is created, the originating manager joins the parent transaction using the `join` method when the managers are different. The `create` method may raise the `UnknownTransactionException` if the parent transaction manger does not recognize the transaction because of an incorrect ID, or the ID has become inactive or has been discarded by the manager.

Jini Smart Device

A Jini-enabled device that usually has its own JVM and is able to consume Jini services.

Jini Transaction Creation

A process where a Jini transaction is created.

To create a transaction, it is necessary for the client to use a lookup service (or a similar directory services metaphor) in order to reference a `TransactionManager` object. A new transaction may be initiated using the `create()` method and by specifying a `leaseFor` period in milliseconds. The `leaseFor` duration is adequate for the transaction to complete, and the `TransactionManager` may forbid the lease request by throwing the

LeaseDeniedException. Expiration of a lease before a Transaction-Participant's vote with a commit or abort leads the TransactionManager to abort the transaction.

Constants provide the currency of the described communications between TransactionParticipants and TransactionManagers and are defined in the TransactionConstants interface:

```
package net.jini.core.transaction.server;

/** Constants common to transaction managers and
      participants. */
public interface TransactionConstants {
  /** Transaction is currently active */
  final int ACTIVE = 1;
  /** Transaction is determining if it may be committed
    */
  final int VOTING = 2;
  /** Transaction has been prepared but not yet
    committed */
  final int PREPARED = 3;
  /** Transaction has been prepared with nothing to
    commit */
  final int NOTCHANGED = 4;
  /** Transaction has been committed */
  final int COMMITTED = 5;
  /** Transaction has been aborted */
  final int ABORTED = 6;
}
```

Jini Transaction Join

A transaction join is a process where a participant joins a transaction.

To join a transaction, a participant invokes the join method held by the transaction manager using an object implementation of the Transaction-Participant interface. The TransactionManager uses the object to communicate with the TransactionParticipant. If the join method results in the RemoteException being thrown, the TransactionParticipant relays the same (or another exception) to its client. The join method has a crash count parameter that holds the TransactionParticipant's storage version that contains the transaction state. The crash count is changed following each loss or corruption of its storage. When a TransactionManager receives a join request, it reads the crash count to determine if the `TransactionParticipant` is already joined. If the crash count is the same as that passed in the original join, the join method is not executed. If it is different, however, the `TransactionManager` causes the transaction to abort by throwing the `CrashCountException`.

The two-phase commit protocol is implemented using the primary types:

1. `TransactionManager` that creates and coordinates transactions
2. `NestableTransactionManager` that accommodates nested transactions or subtransactions
3. `TransactionParticipant` that allows transactions to be joined by participants

The two-phase commit protocol coordinates the changes made to systems resources that result from transactions. It tests for their successful implementation, in which case they are committed. If not, and any one fails, they are each rolled back. In transaction processing (TP), this is left to a transaction coordinator whose function is integrated in the `TransactionManager` using Jini.

The `TransactionManager` is key to the two-phase commit protocol. This requires that all `TransactionParticipants` vote in order to indicate their state. The vote may be prepared (or ready to commit), not changed (or read only), or aborted (when it is necessary to abort the transaction). Having received information of the readiness to commit through a prepared vote, the `TransactionManager` signals `TransactionParticipants` to roll forward and commit the changes resulting from the transaction. `TransactionParticipants` that vote abortedare signaled to roll back by the `TransactionManager`.

Jini API

Jini is made of numerous Jini packages that combine to make the collective API that implements the many required interfaces. These include the methods and exceptions that make light work of developing Jini applications and include the shown `TransactionManager` interface.

```
package net.jini.core.transaction.server;

import net.jini.core.transaction.*;
import net.jini.core.lease.Lease;
import net.jini.core.lease.LeaseDeniedException;
import java.rmi.Remote;
import java.rmi.RemoteException;
```

```
public interface TransactionManager
extends Remote, TransactionConstants {
  /** Class that holds return values from create
    methods. */
  public static class Created implements
    java.io.Serializable {
    static final long serialVersionUID =
    -4233846033773471113L;
  public final long id;
  public final Lease lease;
  public Created(long id, Lease lease) {
    this.id = id; this.lease = lease;
  }
}
Created create(long lease) throws
  LeaseDeniedException, RemoteException;
void join(long id, TransactionParticipant part, long
  crashCount)
  throws UnknownTransactionException,
  CannotJoinException,
    CrashCountException, RemoteException;
int getState(long id) throws
  UnknownTransactionException, RemoteException;
void commit(long id)
  throws UnknownTransactionException,
  CannotCommitException,
    RemoteException;
void commit(long id, long waitFor)
  throws UnknownTransactionException,
  CannotCommitException,
    TimeoutExpiredException, RemoteException;
void abort(long id)
  throws UnknownTransactionException,
  CannotAbortException,
    RemoteException;
void abort(long id, long waitFor)
```

```
throws UnknownTransactionException,
CannotAbortException,
  TimeoutExpiredException, RemoteException;
}
```

Computers/Software

JavaSoft™

A company formed by Sun Microsystems in 1995 and maker of Java language products and technologies that include Jini.

JavaWorld

An on-line magazine dedicated to Java.

Jaz Drive

A removable storage device manufactured by Iomega. Jaz disks offer 1Gbyte data storage capacity.

JBuilder

A Java-based development software suite from Inprise.

JDBC (Java Database Connectivity)

A Java package that allows Java programs to interact with compliant databases, and also includes a bridge so as to provide backward compatibility with ODBC databases.

JDK (Java Development Kit)

Sun Microsystems's development tool for creating Java applets and applications. It is freely available from JavaSoft (www.javasoft.com), and includes:

- Appletviewer for viewing Java source code listings

- Jar for compressing packaging applications
- Java for executing applications
- Javadoc for documenting Java programs
- Javac for compiling Java programs.

Jetsend

A wireless connection technology for home appliances.

JIT (Just-in-Time) Compiler for Java

A compiler that converts OS-independent Java bytecode and optimizes it for execution on the target OS. The conversion naturally takes place on the client side.

JIT Debugging

A method of detecting bugs in a running program, and responding by running an appropriate debugging process.

Jobs, Steve

A co-founder of Apple Computer, who later founded NeXT. His most significant achievements were those at the early years of Apple Computer, an era when a clutch of American companies largely run and owned by college dropouts (including Bill Gates) revolutionized the computer industry by designing affordable microcomputers and accompanying software. Steve Jobs and Steve Wozniak revolutionized the world of computing by mass-producing one of the world's most affordable PCs, known simply as the Apple, and later the Apple II. It was designed by Steve Wozniak, whose dream was always to own a computer, once saying "I don't care if I live in the smallest house, just so long as I have my very own computer." This was a dream that he almost single-handedly made a reality for himself and for millions of people around the world.

Join

A process of combining records from different tables and/or files in a relational database management system (RDBMS).

Just-in-Time (JIT) Compiler

A Java runtime environment that is encapsulated by the Java Developer Kit and permits redistribution of the runtime environment. JRE consists of the JVM, core classes, and miscellaneous files.

JVM

A JVM (Java Virtual Machine) yields an environment for running Java applets. Browsers such as Netscape Navigator and Microsoft Internet Explorer feature Java Virtual Machines. JVMs may be run on clients such as desktop and notebook systems or even consumer appliances so that they connect with Jini networks and consume compliant services.

Miscellaneous

JPEG

JPEG (Joint Photographics Experts Group) is an internationally agreed standard for still-image compression and decompression that was devised by the JPEG—a specialist group set up by the ISO and CCITT. It is a symmetrical algorithm in that the processes required for compression mirror those of decompression. The processes include forward and reverse DCTs (Discrete Cosine Transformations). It may be used to compress 8bit, 16bit, and 24bit graphics. Motion-JPEG (M-JPEG) video uses individual frames compressed according to the JPEG algorithm, giving full-frame updates as opposed to the predominantly partial frame updates of standard MPEG video. The JPEG standard compression scheme for still photographic quality images began development in 1986. Compression and encoding techniques were evaluated during 1987 and 1988, until eventually the components of the symmetrical compression cycle were agreed, with DCT (Direct Cosine Transform) proving a central theme of the JPEG design.

Acronyms

J2ME	Java 2 Micro Edition
J2ME	MIDP J2ME Mobile Information Device Profile
JAE	Java Application Environment
JDBC	Java Database Connectivity

JIT	Just-in-Time
JPEG	Joint Photographics Experts Group
JTC	Joint technical committee
JVM	Java Virtual Machine

L

Layers

Cdma2000 Upper Layers

The upper layers shown in Fig. C-3 address voice, signaling, and end user data-bearing services. Voice services include voiced telephony, IP telephony, PSTN access, and mobile-to-mobile. End user data-bearing services deliver packet data such as Internet protocol/TCP services, circuit data services such as B-ISDN emulation services, and short message service. Packet data services may use IP, TCP, UDP, and ISO/OSI connectionless internetworking protocol (CLIP). Circuit data services include asynchronous dial-up access, fax, V.120 rate-adapted ISDN, and B-ISDN. Signaling services control the operation of the mobile.

Lower Layers (LAC and MAC)—CDMA

The link layer provides features that drive the QoS for upper-layer services, providing functions used to map the data transport to the upper layers. The Link layer includes the Link Access Control (LAC) and Media Access Control (MAC) sublayers. The LAC sublayer addresses point-to-point communication for upper-layer entities, and has a framework supporting end-to-end link layer protocols.

The MAC sublayer holds multiple instances of an advanced-state machine, and delivers 3G media-rich services with QoS management per service, and provides three basic functions:

- Media access control state: Controls access of services to the physical layer, and also handles contention between users and services.
- Best-effort delivery: Provides delivery guarantee over the radio link with protocol (RLP).
- Multiplexing and QoS control: Enforces QoS by prioritizing access requests, and by mediating conflicting requests from services.

Physical Layer—CDMA

The physical layer is a source of coding and modulation services for logical channels that are used by the PLDCF MUX and QoS layer. Physical channels may be F/R-DPHCH (forward/reverse dedicated physical channels) where they carry information in a point-to-point model between the base station and a mobile user. They can also be F/R-CPHCH (forward/reverse common physical channels, or a number of physical channels carrying information in a point-to-multipoint model between the base station and multiple mobile stations.

Operations of a UMTS Transmitter at the Physical Layer

The transport channel data from layer 2 and above are arranged in blocks depending on the type of data. The blocks are cyclically redundancy coded (CRC) for error detection at the receiver.

The data is segmented into blocks and channel coding ensues. The coding may be convolutional or turbo. Sometimes channel coding is not used.

Data is interleaved to decrease the memory of the radio channel and thereby render the channel more Gaussian-like. The interleaved data are then segmented into frames compatible with the requirements of the UTRA interface.

Rate matching is performed next. This uses code-puncturing and call data repetition, where appropriate, so that after transport channel multiplexing the data rate is matched to the channel rate of the dedicated physical channels.

A second stage of bit FI interleaving is executed, and the data are then mapped to the radio interface frame structure.

Suffice to say at this point there are different types of physical channels, namely:

- Pilot channels that provide a demodulation reference for other channels
- Synchronization channels that provide synchronization to all UEs within a cell
- Common channels that carry 0.2 information to and from any user equipment (UE)
- Dedicated channels that carry information to and from specific UEs

The physical layer procedures include:

- Cell search for the initial synchronization of a UE with a nearby cell
- Cell reselection, which involves a UE changing cells
- Access procedure that allows a UE to initially access a cell
- Power control to ensure that a UE and a BS transmit at optimum power levels

- Handover—the mechanism that switches a serving cell to another cell during a call

3G Layers

A 3G network consists of layers dedicated to:

- Transport: Carries data (bits) over the IP backbone and wireless access network that may be ATM, SONET, or an alternative.
- Control: Controls calls, authenticates calls, manages mobility, manages sessions, and is accommodated in the network nodes that include RNC/BSC, MSC, SGSN, and GGSN.
- Applications/services: Hosts applications and services and is otherwise known as the *service network*.

  See Fig. L-1.

Link Types (Bluetooth)

The delivery of different traffic like voice and bursty traffic is accommodated by two link types:

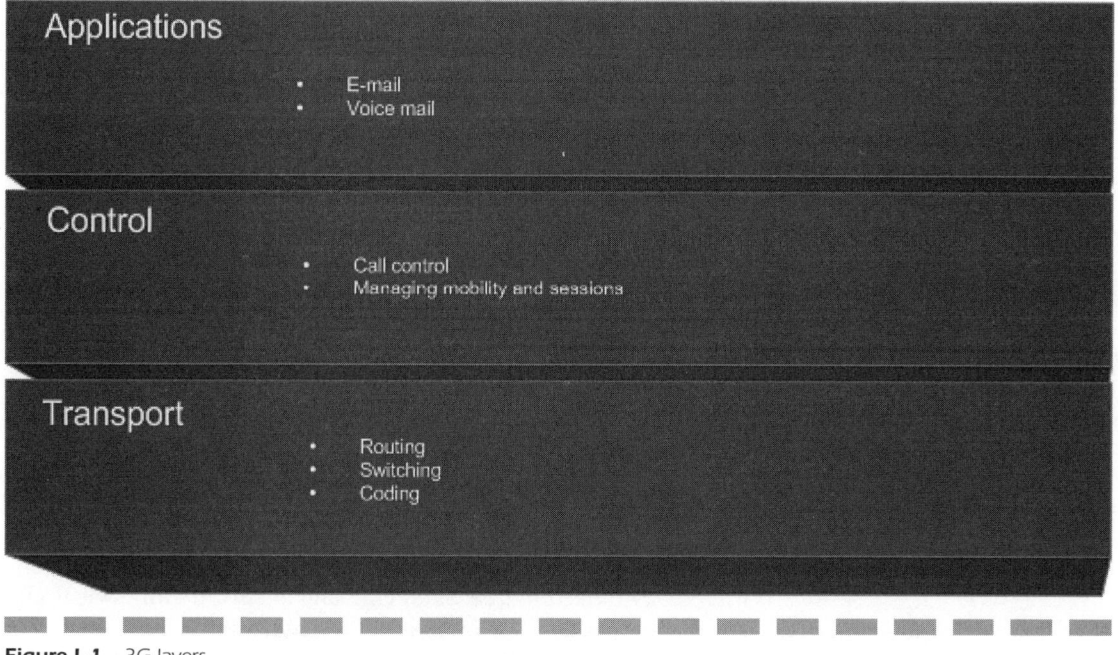

Figure L-1 3G layers.

TABLE L-1

Bluetooth Bit Rates

Type	Symmetric (kbps)	Asymmetric Downlink (kbps)	Asymmetric Uplink (kbps)
DM1	108.8	108.8	108.8
DH1	172.8	172.8	172.8
DM3	258.1	387.2	54.4
DH3	390.4	585.6	86.4
DM5	286.7	477.8	36.3
DH5	433.9	721	57.6

- SCO (Synchronous Connection-Oriented) is applicable using CS services with short latencies and high levels of QoS. This offers symmetric channels that are synchronous, and is used mainly for voice, though data is possible over the 64kbps channel.
- ACL (Asynchronous Connection-Less) is used for data transfer and other asynchronous services, and offers PS channels. Transmission slots are not reserved but are granted using a polling access scheme.

A connection may have several links of ACL and/or SCO, and when using a piconet a master may have SCO and ACL links with several slaves, but there is a three-voice call limit.

Bluetooth analog voice may be digitally encoded using PCM (Pulse Code Modulation) or CSVD (Continuously Variable Slope Delta). PCM uses eight-bit samples at 8kHz, and so requires 64kbps in order to transmit the digitally encoded signal. CSVD uses 1-bit samples that indicate whether the slope of the voice signal is increasing or decreasing. Bluetooth voice calls are ISDN quality using 64kbps, and this is better than many 2G handsets that offer 8kbps.

Bluetooth packets can be sent one slot at a time, or using multislot where a large packet is sent over several slots and the frequency remains the same. Using multislot bit, rates may reach 721kbps but the packet has no protective coding.

See Table L-1.

Leasing

Leasing is an important part of load balancing and ensuring that devices and resources are made available for as long as possible, and as frequently as possible. Leasing is used in the Jini architecture that is used in the following sections to explain the concept.

Leasing Jini

Leasing is the time-sharing policy that Jini uses to allow resources, entities, services, and entries to interact and coexist in an economical and logical way. Surrounding it are simple rules that to most people are rather obvious, but we'll go through them all the same:

- A request for a lease period may or may not be granted, and a shorter lease or no lease at all may be offered if the resource to be leased is unavailable.
- A lease may be terminated by its holding entity, leaving the grantor to restore any lease-related conditions to their preleasing state.
- A lease may be renewed by the holder, or extended, or reduced, provided the grantor obliges.
- A lease may expire without any request for renewal from the holder, and leaves the grantor to return resources associated with the lease to the prelease states.

The `Lease` interface defines a lease object type that may be granted and, among other things, constants that include FOREVER which is an unconditional lease and requires the holder to free the lease resource when no longer required. Alternatively, the constant ANY allows the grantor to specify an appropriate length of lease. A serialized format is also possible, and includes DURATION and ABSOLUTE, where the former converts the time of the lease's expiry into a duration that is expressed in milliseconds. It follows then that this is an appropriate policy when writing a leased object from one address space to the next, such as when you use an RMI call, because obviously there may be time differences. Of course none of this is relevant with ABSOLUTE since this is used simply to specify a lease value in milliseconds.

The Lease Interface The `Lease` interface defines methods that relate to lease objects, and the list below endeavors to explain their usage, and when and where they may be used.

- `GetExpiration:` A method that returns a long value, holding the time that a specified lease will expire. The current time, of course, is obtainable using the java.lang.System.currentTimeMillis.
- `Cancel:` A method used by the holder to cancel a lease and to detach itself from the lease's required resources. It may throw the exception `UnknownLeaseException` when the grantor does not know of the lease.

A RemoteException is thrown if the method's implementation calls the remote lease-holding object.

- Renew: A method that naturally renews a lease, by accepting a period defined in milliseconds, and discarding what remains of the existing lease. A grantor may throw the exception RenewFailedException if it is incapable of renewing the lease, and the existing lease remains. A RemoteException is thrown if the method's implementation calls the remote lease-holding object.
- SetSerialFormat: A method that may be used to declare a lease in terms of duration in milliseconds, or to declare a lease expiration time that is relative to the clock.
- SetSerialForm: A method that returns an integer that indicates the lease format as described above in terms of the setSerialFormat.
- CreateLeaseMap: A method that creates a map object-holding leases with batched renewals and cancellations.
- BatchWith(Lease, lease): A method that shows whether a lease given as an argument may be batched, and it uses a boolean to do so.

The LeaseMap interface lets you renew or cancel groups of lease objects as a batch, rather than use multiple methods or operations, and is an extension of the java.util.Map class, and Longs are used to hold the durations of lease objects.

The LeaseMap Interface

- canContainKey: A method that returns a boolean indicating whether or not a lease object may be added to a map.
- renewal: A method that renews all leases of objects in the map.
- cancelAll: A method that cancels all leases in the map.

Local Multipoint Distribution Service (LMDS)

LMDS is wireless broadband access technology using a 150 or 1150MHz bandwidth in the 28 to 31GHz range. It is among the "last-mile" access technologies that include DSLs, TI, T3, and Frame Relay. Because LMDS uses very high frequencies and short wavelengths, resulting signals cannot negotiate physical obstacles as FM signals can, so it requires a line of sight between the transmitter and receiver using fixed antennas. LMDS may carry voice, data, and video, over a wireless local loop operating in the 28- to 31GHz band. Beginning in 1998 the U.S. Federal Communications Commission (FCC) auctioned

LMDS licenses for local regions or *basic trading areas* (ETAs), and the highest bids were more than $45 million. LMDS licenses are separated into *A block* and *B block,* which are required by each ETA, and the first LMDS licenses were issued for a period of 10 years. See Fig. L-2.

Architecture

LMDS requires a central antenna to serve a cell (see Fig. L-3) that is as small as 15km owing to the high frequency and short wavelength. Modulation and climatic conditions also dictate cell sizes: Efficient modulation reduces the cell size, and rainfall does the same. For wireless cable operators, the base station or LMDS node may connect with an earth station receiving satellite TV feeds. The nodes are controlled by a network operations center (NOC). An LMDS operator may use TDMA, FDMA, and CDMA.

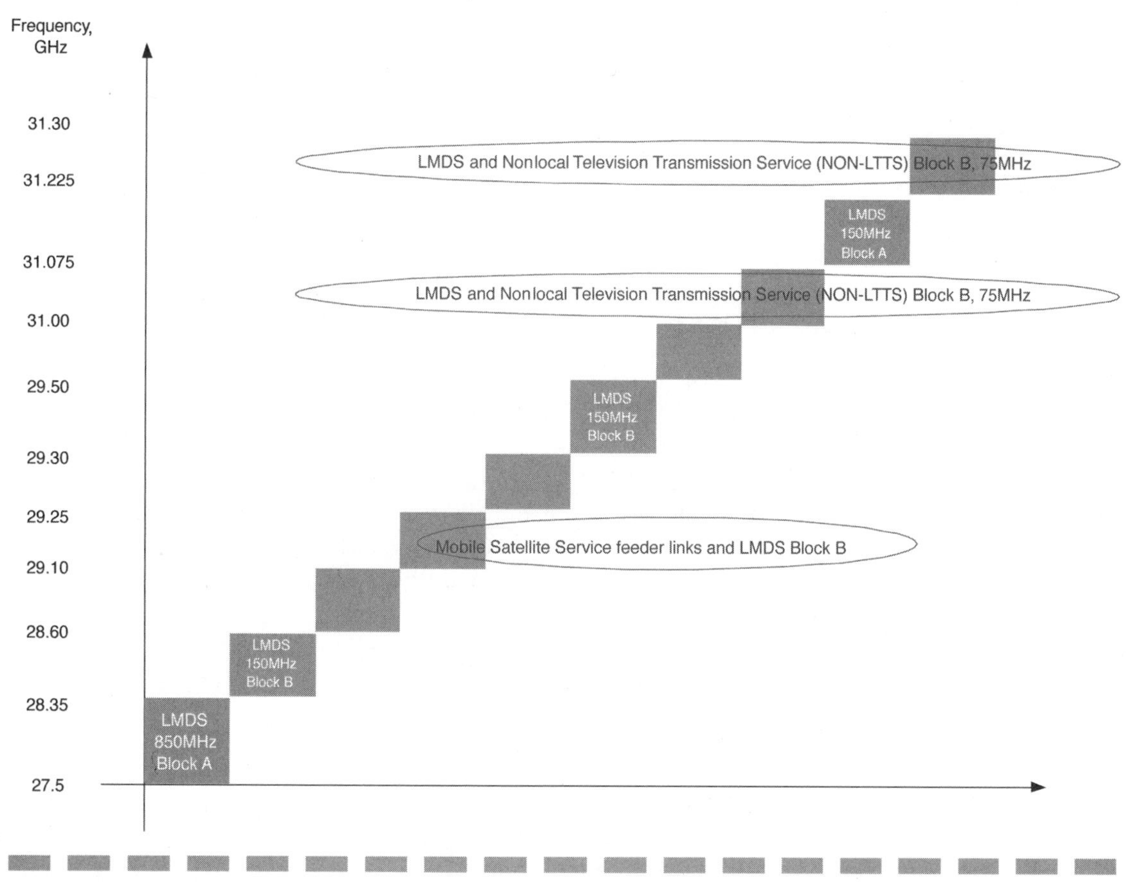

Figure L-2 LMDS frequency allocation.

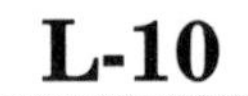

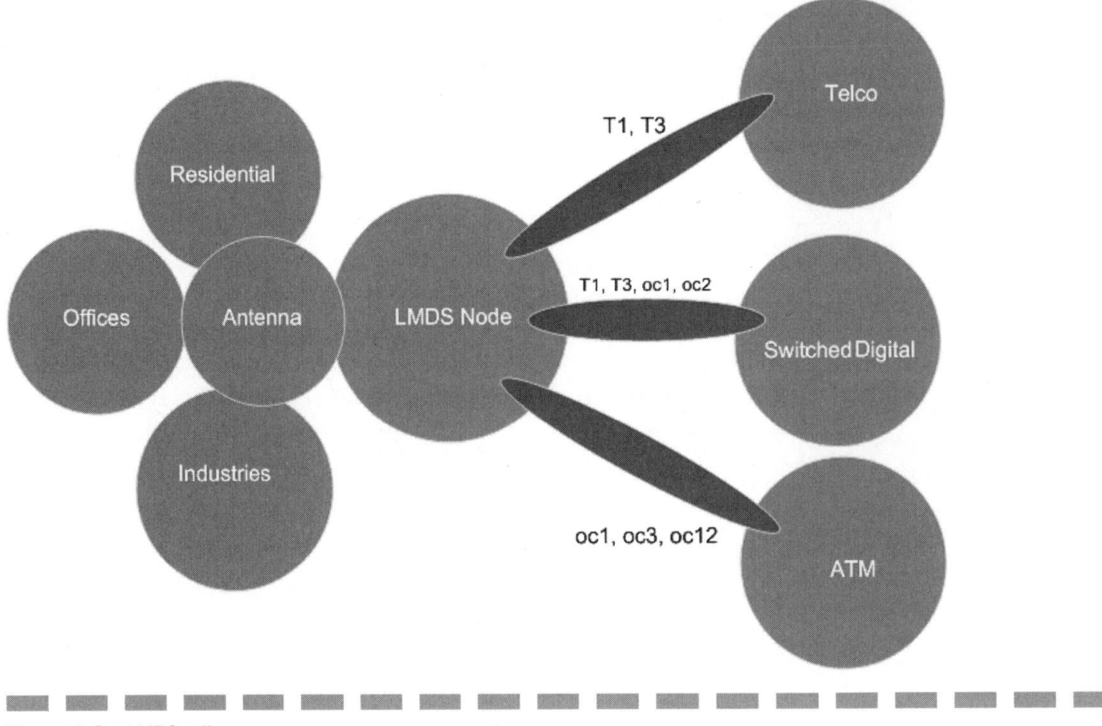

Figure L-3 LMDS cell.

An LMDS base station or node interfaces with the network and outdoor microwave antenna. Typically a typical transmit and receive sector will provide a 15-, 30-, 45-, or 90-degree beam. The subscriber or LMDS customer requires a network interface unit (NIU) that acts as a gateway. The NIU has an RF modem that supports quadrature amplitude modulation (QAM) or phase-shift keying (PSK) modulation, and can support FDMA, TDMA, or CDMA. It may also include connectivity to multiple network protocols such as TI, El, 10BASE-T, 100BASE-T, Gigabit Ethernet, and ATM. A subscriber's NIU may also use an all-TDMA or an all-FDMA for uplink and downlink connections.

Modulation

RF modulation methods include amplitude modulation (AM) and phase-shift keying (PSK), binary phase-shift keying (BPSK), quaternary phase-shift keying (QPSK), differential QPSK (DQPSK), and octal phase-shift keying (8PSK). Table L-2 shows some FDMA access modulation methods.

TABLE L-2

FDMA Access Modulation Methods

Modulation Method	MHz/Mbps CBR Connection
BPSK (binary phase-shift keying)	1.4MHz
QPSK (quaternary phase-shift keying)	0.7MHz
DQPSK (differential QPSK)	0.7MHz
8PSK (octal phase-shift keying)	0.4MHz
4-QAM (quadrature amplitude modulation, 4 states)	0.7MHz
16-QAM (quadrature amplitude modulation, 16 states)	0.3MHz

Location Services

Location services (LCS) are applications based on the mobile station's location, time stamp, and location error estimation. Early applications of LCS include the U.S. services to locate users making emergency calls. See Fig. L-4.

An LCS client requests the location information of a target UE, and may be internal or external to the PLMN, or in the target UE. A gateway mobile location center (GMLC) processes these requests, and interrogates the LCS client and the UE, for authorization and for LCS support, respectively. Approved requests are routed to the appropriate radio network subsystem (RNS) that usually implements LCS procedures. Three methods for locating a UE include:

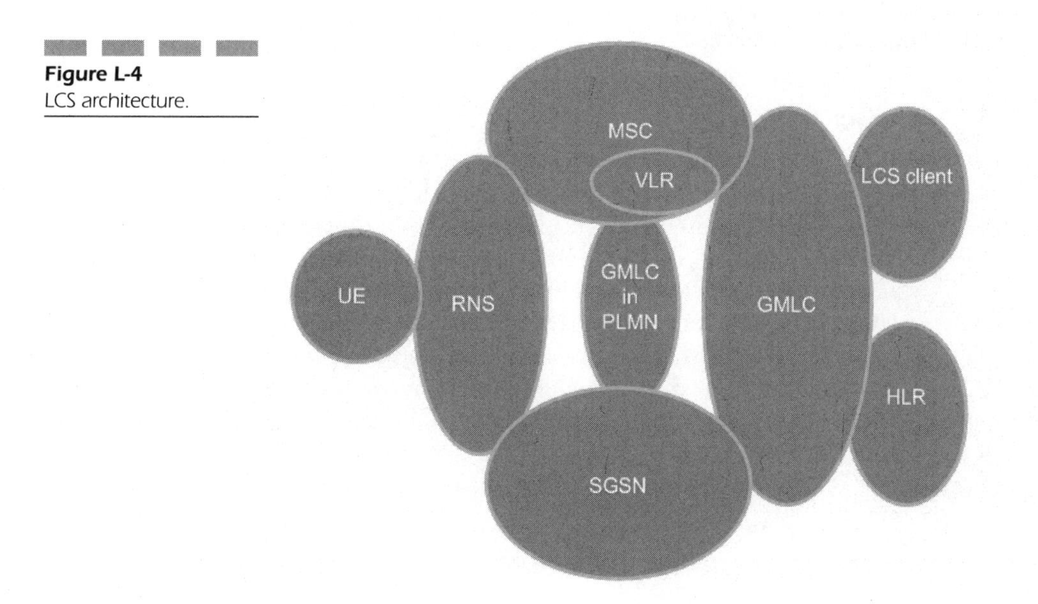

Figure L-4
LCS architecture.

- Cell-coverage-based method
- Observed time difference of arrival (OTDOA)
- Network-assisted GPS

Cell-Coverage-Based Method

The last cell where the UE performed a location-update procedure is acquired using normal signaling. The accuracy threshold spans the entire cell area ranging from hundreds of meters to kilometers. Using this information, operators may implement location-sensitive billing.

Observed Time Difference of Arrival

The UE's measurements using Node's pilot signals provide the observed time difference of arrival (OTDOA). The pilot signals contain scrambling code that identifies cells. The UE's measurements are transmitted to a position calculation function in the serving RNC or in the UE. Determining the UE's location also requires the transmitter locations and the relative time differences (RTDs) of the transmissions that are calculated using:

$$R_{12} = x_2 - x_1$$

x_1 is the pilot signal delay from Node B1.

x_2 is the pilot signal delay from Node B2.

R_{12} provides a hyperbola, and the UE performs many OTDOA measurements for improved accuracy.

GPS (Global Positioning System)

To determine the coordinates of a device and/or user, GPS uses the propagation delays from satellite transmissions. The Interagency GPS Executive Board (IGEB) governs GPS and was set up up by President Clinton in 1996. GPS application by the U.S. military led to the use of encryption, though this reduced the accuracy of the GPS receiver to 100m—but in 2000 President Clinton stopped this jamming/jittering, or selective interference (SA). The inaccuracy caused by SA was remedied by differential GPS (DGPS), but is now largely redundant since 2000, although DGPS beacons are located in coastal areas for maritime use.

DGPS operates as follows:

- GPS receivers are planted in known positions.
- They will receive the GPS signals from satellites and make location calculations.

- The receivers calculate the needed corrections to the received signals.
- These correction values are broadcast repeatedly (or several times per minute) to other GPS receivers in the area.

Other global navigation satellite systems (GNSS) include the Russian GLONASS that is available for civilian use. And the European Union Galileo system aims to reduce the European dependence on U.S. and/or Russian systems.

GPS—Network-Assisted Network-assisted GPS may use the:

- UE-based method when the UE has a full GPS receiver
- UE-assisted method when the UE has a lite GPS receiver

The UTRAN gives the UE an estimate of the GPS signal's time of arrival, and it gives it GPS parameters including the visible satellite list. This information may also be obtained from the GPS though this is slower.

GPS Techniques

- The angle of arrival (AOA) method is used by the BS to estimate the direction of the UE. The RNC uses this information to estimate the location of the UE.
- The observed time of arrival (OTOA) method bases itself on propagation delay measurements made on the time of arrival of signals.
- Using the reference-node-based positioning (OTDOA-RNBP), the network operator can plant equipment in a cell that is difficult for LCS (location services). The network uses signals from the equipment as a reference.
- Using OTDOA positioning elements (OTDOA-PE), elements are placed in known network locations, and they transmit secondary synchronization codes (SSC). The UE then measures the time difference between the arrivals of these signals to estimate the UE's location.

LCS Types LCSes may be commercial or "value added," internal, emergency, or lawful-intercept.

Commercial services are used by service providers that provide applications based on UE location such as Navigation.

Internal LCS enhances the UMTS network perhaps for location-assisted HO, for example.

Emergency LCS locates users whose locations are relayed by the operator to the emergency service. The accuracy according to the FCC is 125m.

Lookup Services

Lookup services are part of the discovery of available services that a device may use or even provide if it is a server. Lookup services are included in Microsoft UPnP and in SunSoft Jini.

Lookup—Jini

In an active Jini environment there are lookup services that hold lists of services, proxy objects that provide communication between clients and services, and attributes that determine behaviors of many kinds, including the presentation of a service in terms of its UI, for example, or even a printer type. And if the service is implemented as a remote object, it may also include an RMI stub, and, collectively, this and all the other related entities that are stored in a lookup service are termed the *service item.* The service items may be located by applications and other services, using their instances in the lookup service, and matches may be based on data types. The lookup service storage model or architecture is flat, but there is scope for the implementation of logical hierarchies based on services items and attributes they hold, and these may then be discovered and browsed with efficiency.

The attributes themselves are in sets that are instances of classes of the Java platform, and service items may hold multiple instances of classes. As an example, you may find that an item may have multiple instances of the `Name` class where it may characterize a service using different descriptions or names based on a context of invocation that reflects the user's requirements. The lookup service environment is part of the overall event-driven architecture, of course, and an administrator is a recipient of its events, indicating the registration of new services, for example. It is the same administrator that may use an applet to add new attributes to a service, which duly responds to the requests by writing changes to its *service item* in the lookup service.

ServiceRegistrar An instance of the `ServiceRegistrar` is constructed by each lookup service and is downloaded to the client as a proxy object, and the methods are predominantly those of RMI. The `ServiceRegistrar` provides a number of methods, none more important than those used to register service items, receive event notifications, and find service items. The list below has types that are in the `net.jini.core.lookup` package and briefly runs through their application in the lookup service architecture.

- `ServiceID`: An instance of this class may hold a UUID universally unique identifier for each service. A service ID is a 128-bit value and is meant to be created by lookup services rather than by clients.
- `ServiceItem`: An instance of this class is used to store service items.
- `ServiceTemplate`: Instances of this class provide a mechanism for matching service items.

- `ServiceMatches`: A class that provides an output parameter when looking up service items.
- `ServiceRegistration`: An instance of the ServiceRegistrar is constructed by each lookup service and is downloaded to the client as a proxy object, and the methods are predominantly those of RMI. It uses self-explanatory methods like getServiceID, getLease, addAttributes, and modifyAttributes.

Computers/Software

LCD (Liquid Crystal Display)

A form of display measuring a few millimeters in depth. Available in monochrome and color, LCDs are used in a wide variety of appliances including pocket televisions, notebook computers, PDAs, cellular phones, calculators, and portable CD-ROM readers/electronic books. Modern notebooks tend to use TFT and DSTN display technology. The former provides improved image definition.

Lease

A time period assigned to an object, data element, or service that dictates longevity. Services that register on a network may be assigned a lease, while objects in a JavaSpaces service may also be assigned leases placing time limits on their availability.

Legacy

A system, application, or operating system (OS) that is of a past generation, and is usually based on mainframe technology. It may nonetheless be integrated into a modern IT implementation, and coexist with modern client/server architectures.

Level

1. A defined RAID architecture.
2. High- and low-levels languages describe macro and micro features, respectively. In terms of programming languages, low-level languages relate most closely to machine code, such as assembly languages, for instance. High-level languages are those that are a considerable

distance from the machine language (in terms of compilation processes), and include C++ and Visual Basic.

3. A U.S.-defined series of security grades.

LIFO (Last In–First Out)

A type of queue. The order in which items are regurgitated opposes that in which they are deposited. The last item placed in the LIFO queue is the first to be retrieved.

It may be used to store the return addresses, when a subroutine is called. In this guise it is called a stack.

LIM (Lotus-Intel-Microsoft) Memory

An alternative name for EMS (Expanded Memory Specification), this is a method of addressing memory in a PC architecture. Introduced by Lotus, Intel, and Microsoft (LIM) in 1984, it was used in Windows 1.x to cache DOS applications. The specification has evolved and numerous different versions are available. In the early days, many PCs were fitted with EMS-compliant memory cards. However, growing extended memory (XMS) on motherboards and their EMS compliance drove EMS memory cards into obsolescence. Expanded memory is accessed by reading 16K pages from EMS into the memory area between 640K and 1Mb RAM. A device driver such as EMM386 responds to EMS requests.

Lingo

An OOP-based multimedia authoring language that was developed by MacroMedia, and is a most intuitive language even to nonprogrammers. It may be used in conjunction with Macromedia Director and Macromedia Authorware Professional.

Lingo's functionality, syntax, and structure are comparable to those of OpenScript that is a proprietary language included with certain versions of Asymetrix ToolBook. It has become the industry's chosen language for authoring multimedia titles.

Using Lingo, Director movies may be interwoven with interactivity, by coding handlers that respond to events. Messages that result from such events may be defined in the program code.

Lingo Events Director is driven by four key event categories that are associated with:

- Frame
- Keyboard
- Mouse

Lingo Program Form

- Movie

```
on eventOfSomeSort
  go to frame 15
end
```

This simple script operates as follows: If the specified event occurs, the play head is moved to frame 15.

Lingo Messages Events invoke the messages:

```
mouseDown
mouseUp
- - are sent when the mouse button is either pressed
  or released
keyDown
keyUp
- - are sent when is either pressed or released
enterFrame
exitFrame
- - are sent when the playback head passes over
  frames
startMovie
stopMovie
- - are sent when the movie either starts or stops
idle
- - is sent during dormant states
timeOut
- - is sent after a specified period following a
  previous action
```

Lingo If...Then Form Additional messages may be defined.

```
on keyDown
  if the key = ESCAPE then
```

```
    alert "Cue previous video clip"
    beep
  else
    alert "Escape?"
  end if
end keyDown
```

Link

1. A process by which the object (.OBJ) files are linked with libraries that include functions, procedures, and classes.
2. A means of connecting related information in a hypertext model for information storage and retrieval. Hence user-interaction may be given context.

   They provide the user with a means of touring nonlinear paths through information. Such links of association are taken to limits that are imposed by the design's levels of granularity.

   Nodes, representing text or images, are linked to provide a potentially infinite number of meaningful paths. For example, a single node representing a linear structure, such as an article, might be linked to numerous other articles and images.

   Links can naturally exist at a number of levels, either to link complete documents (macro features), or to reference words or phrases (microfeatures) within documents, or complete documents or Web sites that are identified by their URLs.
3. A means by which tables/files may be connected in a relational database management system (RBMS).

   In the case of a multitable form within a relational database, such links can refer to how master tables and detail tables are associated as follows:

   - One-to-one: Each master record is linked to only one detail record at any given time.
   - One-to-many: Each master record is linked to a group of detail records.
   - Many-to-one: A number of master records may be linked to a single detail record.
   - Many-to-many: Each master record is one of a group that may be matched to one of a group of detail records.

   Such RDBMSes make possible multiple table changes and updates using a single form as a data entry interface.

4. A communications path between devices that may be processors as is the case in MPP designs.

List Box

A windows component that may be assumed to provide a means of selecting files. Where the number of files exceeds a certain figure, a vertical scroll bar is provided in order to assist in the process of their selection.

Liv Zempel

A data compression algorithm.

Load Balancing

1. A method of distributing the workload across processes and system resources, in an effort to optimize performance. It is usually applied dynamically in OO distributed systems.
2. A method of distributing the workload in an MPP architecture, so that processors are as close to the heightened states of operation as is possible. It is carried out dynamically, and may be referred to as dynamic load balancing.

Local Bus

A method of connecting video cards, hard disk controllers, and other devices more directly to the processor's data bus, thus overcoming the data transfer bottleneck of ISA.

Theoretically, local bus technology should permit the accommodation of expansion cards running generally at clock speeds equal to that of the processor's external data bus. Local bus standards have emerged including VL-bus and Intel's PCI (Peripheral Component Interconnect). PCI generally performs better than VL-bus.

Local Class

A class that is local to a given block of Java code.

Local Glue

A collection of entities that unite client components, so that they may operate collaboratively. OLE, OpenDoc, ActiveX, and JavaBeans components require local glues so that their running operations may be coordinated.

These common OO component architectures use different local glues:

- OLE uses ODL (Object Definition Language).
- ActiveX uses COM.
- OpenDoc uses CORBA IDL (Interface Definition Language).
- JavaBeans uses a subset of the Java programming language.

Logical Client/Server Model

A process of connecting to, and disconnecting from, a computer, network, remote server, Internet service provider, or Internet service. A login name is required, such as a password.

Log In and Log Off

A model that sees the interaction of components and programs where messages are typically used to request services and data. Software components may act as:

- Servers, providing client components with data
- Clients that request data from servers
- Server and clients

Typically it is a distributed OO software architecture platformed on a physical client/server system.

Lomem

Lowest user memory address in system.

Look-and-Feel

A term which broadly describes the user interface, or presentation element, of an application.

Lookup Service

A directory that stores links or objects relating to available or registered network services.

Loss

1. A level of attenuation that a signal is subjected to while passing through media that may be physical or wireless. Optic fiber signal losses are caused by impurities in the silica core, and by fiber couplings.
2. A measure of the number of lost telephone calls or connections due to congestion.

Lossless Compression

A compression technique that does not rely on the omission of pixel information from a video or image file. Authentic lossless compression should result in video or image quality that is equal to that provided by the uncompressed files. However, it may be assumed that attainable compression ratios are lower than those of lossy compression algorithms.

Lotus Notes

A groupware implementation, and remembered as the first commercially successful variant. An evolving solution, it provides network services such as email and document publishing, and provides easy migration of resulting Notes applications to the Web.

Low-Level Language

A programming language that provides access to the low-level elements of a computer such as memory locations and processor registers. Assembly language is considered to be a low-level language. Assembly languages are indigenous to the processor type. The language consists of mnemonics that replace, and translate into, hexadecimal processor instructions.

Miscellaneous

L Band

A section of the electromagnetic spectrum that ranges from 1.53 to 1.66GHz, and is applied in satellite and microwave communications.

LAN (Local Area Network)

A number of computers connected physically or wirelessly so that users may share directories, applications, services, and resources such as printers and fax/modems. LANs are commonly used to connect all computers in a department, while Wide Area Networks (WANs) may be used to connect multiple sites.

LAN data transfer rates over approximate speeds between 10 and 100Mbps though speeds up to 1000Mbps are possible.

There are a number of internationally agreed standards by which computers forming a LAN are connected. Network topologies include

- Star, where computers are connected using a centralized hub
- Ring, where computers are connected in a chain, and is the chosen method for small networks that are perhaps peer-to-peer configurations that might be based on Windows 98/NT

LAN standards include Ethernet, Token Ring, and, occasionally, Fiber Channel (FC). The latter may be used to implement FC arbitrated loops.

LED

An LED (light-emitting diode) is a semiconductor/optronic device that emits visible light when excited electrically. It may provide a basis for:

- Display technology
- Laser light sources for fiber optic lightwave communications

They are used in all types of consumer electronics and computers, such as power indicators and alphanumeric displays, to clocks and vu meters. Advantages over conventional filament bulbs include near-infinite life span, incredible durability, reliability; they are physically robust, easy to manufacture in different colors, have low power consumption, and are inexpensive. LEDs emit optical radiation, and the 800nm (wavelength) emission is seen as red.

Dividing the speed of light c (3×10^8m/s) by the wavelength (λ) of an (800nm) LED emission results in its frequency f:

$$f = c/\lambda$$

$$f = (3 \times 10^8)/(8.00 \times 10^{-7})$$

$$f = 3.75 \times 10^{14}\text{Hz}$$

The emission from an LED is commonly defined in terms of the external stimulation of electrons. Electrons exist in bands that surround a nucleus. The outer or valence band may share its electrons with other atoms that collectively form molecules. The valence band's electrons are stimulated to a higher energy state called the conduction band. This condition occurs when electrons are:

- Passed through a pn junction diode
- Stimulated by a high voltage
- Stimulated by light

The difference that exists between the valence and the conduction bands is described as the forbidden gap. The type of semiconductor determines the behavior of the bandgap, which may be:

- *Indirect*, such as silicon, where electrons may occupy intermediate levels as they pass from the conduction band to the valence band
- *Direct*, where electrons move directly through the forbidden gap, and provide the best results

Photon emissions take place as electrons are displaced from the conduction band, and united with holes in the valence band. The resulting emission of light from an LED might be referred to as *pn junction electroluminescence* or *recombination radiation*. The latter refers to the combination of electrons with holes. The resulting wavelength λ (in micrometers) is calculated thus:

$$\lambda = hc/E$$

where h is Planck's constant (6.63×10^{-34}J-s), c is the speed of light ($3 \times 10^{14}\mu$m/s), and E is the energy difference between the valence and conduction bands.

Line Speed

A data transmission rate over media that may be physical or wireless. The unit measurement is typically in kbps or Mbps.

Linux

The Linux operating system is based on Unix, and includes Unix commands and concepts. The Linux repository is the CVS (Concurrent Versioning System) that permits multiple users to work on the same files in a collaborative team environment. Files may be checked out and returned to the repository where their contents may be merged with other versions of the same file. For Windows systems access to the CVS is provided by WinCVS, or, alternatively, a Telnet and ftp session may be established between the Windows system and the system hosting the CVS.

Consider a scenario where you want to write files from a Windows system to the CVS:

Establish a Telnet session with the network system (such as lin3) holding the CVS. This requires Telnet software on the Windows machine, where the command used would logically follow the pattern:

```
Telnet lin3
```

Respond the prompts:

```
username:
password:
```

Open a DOS window and change to the directory holding the file(s) you wish to write to the CVS. To do this use the standard cd and cd/ commands.

Establish an ftp session with the network system (such as lin3) holding the CVS. Again this requires software on the Windows system. Typically you would type in a DOS window:

```
ftp lin7
```

In the DOS window change to the directory on lin3 where the files are to be copied.

By default Linux is now primed for an ASCII transfer, but if the files are to be read using an intranet, and are perhaps PDFs, they will be corrupt, and the intranet will not work for local and for remote workers. A binary transfer is therefore necessary, and this is achieved simply by typing "binary."

Copy the file(s) by typing:

```
put
```

Or

```
mput outerwall.*
```

Type:

```
cvs commit
```

Finally, in the Telnet window, type:

```
cvs update
```

Acronyms

L2CAP	Logical link control and adaptation protocol
LA	Linear amplifier
LAC	Link access control
LAN	Local area network
LAPD	Link access procedure on the D-channel
LCP	Link control protocol
LD-CELP	Low-delay codebook excited prediction
LIU	Line interface unit
LLA	Logical layer architecture
LLC	Logical link control
LLC	Link layer control
LMDS	Local multipoint distribution system
LMP	Link manager protocol
LOS	Line of sight
LP	Linear predictor
LPAS	Linear-prediction-based analysis-by-synthesis
LSPD	Low-speed data packet
LT	Long-term

M

Macromedia Flash Player 4 for Pocket PC

See Macromedia Flash Player 4 for Pocket PC under letter G.

Microwave Radio

Microwave radio frequencies are greater than 1000MHz (1GHz) with a wavelength measuring a few centimeters. Microwave radio is suitable for high-capacity, point-to-point transmission. Microwave radio may be used by corporations to toll bypass the large network providers.

Operation is a line-of-sight for 40–50km, depending on the radio frequency and the radio propagation conditions. Distances greater than 40–50km are achieved over multilink paths that use radio repeater stations arranged in sawtooth topology. Each transmission path through the repeater stations requires dish antennas facing previous and next sequential station. The atmosphere causes refraction of the radio waves as they adopt curved paths, which influence antenna alignment.

Microwave systems are susceptible to atmospheric interference and attenuation caused by weather conditions. Reflections off buildings and structures cause more serious multipath interference, but this may be remedied by network planning. A microwave oscillator generates a carrier signal that is modulated using a mixer, and is amplified using a traveling wave tube (TWT) or a very high-gain amplifier. The signal is usually emitted using a parabolic reflector antenna.

Tropospheric Scatter

Tropospheric scatter permits over-the-horizon communication using radio waves that are reflected by the earth's atmosphere. Such radio-wave reflection, or *scatter,* is caused by the tropospheric part of the atmosphere. The scatter angle is the angle of refraction (or the *reflection angle*) and permits transmission path curvature.

Frequency bands allocated by ITU-R for point-to-point and point-to-multipoint are as follows:

- 1GHz worldwide cellular mobile radio
- 2GHz worldwide cellular mobile radio DECT some point-to-point microwave
- 3GHz worldwide
- 3.5GHz rural radio
- 4GHz worldwide high-capacity point-to-point (e.g., STM-I) satellite communications
- 6GHz worldwide satellite communications
- 7–8GHz worldwide long-haul point-to-point microwave (up to 50km)
- 10GHz US: cellular base station backhaul
- 11GHz worldwide medium capacity (E3, T3)
- 12, 14GHz satellite communication direct broadcast by satellite

- 13, 15GHz worldwide short- to medium-haul point-to-point (25km)
- 18, 23, 26GHz worldwide short-haul point-to-point (10–15km)
- 28GHz worldwide point-to-multipoint, multimedia applications
- 38GHz worldwide short-haul (5–7km)
- 40GHz worldwide extremely short-haul (1–3km)

Multichannel Multipoint Distribution System

A multichannel multipoint distribution service, or *microwave multipoint distribution service,* operates in the 2.5 to 2.7GHz frequency range. MMDS provides data rates up to Mbps and may deliver broadband services where users may receive data rates of 2Mbps. MMDS operates at a lower frequency spectrum than LMDS, and is therefore not limited to line-of-sight paths and is less susceptible to interference. It also provides an alternative to fiber optic, DSL, cable, or other access technology that may be absent from certain buildings and/or developments. MMDS requires an antenna and radio receiver at the subscriber's premises and a coaxial cable to customer premises equipment (CPE).

MMDS signals may be transmitted for distances of up to 30 mi, and are deflected by objects and reassembled at the receiver using various techniques that include vector orthogonal frequency-division multiplexing (VOFDM), which supports data rates up to 50Mbps. Figure M-1 shows video-on-demand application of MMDS, including a video distribution station using a satellite receive-only (RO) antenna to receive TV signals that are downconverted to VHF frequencies, and upconverted and transmitted to the subscriber.

Low-gigahertz operation means that subscribers require perhaps a 60cm-high Yagi, a partial parabolic, or flat-array antenna.

iSpeed is a commercial MMDS used for Internet access in the San Francisco Bay area, and customers may choose between iSpeed Singlepath and iSpeed Dualpath, which is a two-way service. iSpeed Singlepath uses standard telephone connections for uplink communications and broadly equates to T1. Dualpath operates at 1.5Mbps with bursts as high as 10Mbps.

Local Multipoint Distribution Service (LMDS)

LMDS is wireless broadband access technology using a 150 or 1150MHz bandwidth in the 28 to 31GHz range. It is among the "last mile" access technologies that include DSLs, TI, T3, and Frame Relay. Because LMDS uses very high frequencies and short wavelengths, resulting signals cannot negotiate physical obstacles as might FM signals, so it requires a line of sight between the transmitter and receiver using fixed antennas. LMDS may carry voice, data, and video over a wireless local loop operating in the 28 to 31GHz band.

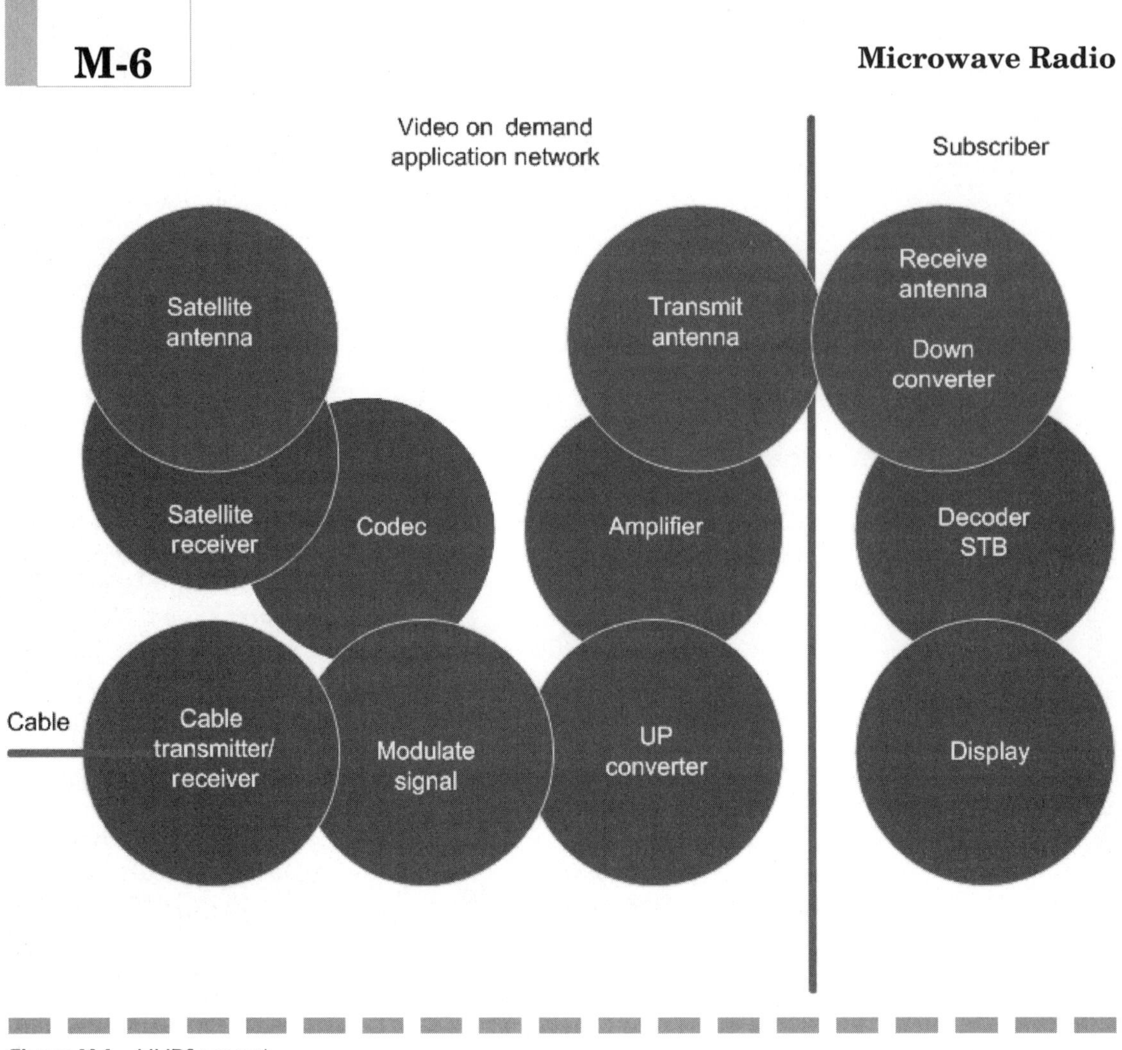

Figure M-1 MMDS network.

Beginning in 1998 the U.S. Federal Communications Commission (FCC) auctioned LMDS licenses for local regions or *basic trading areas* (ETAs), and the highest bids were more than $45 million. LMDS licenses are separated into *A block* and *B block,* which are required by each ETA, and the first LMDS licenses were issued for a period of 10 years.

Architecture LMDS requires a central antenna to serve a cell (see Fig. M-2) that is as small as 15 km owing to the high frequency and short wavelength. Modulation and climatic conditions also dictate cell sizes: Efficient modulation and rainfall reduce the cell size. For wireless cable operators, the base station or LMDS node may connect with an earth station receiving satellite TV feeds. The nodes are controlled by a network operations center (NOC). An LMDS operator may use TDMA, FDMA, and CDMA.

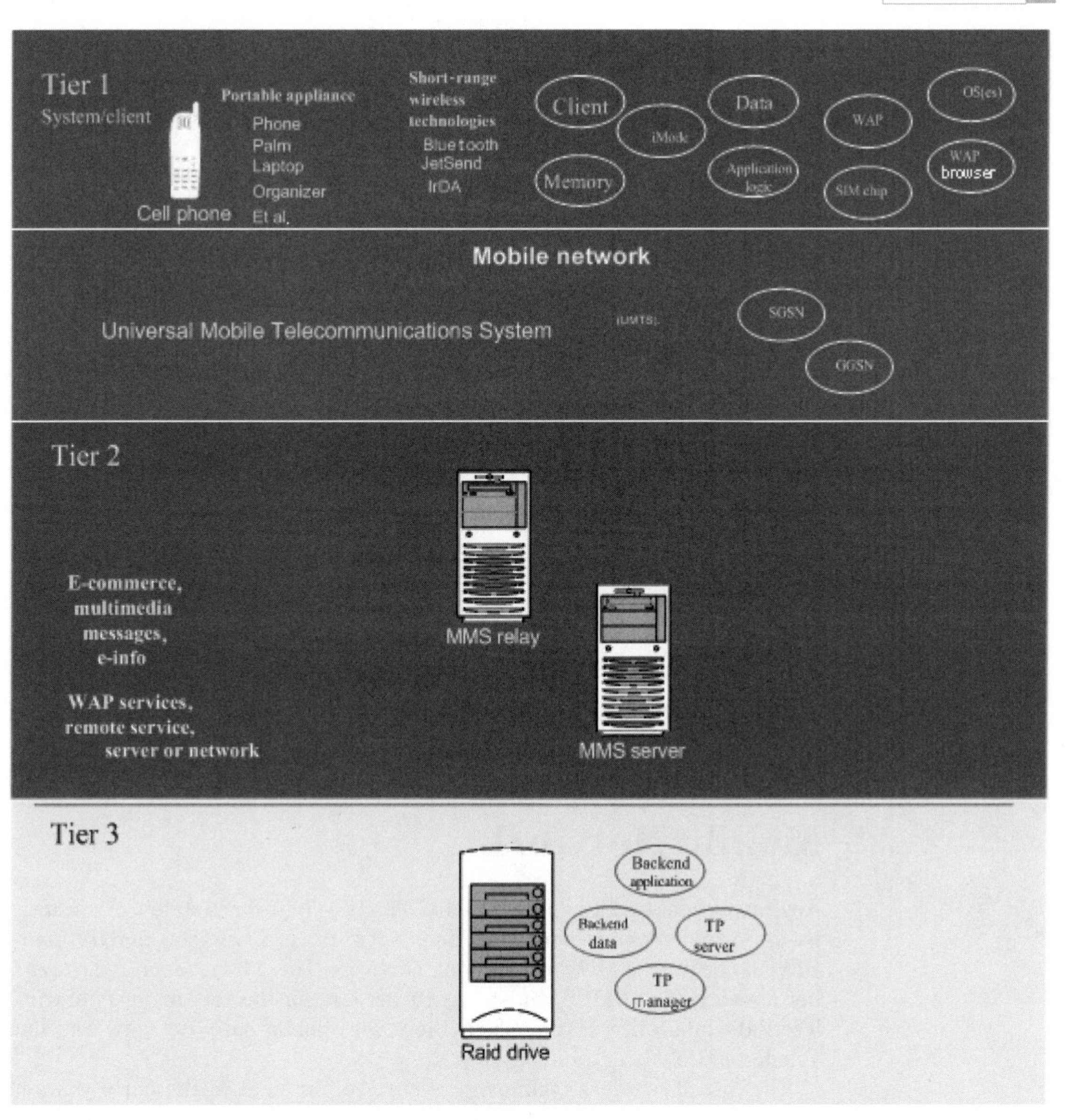

Figure M-2 MMS network.

An LMDS base station or node interfaces with the network and outdoor microwave antenna. Typically a typical transmit and receive sector will provide a 15-, 30-, 45-, or 90-degree beam. The subscriber or LMDS customer requires a network interface unit (NIU) that acts as a gateway. The NIU has an RF modem that supports quadrature amplitude modulation (QAM), or

TABLE M-1

FDMA Access Modulation Methods

Modulation Method	MHz/Mbps CBR Connection
BPSK (binary phase-shift keying)	1.4MHz
QPSK (quaternary phase-shift keying)	0.7MHz
DQPSK (differential QPSK)	0.7MHz
8PSK (octal phase-shift keying)	0.4MHz
4-QAM (quadrature amplitude modulation, 4 states)	0.7MHz
16-QAM (quadrature amplitude modulation, 16 states)	0.3MHz

phase-shift keying (PSK) modulation, and can support FDMA, TDMA, or CDMA. It may also include connectivity to multiple network protocols such as TI, El, 10BASE-T, 100BASE-T, Gigabit Ethernet, and ATM. A subscriber's NIU may also use an all-TDMA or an all-FDMA for uplink and downlink connections.

Modulation RF modulation methods include amplitude modulation (AM) and phase-shift keying (PSK), binary phase-shift keying (BPSK), quaternary phase-shift keying (QPSK), differential QPSK (DQPSK), and octal phase-shift keying (8PSK). Table M-1 shows some FDMA access modulation methods.

Mobile Network

Key mobile networks include GSM (Global System for Mobile Communications), PCN (Personal Communications Network), or DCS-1800, and the many UMTSs (Universal Mobile Telephone Services). These have displaced the earlier networks like AMPS, and currently account for the vast majority of wireless data and voice traffic. Then there are mobile satellite networks like Teledesic, Globalstar, and Odyssey.

The cost of wireless technology is sure to plummet in the early years of the new millennium. For telcos the implications are as serious as those when the copper networks of the 1980s began to be replaced by the then revolutionary optic-fiber links. The level and rate of displacement of physical networks and media then carries with it important questions for analysts and others attempting to forecast the growth of these new deregulated network infrastructures which coexist with those of the telcos, but at the same time also compete with them.

As an example, voice telecommunications has bypassed the comparatively regulated international telcos since 1995 when the first Internet Phones were introduced, and clearly illustrated that IP networks could provide services other than Web applications. By 1998 growing numbers of voice calls migrated from the switched networks that were once the preserve of international telcos like AT&T, Cable & Wireless, BT (UK), and Telstra (Australia), to the largely unregulated packet-switched (PS) networks using what became known as IP or Internet telephony. With appropriate IP telephony gateways (that are now standard), virtually anyone can set up as a carrier, and corporate data networks have done the same with their own IP telephony gateways.

(*See* GPRS under letter G, GSM under letter G, UMTS under letter U.)

Mobile Satellite Service (MSS)

Mobile satellite services or satellite-based mobile-telecommunication services have existed for some time, and the basic technology has been used for various purposes including television broadcasting. MSS mobile-telecommunication services provide links directly to mobile users that theoretically provide large geographic and cell coverage and global roaming. Comparative disadvantages include system capacity thresholds, high investment requirements, signal delays, and low data-transmission capacity. Also significant is the survivability of such networks at times of war and terrorist activity, as such satellites can quite easily be destroyed. See Figs. M-3 and M-4.

Satellites generally orbit at four altitude bands: LEO (low earth orbit), MEO (medium earth orbit), GEO (geostationary orbit), and HEO (highly elliptical orbit).

MSS often uses MEO or LEO that are as low as between 200 and 1400km, as shown in Fig. M-5. Prohibited orbits include the inner and outer Van Allen belts that have high-energy charged particles. The inner and stronger belt is 300km high, and the outer belt is 16,000km. The advantages of using LEOs include low propagation delay, low transmitter power, and higher elevation angles. Disadvantages include frequent handovers and the requirement of a greater number of satellites. For example, the high velocity of an Iridium cell requires one HO per minute.

MEO satellites often have a 10,000km altitude and signal a line-of-sight, and are therefore used for outdoor applications. Advantages of MEOs include the small number of satellites, which is a fraction of LEO networks. For example, ICO will use 10 satellites in two orbits for its global network. MEO satellite longevity is typically 10 to 15 years, whereas LEOs survive for between 5 and 8 years for the LEO bird.

GEO systems orbit at 35,786km height that creates a fixed coordinate relative to the earth, and their velocity equates to their altitude, that is, 35,768km/h. GEO is suitable for broadcast services, and as few as three to four satellites may provide a global network. Comparative disadvantages of GEO systems for mobile-communication services include signal-propagation

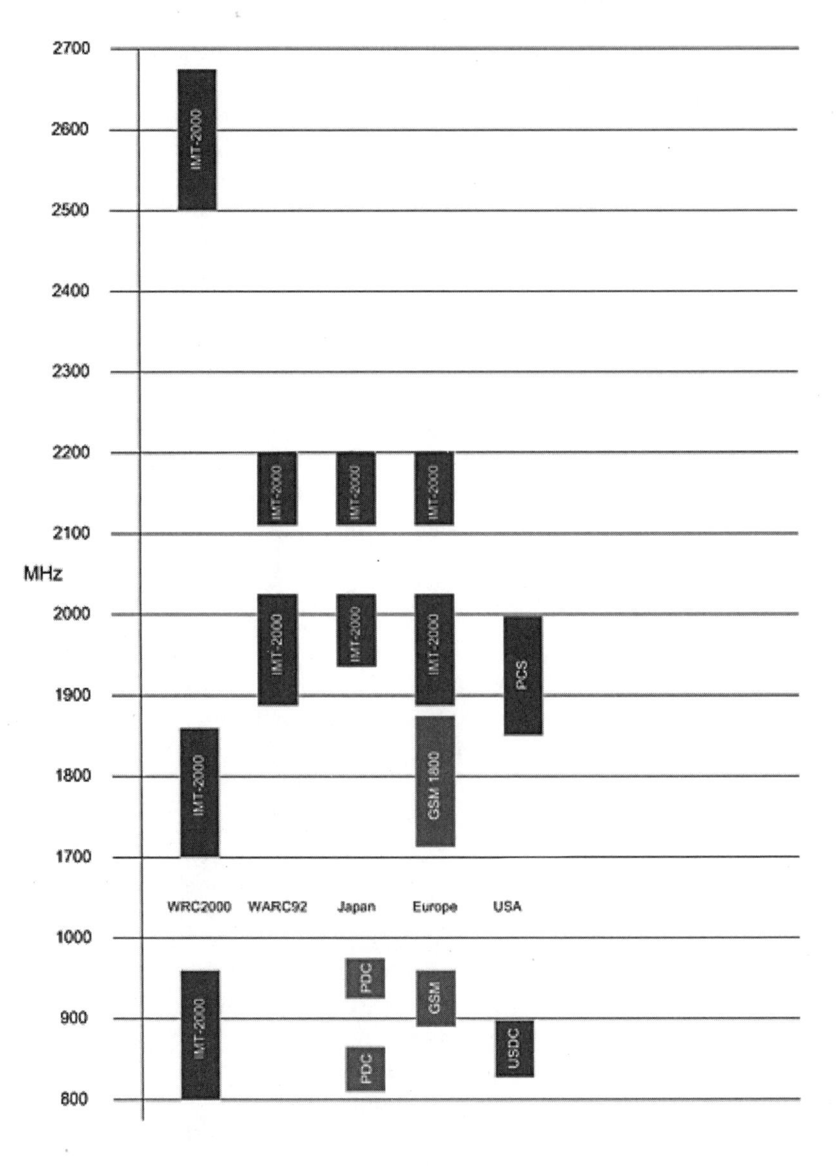

Figure M-3
IMT-2000 allocations.

delays of 250ms, high transmitter powers and large antennas for mobile terminals, and inaccessibility of the earth's poles because GEO satellites are above the equator. Also, the cost of deployment is obviously higher than with lower-orbit satellites. Inmarsat is one of the oldest global networks based on GEO, and it uses large portable terminals.

HEO satellites are deployed in an elliptical orbit, at an altitude of between 1000 and 39,400km. They pass through the Van Allen belt and are immersed in radiation as a result. (See Fig. M-5.) Their lowest altitude point is the *perigee* and their highest altitude point is the *apogee,* which is when it should be used.

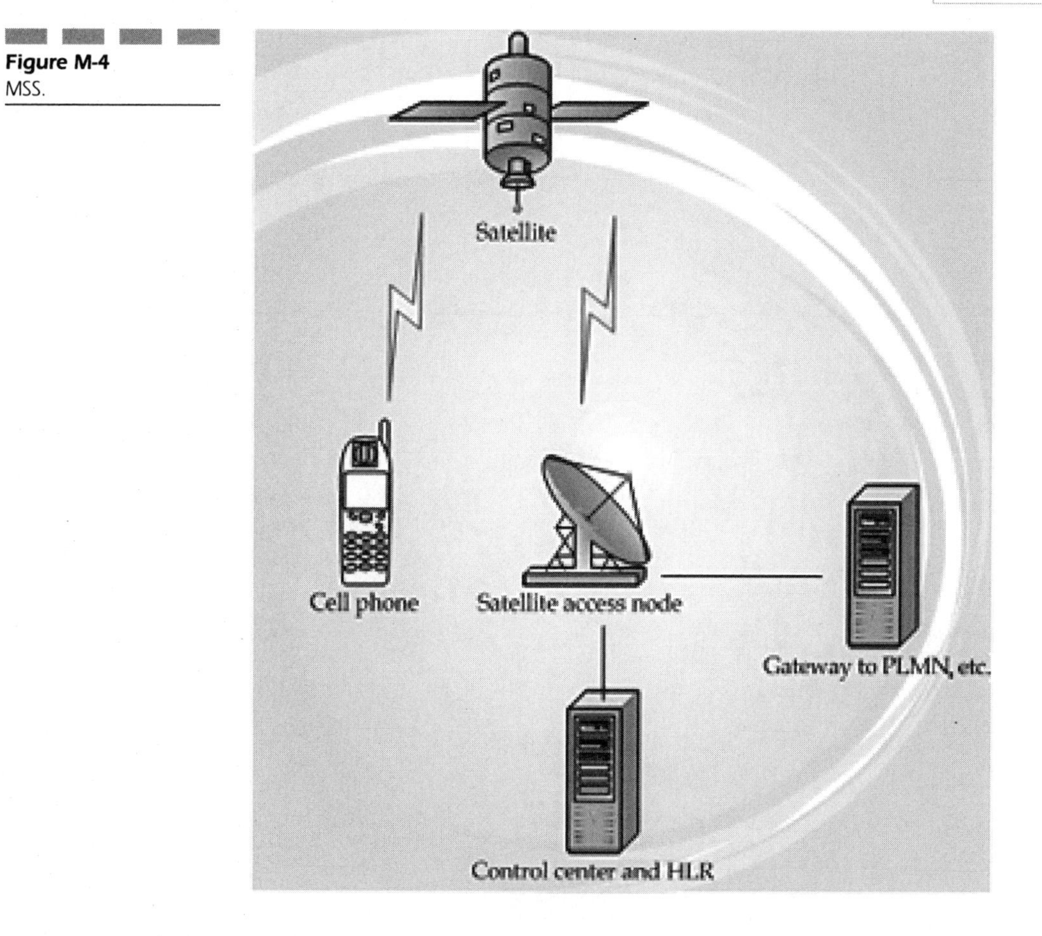

Figure M-4
MSS.

Iridium

Iridium no longer exists but was an early network containing 66 satellites and *not* 77, which is the number of electrons in an iridium atom. The satellites were at 780km and used six polar orbits of 11 satellites at a height of 780km, and each satellite gave 48 cells or *spot beams.* The Iridium carrier had four duplex channels, and a TDMA frame lasted 90ms. Signals were relayed over intersatellite links.

Globalstar

Globalstar uses 48 satellites in eight orbits at 1414km, but it is not a global system. The system coverage is not global and excludes polar areas. Each satellite covers 16 cells, and air interface uses CDMA, and the carrier bandwidth is

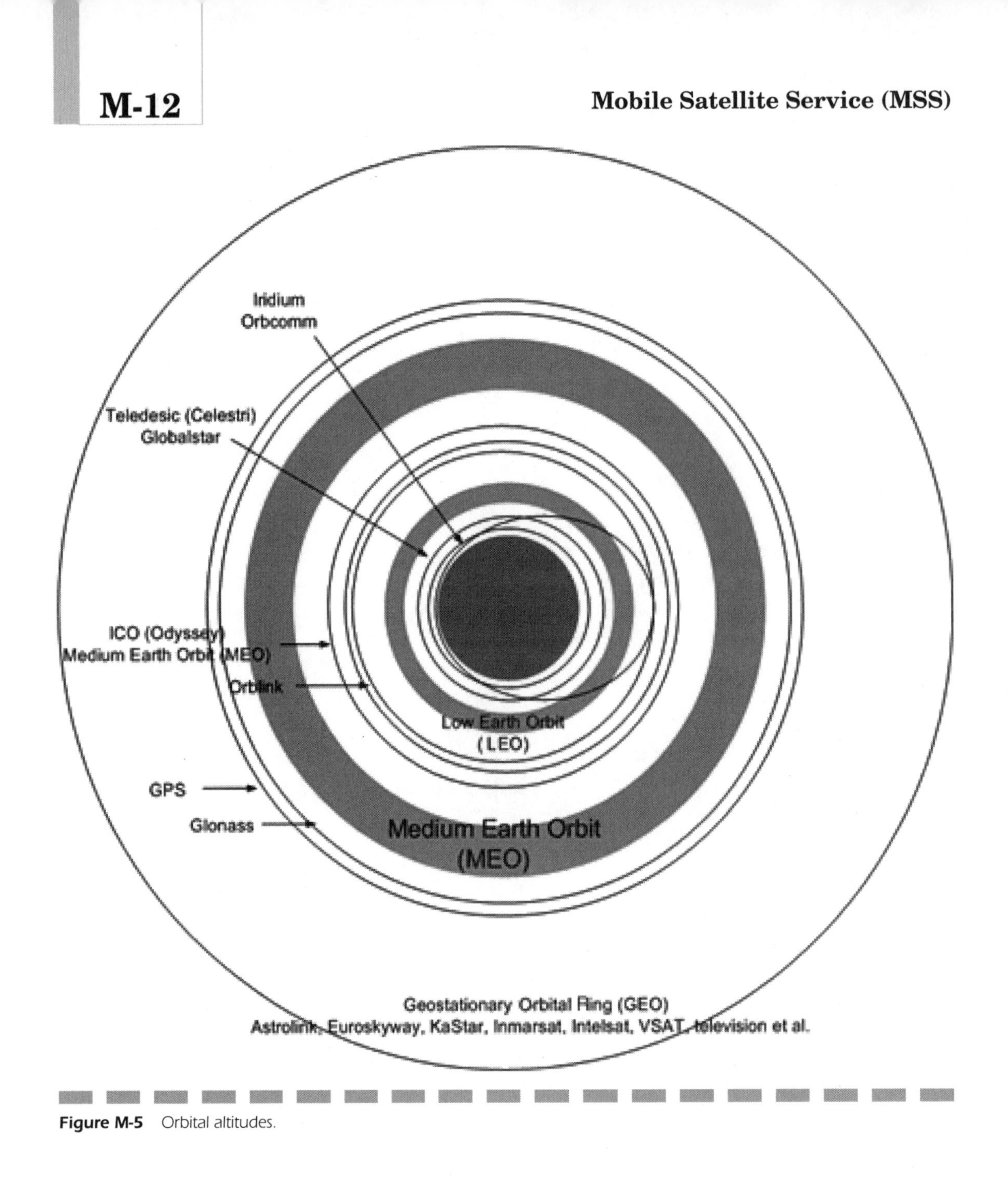

Figure M-5 Orbital altitudes.

1.25MHz, with the uplink using 1610 to 1621.35MHz and the downlink using 2483.5 to 2500MHz. Intersatellite links are not used, so earth stations are harnessed.

Intermediate Circular Orbit

ICO has 12 satellites (including two spares) in two orbits at 10,355km, 12 earth stations or satellite access nodes (SANs), and provides global coverage. Intersatellite connections are not used, and calls are routed through SANs. The air interface technology is a GSM variant, and each satellite serves 163 cells. ICO's bandwidth allocation is 1980 to 2010MHz (uplink) and 2170 to 2200MHz (downlink), and the channel bandwidth is 25.4kHz.

Odyssey

Odyssey was developed by TRW that later opted for ICO instead. Odyssey was intended to be an MEO system with 12 satellites in three orbits, and with a CDMA radio interface.

Teledesic

Teledesic includes a satellite constellation that is as large as 840 active satellites, although fewer will be used, and is backed by Bill Gates, Craig McCaw, and others. It addresses broadband data traffic with up to 64Mbps downlink and 2Mbps uplink) broadband data services. Teledesic's data-transmission is significantly greater than LEO/MEO narrowband systems.

The Teledesic system has 288 small satellites in 12 orbits, covering 95 percent of the globe. Each satellite serves 576 cells that are given 1800 channels offering 16kbps. The radio interface between the user terminal and the satellite for the uplink direction uses multifrequency time-division multiple access (MF-TDMA); the downlink uses asynchronous TDMA (ATDMA).

Orbcomm

Orbcomm uses 35 satellites at 775 km, and does not use intersatellite links. The radio interface technology is CDMA and delivers narrowband data rates suitable for paging-type messages, perhaps for tracking applications where satellites receive data packets from sensors in vehicles, that are then sent to earth stations. The system uses the 137–138MHz and 400MHz frequency bands for transmissions down to mobiles, and 148–150MHz frequencies up to the satellites.

Skybridge

Skybridge offers broadband data rates for two terminal types, including residential terminal with 20Mbps on the forward link and 2Mbps on the return link. There is also a professional terminal with rates of 100Mbps on the forward link and 10Mbps on the return link.

Location in Satellite Systems

Stationary cells may be used to determine location: Satellite beams do not move in unison with satellites, rather they are directed to fixed areas or supercells. The satellite beam tracks the same supercell on the ground for some time, until it is time to track another—or at such time as a beam from another satellite takes over.

Teledesic divides the earth into 20,000 supercells that are each divided into nine cells, which are scanned by a satellite beam.

MPEG

The early days of digital video were plagued by the problem of just how digital video data should be compressed, thus illuminating the need for international standards for the digital storage and retrieval of video data. Sponsored by the then ISO (International Standards Organization) and CCITT (Committee Consultitif International Telegraphique et Telephonique), the Motion Picture Experts Group (MPEG) was given the task of developing a standard coding technique for moving pictures and associated audio.

The group was separated into six specialized subgroups, including:

- Video Group
- Audio Group
- Systems Group
- VLSI Group
- Subjective Tests Group
- DSM (Digital Storage Media) Group

MPEG participants included leaders in computer manufacture (Apple Computer, DEC, IBM, Sun, and Commodore); consumer electronics; audio visual equipment manufacture; professional equipment manufacture; telecom equipment manufacture; broadcasting; telecommunications; and VLSI manufacture. University and research establishments also played an important role.

It provided a basis for the development of the Video CD, which was specified publicly by Philips in late 1993. This is an interchangeable format that may be played using both PCs fitted with an appropriate MPEG video cards and compatible CD-ROM drives, as well as Philips CD-I players fitted with Digital Video cartridges. It is constantly in development so as to accommodate the increasing data transfer rates of both DSMs and other video distribution transports.

The first phase of MPEG work—MPEG-1 compression—is optimized for DSMs with up to 1.5Mbps transfer rates, for storage and retrieval, advanced Videotex and Teletext, and telecommunications. MPEG-1 was finally agreed on, developed, and announced as long ago as December 1991. The second phase, MPEG-2, addressed DSMs with up to 10Mbps transfer rates for digital television broadcasting and telecommunications networks. This phase would cling to the existing CCIR 601 digital video resolution, with audio transfer rates up to 128kbps. MPEG-4 video compression is designed to transmit video over standard telephone lines.

MPEG-1

MPEG-1 is an established and internationally agreed-upon digital video compression standard used widely in fixed and mobile networks. It is used widely for local playback and for streaming multimedia over the Internet and other IP and multimedia networks.

MPEG-1 video production has also become increasingly popular using comparatively inexpensive video capture hardware and encoding software. An alternative to such video production is to use the services of a fully equipped bureau. The decision as to whether the services of a bureau should be used is driven by a number of obvious key factors that include the amount of encoding you require, for which you will be charged on a per-minute basis.

The production of MPEG-1 video using encoding software begins with the capture of a video sequence from a source recording that might be in the VHS or S-VHS formats. Film studios and production companies might rely on professional- and broadcast-quality formats such as Digital Betacam or D1 for their source recordings.

MPEG-1 was developed for narrow-bandwidth media, such as the original single-speed CD drive variants that offered average data transfer rates of approximately 150kbps or 1.2Mbps.

With regard to careful adjustment of the MPEG compression parameters, there is not much you can do if the MPEG encoding software provides no control over them. If it does, you can assume that a greater number of I frames can improve the quality slightly, though this will introduce an overhead in terms of lowering the compression ratio.

It is this single operation that limits the quality of video that can be achieved using MPEG-1, although it has to be implemented in order to confine the video stream to the narrow bandwidth of about 1.5Mbps. If you are unable to capture video at 25 frames/s, you can increase the frame rate following video capture using Video for Windows VidEdit or an equivalent digital video editing program. The increased frame rate is achieved merely by duplicating frames, but it does mean that the finally encoded MPEG video stream will at least be an authentic one.

MPEG-2

An improved version of MPEG-1 video compression, MPEG-2, is supported by DVD technology. It was developed for media and networks able to deliver 10Mbps data transfer rates.

MPEG-2 video may contain considerably more audio and video information than MPEG-1. The most noticeable improvement is the higher playback screen resolutions that are possible, making possible D1 or CCIR 601 quality. DCT is key to MPEG-2, as it is to MPEG-1 and JPEG (or even M-JPEG). As is the case with MPEG-1, MPEG-2 requires decoding solutions that may be hardware-based such as set-top boxes (STBs), or equivalent hardware implementations integrated in computers. Applications of MPEG-2 video include Video-on-demand, multimedia, videoconferencing, etc. It may also be stored and delivered using DVD variants.

MPEG-4

MPEG, a working group in ISO/IEC, produced MPEG-l, MPEG-2, and the MPEG-4 standard that was initiated in Seoul (1993), with completion scheduled for 1998. MPEG-4 was evolved by the ITU-T that addresses teleco standards, and includes the Experts Group on Very Low Bit-Rate Visual Telephony (LBC). MPEG-4 was designed to provide standard, interchangeable very high compression ratio coders, with data rates less than 64kbps. MPEG-4 is optimized for data compression ratios that H.261 or MPEG-1 are unable to deliver efficiently. To obtain this the MPEG-4 algorithm includes image and video analysis and/or synthesis, modeling, fractaling, morphing, and more.

Though much is written about MPEG-4, other video compression standards are also significant including Microsoft AVI as well as the established MPEG-1. The AVI file format is used for storing interleaved audio and video. Using many video editing and video capture tools, the interleave ratio may be varied. The ratio may be specified as a single figure where, for instance, an interleave ratio of 7 indicates that seven video frames separate each audio chunk. Using Microsoft VidEdit, the statistics of a video file may be shown where the interleave ratio is displayed alongside the phrase Interleave Every.

The interleave ratio is expressed as the number of video blocks that separate audio blocks. Generally high interleave ratios are applicable to video stored on hard disk, whereas AVI video stored on a CD variant is optimized using lower interleave ratios, which often equate to one video frame for every audio chunk. Sound track quality commonly found in AVI files ranges from mono 8bit recordings digitized at 11kHz, to 16bit stereo recordings digitized at 44.1kHz.

MPEG-1 Quality

The quality of MPEG video depends on a number of factors ranging from the source video recording quality to the use of important MPEG parameters that affect the overall compression ratio achieved. Probably the most obvious elements that influence MPEG video quality include the analog or digital source recording, the video source recording format, and the video source device specification.

This resolution equates to the MPEG Source Input Format (SIF), which is achieved by omitting odd or even lines from a standard interlaced PAL (Phase Alternating Line) signal. A SIF frame has an MPEG-1 frame resolution of 360 × 240 pixels for NTSC, and 360 × 288 for PAL. This is an exceptionally lossy procedure, omitting a great deal of picture information and losing video quality. It is this single operation that limits the quality of video that can be achieved using MPEG-1, although it has to be implemented in order to confine the video stream to the narrow bandwidth of about 1.5Mbps.

The MPEG claim that this is VHS quality is an area of debate. We can assume that the higher resolution S-VHS format will provide slightly better results than VHS, but there will not be a dramatic improvement in resolution because the MPEG SIF is standardized at 360 × 288 pixels for PAL.

Contrary to popular belief, the logical operations that provide a basis for obtaining high-quality MPEG video are by no means the preserve of expensive video production bureau. Equipped with a reasonably specified PC and a basic understanding of MPEG video, there is nothing to stop you from producing good-quality White Book–compatible video on your desktop.

Video Capture

When capturing a video file so that it may eventually be compressed, it is important to choose an appropriate capture frame rate, capture frame size, and image depth. The capture frame rate should be set for 25 frames/s for PAL and 30 frames/s for NTSC. Frame rates that differ from these will cause the MPEG video sequence to run at the wrong speed, and it will not be White Book–compliant. The capture frame size should correspond with the MPEG-1 SIF, which is 360 × 288 pixels for PAL and 360 × 240 pixels for NTSC. Authentic MPEG requires a truecolor image depth of 24bits per pixel giving a total of over 16.7 million colors that are generated by combining 256 shades of red, green, and blue.

The video source recording might be analog or digital when capturing video for transmission over wireless networks. The latter requires that the video capture card incorporate an appropriate input. The three general types of video capture include

- *Real-time video capture* involves digitizing the incoming video source signal on the fly, and the video source device is not stopped or paused at any moment during capture.
- *Automatic step-frame capture* requires that the source device is stopped, paused, and even rewound so as to digitize a greater amount of the source recording. It offers certain advantages, namely, it is possible to achieve a greater number of colors (or greater image depth), higher capture frame rates, and larger capture frame resolutions than would normally be possible using the same video capture hardware and software configuration to record video in real time.
- *Manual step-frame capture* usually depends on the operator clicking a button on screen in order to capture selected video frames.

Available color depths using fully specified video capture card and capture program partnerships include 8bit, 16bit, and 24bit. The 8bit format gives a maximum of 256 colors stored in the form of color palette that can be edited using programs such as PalEdit. 16bit and 24bit formats are described as truecolor, giving a maximum of 65,536 and 16,777,216 colors, respectively, and when using appropriately specified video capture hardware and software they can produce impressive results.

Using many video capture systems, the data throughput required to capture 16bit and particularly 24bit video in real time limits both the capture frame rate and frame size. The frame dimensions chosen hinge largely on the specification of the capture card, though the image depth chosen is also influential as is the capture frame rate.

Though the video frame dimensions can be scaled using video editing programs and even multimedia authoring tools, enlargement can result in a blocking effect as the individual pixels are enlarged. However, certain graphics cards, particularly those that enlarge Video for Windows video sequences, will apply a smoothing algorithm during playback in an attempt to minimize the blocking effect.

Video editing techniques also can be used to increase the playback frame rate (through frame duplication). This, along with other digital video editing techniques and hardware/software features of the playback system, can help improve the quality of video playback.

However, capturing and compressing optimum quality digital video relevant to the intended playback platform remains the most important process. There are limitations in what can be achieved through digital video editing, and through playback hardware that enhances digital video playback.

Generally editing compressed MPEG video is difficult owing to the paucity of authentic access points. However, editable MPEG files exist, one of which is backed by Microsoft. Additionally an MPEG video stream composed entirely of I frames lends itself to nonlinear editing.

The original video sequence may be enhanced, even enlarged through duplication, but it cannot be used to play video information present in the source recording that it simply does not contain. Even though numerous algorithms can enhance digital video, and numerous other will emerge, it is reasonable to assume that if the video file does not contain a particular frame then that frame cannot be played.

The quality levels available using wave audio recorders together with mainstream sound cards also can be achieved through fully specified Windows video capture programs. 8bit or 16bit sample sizes are available, recorded at frequencies of 11.025kHz, 22.05kHz, and 44.1kHz in mono or in stereo. The size of the sound track, which increases in relation to the recording quality chosen, can be monitored using many video editing programs.

Audio Capture

If you are digitizing the soundtrack of the source video recording also, you will probably obtain the best results with camcorders and VCRs that offer hi-fi quality stereo sound.

The quality of captured audio that is used as an input audio stream obviously depends on the sample size, recording frequency, and whether mono or stereo is chosen. You can assume that your wave audio recorder or video capture program will provide sampling rates of 11kHz, 22kHz, and 44.1kHz, and sample sizes of 8bit and 16bit. While higher sampling rates and larger sample sizes yield improved audio quality, the resultant audio stream can consume an unacceptably large portion of the available MPEG-1 bandwidth.

A video editing program can be used to synchronize audio and video streams, usually by introducing a time offset for the audio track. By then separating the file into video and wave audio files, once again using the video editing program, their play times should become equal.

Some MPEG encoders will automatically alter the length of the input audio file so that it matches the length of the input video file. This does not guarantee that the audio and video material is synchronized correctly when multiplexed. It should be added that the synchronization of audio and video information can also be carried out at the decoding stage.

MPEG Frames

An MPEG video stream generally consists of three frame types:

- Intra (I) frames
- Predicted (P) frames
- Bidirectional (B) frames

Central to MPEG encoding is the use of reference or I frames that are complete frames and exist intermittently in an MPEG video sequence. I frames are compressed in a similar way to JPEG (Joint Photographic Experts Group) images and do not rely on image data from other frames. They exist intermittently, perhaps between 9 and 30 frames, and provide nonlinear entry points. The video information sandwiched between intra frames consists of that which does not exist in the intra frames. Information that is found to exist in the intra frames is discarded or *lossed*. Intra frames can act as key frames when editing or playing MPEG video as they consist of a complete frame.

Increasing the frequency of I frames provides a greater number of valid entry points, but the compression ratio of the overall file diminishes proportionately. Realistically the compression ratios achieved using MPEG may be assumed to be around 50:1. Higher compression ratios lead to an unacceptable loss of quality, and it is wise to forget that 200:1 ratio which MPEG is supposedly capable of producing. Normally this is achieved through a pretreatment process that dramatically reduces the number of frame pixels.

I frames and the following P and B frames are termed Groups of Pictures (GOPs), and the occurrence of each frame might be predefined through the careful adjustment of MPEG parameters prior to encoding. However, this fine level of control over compression parameters may not be provided by low-cost MPEG encoding programs.

Multimedia Messaging Service

Multimedia messaging service (MMS) is a significant application of UMTS and provides services with voice, text, audio, video, images, and sound synthesis. An MSS message may have multiple components (or message elements) that are combined at the user interface to produce a multimedia presentation. This introduces a non-real-time facet where network latencies may cause no significant QoS degradation. This means that dropped packets, packet retransmission, and long interleaving do little to lower the quality of service. 3GPP MMS specifications cover the technology required to distribute multimedia over mobile networks. Applications of MMS include e-cards, e-business cards, multimedia news, traffic information, music on demand, POI, e-business, and many others.

MMS Architecture Figure B-1 shows some of the components used in a multimedia messaging service environment (MMSE) and could conceivably be implemented using 2G and 2.5G networks like GPRS, EGPRS, and EDGE. The MMS user agent is part of the UE, and functions in the application layer providing MMS message composition, presentation, retrieval, and notifications.

TABLE M-2

Multimedia Elements Supported by the MSS User Agent

Text	Character Sets That Have a Subset of the Logical Unicode Characters
Audio	AMR/EFR speech
	MP3
	MIDI
	WAV
Image	JPEG
	GIF89a
Video	MPEG4
	ITU-t H.263
	Quicktime

The MMS server stores and processes messages, may use a separate database for message storage, and may be dedicated to a message type. The MMS relay is an intermediate and control mechanism that is located between the user agent and the MMS server and services.

3GPP MMS relay specified functions:

- Check terminal availability
- Convert media format
- Convert media types
- Enable/disable MMS functions
- Erase MMS messages based on user profile
- Forward messages
- Generate CDRs
- Message notification
- Negotiate terminal capabilities
- Personalize MMS using user profiles
- Receive/send MMS messages
- Retrieve message content
- Screen MMS messages
- Translate addresses

The MMS user database stores:

- MMS subscriptions

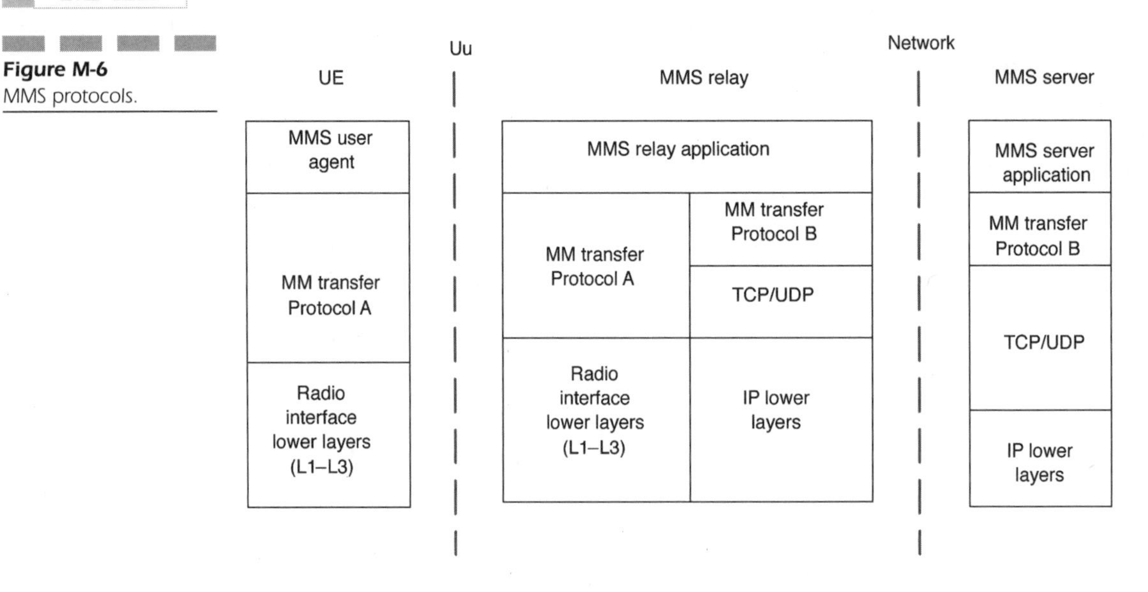

Figure M-6 MMS protocols.

- Access control to MMS
- Data to control the extent of services
- Rules that determine handling of incoming messages and their delivery
- Data about user terminal capabilities

MMS Protocols Figure M-6 illustrates the MMS protocol framework where the MMS transfer protocol A may include WAP and TCP/IP. WAP is now a rather established protocol for deploying wireless applications, and is still being used by 3G users where fast access is provided to Internet sites. The faster data transfer rates provided by UMTS will evolve WAP, and drive the inclusion of additional features that are not possible using GSM and other 2G networks.

For more information, refer to the Web site www.wap.org.

The other technology that fits into the MMS protocol framework is Java and the use of Java virtual machines (JVMs) on the user's device. This gives true device independence, and renders a service universally available. The mobile device may be extremely light in terms of processing and memory because the JVM of software processors does have excessive hardware requirements. This technology model is a most challenging one, as it defies the notion of having a single browser on the mobile device.

For example, PalmPC, WAP, and Windows CE are fixed user interfaces. The JVM concept means that a UI object may be downloaded to the mobile device, so each service could have a suitable user interface. The Java framework that realizes this concept is called Jini. The MMS message transfer protocol layer may include SMTP, POP3, IMAP4, HTTP, or SMPP.

[*See* Discovery (Jini) under letter D.]

Computers/Software

Macro

A macro is a short program or series of instructions. Macros are useful for automating processes, or for performing tasks that would otherwise take a great deal of time to implement.

Typically macros are written when the user interface is restrictive for a given task, or for particular usage habits. They are also written when there is no predefined macro that will perform a desired task. Sometimes certain predefined macros may be improved upon or edited so as to perform different tasks.

Previously sophisticated word processors such as Microsoft Word and Word Pro had indigenous macro languages, as today some Standard macro languages include Visual Basic for applications.

Mainframe

A mainframe is a powerful computer that typically offers centralized processing, serving a number of connected dumb terminals. In terms of its positive characteristics, a modern mainframe may:

- Process data at speeds beyond those attainable on desktop systems, and those based on conventional 32bit and 64bit processors that might be CISC or RISC.
- Provide long-term archiving of data
- Be a massively parallel processing (MPP) architecture, where processes are run concurrently, and offering efficient scaleable processing
- Offer industrial strength operation through robust operating systems and applications
- Provide easy diagnosis of faults as they are isolated to the network or to the mainframe itself—though the mainframe with all its electronics and mass storage remains a complex fault diagnosis domain
- Provide effective migration paths to client/server architectures
- Prove a more durable IT solution in terms of longevity, because mainframe technology advances more slowly than microcomputer technology.

Key disadvantages of mainframe computers revolve around:

- High initial cost.
- High cost of ownership brought about by comparatively high maintenance and servicing bills.

- Fault tolerance is at a low threshold, because a mainframe fault may render an entire IT solution inoperable. However, the fault tolerance of the connected mass storage (which may be shared) might be high.
- The dumb terminals are typically green-screen, but there is scope for renovation.
- Mainframe languages tend to be old-fashioned, like COBOL, although this is a changing situation.

Mapi (Messaging Application Program Interface)

MAPI is a standard that permits e-mail messages to be sent from any application. Originally developed by Microsoft, it is a DLL containing C functions, and allows developers to exploit Windows messaging. Calls to the DLL may allow applications to be given e-mail functions.

Master/Slave Processing

A master/slave architecture where a master computer is connected to slave (intelligent) computers that are connected to dumb terminals. Processing is distributed from the master computer that may be assumed to be a mainframe, to the slave systems.

Media Player

Media player is a Windows program able to play a variety of different media files. Media with the appropriate driver selected and installed using the Control Panel it may be used to play various different media types, including:

- CD-DA
- Midi files
- Wave (.WAV) files
- Video files

When launched it shows controls common with typical audio/visual appliances, including Play, Pause, Stop, and Eject. Finer control over playing various media files and tracks is provided by a horizontal scroll bar.

Merchant Server

A Windows NT-based server that is part of the MCIS, and permits the construction of virtual shopping sites.

The server consists of a:

- Controller that is used to define language, currency, date, and other preconfigurable parameters.
- Router that is an ISAPI (Internet Service Application Programming Interface) DLL. This routes requests from the client to relevant parts of the store server, and routes responses to those requests back to the client browser.
- Store server that is the system's backbone, and functions to implement tasks such as order requests, and to interact with the backend database.

Merchant Server may be used to implement sites that allow customers to:

- Peruse product databases
- Purchase items using a shopping cart metaphor
- Receive e-mail confirmation of orders placed

Merchant Server permits the vendor to:

- Query customer details, and purchase habits
- Conduct promotions
- Conduct marketing campaigns
- Create membership accounts using IDs and passwords
- Offer membership discounts
- Integrate ActiveX, OLE, and COM components into the server
- Use ODBC-compliant databases
- Secure credit card transactions using the SET (Secure Electronics Transfer) protocol together with Verifone's vPOS application

Merchant Services

A bank department that processes information for merchants.

Message

A request sent from one object or component to another, commonly used in OO systems. The message will be of a standard or proprietary format, with address information, and appropriate data.

The messages might require an acknowledge message, before the originating component may continue processing. OO client/server architectures use messages and underlying protocols as their collective glues.

Message Authentication

A process or usually subprocess that verifies that a message is received from the appropriate or legal sender.

Message Wrapper

A top-level data structure that conveys information to message recipients.

Meta Data

A term used to describe data that indicates the information types and subjects. The data may be stored in an information storage and retrieval system.

In the context of the Web, meta data such as indexes and URLs are gathered and stored by search engine implementations. This provides clients with the ability to search and retrieve documents from the Web.

Metadata

A code fragment that is able to perform an operation of a certain type, like writing or reading data, or making a connection with a remote server application, or simply sending a message, and so on.

Methods may be defined in a set of APIs and they may define objects' behaviors in terms of how they respond to an expected event such as a mouse click, and to other stimuli. Other events might be the reception of messages from other objects, and the underlying methods might interpret them, and initiate an appropriate response. The response might be an acknowledge message, or a return value such as the contents of a variable.

MHz (Megahertz)

A measurement that equates to 1 million cycles or pulses per second. It is commonly used to describe the clock speed of computers, thus providing indication of speed of operation. A 50MHz machine will therefore yield 50 million clock cycles per second, and a single clock cycle will have a duration of 1/50,000,000s.

MicroJava

A processor from Sun Microsystems that is optimized for the Java programming language. It is used in network devices, telecommunications hardware, and consumer games.

Microsoft

A large software producer and vendor that was founded jointly by Bill Gates and Paul Allen. Microsoft is a leading computer software company targeting mainly the PC platform.

Its best-known products are Windows, Microsoft Office, and the MS-DOS operating system, among others. It also produces multimedia titles, and has recently extended its operations to the Internet through the Microsoft Network (MSN), and numerous related ventures.

Microsoft ActiveX SDK

An SDK dedicated to the creation of ActiveX controls, and is compatible with Visual C++ 4.2 (or higher).

Microsoft Design-Time Control SDK

An SDK that is used to create Design-Time ActiveX Controls that, as their name suggests, is active only during design. Resulting controls may be used with FrontPage, InterDev, Visual C++, Visual Basic, etc.

Microsoft DirectX SDK

A toolset that is used to develop multimedia elements, and includes:

- Direct3D for three-dimensional graphics
- DirectDraw for 2-D graphics
- DirectInput for connectivity to input devices such as joysticks
- DirectSound for exploiting sound card/software capabilities
- DirectPlay for connecting to remote applications

Microsoft Forms 2.0 ActiveX Control

A suite of ActiveX Controls included in Visual Basic Control Edition.

Microsoft NetShow

A streaming technology server that may be integrated into a Web site and/or application. Its inclusion results in the ability to serve client browsers with streaming audio, video, and multimedia.

Web site and Web application developers may integrate it into IIS-based Web application solutions.

Microsoft NetShow Theater Server

A streaming MPEG video media server that extends Windows NT Server NetShow Services, so as to deliver higher-quality video, including MPEG1 and MPEG2 video streams from 500kbps to 8Mbps.

- Scalability of up to thousands of video streams
- A distributed, fault-tolerant PC architecture for mission-critical applications

Microsoft Proxy Server

A server implementation that may be used to deliver Internet access across an enterprise. The Internet Service Manager is used to manage the Proxy Server, as it can for Chat and Mail servers.

The Microsoft Proxy Server:

- Is compatible with Intel and Risc platforms
- Uses caching algorithms to optimize access to LAN data
- Includes an autodial feature that connects the user with the ISP, if the user's requested data does not reside in the cache
- Assign users with access rights to specified Web sites

Microsoft SDK for Java

A superset of the JDK, it includes Microsoft class libraries, the JIT compiler, and the Microsoft Virtual Machine for Java.

Microsoft SQL Server

A relational database management system (RDBMS) that provides multi- and concurrent-user access to enterprise data. The Microsoft SQL Server's utilities include the:

- SQL Enterprise Manager that provides management features
- SQL Service Manager that provides start and stop functions
- Interactive SQL for Windows that permits sessions with multiple SQL servers
- SQL Security Manager that provides access to security features
- SQL Setup that may be used to upgrade MS SQL Server, as well as to change default settings
- SQL Client Configuration Utility that is used to manage SQL Server client software configurations
- SQL Performance Monitor that offers performance readings
- SQL Server Web Assistant that permits the generation of Web pages that use SQL Server data
- SQL Trace that is used to track SQL Server user habits

Microsoft Transaction Server

A transaction manager. A Microsoft solution for integrating transaction processing in Web applications. Its component architecture includes the:

- Transaction Server Explorer that is used for administration and management purposes
- Transaction Server Executive that is a DLL providing functions used by the application's server components
- ActiveX Server Components that is used to deploy ActiveX server components
- Server Process that hosts the application's components
- ODBC Resource Dispenser that manages database connectivity
- Shared Property Manager that gives access to a Web application's properties
- Microsoft Distributed Transaction Coordinator that coordinates transactions, and is integrated in Microsoft SQL Server 6.5

Other transaction managers include CICs and Encina.

Microsoft Windows

An industry standard operating system and graphical user interface (GUI) for the PC. It uses the windows metaphor as a means to contain documents and applications. Up until late 1995 when Windows 95 appeared, its foundation was considered to be Program Manager, a main window that contained program group windows. The group windows contain selectable icons of related applications.

Windows 95 offers a replacement for Program Manager by way of task bar. By default it underscores all applications, providing buttons to select open applications, and it anchors the all-important start button which invokes the start menu.

This bears options that lead to programs as well as to submenus that replace the group windows of Program Manager. Once invoked the menu system may be navigated by dragging the mouse rather than by clicking menu items, and programs are opened through a single mouse click. Application menu systems echo its operation.

Microsoft Windows CE (Compact Edition)

A version of the Windows OS designed for palmtops, organizers, and other small-scale system solution including those targeting the consumer market. It also supports UpnP.

Microwave Radio

Short wavelength radio waves that have a frequency above 1000MHz or 1GHz.

Middleware

1. A software implementation or glue that exists between the client and the server. It makes the network protocols and other server workings transparent to the client. Middleware implementations include those based on the OMG's CORBA-based Notification Service that supports push-and-pull style communications of an asynchronous nature. There is no real-time synchronization between client (consumer) and server (supplier) applications; rather the client may invoke operations even when a supplier application or complete server is occupied.

   Both consumers and suppliers invoke operations in the API that includes modules or files written in the CORBA Interface Definition Language that is loosely based on C++. Within the files are defined interfaces, operations (or methods), exceptions, and certain error codes. The key conduit in such implementations is the event channel, and all

2. Database middleware connects client applications with back-end applications, and consists of:
   - An application programming interface (API)
   - Network and database translators

MIDI (Musical Instrument Digital Interface)

An industry standard file format and specification for producing and playing electronic music using computers and compatible devices such as Midi keyboards and Midi guitar interfaces. It covers hardware, cables, connectors, and data protocols (MIDI messages) and file formats.

The single most significant advantage of Midi is the compactness of resultant so-called Midi song files. These consume a fraction of the data capacity required by digitized waveform audio such as .WAV files. A one-hour stereo Midi file may consume around half a MByte. Even using compression techniques, an equivalent .WAV file would consume literally hundreds of MBytes.

Mirror Site

An Internet site that duplicates the functionality of another site. Mirror sites help provide an improved service for users by lowering usage demands on individual sites.

Mirroring

A function of hard disk controller that writes data to more than one disk drive simultaneously.

Modal

A term used to describe interaction where the user moves between different modes of program operation. The multimedia authoring tool Asymetrix ToolBook is modal, in that the user switches between Read and Design modes.

Modem (Modulator/Demodulator)

A hardware device used for modulating and demodulating data normally received and transmitted over voice-grade communications systems.

It may be an:

- Internal modem that consists of an expansion card that plugs into the expansion bus.
- External modem connects with the serial port of a computer. It typically measures about 15cm × 10cm × 2.5cm.
- External PCMCIA (Personal Computer Memory Card International Association) modem that is little bigger than a credit card.

A 56.6kbps standard analog modem exceeds the proven bandwidth limit calculated using Shannon's theorem. The higher speed is achieved using PCM, and a digital link between the telephone company and the ISP.

56.6kbps modems are asymmetrical, offering wider downstream bandwidths, thus downloading times are shorter than those of uploading.

The ITU has attempted to amalgamate two of the industry standards:

- X2
- K56flex

The resulting V.90 standard was specified provisionally and finally released in 1998.

Moderator

A person who checks all contributions to newsgroups before posting them.

MOLAP

An OLAP implementation that supports Multidimensional Database Management Systems (MDBMS) that may be assumed to use proprietary data storage techniques.

Monitor

A display device used with computers, multimedia, and digital video playback systems. Desktop systems may be assumed to include CRT (Cathode Ray Tube) displays, but increasingly flat-screen TFT displays are being used.

Notebooks and other portable systems may be assumed to integrate LCD (Liquid Crystal Displays), TFT, or DSTN display technology.

Principal technical factors that dictate a monitor's specification are its:

- Screen size
- Supported resolutions
- Noninterlaced and interlaced screen refresh rates (in the case of CRT-based designs
- Supported number of colors that is irrelevant with CRT-based designs

MOTO (Mail Order/Telephone Order)

A transaction that emanates from a card-not-present scenario, and may take place using voice communications or most often using an e-commerce Web site.

Mouse

A hand-held input device. By dragging it on a flat surface it provides a means of moving a screen pointer/cursor in both x and y directions. It is typically connected to the serial port, but may also be wireless.

It typically includes two or three pushbuttons that are used to make selections either by pressing a button once (or single-clicking) or by pressing a button twice in succession (or double-clicking).

The mouse is also used for dragging (or moving) objects so as to move them from one point to another, or for resizing windows by dragging their borders. Dragging is carried out by holding down the left mouse button above an object or window border, and then moving the mouse appropriately.

Modern notebook systems use mechanism-free touchpads instead of the traditional mouse.

MPP (Massively Parallel Processing)

A computer that has multiple processors that may operate independently and concurrently, as well as interact with one another through interprocess communications.

The strict definition of MPP is a system that offers scalability where resulting processor gains increase in multiples that equate to the processing power of single unit processor. For example, the collective processing power of an MPP system with n processors, should increase by x MIPs per added processor(s).

The processors may have their own memory and I/O capabilities, and constitute complete computers, or use shared memory. The processors also exhibit channels of interconnection between other processors. These connections constitute the network, and its bandwidth naturally influences the collective processor power of the system.

The network is not to be confused with external, industry standard networks such as IP and Ethernet. An MPP network is internal, with the rationale of optimizing system performance by permitting the processors to communicate as quickly as possible.

Typical network topologies include ring, two-dimensional mesh, three-dimensional mesh, and hypercube. The resulting MPP interconnection network may be specified in terms of its:

- Link bandwidth, or the rate at which data may be sent via a direct link, and is a function of clock speed, and data bus width.
- Switching latency that might be defined as the period between a processor data request and the reception of that request. This is a function of clock speed, the network topology, and the physical location of the serving processor in the network; the farther away it is, then the switching latency is extended.

The processing power of an MPP may be measured in:

- Millions of floating point operations per second (MFLOPS)
- Billions of FLOPS (gigaFLOPS or GFLOPS)
- Trillions of FLOPS (teraFLOPS or TFLOPS)—in future
- Millions of instructions per second (MIPS)
- SPECmarks

The optimum processing yield depends on distributing processes evenly across the processor array, matrix, or network. Program algorithms may perform this function of dynamic load balancing that is carried out in real time.

A common denominator in current networks is that not all processors are connected directly.

MPP architectures are divided between:

- Multiple Instruction Multiple Data (MIMD)
- Single Instruction Multiple Data (SIMD)

MIMD architectures feature memory that may be distributed or shared.

The SIMD architecture has a single controller driving multiple slave processors, each with independent storage. The distributed memory DM MIMD architecture has a multiplicity of such processors, and controllers, too.

An MPP architecture variant may be explained in terms of its electronic storage, controller(s), and processor(s).

Leading MPP manufacturers include Cray, Thinking Machines, Intel, and nCube.

Concurrent programming languages include Occam that has its origins in Inmos (U.K.) where it was developed as part of the Transputer parallel processor. Java is the first mainstream programming language that supports the parallel programming model. Languages that are optimized for parallel processing systems offer authentic concurrency.

One of the earliest transputer-based supercomputers was developed by Meiko through the Computing Surface. This was used in the development of DVI, and modern transputer-based implementations are used as Video-on-demand servers. Among the advantages of such parallel processing systems is scalability, where, for example, growing numbers of subscribers to a Vod service may be accommodated through additional processors, and even complete servers.

MReply

An e-mail autoresponder, with added features including the ability to send multiple messages to recpients on a mailing list.

MTBF (Mean Time between Failures)

An average period of time that indicates the frequency at which a device, component, subsystem, or complete system will fail.

MTTF (Mean Time to Restore)

An average period required to return a failed system to its fully operational state.

Multimedia

A broad term that may be applied to a system or process which embodies and combines various different media. Modern (digital) multimedia may comprise computer animations, text, still images, digital audio, synthesized sound, digital video, and interactivity.

Combining still and moving images, sound, audio, text, and interactivity, multimedia has initially culminated in a reevaluation media.

Unlike linear, noninteractive media, such as broadcast television, it provides users with a choice of numerous meaningful paths.

The underlying technology has spawned offshoots, of which the most notable will be video-telecommunications and videoconferencing.

Distribution media disc-based multimedia include Compact Disc—Read Only Memory (CD-ROM). Earlier, and less well-known distribution media include Compact Disc Interactive (CDI), and CD-ROM XA (Extended Architecture) discs.

The 12cm-diameter CD-ROM and CD-I discs typically support up to about 660Mbytes data storage capacity. A single-sided, single-layer DVD-ROM disc supports 4.7Gbytes, and supports MPEG-2 video playback.

Increasingly, however, multimedia networks are being used, and the most significant of these is the ubiquitous Internet.

Multimedia Authoring Tool

A software tool intended for the development of multimedia. Many require no programming skills.

Multimedia Presentation

A multimedia-based presentation that might combine audio, midi, video, text, animations, and graphics. It might be presented on a desktop or even a notebook computer using their attached displays. Or it might be presented using an LCD projector.

Popular multimedia presentation programs include Microsoft PowerPoint that is included with Microsoft Office. Multimedia authoring tools such as ToolBook, IconAuthor, and Authorware may also be used, but these are not dedicated to the production of presentations.

Multimedia Streaming

A real-time delivery and playback of multimedia that may be local or remote. Typically, it takes place over the Web or Internet, and requires a server and client. Web applications include real-time monitoring or surveillance of remote locations, WebTV, and video playback.

Multiple Inheritance

A concept where subclasses inherit methods and data from more than one superclass. It defines a class of objects that inherits attributes and behavior from multiple superclasses.

Multisync Monitor

A monitor that may synchronize itself with various incoming signals. There are many technical implementations of the "multisync monitor," the simplest of which will automatically synchronize with perhaps two or three vertical frequencies. The term *multisync* was coined by NEC and is registered.

Professional versions are able to automatically synchronize with a range of horizontal and vertical frequencies. This is called the scanning range, and the greater it is, the greater the number of acceptable signal sources. Yet higher specification monitors economize on scan range, thus concentrating on the narrow band of professional graphic controllers beginning with VGA. Such monitors may be considered nonproprietary.

Multithreading

A process by which multiple processes within the same application are executed concurrently, or what is perceived to be concurrently.

Miscellaneous

Mbone (Multicast Backbone)

Mbone is a virtual infrastructure for delivering multicast packets over the Internet. It is composed of tunnels, and provides limited bandwidth, but enough for audio/video data.

A restriction mechanism integrated in MBONE routers or mrouters drops packets over tunnels where a predefined threshold rate is exceeded. Mrouters forward multicast packets to specified destinations.

MIME (Multipurpose Internet Mail Extensions)

MME is a standard specification that permits e-mail messages to include multimedia elements. It supports:

- ASCII alternatives such as foreign language character sets
- Images
- Multiple objects
- Audio

- Video
- Postscript

Included in served files is a MIME code that has a type and subtype, denoting the media included. Types of media, such as HTML and GIF, may obviously be displayed by any browser. Others require helper programs, and include MPEG video.

MIME was developed by Nathaniel Bernstein of Bellcore, and by Ned Freed of Innosoft.

M-JPEG (Motion-Joint Photographics Experts Group)

M-JPEG is a type of video that uses individual frames compressed according to the JPEG algorithm. It gives full frame updates as opposed to the predominantly partial frame updates of MPEG-1 video. M-JPEG video, therefore, provides random access points and lends itself to nonlinear editing.

In this respect it is more flexible on playback because applications can simply show any frozen frame of an M-JPEG sequence or play any selected frames of a sequence either backward or forward.

Other advantages of M-JPEG is that it may be compressed into other formats, including MPEG-1/2. A principal disadvantage of M-JPEG, however, is its comparatively low overall compression ratio.

MP3

MP3 is a compressed stream of digital audio created according to level 3 of the MPEG-1 audio/video specification that dates back to 1990. It is a popular format for distributing audio using the World Wide Web, and numerous MP3 Web sites exist from which such audio may be downloaded free of charge. An MP3 player application such as WinAmp or even a consumer appliance may be used to play resulting files, and they may be recorded or created using an appropriate MP3 recorder.

MP3 file quality is determined by the:

- Source recording quality
- Source format (which may be analog cassette, CD, DAT, etc.)
- Source playback device
- Sample rate (which may be 11.025kHz, 22kHz, or 44.1kHz)
- Sample size (which may be 4bit, 8bit, or 16bit)

Multiplexing

1. A process where an MPEG video stream is mixed with an MPEG audio stream to form an MPEG system stream.
2. A process where multiple signals may be communicated along a single transmission path that may be serial or parallel.

The Integrated Services Digital Network (ISDN) standard uses multiplexing and involves creating a data stream consisting of 8bit PCM blocks. The blocks are created every 125μs. By interleaving the blocks with those from other encoders, the result is time division multiplexing (TDM).

In North America ISDN typically interleaves data from twenty-four 64kbits/s sources or channels. This results in connections that provide 1.536Mbits/s. Although in actual fact the connection has a bandwidth of 1.544Mbits/s, because each channel's frame has a marker bit "F," adding 8kbytes/s.

Europe sees ISDN that typically interleave thirty 64kbits/s channels, giving 2.048 Mbits/s. This and the 1.544Mbits/s connection are known as primary rate multiplexes.

Further interleaving of primary rate multiplexes sees:

- 6, 45, 274Mbits/s in North America
- 8, 34, 139, and 560Mbits/s in Europe

PCM was conceived in 1937 by Alec Reeves but was not applied widely for many years.

Acronyms

M3UA	MTP3-user adaptation layer
MAC	Media access control
MAI	Multiple access interference
MAP	Maximum a posterior
MASHO	Mobile-assisted soft-handoff
MBS	Mobile broadband systems
MC	Multicarrier
MCTD	Multicarrier transmit diversity
MCU	Main control unit
MD	Meditation device
MD-IS	Mobile data intermediate system
MDM	Mobile diagnostic monitor

MDR	Medium data rate
MExE	Mobile station execution environment
MFA	Management functional area
MIB	Management information base
MIN	Mobile identification number
MIPS	Millions of instructions per second
MIT	Management information tree
MM	Mobility management
MMSE	Minimum mean-squared error
MNRP	Mobile network registration protocol
MO	Managed object
MOC	Managed object class
MOl	Managed object instance
MOS	Mean opinion score
MPT	Ministry of Posts and Telecommunications
MRC	Maximal ratio-combining
MSE	Mean-squared error

N

Network Design and Deployment

Designing and upgrading mobile networks have become frequent exercises for operators that historically did little more than maintain network infrastructures, install consumer appliances, and bill subscribers. Today's operators rely heavily on technical departments and network architects to evolve and develop new networks such as GPRS, EDGE, and 3G. This section addresses the design process in terms of key technical and business considerations that range from population densities to network core components. The design process begins with certain milestone decisions that rely on the following common vernacular:

- *Cell loading* is the occupancy of the cell, as a percentage of the maximum capacity.
- *Erlang/km*^{2} is the number of calls per square km and is suitable for CS voice calls. For data services, traffic may be measured in Mbps/km^2.
- One erlang is a unit measure equating to a 1-h call, so traffic intensity is erlang(s) = (calls/hour $\times$ average call duration in seconds)/3600; mErlangs (= 0.001 erlang) are used if the result is less than 1.
- *Loading factor* is cell interference caused by surrounding cells, and is the ratio (power received by BS from other cells):(power received from mobiles in cell)
- *Outage* is the probability that a QoS level won't be met.
- *Spectral efficiency* is the traffic capacity delivered at a given bandwidth and in a specified area: traffic intensity (erlang)/(bandwidth $\times$ area) = bps/(MHz $\times$ km^2).

Classification of Environment

A mobile network is typically deployed over numerous environments including dense metropolitan, urban, suburban, and rural. Dense metropolitan districts include tall buildings with 10 or more stories with a population density of 20,000 subscribers per square mile. Urban areas include residential and office areas with buildings of 5 to 10 stories high, and with a population density of 7500 to 20,000 subscribers per square mile. Suburban areas have housing with one to two stories that are 50m apart, shops and offices that are two to five stories, and population densities between 500 and 7500 subscribers per square mile. Rural areas include farming communities with a population density as small as 500 subscribers per square mile.

Cell Architectures and Frequencies

GSM network design requires frequency planning because all cells cannot share the same frequency, and it optimizes network capacity. Adjacent cells may not share frequencies because of interference levels that would peak at cell borders, and communications would fail where cells overlap. Cell dimensions are driven by transmitter and receiver performances, and general signal losses, and hundreds or thousands of cells require frequency reuse where some cells share the same carrier frequencies. See Figs. N-1, N-2, and N-3.

GSM cells are circular in reality, but are considered to be hexagonal for design purposes, arranged in consistent clusters that have symmetric geometries with 3, 4, 7, 12, 15, or 21 or more cells. Small clusters allow more reuse of the same frequency, but increases cochannel interference caused by cells operating on the same frequency. As such efforts have to be made to architect a network with an appropriate cluster size.

Cells may be divided into *i* sectors or subcells, and hierarchical cell structures may be used where traffic is partitioned among cells. Macrocells may handle fast mobile stations for fewer handovers (HOs) in underlying smaller cells. Macrocells have a diameter of a few hundred meters diameter and are used to increase capacity, and target city pedestrians. Picocells target traffic hot spots, and measure only tens of meters, and prevail in indoor environments. Picocells are unsuitable for frequent handovers such as those caused by mobile users in transport. GSM cochannel interference may be combated using frequency hopping, power control, discontinuous transmission (DTX), multi-user detection (MUD), and cell sectorization. The latter sees increasingly reduced cochannel interference as more cell sectors are used.

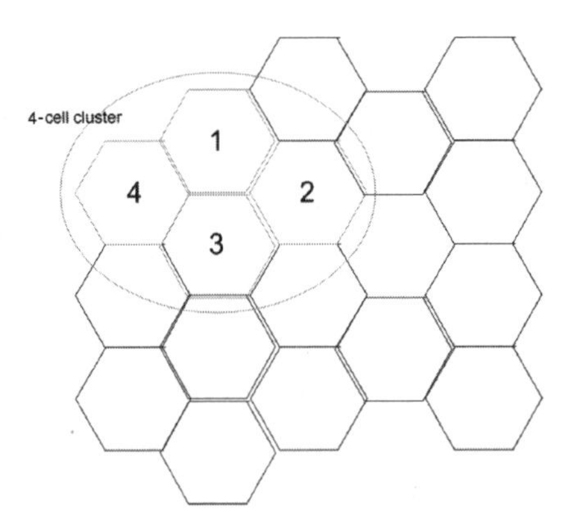

Figure N-1 GSM (TDMA) 4-cell cluster.

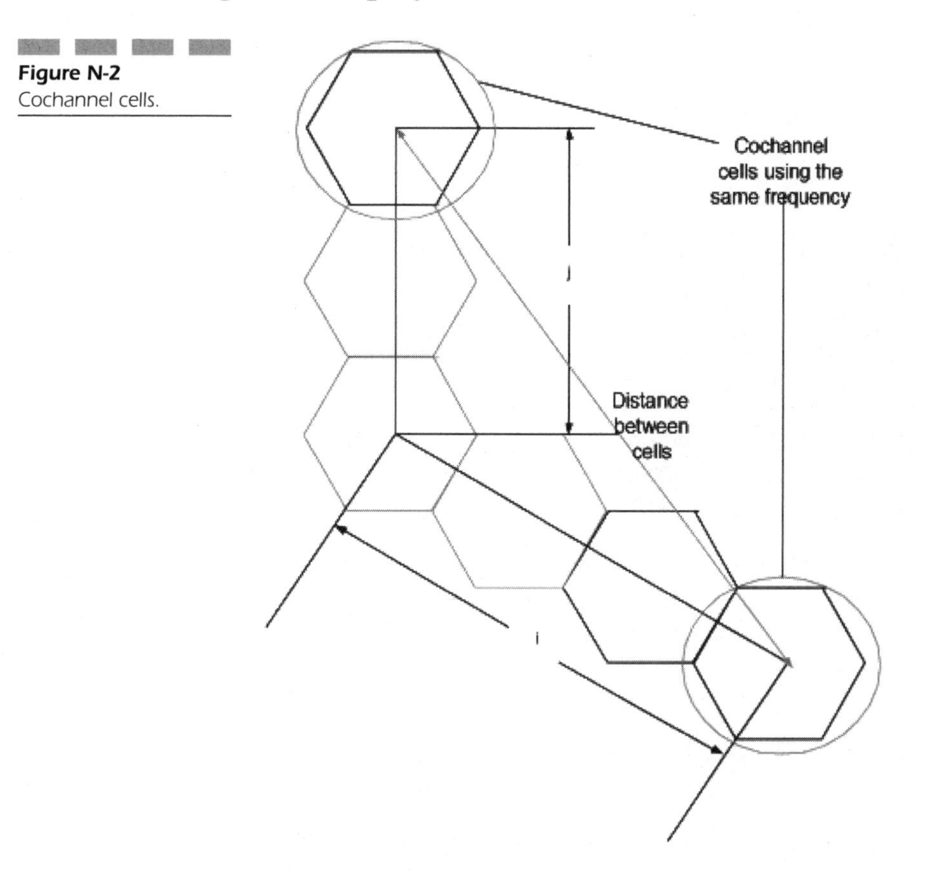

Figure N-2
Cochannel cells.

CDMA network cells use an alike frequency, and a typical WCDMA operator has a 15MHz frequency allocation for 3 × 5MHz WCDMA frequency channels. The UMTS Forum recommends a minimum license allocation of 2 × 15MHz (FDD) + 5MHz (TDD). WCDMA cell sizes are variable, and may be "pulsating"

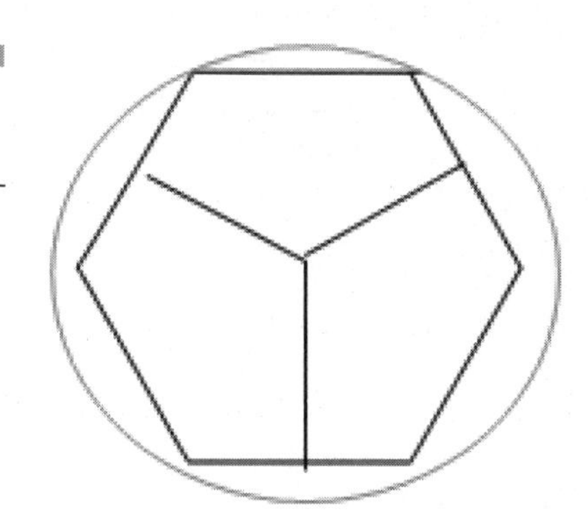

Figure N-3
Antenna pattern for three-sector cell.

or "breathing" and alter with capacity. It can be generally assumed that new BSs are needed to increase capacity.

Radio Network

Obtaining cell sites may be difficult when the land is unobtainable or unsuitable. The electromagnetic emission from BSs also continues to be an issue. Reducing the number of HOs and subsequent signaling traffic may be achieved using large macrocells. The following procedures should be implemented:

- Control channel power planning
- Detailed study of radio environment
- Interfrequency (HO) planning
- Iterative network coverage analysis
- Radio-network testing
- Soft handover (SHO) planning

CDMA Signal Strength

Signal strength is the power received by a copolarized dipole antenna into a 50Ω load. ERP (effective radiated power) and path loss drives signal strength, where the loss is a culmination of propagation characteristics, distance, and obstructions. The sensitivity of the mobile user is the signal strength required in an open area under normal atmospheric and environmental conditions like a rural area where less than 10 to 12dB attenuation allowance is adequate. Densely populated metropolitan areas, on the other hand, may require 15 to 20dB.

Cell Configurations The cell configuration depends on the classification of environment, and in rural areas 300ft towers with high-gain antennas may be appropriate. In suburban areas 100 to 200ft towers may be used with low-gain antennas. Dense metropolitan districts include tall buildings with 10 or more stories with a population density of 20,000 subscribers per square mile. Urban areas include residential and office areas with buildings about 5 to 10 stories high, and with a population density of 7500 to 20,000 subscribers per square mile. Suburban areas have housing with one to two stories that are 50m apart, shops and offices that are two to five stories, and population densities between 500 and 7500 subscribers per square mile. Rural areas include farming communities with a population density as small as 500 subscribers per square mile.

Cell Overhead Channels CDMA cells transmit pilot, synchronization, and paging signals. Power allocations for the forward link as a percentage of total transmitted power are:

- Paging channels: 6–7 percent
- Pilot channel: 15–20 percent
- Synchronization channel: 2 percent
- Traffic channels: 71–77 percent

(*See* cdmaOne Radio Interface under letter C.)

Power Allocation Base station output power calibration ensures that when the BCR and digital gains (dg) are established, the antenna receives the required output power. When the transmit path is calibrated, the power is adjusted electronically, so when the BCR attenuation falls, more power is given to the traffic channel. And when attenuation increases, traffic channels are allocated less power. The transmit amplifier is not overdriven because maximum output power controls forward link overload control. See Fig. C-2.

Power allocations for the pilot, paging, sync, and traffic channels may be calculated using the formula:

$$\text{Power} = \text{scale} \times 10^{-(\text{bcr})/10} \times \left[v_{pl} G^2_{pilot} + v_{pg} G^2_{page} + v_{sy} G^2_{sync} + v_{tr} G^2_{traffic,m} \right]$$

G_{pilot} and v_{pl}	Pilot channel gain and channel activity
G_{sync} and v_{sy}	Sync channel gain and channel activity
G_{page} and v_{pg}	Paging channel gain and channel activity
$G_{traffic,m}$ and v_{tr}	Traffic channel gain and channel activity

Cdma2000

Cdma2000 uses a wideband, spread spectrum CDMA radio interface that accommodates 3G requirements and those specified by the ITU and by the IMT-2000 (International Mobile Telephony), and it is also compatible with cdmaOne standards. Cdma200 may be deployed in indoor/outdoor environments, wireless local loops (WLLs), vehicles, and hybrid vehicle/indoor/outdoor scenarios. Outdoor megacells are less than 35km in radius, and outdoor macrocells may be 1km to 35km in radius. Indoor/outdoor microcells are up to 1km in radius, and indoor/outdoor picocells are less than 50m radius.

Cdma2000 may be summarized as supporting:

- Data rates from a TIA/EIA-95B compatible rate of 9.6kbps to greater than 2Mbps

- Circuit-switched and packet switched
- Channel sizes of 1, 3, 6, 9, and 12 × 1.25MHz
- Modern antennas
- Large cell sizes to reduce cell numbers per network
- Higher data rates for all channels
- High-speed circuit data, B-ISDN, or H.224/223 teleservices
- Mobility for speeds up to 300mph

Cdma2000 reuses the following standards:

- IS-127 enhanced variable rate codec
- IS-634A
- IS-637 short message service (SMS)
- IS-638 over-the-air activation and parameter administration
- IS-707 data services
- IS-733 13kbps speech coder
- IS-97 and IS-98
- TIA/EIA-41D
- TIA/EIA-95B channel structure
- TIA/EIA-95B extensions
- TIA/EIA-95B mobile station and radio interface specifications

Network Reference Model

IS-657 specifies a network reference model for packet data services and protocol options. See Fig. N-4.

- Base station (BS): Equipment on the land side of the *Um* interface, including radio processing and management and protocol processing and management.

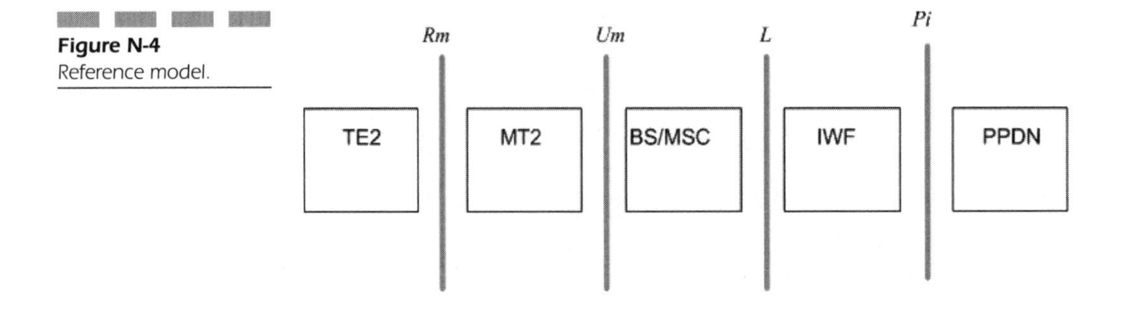

Figure N-4
Reference model.

- Interworking function (IWF): Functions needed for equipment connected to mobile terminations to interface with networks like PSTN or CDPD.
- Mobile switching center (MSC): Functions of the cellular switch.
- Mobile termination 2 (MT2): Provides a non-18DN (*Rm*) user interface.
- Mobile termination O (MTO): A self-contained mobile termination that does not support an external interface.
- Public packet data network (PPDN): A public packet data network is a transport for packet data between elements.
- Terminal equipment 2 (TE2): A data terminal device that has a non-15DN user network interface.

Reference Points

- *Rm*: Physical interface between TE2 to an *MT2*
- *Um*: Physical interface connecting an MTO or MT2 to a IS/MSC
- *L*: Physical interface connecting a BS/MSC to an IWF
- *Pi*: Physical interface connecting IWF to a PPDN

Networks

1G (First-Generation) Networks

See 1G (First-Generation) Networks in Numerals chapter.

2G (Second-Generation) Networks

See 2G (Second-Generation) Networks in Numerals chapter.

GSM Network Operation

See GSM Network Operation in Numerals chapter.

2G Origins

See 2G Origins in Numerals chapter.

2.5G Networks

See 2.5G Networks in Numerals chapter, 2.5G GPRS in Numerals chapter.

Terminal Classes

See GPRS Terminal Classes in Numerals chapter.

GPRS Architecture

See GGSN, GPRS Mobility Management, GPRS Network, SGSN, all in Numerals chapter.

3G (Third-Generation) Networks

See 3G (Third Generation) in Numerals chapter, 3G Layers in Numerals chapter.

3G Origins WAP and I-mode have been key to mass-market mobile wireless applications that converge on the Internet, giving limited access to Web content. The precursor to WAP is, of course, SMS (Short Message Service) where text is sent to mobile users' handsets, and while this is considered the trailing edge of mobile applications, such solutions remain a practical industry for many WASPs globally. WASPs such as these are (generally) positioned close to the trailing edge, as opposed to being leading-edge enterprises that may be engaged in developing core software solutions.

The European RACE 1043 project began with the aim of identifying services that Y2K 3G services would deliver, and evaluating how the mobile network infrastructure would evolve in the mass market telecommunications sector. The project's forecasts include a displacement theory where TACS (Total Access Communications System) would be displaced by GSM that would then be displaced by UMTS (Universal Mobile Telecommunications System)–a term that was coined by the project.

However the intermediate transitions from 2G GSM to 3G were not foreseen, including the incremental advancements of GPRS (General Packet Radio Service) and EDGE.

The 3GPP (Third-Generation Partnership Project) was assigned the task of specifying a 3G system based on underlying GSM network. 3GPP consisted of:

- Standards organizations: ARIB (Japan), CWTS (China), ETSI (Europe), TI (USA), TTA (Korea), and TTC (Japan)
- Market representation partners: Global Mobile Suppliers Association (OSA), the OSM Association, the UMTS Forum, the Universal Wireless Communications Consortium (UWCC), and the IPv6 Forum
- Observers: TIA (USA) and TSACC (Canada)

Satellite Networks

See Mobile Satellite Service (MSS) under letter M.

Networking Approaches

Satellite constellations providing broadband services may be:

- Ground-based with terrestrial network functions
- Space-based with space segment network functions

Ground-Based Constellation Network Ground-based constellations have satellites that are space-based retransmitters. These use "bent-pipe" frequency shifting and amplification, or signal regeneration with baseband digital signal processing (DSP). The satellite provide a "last hop" to the ground network, and commercial examples of this system include *Globalstar* and *Skybridge*. The ground-based network topology may take many forms and is shaped by geography, politics, and legislation. Satellite telemetry, tracking, and control (TT&C) ground stations ware networked to share the constellation's information.

Space-Based Constellation Network Space-based constellation satellites have on-board processing (OBP), and communicate with neighboring satellites using high-frequency radio or laser intersatellite links (ISLs). Using ISLs the space segment holds the network layer, and satellites have routing and switching functions. Commercial and proposed ISL systems include the *Teledesic* constellation and Hughes' GEO. See Fig. N-5.

Internetworking with Satellite Constellations Fixed "fore" (ahead) and "aft" (behind) intersatellite link equipment for intraplane communication between satellites is possible with circular orbits. This is not possible with interplane communication between satellites in different orbits because of:

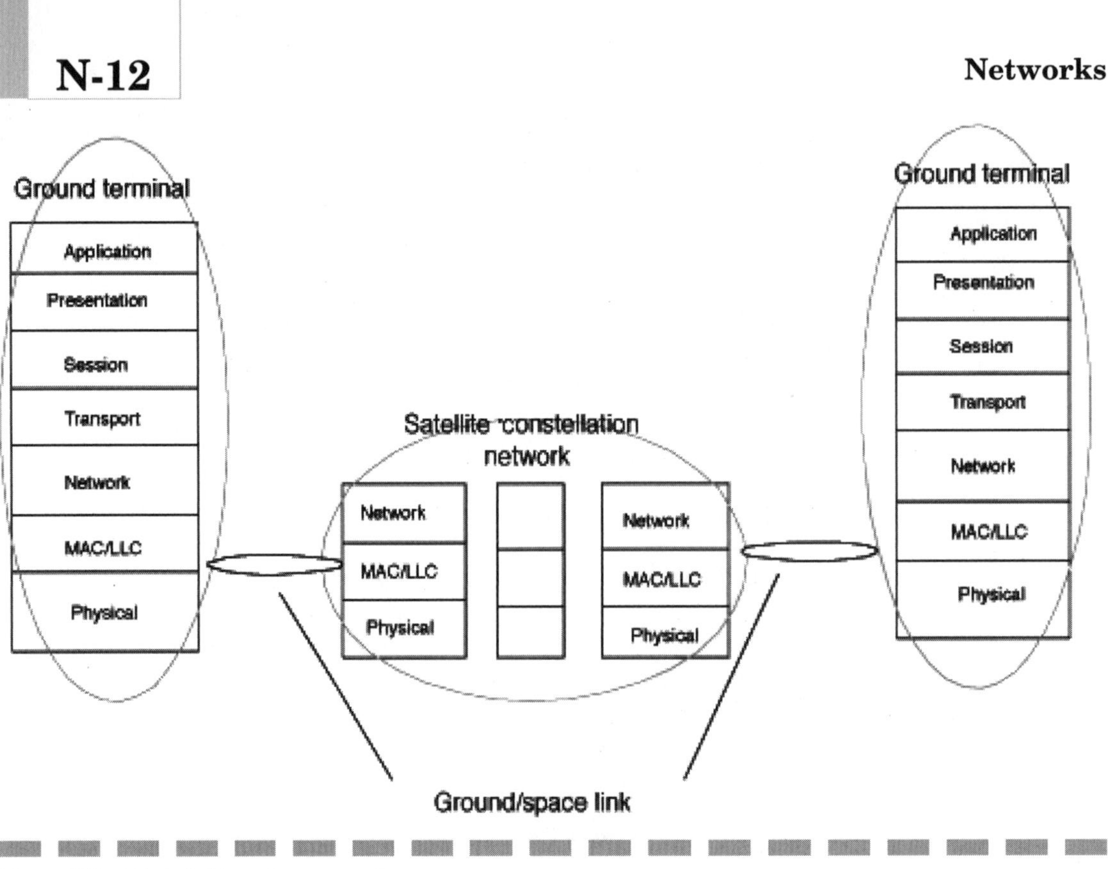

Figure N-5 Teledesic ISL routing approach.

- High relative velocities between satellites
- Complexity of tracking control as antennas slew around
- Doppler shift with changing distances

UMTS Acronyms

See UMTS Acronyms in Numerals chapter.

UMTS Network Architecture

See UMTS Network Architecture in Numerals chapter.

UMTS Terrestrial Radio Interface

See UMTS Terrestrial Radio Interface in Numerals chapter.

Nonaccess Stratum

In the UMTS architecture, UTRAN sees the air interface isolated from the MM (mobility management) and the connection management layers. This notion takes the form of the access stratum (AS) and the nonaccess stratum (NAS). The AS includes radio access protocols that terminate in the UTRAN, while the NAS has core network protocols between the UE and CN that are not terminated in the UTRAN that is transparent to the NAS.

The NAS attempts to be independent of the radio interface, as do the protocols MM, CM, GMM, and SM. These protocols are theoretically constant in terms of the radio access specification that carries them, so any 3G radio access network (RAN) should connect to any 3G CN. The GSM's MM and CM layers are almost the same as 3G NAS, and NAS layers will compare with future GSM MM and CM layers according to the phase 2+ specification.

Lower layers from the AS are different from GSM where the radio access technology (RAT) is TDMA, as opposed to CDMA for UTRAN. The protocols used therefore are also radically different, as is the packet-based GPRS protocol stack. So this reveals an obstacle in the so-called smooth migration path from 2.5G GPRS to 3G. However, the GPRS CN components can be reused when renovating the system to become a 3G solution.

The AS has three protocol layers that include sublayers:

Layer	Sublayers
Physical layer (L1)	
Data link layer (L2)	Medium Access Control (MAC) Radio link control (RLC) Broadcast/multicast control (BMC) Packet data convergence protocol (PDCP)
Network layer (L3)	Radio resource control (RRC)

The two vertical planes include the control (C-) plane and the user (U-) plane, and both have the MAC and RLC layers, while the RRC layer is present only in the C-plane.

Computers/Software

NCSA Mosaic

A browser developed at the National Center for Supercomputing Applications at the University of Illinois at Urbana-Champaign. This browser is distributed under a licensing agreement with Spyglass, Inc., and contains security software licensed from RSA Data Security Inc. Portions of this software are based in part on the work of the Independent JPEG Group and also contains SOCKS client software licensed from Hummingbird Communications Ltd.

Nested

1. *Nested transaction:* A technique that sees the integration of subtransactions within transactions. The subtransactions are said to be nested. *See* Server, Transaction.
2. *Nest loop:* A loop in a computer program that is encapsulated within another.

Nested Top-Level Class

A static member of an enclosing top-level class.

NetMeeting

A Microsoft technology that permits a multiplicity of communication and information exchange types over the Internet and over compatible IP networks such as intranets. The communications types supported include:

- Internet telephony
- Whiteboards
- Application sharing
- File transfer
- Chat
- Multiple participant conferences

NetMeeting SDK (Software Development Kit)

A software suite that may be used to integrate conferencing features into applications.

Netscape Navigator

A Web browser produced by Netscape. Its functionality is improved through the addition of plug-ins. Plug-ins for streaming audio and video are available. Like many other browsers, Navigator may be used to send e-mail messages, but it is not an e-mail application. A bookmarks window assists users to list and revisit Web sites that are of interest. Images that are shown in the client area may be saved to disk by right-clicking on them.

Netware

A network operating system (NOS).

Network

1. A physical or wireless entity that unites computer systems.

2. A physical entity that interconnects processors in an MPP.

3. An interconnecting scheme for neural networks.

Network Computing

A broad term denoting the use of systems that are connected to physical networks.

NFS (Network File System)

An NFS server allows users to share files on other hosts.

NNTP (Network News Transfer Protocol)

A protocol for transferring Usenet news between servers, clients, and a central server.

Noninterlaced

A mode of CRT-based monitor/display operation in which the screen image is generated by scanning all lines in a single scan field. The rate at which all lines are scanned is termed the refresh rate or the vertical scan rate. The frequency that lines are scanned is termed the horizontal frequency.

Nonpreemptive Multitasking

A type of multitasking in which the operating system does not interrupt applications. It is less seamless than preemptive multitasking in that a reasonable degree of concurrency is not achieved. Windows 3.1 (and earlier) and Windows for Workgroups 3.1x offer nonpreemptive multitasking.

Notification Services

An extension to the OMG Event Service that defines interfaces and operations for transmitting events between client and server applications.

n-Tier Client/Server Architecture

A client/server architecture that sees multiple divisions of application logic and data. The divisions are distributed across four or more systems that represents the number of tiers (n).

NUMA (Nonuniform Memory Architecture)

A variation of the SMP system architecture that attempts to solve the bottleneck of using a single shared bus to interconnect processors. Instead, a number of internal buses are introduced, thus promoting processor scalability.

Numa systems have modules called quads that include processors, memory, and I/O devices that share an internal bus. The modules are called quads and interconnect via a main bus.

The Numa architecture permits processors to access:

- Local and external caches
- Local and external memory

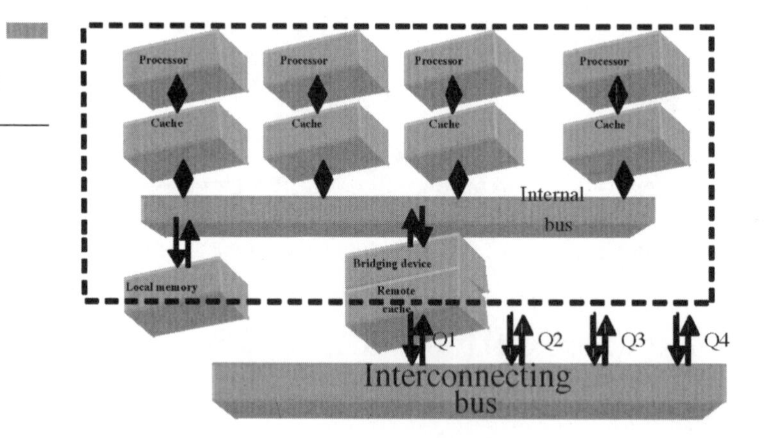

Figure N-6 Numa quad-based architecture.

Its nonuniform characteristic hinges on the varying access time, between local and remote memory access.

Though Numa is architecturally superior to SMP, it is not classifiable as an MPP system. See Fig. N-6.

Miscellaneous

Network Interface Card

A device used to connect a system to a network. It may be one (or even more) of a number of standard and proprietary variants, including:

- Ethernet
- Token Ring
- ISDN interface
- Fibre channel interface (arbitrated loop)
- Modem (such as an analog or cable variant)

NDIS (Network Driver Interface Specification)

A standard specification for network interface cards (NICs). It provides functions collectively referred to as a wrapper that may be used by TCP/IP drivers. It was developed by Microsoft and 3COM. Its implementations for Windows may be assumed to be proprietary.

Nyquist Relationship

Harry Nyquist determined the relationship between a channel's bandwidth (W Hz) and signaling rate (baud rate B) as:

$$B = 2W$$

Acronyms

MSC	mobile switching center
MS-GSN	mobile-to-GSN
MT	mobile terminal

MT	mobile termination
MTO	mobile termination
MTP3-B	message transfer part 3-B
MUD	multiuser detection
MUX	multiplex, or multiplexer
MWIF	Mobile Wireless Internet Forum
NAK	negative acknowledgment
NBAP	Node B application part
NCAS	non-call-associated signaling
NE	network element
NEI	network entity identifier
NEL	network element layer
NIC	network interface card
NIU	network interface unit
NLUM	neighbor list update message
NM	network management
NMF	Network Management Forum
NML	network management layer
NOS	network operating system
N-PDU	network layer packet data unit
NRT	non-real-time
NSM	network and services management

O

O-2

OFDM

Orthogonal frequency-division multiplexing (OFDM) is a transmission method used in MMDS, and uses multiple carriers or discrete tones that are orthogonal or independent. The independence leaves guard bands needed only for groups of tones. Typically there are as many as 512 tones, each representing a carrier using QPSK modulation. OFDM systems transmit data in bursts to help lower intersymbol interference. The bursts have a cyclic prefix that absorbs transients as a result of multipath signals.

Developed by Cisco, the VOFDM (vector orthogonal frequency-division multiplexing) method is also used by MMDS, and adds spatial diversity to OFDM. This method relies on multiple antennas that add up to −10dB over MMDS, and provide a different set of multipath signals for each antenna. VOFDM gives broadband performance and spectral efficiency.

Multichannel Multipoint Distribution System

A multichannel multipoint distribution service or *microwave multipoint distribution service,* operates in the 2.5 to 2.7GHz frequency range. MMDS provides data rates up to Mbps and may deliver broadband services where users may receive data rates of 2Mbps. MMDS operates at a lower frequency spectrum than LMDS, and is therefore not limited to line-of-sight paths and is less susceptible to interference. It also provides an alternative to fiber optic, DSL, cable, or other access technology that may be absent from certain buildings and/or developments. MMDS requires an antenna and radio receiver at the subscriber's premises and a coaxial cable to customer premise equipment (CPE).

MMDS signals may be transmitted for distances of up to 30mi, and are deflected by objects and reassembled at the receiver using various techniques that include vector orthogonal frequency-division multiplexing (VOFDM), which supports data rates up to 50Mbps. Figure M-1 shows video-on-demand application of MMDS, including a video distribution station using a satellite receive-only (RO) antenna to receive TV signals that are downconverted to VHF frequencies, and upconverted and transmitted to the subscriber. Low-gigahertz operation means that subscribers require perhaps a 60cm-high Yagi, a partial parabolic, or flat-array antenna.

iSpeed is a commercial MMDS used for Internet access in the San Francisco Bay area, and customers may choose between iSpeed Singlepath and iSpeed Dualpath which is a two-way service. iSpeed Singlepath uses standard telephone connections for uplink communications and broadly equates to T1. Dualpath operates at 1.5Mbps with bursts as high as 10Mbps.

Offset Quadrature Phase-Shift Keying

QPSK has two orthogonal binary phase-shift keyed (BPSK) channels. There are even (I) and odd (Q) bitstreams, where the former modulates the cosωt term, and the latter modulates the sinωt term. The carrier phase may change every $2T$ seconds, and if neither bitstream changes sign, it is unchanged. A change of sign for I and Q results in a π or 180-degree phase shift. When this is filtered by the transmitter and receiver filters, it changes the amplitude of the detected signal and causes errors.

When I and Q are offset by a half-bit interval, amplitude changes are reduced because there is no 180-degree phase shift. This is offset quadrature phase-shift keying (OQPSK) modulation, and is QPSK with the odd bitstream Q *delayed* by a half-bit interval relative to I. The resulting phase transition is only 0 and 90 degrees, and the 180-degree phase change at the end of the bit interval is reduced to 90. Phase changes still give small amplitude fluctuations in the transmitter and receiver, however.

Open Loop Power Control

Taking a CDMA network as an example, uplink signals arriving at the BS receivers should be equal strength, irrespective of the mobile stations' distances. This is the so called near-and-far effect that is remedied by mobile stations transmitting with varying power levels.

Downlink signals from a single base station will not interfere with each other if fully orthogonal, but this is difficult to achieve. For example, signals from other BSs will present interference, and signal reflections also cause nonorthogonal interference. Ideally these signals should be transmitted with the lowest power level.

Open-loop power control requires the estimate of interference based on the received signal, which often uses a different frequency to the transmitted signal. The UTRA time-division duplex (TDD) mode uses the same frequency for uplink and downlink, so open-loop power control is fairly accurate.

A constant BS transmission power allows the mobile station to measure the received power level and adjust transmission power (see Fig. O-1). If there is

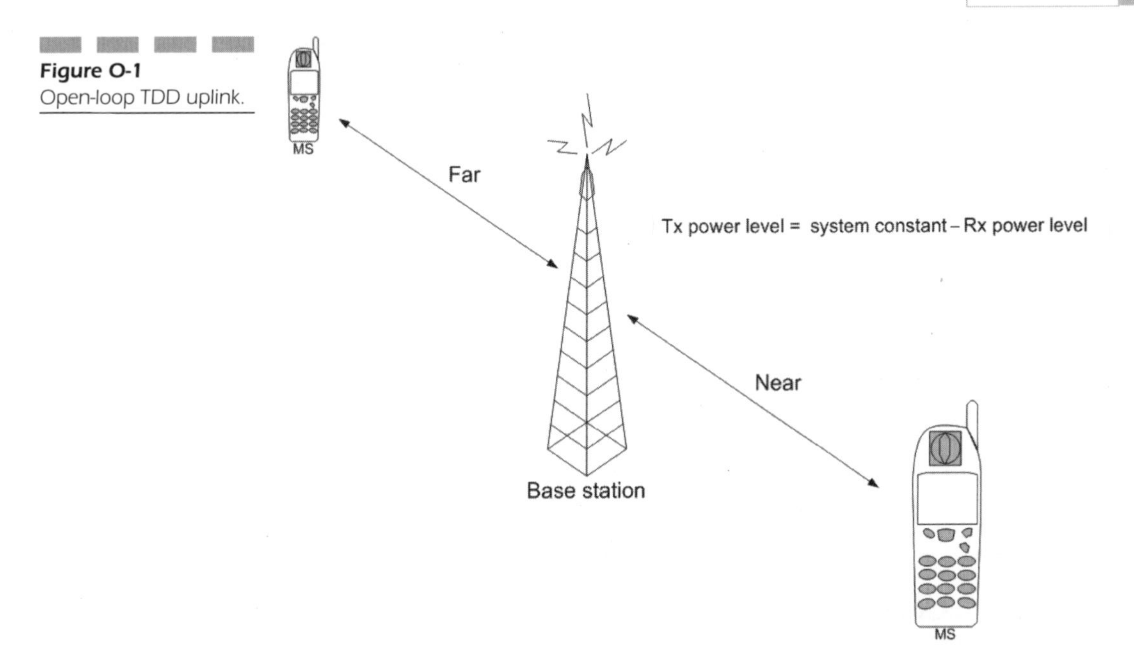

Figure O-1
Open-loop TDD uplink.

not a constant transmission, the MS monitors some other channel or obtains the transmission power of the BS, perhaps using a sync channel.

Closed-Loop Power Control

Closed-loop power control relies on quality measurements at the BS that are sent to the mobile station, and is more accurate than open-loop power control, but is not responsive to channel changes. See Fig. O-2.

Closed-loop power control is used by UTRAN FDD where the received signal-to-interference ratio (SIR) is measured over a timeslot (or 667μs). This information drives automated control to increase or decrease transmission power.

Operating Systems

Client operating systems (OSs) are of various levels of sophistication and include EPOC, Windows CE, and Palm OS. There are also OS-independent environments from Sunsoft. Most of the mobile OS solutions are lite and require variously thin hardware requirements. 3G applications that exploit bandwidth with video require relatively fat mobile clients with displays and browsers able to deliver video that may be processed by an onboard codec, or displayed using perhaps the Flash plug-in for Windows CE.

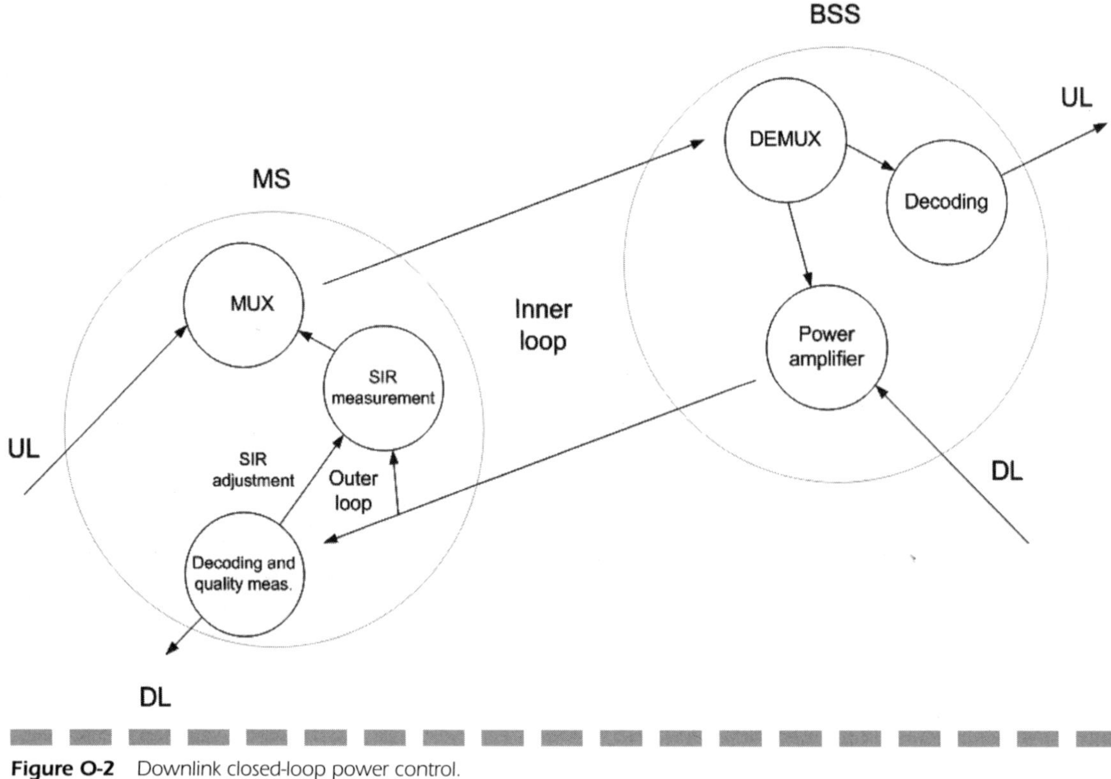

Figure O-2 Downlink closed-loop power control.

Palm OS

The Palm Pilot from Palm Computing drove the development of Palm OS that had an 80 percent market share of the PDA market (2000). Licensees of Palm OS helped evolve it, with Handspring adding the Springboard module for plug-ins like MP3 players. The Palm VII uses the Mobitex packet-based mobile system in a proprietary Palm.net service that provides *Palm Query Applications* (PQA). Each PQA uses *Hypertext Markup Language* (HTML) and a proprietary Wireless access protocol called WebClipping.

Windows CE

Microsoft Windows CE is a thin "consumer device" version of the desktop or mobile version. Windows CE 3.0 is the basis for Pocket PC devices that offer multimedia through Media Player and Flash. It has been successful with A/V appliances like set-top boxes (STBs). Windows CE uses Wm32 APIs, and uses multithreading and multitasking.

EPOC

Symbian's EPOC is aimed at smartphones and communicators, and three variants are available addressing different devices: Pearl, Crystal, and Quartz. Pearl is a smartphone, Crystal is the data-centric design, and Quartz is midway between the two. EPOC development is mostly in C++, and to a lesser extent, Java.

Optic Fiber

Silica-based fiber propagates light signals while inducing minimal losses. Light propagation is driven by total internal reflection that is made possible using a core fiber and surrounding cladding of different refractive indexes. The light source must emit light into the cable at the critical angle in order to achieve total internal reflection. Numerous different types of optic fiber exist, including graded and step index.

A step index has a core silica of one refractive index and a cladding of another. A graded fiber consists of a core fiber that is coated with a number of grades of silica of differing refractive index. The advantages of optic fiber include:

- Light and easy to install
- Immunity to electrical and reasonable levels of electromagnetic interference
- Exceptionally wide bandwidth when compared to electrical conductors
- Cost effective

The operation relies upon total internal reflection, given by reflecting injected rays in the cladding. The core and the cladding, therefore, have a different refractive index. The angle at which rays are injected into a fiber is critical, in order to achieve total internal reflection, and to propagate the ray appropriately. The numerical aperture (NA) of a fiber is a measure of the size of its acceptance cone, or the range of angles at which rays must be injected.

Propagated rays may be:

- Meridional that repeatedly intersect the core's axis.
- Skew that spiral through the core without ever intersecting the axis. Their launch angle tends to be greater than that of meridional rays.

The light source must be an Led or laser device, which *lases* at an appropriate wavelength. A multimode step-index fiber may have a core diameter of between 125 and 500μm, and an NA of the order of 0.15 to 0.4. They are able to propagate a substantial amount of emitted light from an Led. Injected light is dispersed into many thousands of paths called nodes.

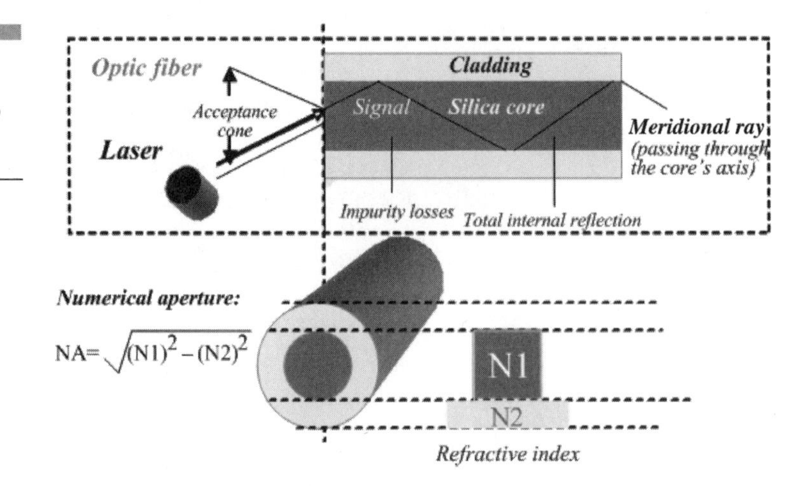

Figure O-3 Optic fiber—monomode. (See also 1979 in Numerals chapter.)

Optic fiber displaced many media including a small portion of satellite communications and particularly transoceanic telephone systems. Fiber capacities are in the Gbps range, and wavelength division multiplexing (WDM) allows a single fiber to carry 10Gbps channels. See Fig. O-3.

Origins

On February 19, 1880, wireless telecommunications were invented by Alexander Graham Bell, as he and his assistant Sumner Tainter transmitted a beam of light that was modulated with a voice signal and was successfully decoded. The transmission over 213m took place in Washington, D.C., from the roof of Franklin School House to Bell's Laboratory at 1325 L Street. Bell concluded that it was his most important invention. This momentous

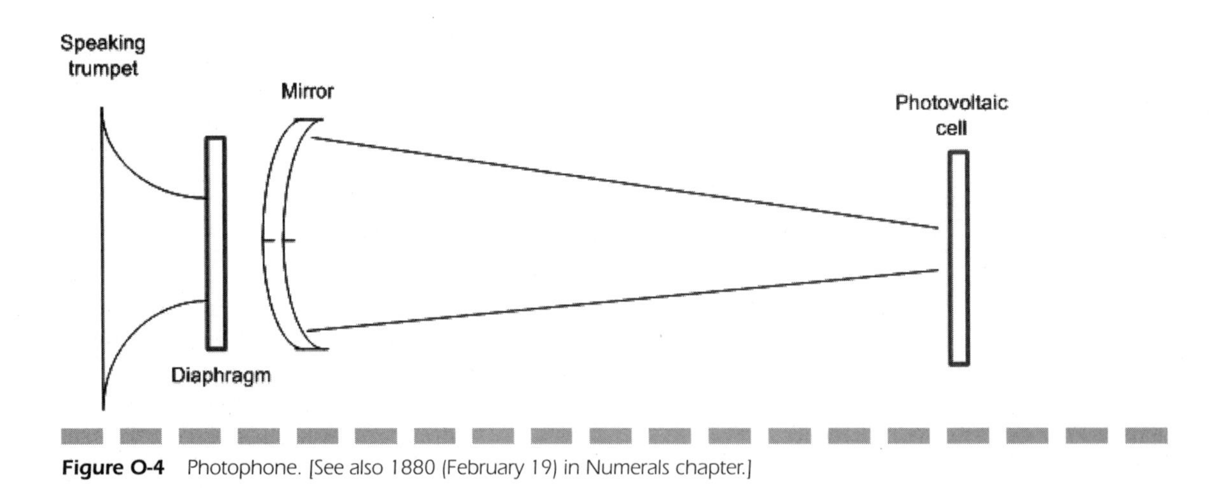

Figure O-4 Photophone. [See also 1880 (February 19) in Numerals chapter.]

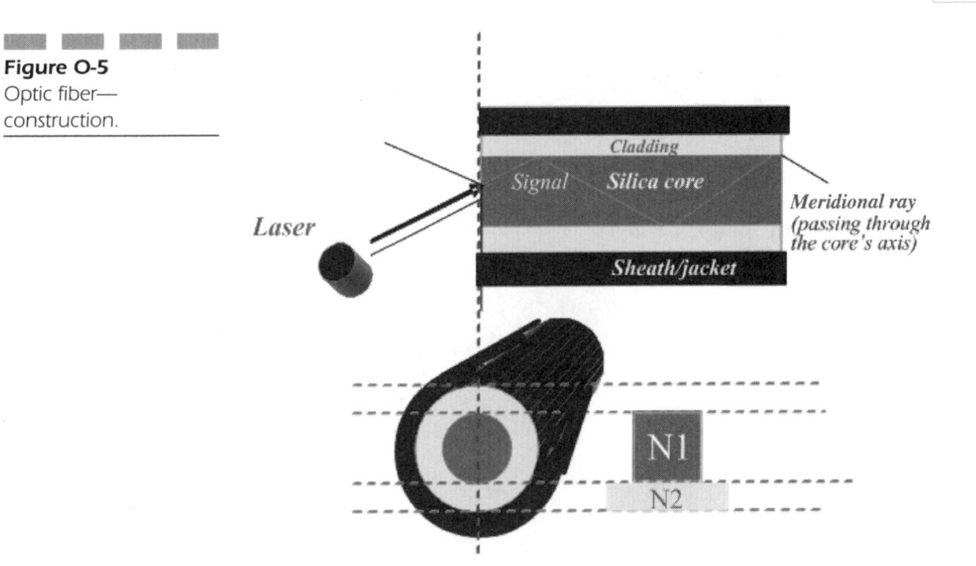

Figure O-5
Optic fiber—construction.

occasion was, of course, also the invention of lightwave communication. This invention was mocked in the *Washington Post* where an anonymous person wrote: "Does Professor Bell intend to connect Boston and Cambridge with a line of sunbeams hung on telegraph posts, and if so, what diameter are the sunbeams to be, and how is he to obtain the required size?"

Irony surrounds this cynicism because it has proved an accurate and prophetic statement, when considering modern fiber-optic networks. See Figs. O-4, O-5, and O-6.

Japan's Nippon Telegraph and Telephone Public Corporation

In 1979 Japan's Nippon Telegraph and Telephone Public Corporation developed a silica fiber with losses of just 0.2dB per kilometer. Until this point losses

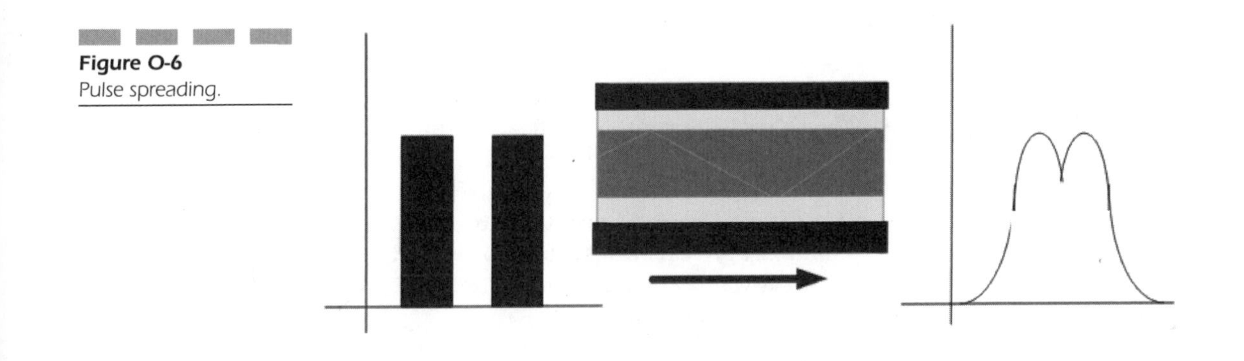

Figure O-6
Pulse spreading.

in the fiber's core that were caused by an unacceptable density of impurities meant that optic fibers were only practical over short distances because of the attenuation.

The resulting low fiber core losses meant the optic fibers became a practical lightwave communications medium, and a viable replacement for copper and aluminum conductors. The advantages of silica-based optic fiber became apparent, and included:

- Ease of installation due to their light weight, compactness, and flexible construction
- Wide bandwidth offering large scale multiplexing
- Immunity to corrosion
- Immunity to electrical and electromagnetic interference
- Cost effective—in certain instances, the replaced copper scrap value helped cover the cost of optic fiber installation
- Durable, reliable, offering longevity
- Inexpensive

Operation

Snell's Law Snell's Law explains the operation of an optic fiber and is based on the refraction of light as it passes from one medium to the next. It states that the ratio of the sine of the angle of incidence to the sine of the angle of refraction is equal to the ratio of the propagation velocities of the wave in the two respective media:

$$\frac{\sin A_1}{\sin A_2} = \frac{V_1}{V_2} = K = \frac{n_2}{n_1}$$

where A_1 = angle of incidence, A_2 = angle of refraction, V_1 = light velocity in n_1, V_2 = light velocity in n_2, and n_2 and n_1 are the refractive indexes of core and cladding.

Construction

The fiber core is a dielectric that has cladding, and a protective sheath or jacket. The refractive index is highest at the core material, and lowest at the cladding's outer parts. A graded index fiber has a refractive index that gradually decreases as light moves from the core's center. A step-index fiber has a refractive index that abruptly changes from that of the core to that of the cladding. A typical specification for a multimode fiber might be a core with a refractive index of 1.5; a cladding index of 1.485 ($.99 \times 1.5$); and a core diameter of 50, 62.5, or $100\mu m$.

Multimode light signals interfere with one another as they exit the core, causing *modal delay spreading.* This spreads or broadens the digital pulses and may confuse the optical receiver.

Attenuation

Attenuation or power loss occurs because of scattering, absorption, fiber connections or splices, and fiber bends. Scattering is the result of the core's microscopic imperfections in the fiber. The glass core's molecular structure places a minimum threshold for scattering. This *Rayleigh scattering limit* depends on wavelength, and, when increased, the optical loss is reduced.

Absorption Losses Absorption is power conversion into light in a medium that has opaque properties. This prevents light from escaping via the fiber's jacket, and cause losses in the core.

Connection Losses Connection loss accounts for a large percentage of overall loss. A splicing technique using a mounting fixture and small electric heater may fuse together single-mode fibers with less than a 10μm diameter. A resulting splice loss may be approximately 0.15dB. Alternatively, connectors may be used, providing a connection loss of between 0.25 to 1.5dB.

Bending Light traveling on the longest paths of curvature has to increase speed to stay in phase. Extreme paths of curvature lead to a situation where the light speed reaches its threshold, and is therefore lost.

OSI (Open System Interconnection)

OSI is a seven-layer industry standard reference model that is applied extensively to client/server architectures, and was introduced in 1984 by the ISO (International Standards Organization). It provides a standard infrastructure for the applications, glues, and communications required of modern client/server implementations.

The seven layers include:

- *Application* that encompasses client- and server-side programs, such as e-mail clients and browser at the front end.
- *Presentation* that is the formatting layer, delivering such operations as protocol conversion and compression. A typical application sees clients' SQL requests converted to a format that complies with the SQL server.
- *Session* that permits a conversation between programs, objects, or processes.

- *Transport* that provides error detection and correction operations for communicated data, and adds a transport-layer ID.
- *Network* that operates to break down transmitted data into packets (with sequence numbers), and to reassemble them into a readable message on reception. It may be assumed to route packets to an appropriate node.
- *Data-link,* which receives packets from the network layer, and adds control information to their headers and trailing regions. The resulting frames are passed to the physical layer when appropriate access is detected.
- *Physical* that converts frames into binary data so that it may be transmitted, and returns this data to frames upon reception at its intended destination.

Computers/Software

Object

An object is an entity in a running OO program, and provides a link between data and methods that read and operate on that data. An object is stimulated by messages that are sent to it, and its responses to messages are defined by its repertoire of methods.

Object Implementation

A coded solution that dictates an object's behavior and response to events. The code represents the object's methods.

Object Interface

An object's outer layer that intercepts messages, and directs them appropriately. The layer is sometimes referred to as a shell. It is the first entity that an inbound message meets. The message may then be processed internally by the object's methods.

Object Schema

A structure that defines the interactions and relationship of objects in an OO system.

Object Scraping

A method of mapping data from a server to objects. The objects are used to perform transactions or other types of processing. It may be applied in an application renovation solution.

Objective-C

An authentic object-oriented programming language that is based on ANSI-C, and adds constructs for classes, messages, and inheritance. The language supports polymorphism and dynamic binding where the runtime system determines what code is executed based on an object type. Objective-C is intuitive for C programmers and for OOP programmers using Java. It executes faster than Java because it is compiled like C, and the Objective-C OO extensions are compiled to C calls to the Objective-C runtime library (libobjc).

Objective-C programs have the main function:

```
#include <stdio.h>

int main (void)

{

  printf ("Hello my friend\n");

  return 0;

}
```

The example is also a C program as it holds none of the Objective-C constructs.

Objective-C Messages An object responds to messages that stimulate objects to respond in some way. Using Objective-C a message is sent to an object (or receiver) using the expression:

```
[receiver message_name];
```

Receiver is the object and message_name is the message or method name that is to be invoked—and may be referred to as a method selector.

To invoke the display method of the mySquare object:

```
[mySquare display];
```

Labels describing arguments precede colons:

```
[myRect setWidth:20.0 height:22.0];
```

Messages with a variable number of arguments:

```
[receiver makeList:list,argOne,argTwo,argThree];
```

OCX

A control or object that was a forerunner to ActiveX. OCX controls may be integrated into compatible applications, yielding functionality gains that may take the form of complete applications such as grammar checkers. OCX controls may be written using Visual C++.

ODBC 3.0 SDK

A set of tools, libraries, and headers that may be used to integrate ODBC 3.0 connectivity access in Web sites.

OLAP (On-Line Analytical Processing)

A data analysis technique used predominantly in the client/server computing environment. It is a decision-making support technique that may be applied to interrogate data from disparate sources. Resulting data may also be analyzed.

OLAP implementations may be assumed to embody multidimensional data analysis techniques, and integrate:

- OLAP GUI for user communication
- OLAP analytical processing logic
- OLAP data processing logic

OLAP empowers users to generate query data in order to answer complex questions based on what-if scanarios, or on current and historical data. It is an advancement of the primitive querying techniques harnessed in RDBMS designs. These include Borland (now Inprise) QBE, and to even query languages such as the industry standard SQL.

OLE (Object Linking and Embedding)

An object architecture. It is a method by which one application may be linked with, or embedded into, another. An OLE server application is the underlying source of an OLE client application. Objects may be video, wave audio, speech synthesis, Midi files, graphics, or text.

The objects may be shown in the client OLE document or application as an icon, and may be launched by double-clicking that icon. OLE may be used to embed Windows Media Player into client applications so as to add voice or video annotations to documents.

Using OLE1-compliant applications, the process of embedding an object is more intensive than that associated with OLE2 applications.

Object embedding is made easier using OLE2-compliant applications, because objects can simply be dragged from one application to another. An increasing number of Window applications are OLE2-compliant.

OLE Client (Object Linking and Embedding Client)

An application into which an object or application from an OLE server application has been embedded.

OLE DB SDK

A Microsoft database access specification that bases itself on OLE and COM object architectures. It complies with SQL and non-SQL databases.

OLE Server (Object Linking and Embedding Server)

An application which provides an object for an OLE client application, providing a means of running that object from within the client application.

OO (Object Oriented)

A prefix used in object-oriented systems, software, and development tools.

OO User Interface

A user interface that uses the object model as its underlying interface components. They are typically graphical user interfaces (GUIs or "gooey"). The Apple Macintosh is remembered as one of the first systems to feature a commercially successful OO UI, followed by NextStep that was founded by Apple Computer's co-founder, Steve Jobs. This was followed by the Microsoft Windows and IBM OS/2 OSes that featured OO user interfaces.

OODBMS (Object-Oriented Database Management System)

A database used to store and retrieve complete objects, including their code and their data. Stored objects may be categorized and stored in compound structures or objects.

OODBMSes are characterized by their ability to:

- Store complex objects
- Be renovated or updated without radical renovation of data table structures associated with RDBMS implementations
- Be extensible, providing a means of defining new data types
- Support OO methodologies and concepts, including encapsulation (where objects' inner workings are hidden) and inheritance (where objects may be granted the methods and data of other objects). Multiple inheritance may also be supported where subclasses inherit methods and data from more than one superclass.

OODL (Object Oriented Dynamic Language)

A programming language that is both object-oriented and dynamic, of which Dylan is a commercial example.

OOL (Object-Oriented Language)

A programming language that adheres to the object oriented programming model.

OOP (Object-Oriented Programming)

A modular programming approach that depends on reusable objects. OOP programming tools include Inprise Delphi, Optima++, and PowerSoft PowerBuilder. OOP languages include C++, Java, and Visual Basic.

In the real world we unconsciously place objects in classes. We know, for example, that cars, holiday chalets, and computers are from different classes, but each time we see a car we don't ask ourselves Which class does a car belong to? Or why is it different from a holiday chalet? We know that it is a member of the class Vehicles because we have learned how it behaves, and that behavior, with all its methods, is in our mind. We do not have to learn or consider an object's behavior each time we come in contact with it. For example, you know that you cannot drive the holiday chalet because of its behavior, and the class to which it belongs. You know these things without having to, repeatedly, decide that a holiday chalet cannot be driven because it has no wheels, no axle, no engine, and so on.

Object-oriented programs are much the same. Classes of objects are carefully defined. Hierarchies form another important part of classes where, once again, as in the real world, classes are subdivided into further classes. This helps distinguish between, say, a sports car and a jeep. The jeep would be a member of the class OffRoadVehicles, which is a member of the class, Cars, which in turn is a member of the class, Vehicles.

This additional information tells us that a sports car cannot be driven up a steep, muddy slope etc. Everything in the real world is a member of a class, of which there are an infinite number.

Some Windows databases come with a number of built-in methods to choose from. These cover standard activities such as opening tables and forms, and even opening the Help window. This type of database building is achieved through a so-called pick-and-build interface.

The OOP model embraces:

- Data hiding
- Encapsulation
- Reuse
- Polymorphism

OOP languages include:

- Java
- C++
- Smalltalk
- Visual Basic

Open Transport

A communications architecture for implementing network protocols and other communication features on computers running the Mac OS.

Operating System (OS)

A generic term used to describe the software elements that manage system resources and so provide an interface between the user and the system, as well as between software and hardware. The shell, user interface, or front end is sensitive to a number of user commands. Popular operating systems include Windows 98/NT, OS/2 Warp, MS-DOS, DR-DOS, OS/2, and Unix.

Oracle

A database development environment produced by a company of the same name.

OSGi (Open Standard Gateway Interface)

A standard for residential networks.

OSPF (Open Shortest Path First)

A protocol used in routers.

Miscellaneous

Odyssey

Odyssey is a satellite constellation and was developed by TRW that later opted for ICO instead. Odyssey was intended to be an MEO system with 12 satellites in three orbits, and with a CDMA radio interface.

OMG Notification Service (NS)

An extension to the OMG Event Service that defines interfaces and operations for transmitting events between client and server applications. In OMG terminology such connected entities are termed consumers and suppliers, and in some NS implementations the terms are interchangeable with consumers and/or receivers and publishers.

The OMG NS is a defacto industry standard messaging architecture, and a mainstay for many client/server implementations that adhere to the push-and-pull models as a means of driving information –on demand. It enhances the OMG Event Service by allowing:

- Clients to specify received events by filtering proxies in a channel
- Event transmission in the form of structured data types, adding to the Anys and Typed-events of the OMG Event Service
- Defined operations (or methods) to provide parameters to suppliers, thus informing them of event types required by consumers on a channel, and providing information about usage habits

- Event types offered by suppliers to an event channel to be discovered by consumers of that channel so that consumers may subscribe to new event types as they become available
- Quality of service (QoS) properties to be set for individual channels, proxies, or structured events (Relevant CORBA operations include `set_qos` and `get_qos`.)
- Optional event-type repositories that facilitate filter constraints by end-users, and make available information about the structured events flow through the channel.

The NS is based on a series of CORBA IDL (Interface Definition Language) modules that define operations and interfaces that may be used by consumers and suppliers that connect over a client/server environment, and include:

- CosNotification (*see* CosNotification under letter C)
- CosNotifyFilter
- CosNotifyComm
- CosNotifyChannelAdmin

Orbcomm

Orbcomm is a satellite constellation with 35 satellites at 775km, and does not use intersatellite links. The radio interface technology is CDMA and delivers narrowband data rates suitable for paging-type messages, perhaps for tracking applications where satellites receive data packets from sensors in vehicles, that are then sent to earth stations. The system uses the 137–138MHz and 400MHz frequency bands for transmissions down to mobiles, and 148–150MHz frequencies up to the satellites.

OTP (One-Time Password)

A password protection security policy to prevent illegal access. Unfortunately, in many instances it does not prevent hackers from monitoring the network and gaining access to information.

OTP variants include:

- Wietse Venema's LogDaemon
- Bellcore's S/KEY Version 1.0
- BellCore's Commercial S/KEY Version 2.0
- United States Research Laboratory's (NRL) One Time Passwords in Everything (OPIE)

Outage

Outage is the probability that a QoS level won't be met.

Acronyms

OAM&P	Operation, administration, maintenance, and planning PC
OBEX	Object exchange PCI
OCCCH	OOMA common control channel
OCNS	Orthogonal channel noise simulator
OFDMA	Orthogonal frequency division multiple access
OHG	Operators Harmonization Group
OMG	Object Management Group
OOCCH	OOMA dedicated control channel
OOCH	Opportunity-driven dedicated channel
OOMA	Opportunity-driven multiple access
OOTCH	OOMA dedicated traffic channel
OQPSK	Offset quadrature phase-shift keying
ORACH	OOMA random access channel
OS	Operating system
OSI	Open system interconnect
OSS	Operations support system
OTASP	Over-the-air service provisioning
OTD	Orthogonal transmit diversity
OVSF	Orthogonal variable spreading factor

P

P-2

Paging (UTRAN)

The UTRAN pages mobiles to indicate incoming call waiting. The idle mode requires paging type 1 messages that may hold several paging requests for mobile users. Using discontinuous reception (DRX) to preserve battery life, the mobile monitors the PCH for such paging messages. Several mobiles may listen for paging, so the UE's RRC layer checks the received paging records for the occurrence of its own paging identities.

Connected-mode paging in the CELL_PCH and URA_PCH states triggers a state change of the UE. When a paging message is received, and there is idle mode, the UE will respond as follows:

1. Compare the "CN UE" identities with its assigned CN UE identities.
2. If an alike identity is located, forward it to the upper-layer entity denoted by the IE "CN domain identity."
3. Do nothing if the IE paging originator is the UTRAN.

The UE responds as follows when in connected mode:

1. Compares identities of type "UTRAN originator" with its allocated U-RNTI—when the IE "paging originator" is the UTRAN.
2. Every match causes the UE to enter CELL_FACH state and implement a cell-update procedure.
3. Do nothing if the IE "paging originator" is the CN.

Dedicated Paging

Dedicated paging is a type-2 paging message sent in CELL_DCH and CELL_FACH connected-mode states, and may establish signaling connections. The RRC forwards the resulting paging identities and causes to the NAS entity determined by the CN domain identity.

RRC Connections

The layered architecture of UMTS sees the separation of the radio connection and the radio bearer (RB), and is analogous of a freeway and cars, where

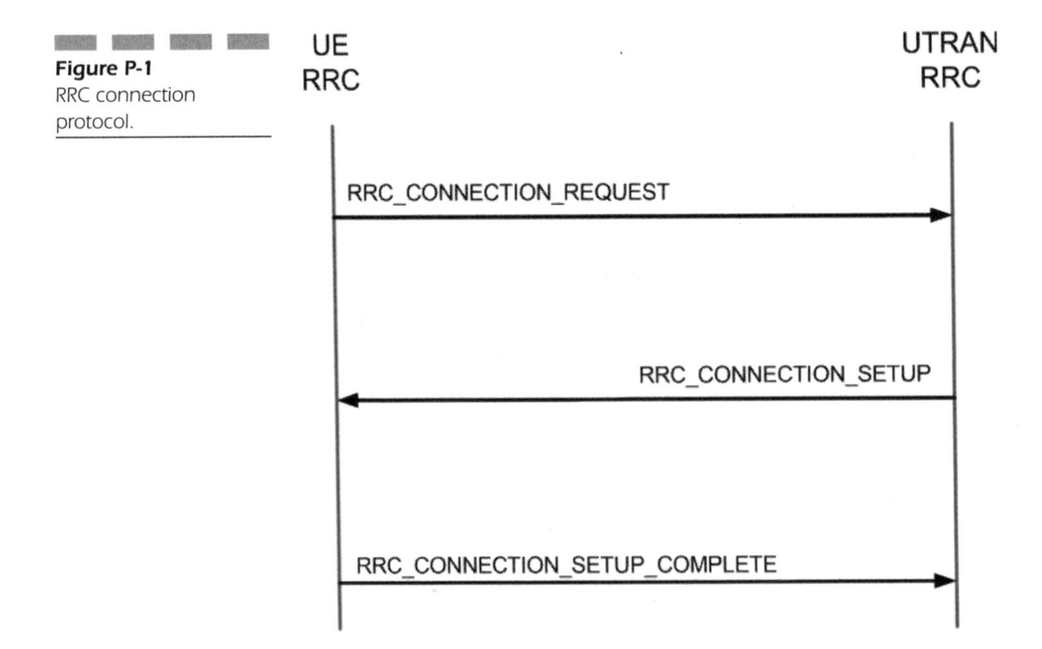

Figure P-1 RRC connection protocol.

the road is the RRC connection and the bearer connection is the traffic (or cars). The RRC stores "initial UE identity" identities that are in the RRC_CONNECTION_REQUEST message, and verifies that the same value is returned in the RRC_CONNECTION_SETUP message. The RRC_CONNECTION_SETUP message holds information used by the L1/L2 configuration, and the final message RRC_CONNECTION_SETUP_COMPLETE (see Fig. P-1) is sent through the new DCH.

PCM

Pulse-code modulation (PCM) for voice coding is prevalent on PSTNs where the analog voice signal is digitized at a central office where it becomes a 64kbps datastream. Eventually the data stream is returned to analog form for reproduction at the subscriber's telephone. The PCM codec creates the datastream by amplifyng the analog signal at 8kHz, eventually resulting in pulse-amplitude modulation (PAM).

A Time Division Multiplexer (TDM) combines 24 PCM digitized-voice signals for the Line Driver that takes a group of twenty-four 8bit samples, and creates a 193-bit frame. Eight thousand such frames are generated per second resulting in a 1.544Mbps data rate. The Line Driver also converts the data stream from unipolar to bipolar with an equal number of positive and negative pulses.

This waveform coding method is the basis of adaptive differential PCM (ADPCM) that is standardized at operating rates of 40, 32, and 24kbps.

Hybrid Coding

Hybrid coding extracts speech parameters from the analog waveform, and the voice sample is synthesized and compared with the original sample. If necessary, differences are reduced by adjusting the speech parameters. Hybrid coders may also use coded speech parameters (in a codebook) that are transmitted. This results in a voice digitization rate of between 5 and 16kbps. Hybrid coders may also use code-excited linear predictive coding (CELP). D-AMPS' vector-sum CELP (or VCELP) uses a 50Hz sampling rate and a 159bit sample size, requiring a bit rate of 7950bps.

ISDN PCM (Pulse Code Modulation)

PCM provides a means of encoding data in digital form, for transmission over a network, or for storage on DSM. Used in the Integrated Services Digital Network (ISDN) standard, multiplexing involves creating a data stream consisting of 8bit PCM blocks. The blocks are created every 125 micro seconds. By interleaving the blocks with those from other encoders, the result is time division multiplexing (TDM).

In North America ISDN typically interleaves data from twenty-four 64Kbits/s sources or channels. This results in connections that provide 1.536Mbits/s. Although in actual fact the connection has a bandwidth of 1.544Mbits/s, because each channel's frame has a marker bit "F," adding 8KBytes/s.

Europe sees ISDN that typically interleaves thirty 64kbits/s channels, giving 2.048 Mbits/s. This and the 1.544Mbits/s connection are known as primary rate multiplexes.

Further interleaving of primary rate multiplexes sees:

- 6, 45, 274Mbits/s in North America
- 8, 34, 139, and 560Mbits/s in Europe

PCM was conceived in 1937 by Alec Reeves but was not applied widely for many years.

Sampling Using ISDN, a 3.4kHz analog signal is sampled at 8kHz. The sampling rate is less than twice the bandwidth of the analog signal, in accordance with Nyquist's sampling theorem, and prevents aliasing.

A sampling frequency in a multiple of 4kHz was used, also because the existing networks used 4kHz carriers, and would cause audible interference in the form of whistles.

Coding The amplitude of each sample is measured and encoded using 12bit values that give +2048 possible values.

Compression The 12bit samples are reduced to 8bits using logarithmic compression that may be:

- "Mu-law" in North America
- A-law in Europe

These compression standards permit the system to be embedded anywhere in an analog network.

Physical Channel—UMTS

The physical channels in UMTS transfer information across the radio interface. A physical channel is defined by its:

- Code and carrier frequency in an FDD version
- Code, carrier frequency, and timeslot in TDD.

TABLE P-1

Physical Channel Names

Name	Physical Channel
F/R-FCH	Forward/Reverse Fundamental Channel
F/R-SCCH	Forward/Reverse Supplemental Coded Channel
F/R-SCH	Forward/Reverse Supplemental Channel
F/R-DCCH	Forward/Reverse Dedicated Control Channel
F-PCH	Forward Paging Channel
R-ACH	Reverse Access Channel
R-EACH	Reverse Enhanced Access Channel
F/R-CCCH	Forward/Reverse Common Control Channel
F-DAPICH	Forward Dedicated Auxiliary Pilot Channel
F-APICH	Forward Auxiliary Pilot Channel
F/R-PICH	Forward/Reverse Pilot Channel
F-SYNC	Forward Synchronization Channel
F-TDPICH	Forward Transmit Diversity Pilot Channel
F-ATDPICH	Forward Auxiliary Transmit Diversity Pilot Channel
F-BCH	Forward Broadcast Channel
F-QPCH	Forward Quick Paging Channel
F-CPCCH	Forward Common Power Control Channel
F-CACH	Forward Common Assignment Channel

Operations of a UMTS Transmitter at the Physical Layer

See Operations of a UMTS Transmitter at the Physical Layer under letter L.

Cdma2000 Channels

A logical channel is designated a three- or four-letter acronym followed by "ch":

First Letter	Second Letter	Third Letter
f–forward—BS to MS	d–dedicated	t–traffic
r–reverse—MS to BS	c–common	m–media access control
		s–signaling

The naming convention for physical channels is the use of uppercase letters, with the first letter indicating direction, except for paging and access channels where direction is implied.

See Tables P-1 and P-2.

Physical Layer The physical layer is a source of coding and modulation services for logical channels that are used by the PLDCF MUX and QoS layer. Physical channels may be F/R-DPHCH (forward/reverse dedicated physical channels) where they carry information in a point-to-point model between the base station and a mobile user. They can also be F/R-CPHCH (forward/reverse common physical channels), or a number of physical channels carrying information in a point-to-multipoint model between the base station and multiple mobile stations.

TABLE P-2

Logical Channels Used by PLICF

Channel	Description
Dedicated traffic channel f/r-dtch	This carries user data and is a point-to-point channel and is leased for the time period of an active data service.
Common traffic channel f/r-ctch	This carries short data bursts of a data service in a dormant/burst substate.
Dedicated media access control channel f/r-dmch_control	This carries media access control messages and is a point-to-point channel. It is leased for the time period of an active data service.
Reverse common media access control channel r-cmch_control	This carries media access control messages, and is used by the mobile while the data service is dormant or idle.
Forward common media access control channel f-cmch_control	This is used by the base station when the data service is dormant or idle.
Dedicated signaling channel Dsch	This carries upper-layer signaling data for a single PLICF instance.
Common signaling channel Csch	This carries upper-layer signaling data shared access.

Protocol Stacks

Bluetooth Protocols

See Bluetooth Core Protocols under letter B.

WAP

WAP permits compatible devices to browse compliant Internet sites using an appropriate browser. A core part of the WAP architecture is the WAP gateway that unites dissimilar networks, providing a bridge between an IP network (namely the Internet) and Mobile Operator network.

Figure P-2 presents a "multiple scenarios" high-level representation of a WAP gateway with all its surrounding software architectures and physical infrastructures. It does not actually use every component part of the shown

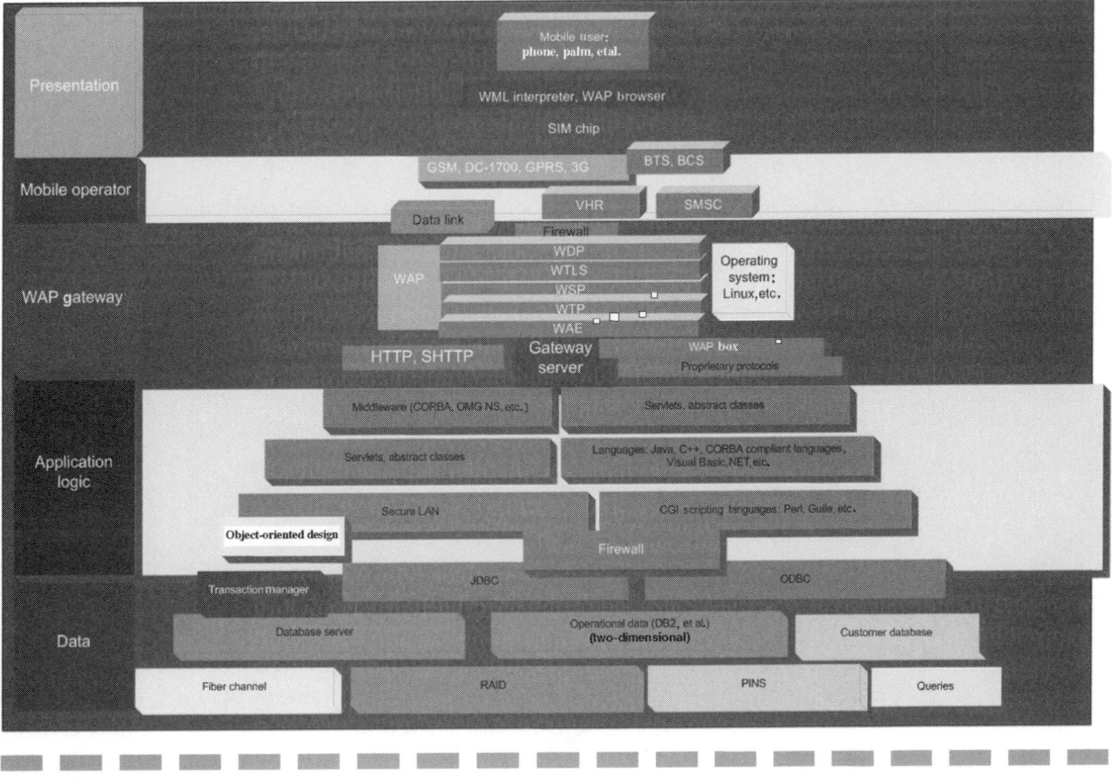

Figure P-2 WAP scenarios.

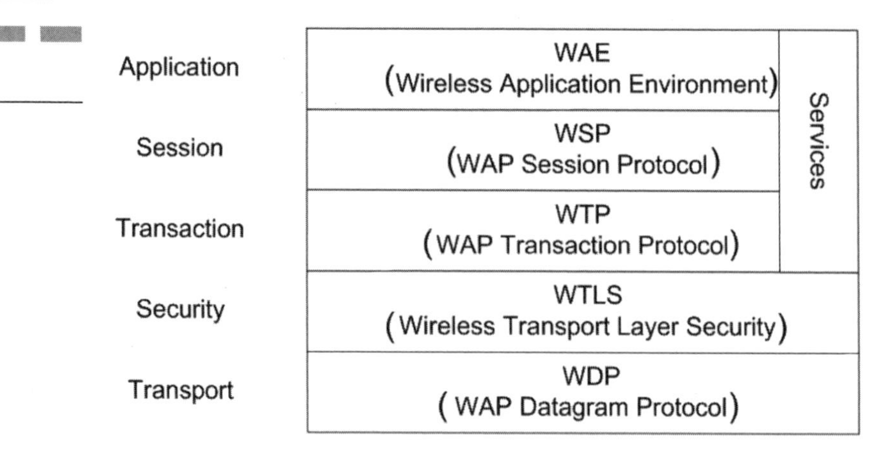

Figure P-3
WAP stack.

architecture since the diagram is intended to show the collective client/server architecture. The illustration attempts to explain the basic strata beginning with the presentation layer through to the back end. A practical implementation of WAP would obviously focus on a single descent through the shown layers and tiers.

A WAP Gateway interfaces the WAP stack with the TCP/IP stack, provides Domain Name Server (DNS) services, and translates WML pages into encoded packets for transmission over wireless networks.

WAP can operate over TDMA networks such as D-AMPS and GSM, and over CDMA networks, and also operates using SMS, and circuit-switched cellular data (CSD). See Fig. P-3.

The WDP is a transport layer for transmitting and receiving messages using one of many bearer network types, including TDMA, GSM, PCS, and CDMA. The WTLS is an optional security transport layer that was not supported by some of the earlier WAP Gateways. It is intended for secure transactions.

The WTP provides transaction TCP and UDP transactions, and reliable and unreliable datagram delivery. The WSP supports data exchange using sessions made up of transactions at the WTP layer.

Access Stratum

In the UMTS architecture, UTRAN sees the air interface isolated from the MM (mobility management) and the connection management layers. This notion takes the form of the access stratum (AS) and the nonaccess stratum (NAS). The AS includes radio access protocols that terminate in the UTRAN, while the NAS has core network protocols between the UE and CN, but they are not terminated in the UTRAN that is transparent to the NAS.

The NAS attempts to be independent of the radio interface, as do the protocols MM, CM, GMM, and SM. These protocols are theoretically constant in terms of the radio access specification that carries them, so any 3G radio access network (RAN) should connect to any 3G CN. The GSM's MM and CM layers are almost the same as 3G NAS, and NAS layers will compare with future GSM MM and CM layers, according to the phase 2+ specification.

Lower layers from the AS are different from GSM where the radio access technology (RAT) is TDMA, as opposed to CDMA for UTRAN. The protocols used, therefore, are also radically different, as is the packet-based GPRS protocol stack. So this reveals an obstacle in the so-called smooth migration path from 2.5G GPRS to 3G. However, the GPRS CN components can be reused when renovating the system to become a 3G solution.

The AS has three protocol layers that include sublayers:

Physical layer (L1)	
Data link layer (L2)	Medium Access Control (MAC)
	Radio link control (RLC)
	Broadcast/multicast control (BMC)
	Packet data convergence protocol (PDCP)
Network layer (L3)	Radio resource control (RRC)

The two vertical planes include the control (C-) plane and the user (U-) plane, and both have the MAC and RLC layers, while the RRC layer is present only in the C-plane. See Fig. P-4.

FMA1

See FMA1 under letter F.

FMA2 (W-CDMA)

See FMA2 under letter F.

IP Mobile Networks

See IP Mobile Networks under letter I.

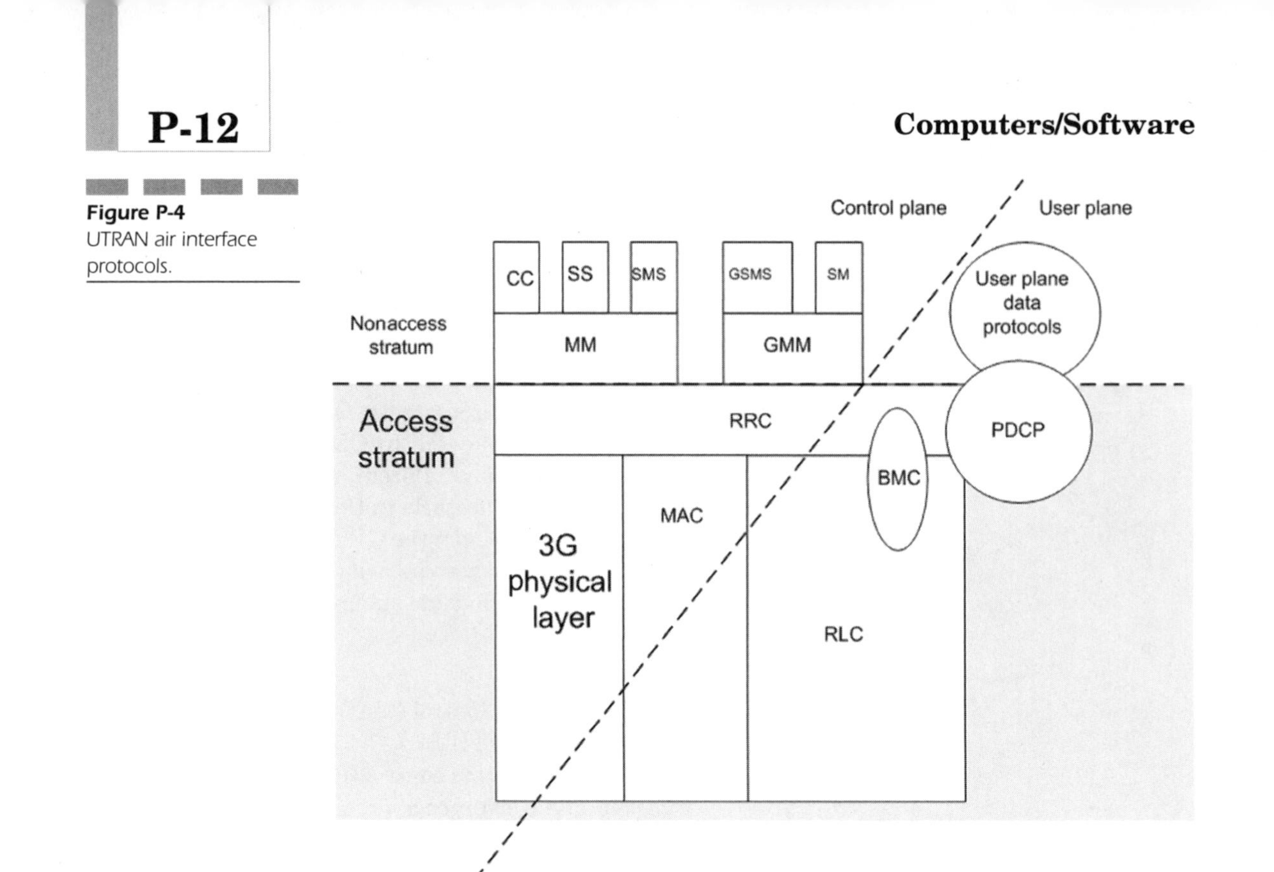

Figure P-4 UTRAN air interface protocols.

Computers/Software

Package

A set of Java classes that address specific functions, where, for instance, `java.io` addresses input and output functions, and `java.net` addresses Internet and network operations.

Palette Editor

A program used to edit the palette of 8bit graphics or video. Palette editors such as Microsoft PalEdit are used to alter the color characteristics of 8bit video sequences. They are also useful for building palettes that work with a number of different 8bit video files.

The importance of this relates to switching between two or more different 8bit sequences on screen that contain different palettes that can result in a flicker. The degree to which the flicker occurs depends on the difference that separates the palettes, as well as on the general video and graphics speed of the playback system.

Building a common palette is easiest if you run multiple instances of the palette editor, provided the program has this capability, or where it is able to open multiple palettes.

Palette Switching

An instant when a color palette is switched from one to another. Palette switching occurs most often when 8bit video is cut from one sequence to another that has a different color palette, or when one 8bit still image is cut to another that has a different palette. Palette switching can result in a brief screen flicker. The screen flicker may be eliminated using a common palette for all bitmaps (images) and video clips. A common palette can easily be achieved using a palette optimizer, or using an editing program such as PalEdit.

Palmtop

A small-scale portable PC that runs a streamlined operating system such as Windows CE.

PARC (Palo Alto Research Center)

A research establishment founded in 1970 by Xerox. It is the birth place of many multimedia-associated technologies and concepts including laser printing, local area networks, the graphical user interface (GUI), and object oriented programming (OOP). The GUI system integrated into the Apple Macintosh launched in 1984 was a direct result of Apple's Steve Jobs visiting PARC. During his visit he saw a GUI platformed on PARC's Alto system.

Parser

A function of a compiler, interpreter, or translator that attaches semantics to tokens that are generated by the lexical analyzer.

Path

A series of characters used for authorization purposes, and that provide access to appropriate services, applications, and hardware.

Path/Trail

1. A path through a series of links in hypertext, hypermedia, or multimedia material.
2. A statement that points through the hierarchy of directories to a file or folder, and may take the form:

C:\jini1_1\source\examples\, and are often included in autoexec.bat files so the user need not type the path of a file or program in order to run it.

PATH_INFO

A CGI variable that holds the URL's suffix or that data which follows the script's name.

Payload

A user data capacity of a packet, block, cell, or frame that forms part of a protocol.

Pay-to-View

An e-commerce site that is created to sell content rather than tangible goods, and may vary from a virtual publication to a TV broadcasting facility or on-line jukebox. Implementing such sites requires password access to selected areas, and forms for gathering credit card or bankcard details.

PC Card or PCMCIA

A slot that connects with almost credit-card-size peripheral devices that may be modems, NICs, interfaces to CD-ROM drives, hard disks, etc. The original PCMCIA was designed for memory cartridges only, but in September 1991 the PCMCIA Type II (PCMCIA 2.0) specification was launched, facilitating hard disks and modem/facsimile devices.

PCX

A bitmap file format developed by ZSoft that features RLE compression.

PEM (Privacy Enhanced Mail)

An encryption standard for e-mail that was created by the IETF.

Pentium

An Intel family of processors.

Perl

A programming language for processing text. It was developed by Larry Wall who once joked that Perl stood for "Pathologically Eclectic Rubbish Lister." He describes Perl as:

> . . .an interpreted language optimized for scanning arbitrary text files, extracting information from those text files, and printing reports based on that information. It's also a good language for many system management tasks. The language is intended to be practical (easy to use, efficient, complete) rather than beautiful (tiny, elegant, minimal). It combines (in the author's opinion, anyway) some of the best features of C, sed, awk, and sh, so people familiar with those languages should have little difficulty with it. (Language historians will also note some vestiges of csh, Pascal, and even BASIC-PLUS.) Expression syntax corresponds quite closely to C expression syntax. . . .

Perl Variables Scalar variables are assigned single data values that may be integer, floating point, or string:

```
$trasform=25;
$response="You did not enter the correct patient
  symptoms.";
```

Numeric variables are incremented using the syntax:

```
$transform=$transform+1;
```

or,

```
$transform++;
```

Numeric variables are decremented using the syntax:

```
$transform--;
```

Subroutines are named using the sytax:

```
&subroutine;
```

Perl Arrays An array of scalars is defined using the syntax:

```
@trans(2,4,5);
@forward (7,8,9);
```

The number "4" in the defined three-element array is addressed thus:

```
$tate=@trans(1);
```

Arrays may be combined using the syntax:

```
@combine=(@trans, @forward);
```

The definition of array variables is accompanied by the generation of scalar variables (of the same name) that have the @# prefix. These store the array size, or, more precisely, the sequence number of their final element.

The array size need not be defined, and it is legal to insert elements into an array at whatever point, thus:

```
@transform = (10,25,35,55);
$tranform[25] = 7;
```

An associative array of scalars may be assigned to a variable (with the "%" prefix), thus:

```
%transform = ("x",100, "y",20, "z", 20);
```

This equates transform to the element strings "x," "y," and "z," whose values are 100, 20, and 20.

Subroutines Subroutines begin with the word *sub* and its code or block is contained within opening and closing braces, thus:

```
sub transformt_every {
  $tate=@trans(1);
}
```

PGP (Pretty Good Privacy)

An asymmetric cryptosystem that is in the public domain, and was invented by Phil Zimmerman. It is used along with RSA by SET (Secure Electronic Transactions) for security and authentication services.

PicoJava

A Chipset from Sun Microelectronics that is optimized for the Java programming language. It is used in cellular phones and computer peripherals.

PIN (Personal Identification Number)

A number assigned to an ATM cardholder.

Ping

A name for ICMP (Internet Control Message Protocol) Reply/Echo. It is also used to describe programs that use ICMP. ICMP is used to test the reliability and connection speed to a remote host. The reply to such a test is called a pong.

Pipe

A network communication channel that provides a means of transferring packets between local or wide area destinations. Pipes have addressable names and may be used to send and receive data (that is typically assembled into packets) to and from a central computer over a WAN.

Plaintext

An input into an encryption algorithm or cipher, and becomes ciphertext. When ciphertext is processed by a decryption algorithm, it is returned to plaintext.

Plug-In

A module that adds functionality to an application. In the case of a browser, a plug-in might take the form of a video playback feature. Numerous plug-ins are available for the Netscape Navigator browser. An alternative technology to Microsoft's ActiveX Controls, plug-ins tend to be a feature of browsers. ActiveX Controls are applied more broadly and cut through boundaries that separate many different software sectors, including client-side and server-side component architectures. The plug-in architecture adds processing capabilities to the client browser, and displaces logic from the server side.

PnP (Plug-and-Play)

A hardware specification that ensures easier installation using Windows 95/NT. PnP hardware devices may be detected and installed by Windows 95/NT.

Polymorphism

An object-oriented concept where two objects respond differently to the same message.

POP3 (Post Office Protocol)

A protocol for sending and receiving e-mail. Compliant e-mail applications are called POP3 agents.

Port

1. A channel through which a computer communicates/drives a peripheral device. Standard PCs include Centronics parallel ports, and serial ports (that are often referred to as COM ports). Typically, parallel ports are used to connect with such peripherals as printers, mass storage devices, and scanners. Serial ports are often used to connect with external modems. Other ports include FireWire and USB (Universal Serial Bus).
2. A method of translating a program from one platform to another, or from one language to another.
3. A port number used by a server.

POS (Point of Sale)

1. An automated credit card or bankcard transaction
2. A method of selling products or services from e-commerce Web sites, or even multimedia booths

PPP (Point-to-Point Protocol)

A standard protocol used with standard access technologies such as POTS and ISDN. Such connections to ISPs see it used with the IP protocol.

Precedence

In C++, arithmetic operators have a precedence value. These indicate the order in which such operators are implemented that is significant with expressions such as:

```
dev = xx + yy * zz + yy;
```

Control over such arithmetic operations is obtained by using parentheses, that is:

```
dev = (xx + yy) * zz;
```

Parentheses may be nested.

Preemptive Multitasking

A type of multitasking in which the operating system interrupts applications running concurrently. It is more seamless than nonpreemptive multitasking in that a higher degree of concurrency is achieved. Windows 95 and OS/2 Warp embody preemptive multitasking.

Private Key Encryption

An encryption model that requires the sender and the receiver of encrypted matter to use a single password key. The size of the key (in bits such as 56bit, etc.) is a function of the encryption techniques harnessed.

Process Flow

A diagram that shows the processes included in a system architecture. It shows how the processes, and their leaf processes, interrelate and interact with entities that might be data or program modules.

Processing Fees

A fee charged by Acquirers and Merchants for using interchange networks or for using Merchant Account services.

Processing Power

A measure of a system's processing performance and may be measured in:

- Millions of floating point operations per second (MFLOPS)
- GigaFLOPS or billions of FLOPS (GFLOPS)
- TeraFLOPS or trillions of FLOPS (TFLOPS)

- The rate at which instructions are executed in millions of instructions per second (MIPS)
- SPECmarks
- Whetstones
- Dhrystones

Processor

A device that embodies the functionality of a CPU (Central Processing Unit). The familiar Intel PC processor continuum broadly equates to: 4004, 8088, 80286, 80386DX, 80836SX, 80486DX, 80486SX, 80486DX2, Pentium, Pentium Pro, Pentium II, Pentium III, and Pentium IV. The generic PC processor continuum is a little more complex with companies such as AMD (Advanced Micro Devices) and Cyrix producing reverse-engineered, and often enhanced, Intel compatible processors.

Programming Tool

An item of software used to develop software.

Properties

A set of attributes associated with an entity that may be a simple font or window, or a complex container that has an embedded application.

Proprietary

A prefix denoting nonstandard hardware or software.

Proxy Object

An object that may be passed to a client so as to provide access to remote objects via an interface. Jini devices consumer Jini services in this matter, where a proxy object is passed to a JVM.

Proxy Server

An intermediate server on the communications path between server applications and data entities, and the client systems and software.

Pseudo-Conversational Communication

A communication regime between two software components or objects that exist only for the duration of interaction.

Public Key Encryption

An encryption technique that requires both private and public keys. A public key is used to encrypt sent data.

Publishing Medium

A medium that may be used to publish information. The Internet, CD-ROM, and DVD-ROM are publishing media.

Pure Transaction

A transaction that occurs under control, and all shared access to resources is coordinated.

Push Technology

A technology with which a user is served requested Web-sourced information. It is sometimes referred to as the push model.

PWS (Personal Web Server)

A downsized implementation of IIS, and is bundled with Microsoft FrontPage. It may be used to:

- Test Web applications
- Build intranets

Miscellaneous

Packet-Switched Network

A data transmission and reception technique where data streams are divided into packets coded with origination and destination information. The packets may be interleaved with different data transmissions. For instance, the packets that may be providing a two-way audio communication link in IP telephony, might be interleaved with other streams such as videoconferencing data.

Packets may follow dissimilar routes over a network, and are directed over what are perceived as the quickest and least congested routes. If available routes or logical channels are congested, then packets are buffered before transmission. The buffer is a FIFO (first in first out) storage, where the first packet placed in the buffer is the first to be retrieved and transmitted when the appropriate virtual channel is available.

The X.25 protocol standard dictates that a packet may contain between 3 and 4100 octets or bytes. (*See* X.25 under letter D.) Up to 4095 logical channels might be accommodated on a single physical link (1997). The logical channel followed by a packet is determined by its header information. There is also error correction, where the receiver might request that a particular packet be retransmitted.

The original packet-switching standard for public data networks is CCITT X.25. This is a multitiered recommendation embodying everything from physical connectors to data formatting and code conversion.

Packet switching is rather like the logistics involved in shipping an automobile part by part. The disassembled parts are sent, and assembled at the factory of destination. Equally if a part is damaged, the factory will request that it is sent again.

The packets may have one of two identities:

- *Multicast* packets (or items of transmitted information) may be delivered to more than one destination.
- *Unicast* packets that have one destination only.

Packet-switched networks (that use IP) are currently displacing switched networks in the telecommunications industry, and is driving the growing use of IP telephony or Internet telephony. Comparative advantages of IP telephony include:

- *Reduced costs, and reduced cost of ownership, for telcos and corporations running IP-compatible networks such as intranets.* The reduced costs are largely brought about by the fact that IP and Internet traffic are unregulated. Corporations and government departments may experience savings in the cost of voice traffic that might reach as much as 80 percent.

- *Flexibility.* IP telephony makes better use of bandwidth. For example, Australian telecommunications giant, Telstra, introduced the virtual second line in the late 1990s. This allows subscribers to its ISP (Telstra Big Pond) to receive incoming calls while connected to the Internet.

Password

A series of alphanumeric characters used to protect a system against unauthorized access. Using TCP/IP, the password file, /etc/password, is used to prevent unauthorized access. Servers like Apache, NCSA, and Netscape integrate facilities for authenticating users that do not require programming. The principal files are the *access file* (.htaaccess) and the *password file* (.htpasswd) that are stored in a secure directory on the server and listed below:

```
#!/usr/bin/perl

$passfile = "/disk/mysite.com/ood/.htpasswd";

require "ctime.pl";
$method=$ENV{"REQUEST_METHOD"};
$type=$ENV{"CONTENT_TYPE"};

%input_values=&break_input;
$username=&normalize_query($input_values{"username"});
$password=&normalize_query($input_values{"password"});

open(PWFILE,"+< $passfile") | | &croak("Can't open
  $passfile: $!");
$salt=reverse time;
seek(PWFILE,0,2);
print PWFILE $username,":",crypt($password,$salt),"\n";
close (PWFILE);
&croak ("All done");
exit;

sub break_input {
local ($i);
read(STDIN,$input,$ENV{'CONTENT_LENGTH'});
@form_names = split('&',$input);
foreach $i(@form_names) {
  ($html_name,$html_value) = split('=',$i);
  $input_values{$html_name} = $html_value;
  }
```

```
return %input_values;
}

sub croak {
local($msg)=@_;
&print_header("System Error");
print $msg;
&print_footer;
}

sub print_header {
local($title) = @_;
print "Content-type: text/html\n\n";
print "<HTML>\n<HEAD>\n<TITLE>$title</TITLE>\n";
print "</HEAD>\n<BODY>\n<H1>$title</H1>\n";
}

sub print_footer {
print "</BODY>\n</HTML>\n";
}

sub normalize_query {
local($value) = @_;
$value =~ tr/+/ /;
$value =~ s/%([a-fA-F0-9][a-fA-F0-
  9])/pack("C",hex($1))/eg;
return $value;
}
```

The access file (.htaccess) holds the following:

- AuthUserFile/disk02/.htpasswd: points to the file containing user names and passwords.
- AuthGroupFile/dev/null: points to groups of names, although this is unusual in a Web context.
- AuthName Name goes_here: specifies the realm.
- AuthType Basic: specifies the user authentication system.
- <Limit GET>: specifies the server method that may also be POST or PUT.
- Require valid-user: ensures that only valid users access the implementation.

Using HTML the passwd.cgi program is integrated as follows:

```
<BODY>
<H1>Input Password</H1>
<FORM ACTION=http://,botto.com/cgi-bin/passwd.cgi"
      METHOD=post>
User name: <INPUT TYPE="text" NAME="username"><P>
Password; <INPUT TYPE="text" NAME="password"><P>
<INPUT TYPE="Submit">
</FORM>
</BODY>
```

PCS-1900

PCS 1900 uses the 1900MHz band and compares with the UK's DCS 1800. The U.S. frequency band allocation for PCS is between 1850 to 1910MHz for the uplink, and 1930 to 1990MHz for the downlink. PCS uses a frequency separation of 80MHz.

PDA (Personal Digital Assistant)

A portable device that may serve a number of functions including that of an e-mail client, Web browser, and organizer.

Peer-to-Peer Network

A network that permits each network user to access the directories and the peripheral devices associated with any connected computer. When computers are linked together so that they can share the resources of one or more computers, a network is formed. You can build a peer-to-peer network using Windows 95/98, simply by adding Ethernet cards to connected systems. Another type of local area networks (LAN) is the server-based variant that permits users to access and share information stored on a powerful computer commonly termed a server.

Protocol

A format used to transmit and to receive data. Examples of industry standard protocols include IP, Ethernet, SMTP, HTTP, etc. Each protocol is optimized for the information it is intended to carry, and for the network over which it is to be used. A protocol often consists of:

- An information field for data
- The destination address
- Error detection and correction codes
- Originating address

All of this information is held together in a single unit that might be a packet, cell, or frame. In IP networks, such as the Internet and intranets, they're called packets, but it's just a new name; they are really all the same thing. The packets are assembled at the point of transmission, and sent over various different paths to their destination. Once received they are checked for errors, and then appropriately assembled.

Network protocols are analogous to the Postal Service; the packets are comparable to envelopes, and they have destination and originating addresses, etc.

Acronyms

PACS	Personal Access Communication System
PACS-UB	Personal Access Communication System unlicensed-version
PAGCH	packet access grant channel
PBCCH	packet broadcast control channel
PC	power control
PCCCH	packet common control channel
PCCH	paging control channel
P-CCPCH	primary common control physical channel
PCG	power control group
PCH	paging channel
PCM	pulse code modulation
PC-P	power-control preamble
PCPCH	physical common packet channel
PCS	personal communication services
PCS HCA	personal communication system high compression algorithm
PCU	packet control unit
PDA	personal digital assistant
PDC	public digital cellular
PDCH	packet data channel

PDrcH	packet data traffic channel
PDSCH	physical downlink shared channel
PDSN	packet data service node
PH	protocol handler
PHS	Personal Handyphone System
PLM	power level measurement
PMRM	power measurement report
PNCH	packet notification channel
POP	post office protocol
POTS	plain old telephone service
PPDN	public packet data network
PPM	pulse-position modulation point-to-point protocol
PRACH	physical random access channel
PRAT	paging channel rate
PS	packet switched
PSMM	pilot strength measurement message
PSPDN	packet-switched public data network
PSTN	public switched telephone network
PSU	packet switching unit
PTCH	packet traffic channel point-to-multipoint
PTM-SC	point-to-multipoint service center
PTP	point-to-point

R

Radio

See Microwave Radio under letter M.

Radio Frequency Spectrum Allocation

Most countries comply with the International Telecommunications Union (ITU) radio frequency allocations, ensuring that different sectors such as aviation and satellite constellations may operate without levels of interference to make them unsafe. The 1934 U.S. Communications Act revision entrusted the U.S. Commerce Department's National Telecommunications and Information Administration (NTIA), and the Federal Communications Commission (FCC) with radio frequency allocation. The FCC controls the frequency spectrum for nonfederal government use, ranging from 9kHz to 300GHz that is divided into 450 bands.

Frequency Bands

See Table R-1.

TABLE R-1 Frequency Bands

Frequency Band	Wavelength (m)
EHF (extremely high frequency)	10^2–10^3
ELF (extremely low frequency)	10^7–10^5
EOF (electro optical frequency)	10^3–10^8
HF (high frequency)	10^2–10^1
LF (low frequency)	10^4–10^3
MF (medium frequency)	10^3–10^2
REF (high-energy frequency)	10^8–10^{13}
SHF (super high frequency)	10^1–10^2
UHF (ultrahigh frequency)	1–10^1
ULF (ultralow frequency)	10^8–10^7
VHF (very high frequency)	10^1–1^{-1}
VLF (very low frequency)	10^5–10^4

TABLE R-2

Band Nomenclature

Frequency Band	Frequency Range
EHF (extremely high frequency)	32–400GHz
HF (high frequency)	3230–28,000kHz
HF/VHF (high frequency/very high frequency)	33–162.0125MHz
LF/MF (low frequency/medium frequency)	130–505kHz
MF (medium frequency)	505–2107kHz
MF/HF (medium frequency/high frequency)	2107–3230kHz
SHF (super high frequency)	3700MHz–27.5GHz
SHF/EHF (super high frequency/extremely high frequency)	27.5–32GHz
UHF (ultrahigh frequency)	322–2655MHz
UHF/SHF (ultrahigh frequency/super high frequency)	2655–3700MHz
VHF/UHF (very high frequency/ultrahigh frequency)	162.0125–322MHz
VLF/LF (very low frequency/low frequency)	0–130kHz

Band Nomenclature

See Table R-2.

Applications

See Table R-3.

Reverse Physical Channels (Cdma2000)

Reverse physical channels transfer data from single mobiles to base stations, and multiple mobiles may do the same over so-called common channels. Six radio configurations are available, including RC1, 2, 3, 4, 5, and 6. See Fig. R-1 and Tables R-4 and R-5.

Reverse Pilot Channel (R-PICH) The pilot channel helps the BS detect mobile station transmissions. The R-PICH is applied for initial acquisition, time tracking, RAKE-receiver coherent reference recovery, and for power control measurements. See Fig. R-2.

TABLE R-3

Applications

Application	Frequency
AM radio	535–1635kHz
Analog cordless telephone	44–49MHz
Cellular	806–890MHz
Digital cordless	900MHz
FM radio	88–108MHz
Large-dish satellite TV	4–6GHz
Nationwide paging	929–932MHz
Personal communications	900–929MHz
	1850–1990MHz
RF wireless modem	800MHz
Satellite telephone downlinks	2483.5–2500MHz
Satellite telephone uplink	1610–1626.5MHz
Small-dish satellite TV	11.7–12.7GHz
Television	54–88MHz
	174–216MHz
	470–806MHz
Wireless cable TV	28–29GHz
Wireless data	700MHz

Reverse Power Control Subchannel At certain times the mobile station may insert a reverse power control subchannel on the R-PICH, and it supports inner and outer power control loops for the forward traffic channel power control. At the rate of 1bit per 1.25ms power control group, the subchannel carries data regarding the forward link's quality.

Reverse Access Channel (R-ACH) The R-ACH initiates communication with the BS, priming it to respond to PCH messages. The ACH transmission is an interleaved and modulated spread spectrum signal, and uses a random access protocol.

Reverse Enhanced Access Channel (EACH) EACH initiates communication with the BS, and has the modes: basic access, power controlled, and

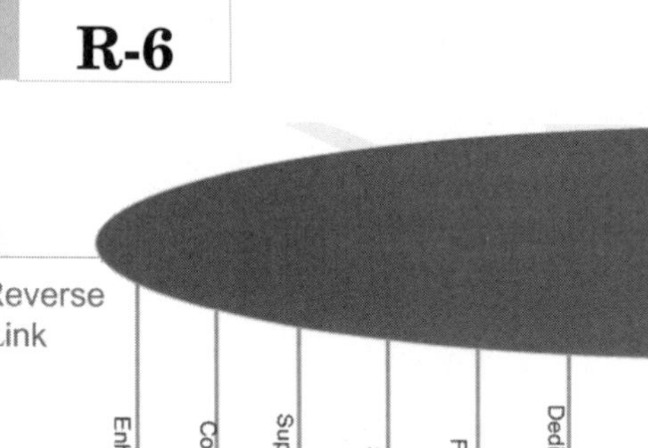

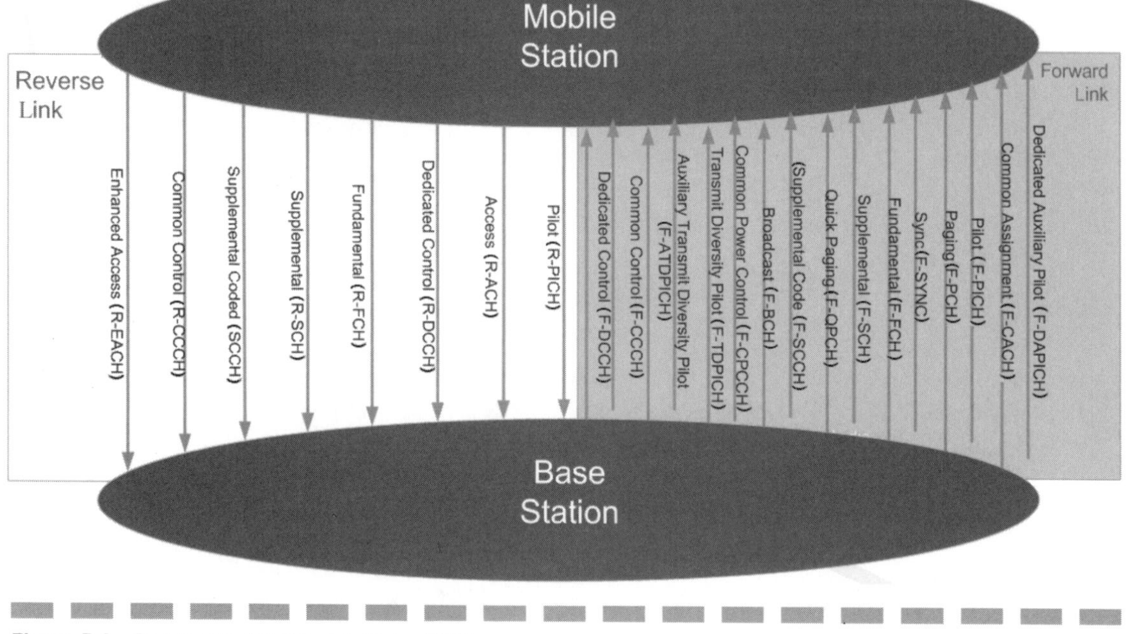

Figure R-1 Reverse physical channels (Cdma2000).

reservation access. The latter two may operate on the same EACH, while "basic access" operates on a separate EACH. The mobile does not transmit the enhanced access header on the EACH when in "basic access" mode. See Fig. R-3.

RF Propagation Models

Propagation models determine the attenuation losses of transmitted signals between transmitters and receivers. They are used during network design so as not to exhaust the link budget, and include Hata, Carey, Elgi, Longley-Rice, Bullington, Lee, and Cost231. Table R-6 lists the attenuation losses that occur in a downlink link.

See Fig. R-4.

Most cellular operators use the Rata model for propagation characterization, while the Carey model is required to submit information to the FCC. Communication Services (PCS) operators use Hata or Cost231, with the latter becoming the 3G model as it may be applied to the spectrum allocations defined by the ITU.

TABLE R-4

Reverse Physical Channels

	Physical Channels	Channel Name
Reverse Common Physical Channel	Reverse Access Channel	R-ACH
	Reverse Enhanced Access Channel	R-EACH
	Reverse Common Control Channel at 9.6kbps	R-CCCH
Reverse Dedicated Physical	Reverse Pilot Channel	R-PICH
	Reverse Dedicated Channel Control Channel	R-DCCH
	Reverse Traffic Channel:	
	Fundamental	R-FCH
	Supplemental	R-SCH
	Supplemental Code	R-SCCH

Propagation modeling may begin with morphologies that are defined in four categories:

- Dense urban may be business districts where buildings are 10 to 20 stories or more.
- Urban may have buildings from 5 to 10 stories in height.
- Suburban may have residential and business-type buildings from one to five stories.
- Rural may have a high percentage of open spaces with few tall structures, and are sparsely populated.

Free Space Free space path loss is the basis for all path-loss models, and is based on a 20dE/decade path loss:

$$Lf = 32.4 + 20 \log (R) + 20 \log (f)$$

where R is in km, f is in MHz, and Lf is in dB.

Hata Hata's empirical model is used widely in cellular, and was developed from the technical report by Okumura:

$$L_H = 69.55 + 26.26 \log (f) - 13.87 \log (h_b) - a (h_m) + (44.9 - 6.55 \log h_b) \log R$$

where f = 150–1500MHz (frequency), h_b = 20–200m (height of base station), h_m = 1–10m (height of receive antenna), R = 1–20km (distance from the site), and L_H is in dB.

TABLE R-5

Radio Configuration Characteristics and Data Rates of Reverse Channels for SR1 and SR3

Channel Type		Data Rate (kbps)	
		SR1	SR3
Enhanced Access Channel	Header	9.6	9.6
	Data	38.4–5, 10, or 20ms frames 19.2–10 or 2ms frames 9.6–20ms frames	38.4–5, 10, or 20ms frames 19.2–10 or 20ms frames 9.6–20ms frames
Access Channel (R-ACH)		4.8	NA
Reverse Control Channel (R-CCCH)		38.4–10 or 20ms frames 19.2–10 or 20ms frames 9.6–20ms frames	38.4–10 or 20ms frames 19.2–10 or 20ms frames 9.6–20ms frames
Reversed Dedicated Control Channel (R-DCCH)	RC3 RC4 RC5 RC6	9.6 14.4–20ms frames 9.6–5ms frames N/A N/A	NA NA 9.6 14.4–20ms frames 9.6–5ms frames
Fundamental Channel (R-FCH)	RC1	9.6, 4.8, 2.4, or 1.2	N/A
	RC2	14.4, 7.2, 3.6, or 1.8	N/A
	RC3	9.6, 4.8, 2.7, or 1.5–20ms frames, or 9.6–5ms frames	N/A
	RC4	14.4, 7.2, 3.6, or 1.8–20ms frames, or 9.6–5ms frames	N/A
	RC5	N/A	9.6, 4.8, 2.7, or 1.5–20ms frames, or 9.6–5ms frames
	RC6	N/A	14.4, 7.2, 3.6, or 1.8–20ms frames, or 9.6–5ms frames
Supplemental Code Channel (SCCH)	RC1 RC2	9.6 14.4	N/A N/A
Supplemental Channel (SCH)	RC3	307.2, 153.6, 76.8, 38.4, 19.2, 9.6, 4.8, 2.7, or 1.5–20ms frames 153.6, 76.8, 38.4, 19.2, 9.6, 4.8, 2.4, or 1.35–40ms frames 76.8, 38.4, 19.2, 9.6, 4.8, 2.4, or 1.2–80ms frames	N/A
	RC4	230.4, 115.2, 57.6, 28.8, 14.4, 7.2, 3.6, or 1.8	N/A

TABLE R-5

Radio Configuration Characteristics and Data Rates of Reverse Channels for SR1 and SR3 (Continued)

Channel Type		Data Rate (kbps)	
		SR1	SR3
	RC5	N/A	614.4, 307.2, 153.6, 76.8, 38.4, 19.2, 9.6, 4.8, 2.7, or 1.5–20ms frames 307.2, 153.6, 76.8, 38.4, 19.2, 9.6, 4.8, 2.4, 1.35–40ms frames 153.6, 76.8, 38.4, 19.2, 9.6, 4.8, 2.4, or 1.2–80ms frames
	RC6	N/A	1036.8, 518.4, 460.8, 259.2, 230.4, 115.2, 57.6, 28.8, 14.4, 7.2, 3.6, or 1.8ms frames

The Rata model should be used when predicting a path loss less than 1km from the cell site, or when the cell site is less than 30m in height. The value h_m is used to correct the mobile antenna height. The Rata model uses three correction factors based on the environmental conditions:

- Urban: 0dB
- Suburban: –9.88dB
- Open: –28.41dB

Cost231-Hata The Cost231 Rata model is pitched at PCS 1900MHz environment:

$$L_{CH} = 46.3 + 33.9 \log (f) - 13.82 \log (h_b) + (44.9 - 6.55 \log h_b) \log d + c$$

where c = 13dB (dense urban)

= 0 (urban)

= –12 (suburban)

= –27 (rural)

Quick The Quick model estimates the general propagation expectations, but is unrefined, although it does provide a starting point. There are two equations:

$$880\text{MHz PL} = 121 + 36 \log (\text{km})$$

$$1900\text{MHz PL} = 130 + 40 \log (\text{km})$$

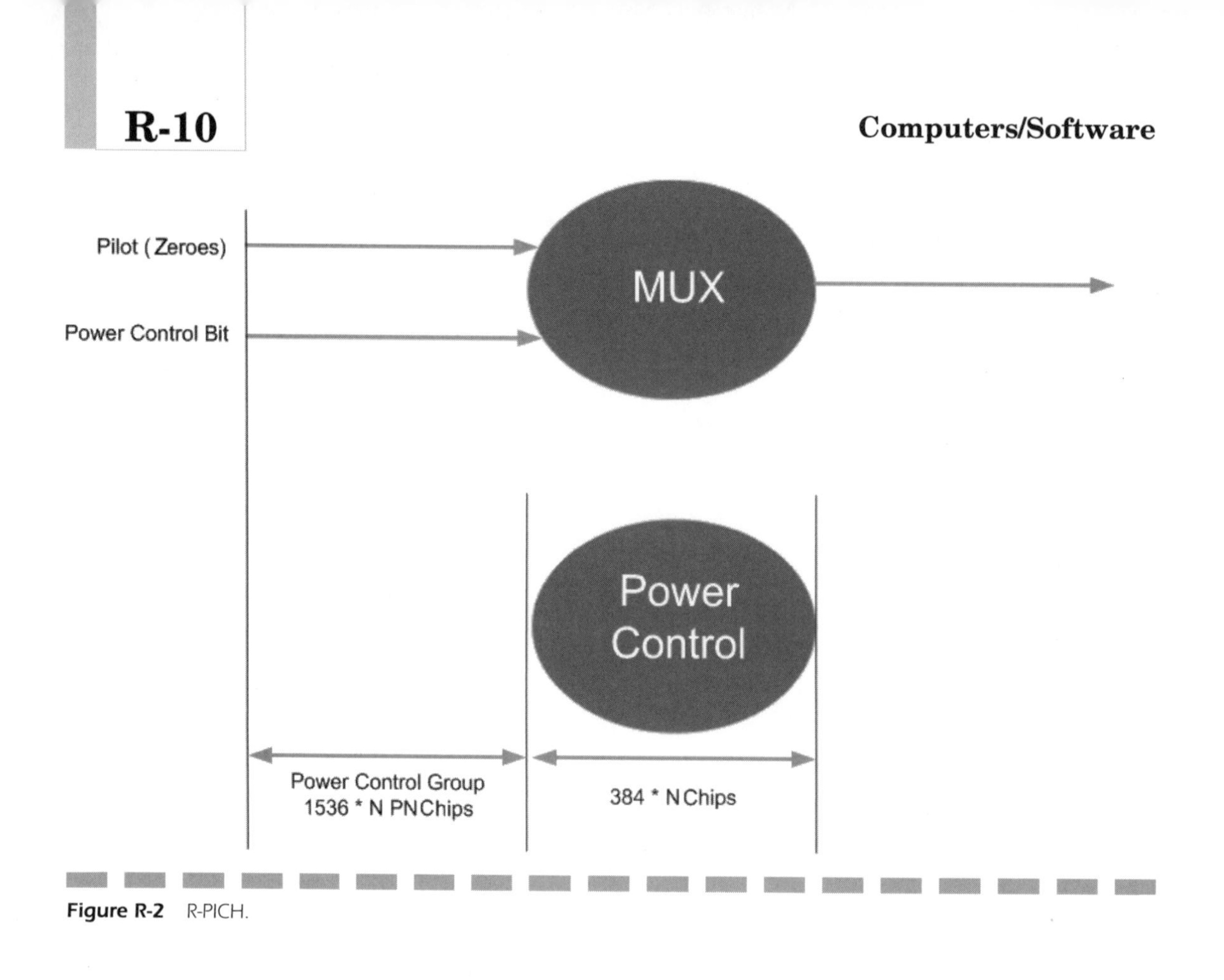

Figure R-2 R-PICH.

Computers/Software

RAD Tool (Rapid Application Development Tool)

A development tool that expedites application development. Its identity hinges on a number of identifying features that may include:

- Authentic Object Oriented Programming (OOP).
- Visual programming methodologies.
- Industry standard component architectures such as ActiveX or JavaBeans.
- Useful program libraries.
- Features that are appropriate to the collaborative team development environment. These may include security features that may be used to provide team members with access rights to objects so that they may be developed.

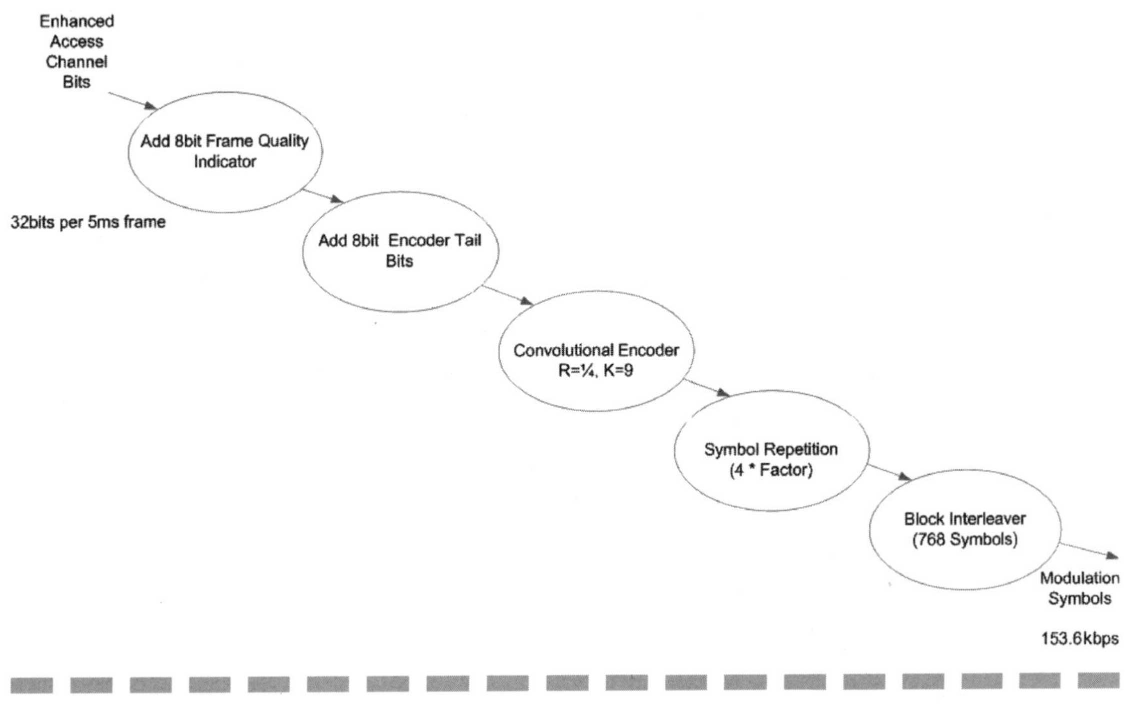

Figure R-3 EACH processing.

Radio Button

A group of buttons where only one may be selected at a time. For example, using HTML, you may add radio buttons using the following form that merely displays four radio buttons labeled $30, $40, $50, and $60:

```
<FORM> NAME="Customer"
       ACTION="http://botto.com/cgibin/form/cgi
       METHOD=get>

<INPUT TYPE="radio" NAME="rad" VALUE="1">

$30

<INPUT TYPE="radio" NAME="rad" VALUE="2">

$40

<INPUT TYPE="radio" NAME="rad" VALUE="3">

$50

<INPUT TYPE="radio" NAME="rad" VALUE="4">

$60

</FORM>
```

TABLE R-6

Generic Downlink Link Budget

Base Station Parameters	
Tx PA Output Power	dBm
Tx Combiner Loss	dB
Tx Duplexer Loss /Filter	dB
Jumper and Connector Loss	dB
Lightening Arrestor Loss	dB
Feedline Loss	dB
Jumper and Connector Loss	dB
Tower Top Amp Tx Gain or Loss	dB
Antenna Gain	dBd or dBi
Total Power Transmitted (ERP/EIRP)	W or dBm
Environmental Margins	
Tx Diversity Gain	dB
Fading Margin	dB
Environmental Attenuation	dB
Cell Overlap	dB
Total Environmental Margin	dB
Subscriber Unit Parameters	
Antenna Gain	dBd or dBi
Rx Diversity Gain	dB
Processing Gain	dB
Antenna Cable Loss	dB
C/L or Eb/No	dB
Rx Sensitivity	dB
Effective Subscriber Sensitivity	dBm

RAM (Random Access Memory)

A form of volatile electronic memory. It is described as random access because the access time is constant.

DRAM is used for system memory and is used mainly in the form of SIMM (Serial In-line Memory Module) electronic assemblies and occasionally SIPP

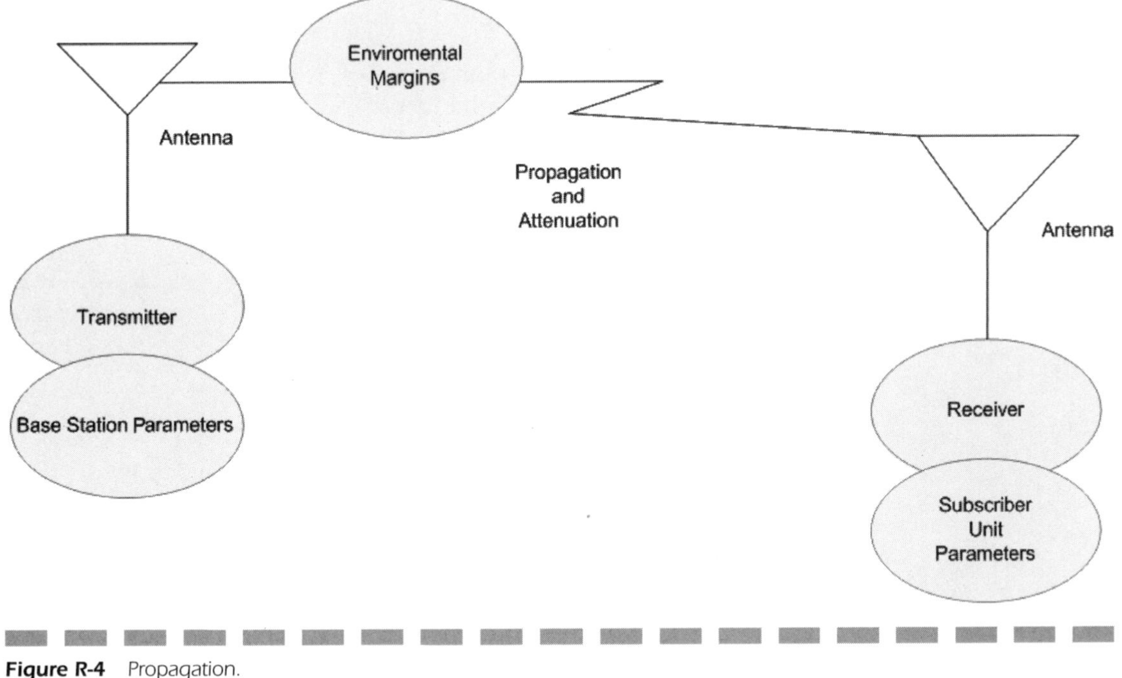

Figure R-4 Propagation.

(Serial In-line Pinned Package) assemblies. The principal advantage of DRAM is its low cost, while its disadvantage is its comparatively slow speed.

Typically, slow RAM will be of the order of 100ns (nanoseconds). Fast RAM will be of the order of 60ns or even 50ns. Enhanced Data Out (EDO) RAM offers higher performance, while Static Dynamic RAM (SDRAM) might offer access times of 10 or 12ns.

Static Ram (SRA) is generally faster than DRAM because it does not require constant cyclic refreshment. Its disadvantage is that of high cost. It is used widely for external processor caches.

RAS (Remote Access Services)

A RAS feature/program permits you to dial in to remote networks and to ISPs. Windows NT features RAS compliance.

RealAudio

A streaming audio technology for deploying real-time audio over the Web.

RealNetworks

A software publisher engaged in the manufacture of media players that include RealPlayer, RealPlayer G2, and RealPlayer Plus.

RealPlayer

A media player that is able to deliver streaming audio and video, as well as play local media files. It provides options to connect to streaming media sites including those associated with radio and broadcasting.

Real-Time

1. A program or system that responds to user interaction instantly.
2. A program or system that captures and/or compresses data at the rate it actually occurs. For example, a live satellite broadcast link is in real time.

Real-Time Compression A technique where an uncompressed video stream is compressed while it is played at full speed.

Record

A row in a database table, or a collection of fields that contain field values.

Recursion

A property of a programming language that enables procedures to be called by their own code. Such compliant languages are termed recursive.

Redirected URL

A page or URL that converts one URL into another.

Referrer Log

A log that tracks the user's visited pages.

Refresh Rate

A measurement of the rate at which all lines on a CRT-based monitor are scanned. It is quoted in Hz.

Relational Database

An information storage and retrieval application. Using a relational database, information is stored in records that are divided into fields of different types, including text, numeric, date, graphic, and even BLOB (Binary Large OBject). The records are stored in tables or files.

Records from one file may be linked to records stored in a separate file or table. Codd's standard text about relational databases published in the 1960s specified different types of relational links. Types of link include one-to-one, one-to-many, and many-to-many.

There are many commercial examples of the relational databases that base their design on the original writings of Codd. Relational databases are formally referred to as RDBMSs (Relational Database Management Systems) whereas flat-file databases are termed simply DBMSs (Database Management Systems). Commercial examples of software products that permit the development of RDBMSs include Borland Paradox for Windows, dBase, Microsoft Access, and Ingress.

Removable Medium

A medium that may be removed from the computer. Examples include floppy disk, CD-ROM disc, DVD disc, and Iomega Zip disks.

Requester Path

A unidirectional path from the client to the server that supports GET requests, and may deliver to the server such information as the client's:

- Domain
- E-mail
- User agent denoting the browser type
- Variables such as a list of file types with which the browser is compatible

The requester path naturally coexists with the unidirectional path from the server to the client which may deliver:

- Web pages
- Streaming media

Resolution

A measurement of the concentration of dots or pixels in a digital image. In display technology, resolution is specified in terms of screen dimensions expressed in pixels, and the dot-pitch expressed as the distance between displayable pixels.

Typical display resolutions of commercially available monitors include 640-by-480 pixels, 800-by-600 pixels, 1024-by-768 pixels, 1280-by-1024 pixels, and 1600-by-1200 pixels.

In terms of printer technology, resolution is expressed in terms of the number of dots per inch (dpi). Generally, low-cost laser printers produce output of 300dpi. More expensive variants offer 600 and 1200dpi resolutions.

Route

1. Noun: A path taken by a packet or message that leads from a sending device to a receiving device. The path might involve the interaction with software components that may form part of an OO distributed system.
2. Verb: An action taken in order to send or forward a packet or message to a receiving device, or even software component.

Miscellaneous

RAID (Redundant Array of Independent Disks)

A mass storage device that has many individual disks. Identifying features of RAID may include:

- High levels of fault tolerance
- Scalability through the addition of hard disks
- Hot-swappable disks, meaning they may be removed and replaced without the need to power down the RAID
- Redundant power supplies for improved fault tolerance
- Shared mass storage, serving disparate computers/networks

- Heterogeneous characteristics, where they may be integrated into environments comprising multiple OSes
- High speed interfaces such as Fiber Channel, Ultra SCSI, etc.

The original RAID specification originated from UC Berkeley in 1987, and was named Redundant Array of "Inexpensive" Disks. The aims of the Berkeley group were threefold:

- Improve fault tolerance of mass storage
- Reduce mass storage costs
- Improve mass storage performance

Realizing the inescapable fact that no single mass storage system could be optimized in all three of the aforementioned areas, the group defined what were to become a number of industry standard solutions.

Achieving its objectives to varying degrees, the Berkeley group defined a series of RAID levels employing several tried and tested data storage techniques. One of these was mirroring where data is written to, and read from, pairs of disks concurrently in order to deliver fault tolerance.

Modern RAID systems may be specified in terms of:

- Maximum data storage capacity that is typically in the Gbyte range for a single RAID unit, and is in the TByte range for multiple connected units
- Average access time measured in milliseconds (ms)
- Average and burst data transfer rates
- Cache size
- Interface type
- Multiplicity of host types that may be connected
- OS compatibility

RAID performance has obvious effects, and high performance echoes performance gains that are felt locally and remotely.

The five levels of RAID defined by the Berkeley group include:

- Level 0 that stripes data across multiple disks, but provides no error correction or redundancy.
- Level 1 that uses duplexing or mirroring, where data is written concurrently to pairs of independent disks, promoting a high degree of fault tolerance.
- Level 2 that stores and reads data by dividing it into bits, and storing them on different drives—otherwise known as striping. It also stores ECC codes on dedicated disks.
- Level 3 that divides data into blocks, storing them on different independent disks. One additional disk contains parity data.

- Level 4 that stripes data blocks across multiple disks. One additional disk contains parity data.
- Level 5 that stripes data blocks across multiple disks, while parity data is stored on multiple disks.

Other RAID configurations include Level 0 and Level 7, neither of which were devised by the Berkeley group. Level 7 offers improved fault tolerance, and is patented by Storage Computer Corporation.

Reliability

A measure of the period of downtime that a system will endure. It may be expressed as a percentage value, indicating the portion of time that the system will be fully or even partially operational. Such a measure of availability (A) may be applied to devices, components, subsystems, systems, networks, etc.

Availability may be calculated using the:

- MTTF (mean time to restore): the average time period required to return a failed system to its fully operational state.
- MTBF (mean time between failures): the average time period that indicates the frequency at which a device, component, subsystem, or complete system will fail.

$$\text{Availability (A)} = \frac{\text{MTBF}}{\text{MTBF} + \text{MTTR}}$$

Collective availability (Ac) of a complete system is equated to the product of the availability for each individual component. For example:

$$\text{Availability (Ac)} = \text{Clients (Au)} \times \text{Server (As)} \times \text{Network (An)} \times \cdots \times \text{Router (Ar)}$$

Risk Exposure (RE)

A product of risk probability (RP) and risk cost (RC):

$$\text{RE} = \text{RP} \times \text{RC}$$

- RP is the probability of attempted attacks on a system leading to a security breach.
- RC is the estimated cost of a particular (or average) security breach.

Router

A device that receives and routes messages between network systems, or between complete networks. The messages may be packets, cells, or frames, depending on the protocol used.

RSA

See RSA under letter E.

RTP (Real-Time Transport Protocol)

A protocol that supports real-time audio/video communications.

Acronyms

RAB	Radio access bearer
RACE	Research in Advanced Communications Equipment
RACH	Random access channel
R-ACH	Reverse access channel
RAM	Random access memory
RANAP	Radio access network application part
RBC	Radio bearer control
RBP	Radio burst protocol
RBS	Radio base station
RC	Radio configuration
RC	Resource control
R-CCCH	Reverse common control channel
RCD	Resource configuration database
RCELP	Relaxed code-excited linear prediction
RCPC	Rate-compatible punctured convolution
R-DCCH	Reverse dedicated control
RES	Radio and equipment systems RF
R-FCH	Reverse fundamental channel
RFU	Radio frequency unit

RILPC	Reverse inner loop power control
RITT	Research Institute of Telecommunications Transmission
RIU	Radio interface unit
RLAC	Radio link access control
RLC	Radio link control
RLP	Radio link protocol
RNC	Radio network controller
RNCP	Radio network control plane
RNG	Random number generator
RNL	Radio network layer
RNS	Radio network subsystem
RNSAP	Radio network subsystem application part
ROLPC	Reverse outer loop power control
ROM	Read only memory
ROPC	Reverse open loop power control
ROSE	Remote operation service element
RPC	Radio port controller
RPC	Remote procedure call
RPCU	Radio port controller unit
R-PICH	Reverse pilot channel
RPOM	Radio port operation and maintenance
R-SCCH	Reverse supplemental code channel
R-SCH	Reverse supplemental channel

S

Satellite Communications

See Mobile Satellite Service (MSS) under letter M.

Networking Approaches

See Networking Approaches under letter N.

Security—UTRAN

The UTRAN initiates Ciphering Control that may start ciphering or change the cipher key. The UTRAN sends a SECURITY_MODE_COMMAND message to the UE holding the new ciphering key and the activation time. The activation time is given as an RLC send sequence number or as a CFN, and when it elapses, the UE starts to use the new configuration and returns a SECURITY_MODE_COMPLETE message.

There are several 3GPP ciphering algorithms, and the algorithm is used in the MAC and in the RLC, but the ciphering key sequence or "COUNT_C" differs. There are three ciphering key sequences, and these are shown in Fig. S-1.

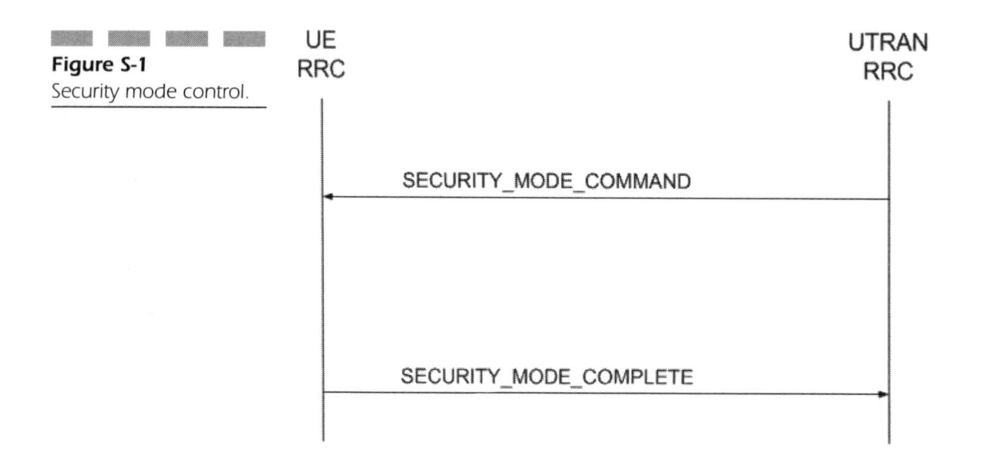

Figure S-1
Security mode control.

Integrity Protection

Integrity protection places a security perimeter around the air-interface signaling, by preventing the tampering of signaling procedures. The integrity-protection process is invoked by the security mode procedure. To start or reconfigure the integrity protection, the UTRAN sends a security mode command message on the downlink DCCH in AM RLC using the present integrity-protection configuration.

The UE uses two integrity-protection hyperframe numbers for the CCCH and for the signaling RB:

1. Uplink HFN
2. Downlink HFN

The message sequence numbers are:

1. Uplink RRC message sequence number
2. Downlink RRC message sequence number

Counter Check

The air interface also uses the "counter check" security procedure to check that the sent data volume is similar in the UE and UTRAN. For each radio bearer, the UE maintains a ciphering sequence number (COUNT -C) that can be queried by the UTRAN in order to detect intruders.

Soft Handoff

Soft handoff (handover) differs from hard handoff (where a definite decision regarding handoff is made), and without the user attempting simultaneous communications with the two or more BSs. A soft handoff results in a conditional decision that depends on changes in pilot signal strength from the two or more BSs. And when the signal from one BS reaches a threshold difference, a handoff is made. So during soft handoff, the user communicates with multiple BSs.

(*See* Handovers under letter H.)

Handoff Categories

Intersector or Softer Handoff The mobile communicates with two sectors in a cell, and a RAKE receiver at the BS amalgamates the best signals from the diversity antennas, creating a single traffic frame.

Intercell or Soft Handoff The mobile communicates with two or three sectors that are not in the same cell. A "primary" BS controls call processing during handoff, and initiates a forward control message. Other BSs are "secondary." Soft handoff ends when either the primary or secondary BS is dropped.

Soft-Softer Handoff The mobile communicates with two sectors in one cell and one sector in another.

Hard Handoff With hard handoffs the connection with the old traffic channel is broken before the connection with the new traffic channel is operational.

Sigtran Protocol

Using the Sigtran protocol suite, it is possible to have a partial SS7 implementation with SS7-based applications at point code 4. The Stream Control Transmission Protocol (SCTP) is a transport protocol used to carry signaling. An Upper Layer Protocol (ULP) is an adaptation layer.

Applications Layers

- *ISDN Q.921-User Adaptation Layer* (IUA) is equivalent to the Q.921 datalink layer that carries Q.931ISDN signaling.
- SS7 *MTP2-User Adaptation Layer* (M2UA) provides an interface between MTF3 and SCTF, so standard MTF3 may be used in the IP network.
- SS7 *MTP3-User Adaptation Layer* (M3UA) provides an interface between SCTF and applications that normally use the services of MTF3. These include ISUP and SCCF.
- SS7 *SCCP-User Adaptation Layer* (SUA) provides an interface between SCCF user applications and SCTF. Applications such as TCAP use the services of SUA in the same way they would use the services of SCCF in the SS7 network.

SMS (Short Message Service)

GSM's short message service (SMS) allows text messages to be sent using any SMS device, and originally permitted 160-character messages. But using data compression the message length may be extended to 239 characters.

Vodafone developed a Huffman-based compression algorithm that later became ETSI 03.42. Because of the short message limitation, an SMS shorthand has developed and mainly consists of mnemonics:

All the best	ATB
Anything	NETHING
Are	R
Are you ok?	RUOK?
Be	B
Be back later	BBL
Be right back	BRB
Be seeing you	BCNU
Before	B4
By the way	BTW
Excellent	XLNT
For your information	FYI
For/four	4
Free to talk	F2T
Great	GR8
Late	LT
Later	LTR
Laughing out loud	LOL
Love	LUV
No one	NO1
Oh I see	OIC
Please	PLS
Please call me	PCM
Regards	RGDS
See	C
Speak	SPK
Talk to you later	TrYL
Thank you	THNQ
Thanks	THX
Today	2DAY
Too/to/two	2
Want to	WAN2
What	WOT

You	U
Your	UR

SMS handsets include SMS-MO and CB, where the former may only send SMS messages. SMS Centers provide the core message services and conversions into facsimile, e-mail messages, and voice-mail notifications. GSM operators may have one or more SMS centers that may connect with e-mail and the PSTN, enabling both inbound and outbound data transfers.

SS7

SS7 is a protocol for Voice over IP (VoIP). The SS7 protocol stack is shown in Fig. S-2, where the MTP transfers messages from the source to the destination signaling point. The ISUP (ISDN User Part) establishes phone calls, as can the SCCP that normally transports higher-layer applications such as the Intelligent Network Application Part (INAP). These generally use the services associated with the Transaction Capabilities Application Part (TCAP) that uses the SCCP services.

SCCP provides global title addressing that permits signaling between entities that may not know the target entities' point codes or signaling addresses; addresses or telephone numbers are mapped to point codes at nodes that may have initiated the message.

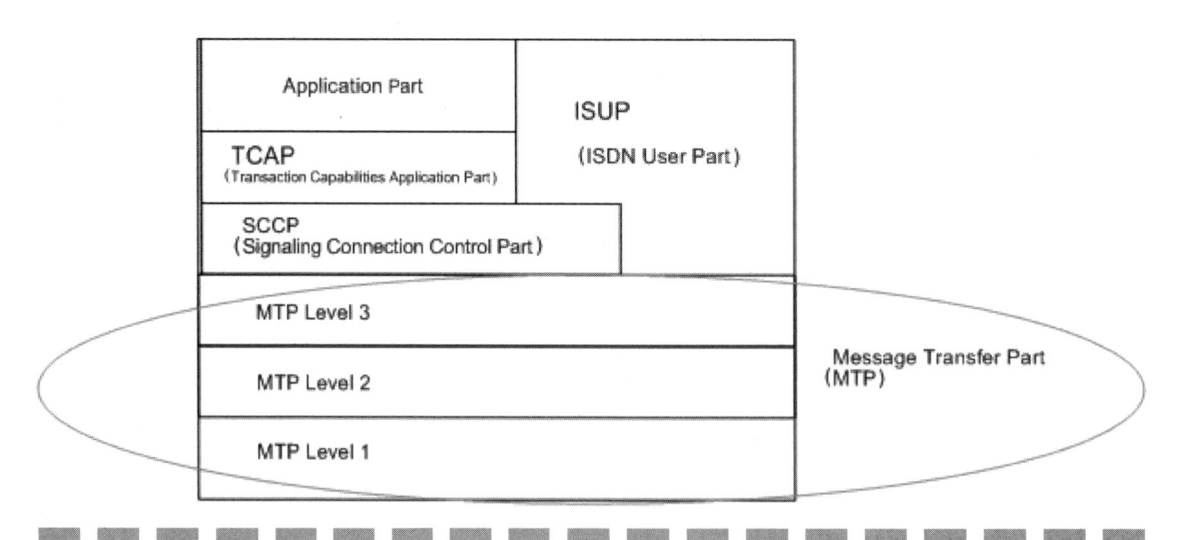

Figure S-2 SS7 protocol.

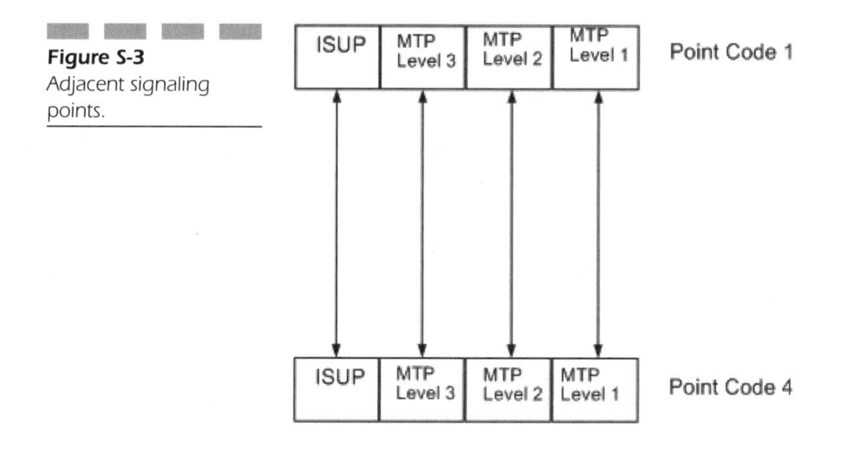

Figure S-3 Adjacent signaling points.

SS7 Communication

Communication between adjacent signaling points makes every layer a peer-to-peer link between entities that may be referred to as point code 1 and point code 4 (as shown in Fig. S-3). Figure S-4 shows nonadjacent signaling points with peer-to-peer links at layers 1 to 3, between point codes 1 and 2, 2 and 3, and 3 and 4. And at the SCCP layer, peer-to-peer links unite point codes 1 and 2 and point codes 2 and 4.

At the TCAP and Application layers, peer-to-peer links unite point codes 1 and 4, so the application at point code 1 is aware of the TCAP layer at point code 1 and the application layer at point code 4. Using the Sigtran protocol suite, it is possible to have a partial SS7 implementation with SS7-based applications at point code 4.

SSL (Secure Sockets Layer) Protocol

A secure channel or protocol that is supported by the TCP transport protocol, and includes the higher-level protocol SSL Handshake Protocol that authenticates the client and server devices, and allows them to decide on an encryption algorithm and keys before data reception or transmission commences. Its ability to allow high-level protocols to sit on top of it is perceived as an advantage. Used alone SSL is not perceived as a complete security solution, although it does present one significant security perimeter in the eyes of many security analysts. It ensures a secure connection by authentication of a peer's connection and uses integrity checks and hashing functions, to secure the channel between applications.

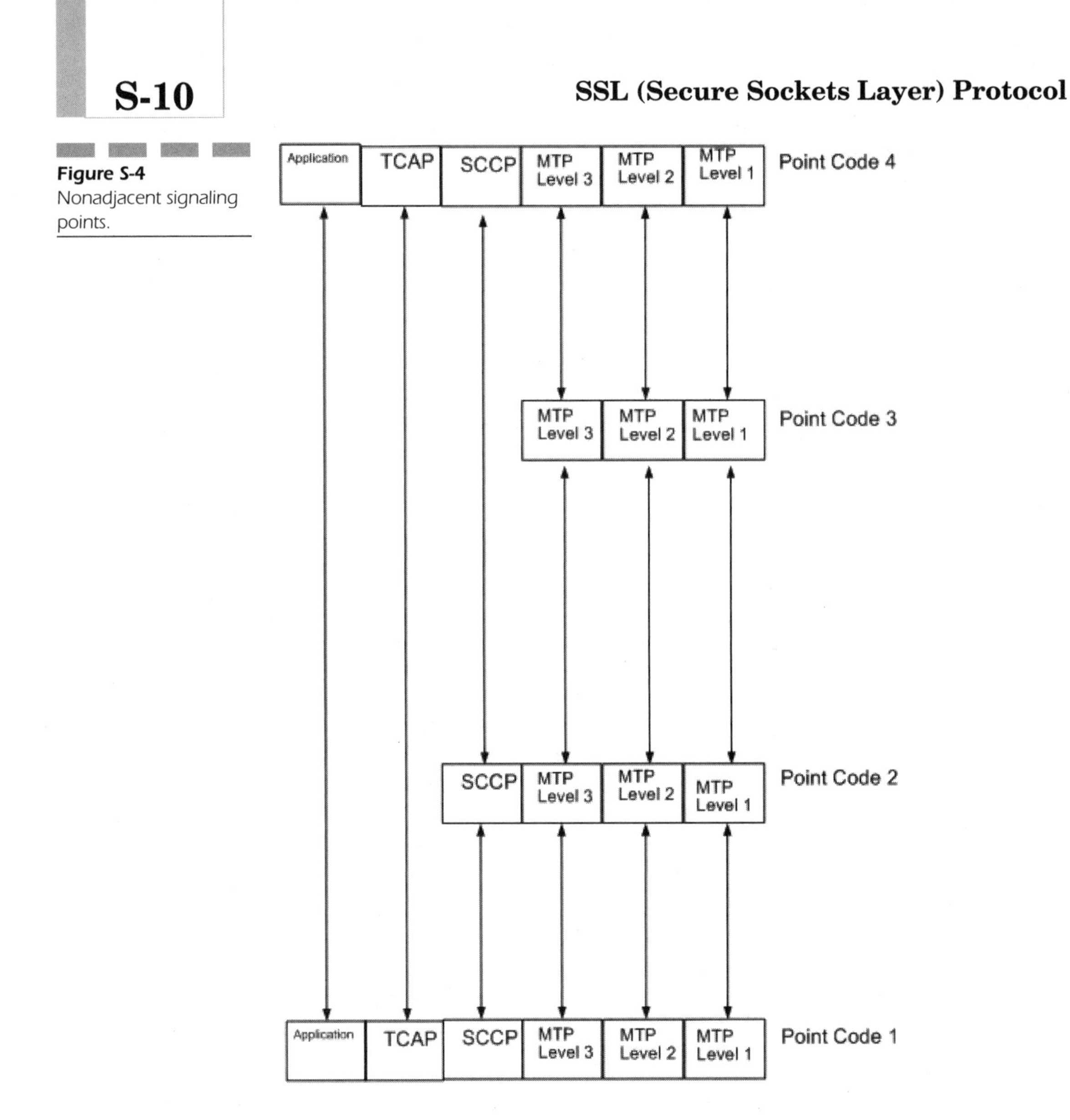

Figure S-4 Nonadjacent signaling points.

It was designed to prevent:

- Information forgery
- Eavesdropping or sniffing
- Data changes

Private or symmetric key is the basis of the SSL's cryptography, and authenticates a peer's identity using asymmetric cryptography.

SLL's flaws are documented widely and include:

- When a browser connects with an SSL server, it receives a copy of its public key wrapped in a certificate that the browser sanctions by checking the signature. The flaw here is that the browser hasn't the means to authenticate the signature, since no verification is performed up the hierarchy because many certificates used by SSL are root certificates.
- It consumes client and server processing resources.
- It has operational difficulties with proxies and filters.
- It has operational difficulties with existing cryptography tokens.
- Its key management tends to be expensive.
- Expertise to build, maintain, and operate secure systems is in short supply (1999).
- It creates network traffic when handshaking.
- The migration path from nonpublic key infrastructures is arduous.
- It requires Certificate Authority with appropriate policies.
- Its communication data does not compress and therefore steals network bandwidth.
- It is subject to certain international import restrictions.

Computers/Software

SAA (Systems Application Architecture)

A strategy initiated by IBM for enterprise computing that defines the three layers:

- Common User Access
- Common Programming Interface
- Common Communications Support
- Sample
- A digital value derived from an analog source

SAP R/3

A client/server development environment that is the successor to SAP R/2. The transition was a response to the shift from the two-tier client/server model to the three-tier client/server model. The product is used globally, and came to prominence through its application in German industry, notably the automotive sector.

Screen Scraper

A client/server software component or function that removes or "scrapes" display information from requested data and formats it so that it may be displayed by the client system. It may also do the same for outgoing traffic from the terminal or client system.

Screened Subnets

A subnet that restricts TCP/IP traffic from entering a secured network. The screening function may be implemented by screening routers.

Screening Router

A router variant able to screen packets that match a predefined criteria, including the:

- Source address
- Destination address
- Protocol type

Script

A series of instructions that may be interpreted by a program. Sometimes referred to as macros. Scripts can sometimes be generated through menu selections or by writing code.

SCRIPT_NAME

A CGI variable that holds the name and path of the CGI script being executed.

Scripting

A scripting language such as VBScript or JScript may also be perceived as a glue, as may be HTML.

SCSI (Small Computer System Interface)

A universal and internationally agreed interface standard backed by ANSI (American National Standards Institute), intended to provide interchangeability between peripherals and computer systems from different manufacturers.

Apple Computer has long since realized the importance of SCSI fitting Macintosh machines with appropriate controllers. The SCSI continuum approximates:

- SCSI-1
- SCSI-2
- Wide SCSI
- Fast Wide SCSI
- Ultra SCSI

Search Engine

1. A site used to retrieve documents from the World Wide Web, and operates using gathered and evolving indexes stored locally. (See Fig. S-5.) The indexes are not common to all search engines, although they share a standard format that dictates that the document's:
   - Heading is enclosed by HTML <TITLE> tags
   - Description consists of the 200 characters that follow the <BODY> tag, or the matter enclosed by <META> tags

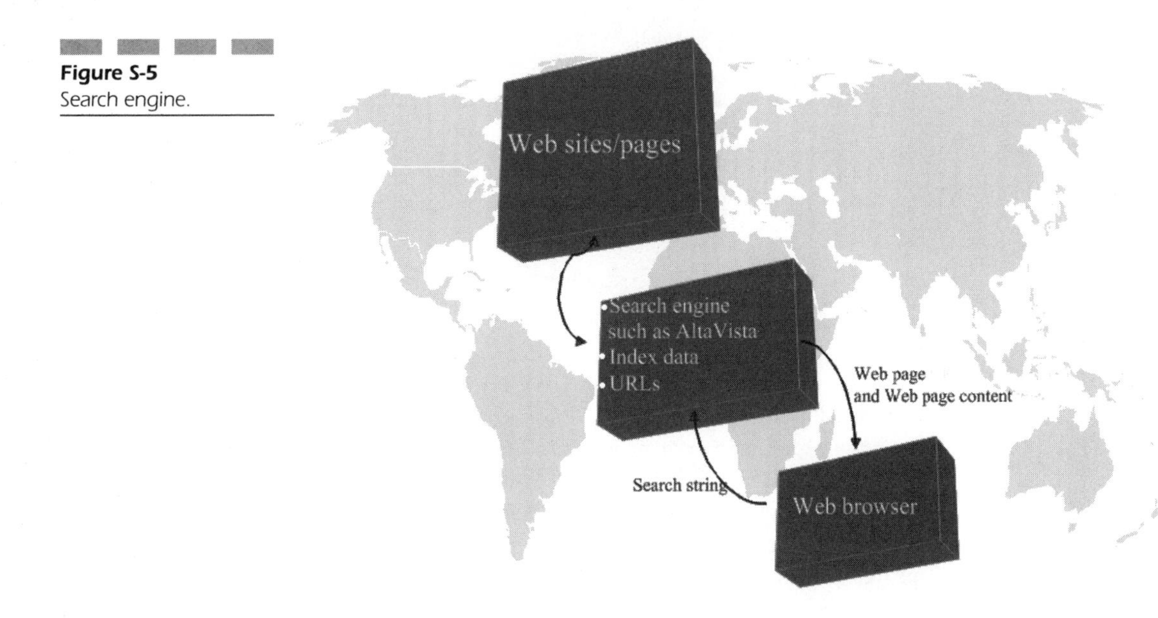

Figure S-5
Search engine.

The interface permits the entry of search words and phrases that may include logical operators. Popular search engines include:

- AltaVista
- Yahoo
- HotBot

An HTML tag may be used to enclose descriptive meta data that may be used by search engines as an alternative to the 200 characters that follow the <BODY> tag.

```
<HEAD>
  <TITLE>Francis Botto home page</TITLE>
  <META name="description" content="IT Research">
  </HEAD>
```

The <META> tag may also be used to add keywords of up to 1000 characters to a Web page, and may be retrieved through appropriate search phrases, such as:

```
<META name="keywords" content="Multimedia, MPEG, DVD">
```

An HTML tag that encloses the Web page title is used as meta data by popular search engines when retrieving Web documents, displaying it as the document's title. Such data is collected by search engines periodically, but may remain transparent to some if your ISP uses a robots.txt file to stop Web robots from indexing Web pages. It is possible to determine if a server has a robots.txt file by entering the Web page's URL (including its domain name and domain category) and including robots.txt as a suffix:

- http://www.FrancisBotto.com/robots.txt

Sending Web page URLs to search engines may cause them to be categorized as available via additional search words and phrases, other than those contained in the Web pages themselves.

(*See* <TITLE> in Numerals chapter, HTML under letter H.)

2. A feature that permits a Web site to be searched.

Search String

A single word, phrase, sentence (or a number of words, phrases, sentences) by which a document or number of documents are searched.

A search engine or retrieval system retrieves documents (or document details) based on the search string.

Search engines and retrieval systems support wild cards and logical operators. For instance, if it were necessary to find documents containing the name van Gogh along with the city Amsterdam, using logical operators you would use the phrase "van Gogh" AND Amsterdam.

Second-Generation Language

A programming language where instructions are represented by concise mnemonics. The language is said to be a second-generation language. Such "assembly languages" are indigenous to processors. Often the terms assembly language and machine code are interchangeable.

Secure Mail

An e-mail that is suitably encrypted.

Secure Systems

A system that has cryptosystems integrated into its design.

Secure Transaction

A transaction that is made secure using cryptography.

Security

A method of restricting access to applications, data, and systems to their intended users.

The term may include automated virus checks on incoming documents, and on executable code such as Java applets and ActiveX controls using security gateways. Firewall technology is key to Web security, as is data encryption and password protection. Security is paramount to organizations that deploy corporate data, and for companies running e-commerce Web sites (such as Amazon, for example).

Security standards include those developed by the U.S. Department of Defense, and named the Trusted Computer Standards Evaluation Criteria that are otherwise known as the Orange Book.

This was introduced in 1985, and was originally aimed at mainframe and mini computers for many years. It is also applicable to databases and to networks, through the Trusted Database Interpretation, and the Trusted Network Interpretation.

The Orange Book is a multitier set of guidelines, including:

- Level D1 that is the lowest level of security, rendering the system untrusted.
- Level C1 that is a discretionary security protection system, requiring a login name and password, and access rights are allocated.

- Level C2 that includes Level D1 and Level C1 security, and integrates additional security features. For instance, this level requires that the system's relevant events be audited.
- Level B1, or Labelled Security Protection, that provides tiered security. Compliance sees object permissions that may not be changed by file owners.
- Level B2, or Structured Protection, requires the labeling of all objects.
- Level B3, or Security Domains level, that requires terminal connections via a trusted path.
- Level A, or Verified Design level, is the highest Orange book security level.

Security Boundary

Also known as security perimeters, such boundaries encapsulate systems, software, objects, etc. They may be implemented in software, hardware using firewalls, passwords, encryption, and the assignment of dongles to users.

Boundaries exist at a number of different levels, including:

- Physical that covers tangible resources such as systems
- Application that cover access rights to applications
- Data that covers access and editing rights to data
- System that dictates who may log on to a network or system

Security Gateway

A security layer that fortifies a network against hostile virus attacks, by screening incoming executables and data. The executables might be Java applets, ActiveX controls, or Plug-ins. Each of these represents a potential threat, not just in terms of viruses, but in terms of what they may do to client-side documents, files, and even system files such as those concerned with the initialization of Windows variants.

For instance, an ActiveX control might take control of a client's Word documents, perform an operation on them, such as convert them to HTML, and then possibly abstract them, for display or even processing on a remote Web site. For many security managers, this is unacceptable behavior.

Additionally, the unregulated Internet means that virtually anyone can deploy applications that may potentially damage clients. It is desirable, therefore, to attempt to check and possibly screen such inbound traffic.

Such a comparative centralization of the antivirus security layer, makes the need for standalone variants on clients redundant, although these may still be employed, particularly if removable media are being used for file transfer.

Such a security gateway may:

- Provide multiple OS support, such as Windows, DOS, Netware, although many are confined to Windows NT
- Support ActiveX, Java, cookies, Plug-ins, JavaScript, JScript, VBScript, and various executables
- Check all Java classes on downloaded applets
- Provide intelligent filtering features

Security Protocols

A protocol that integrates a cryptosystem.

Security Proxy

A Web proxy that integrates security features that are used to authenticate connections to servers.

Semaphore

A communications method, involving physical signaling, that was invented by the French in 1792. Even today the semaphore principle is applied in programming, where flags may be used to relay certain states and events.

In computer terms, semaphore may be applied to coordinate processes.

Sequential Prose

A continuous stream of linear text.

Serial Port

An input/output port through which data are transmitted and received sequentially. RS232 is a standard serial port used to transmit serial digital information over modest distances so as to connect communication devices and other serial peripherals to computers. The PC COM (communications) ports are serial ports.

Server

1. A Web server may also be considered as the software implementation that serves HTML pages, etc., and commercial examples include:
   - Microsoft IIS
   - Apache server
2. A transaction server is allocated the task of transaction processing (TP), and often invokes the application logic necessary to perform database interactions and manipulations. The process(es) invoked directly or indirectly by the client are collectively referred to as the transaction.

   Transaction servers may include UI logic, driving the client UI, relegating the client device as little more than a dumb terminal. Typically, mainframe-based transaction systems might adhere to this model. Alternatively, the UI logic, or presentation, may be distributed to the client.

   The server consists of a TP monitor that performs transaction management and resource management. Transaction management ensures the so-called ACID properties of transactions. These are Atomicity, Consistency, Isolation, and Durability. ACID property compliance is achieved through the two-phase commit protocol.

   Resource management is intended to optimize the use of resources that include memory, mass storage, and processing. It may also be involved with load balancing between resources and between the software processes that may be threads.
3. An entity that serves clients. The services provided might include the implementation of processes and the distribution of data, and may be categorized under:
   - Fax, where the server provides fax reception and transmission facilities for connected client systems.
   - Database, in the client/server configuration that SQL requests from the clients, that performs the necessary data requests.
   - Communications that enable client systems to make remote connections to external networks and servers.
   - Print, where the server is dedicated to printing locally or remotely.
   - File, where a centralized server, perhaps connected to RAID storage devices, is utilized by clients to provide high-volume data storage, and high-performance disk access and data transfer rates.
   - Transaction, where the server updates data that may form part of simple client/server two-dimensional database, or warehouse data that may be multidimensional (in data cubes) or even hypercubes.
4. A Web server is the hardware platform that supports one or more Web sites. Traditionally, Web servers have been based on the Unix or Windows NT OSs.

5. An intranet server may be considered in the same terms as a Web server, but with a security perimeter to prevent public access.
6. A peer-to-peer server is a system on a network in which the resources of any connected system may be shared. While any system on a peer-to-peer network might be a server, typically the most highly specified system performs as a server.
7. A file server provides centralized resources for network users.
8. A database server provides centralized data storage.
9. An object or application that serves an application or object with embedded or linked data. The server might be OLE 2.0–compliant, or may conform to another component architecture.
10. A CD-ROM server dynamically distributes requested information from CD-ROM drives to LAN users.

    If the maximum number of drives per server is exceeded, additional servers are added. The incorporation of additional servers, prior to reaching the network maximum of drives per server, leads to better service for users.

    There are several commercially available CD-ROM network packages, many of which are software-oriented.
11. A video server is a hardware solution that provides the basis for a Video-on-Demand (VoD) service. It may be implemented using MPP. Advantages of such parallel processing systems include scalability, where growing numbers of subscribers to a VoD service may be accommodated through additional processors, and even complete servers.
12. A video, audio, or multimedia server that serves client systems with streaming media.

Server Application

A term used to describe a server-side application that may drive or provide services for client applications and systems. The latter tier is the front end, with which the user interacts.

Between the back- and front-end applications is middleware or glues that exist at a number of levels. These may bind together and coordinate application logic, data, and presentation distributed across the back and front ends.

An application that runs on a server and may be a CGI program.

Server Crash

A state where a Web server is rendered inoperable by a hacker that has in some way overloaded the services it provides, perhaps by:

- Subjecting the server to excessive e-mail traffic
- Using a program that continually attempts to access content files.

Server Object

An object that runs on a server or at some location at tier 3.

SERVER_PORT

A CGI variable that holds the port on which the server is running.

SERVER_PROTOCOL

A CGI variable that contains the protocol version.

SERVER_SOFTWARE

A protocol that holds the name and perhaps the version of the server software.

Service

An entity that may be used to provide functions of some kind that may be relevant to users, devices, applications, or to other services that are typically on a network. Jini services are an example, and may provide access to communications applications, printers, remote residential networks, etc.

SET (Secure Electronic Transactions)

A standard means of securing payment transactions made to on-line merchants. By integrating cryptosystem techniques, it is perceived as a credit card security system, and was initiated by Visa and Microsoft. SET implementations are an amalgam of cryptosystems, protocols, secure protocols, and techniques.

SET Application

An application that uses the SET internationally agreed technologies and methodologies.

SET ASN.1 (Abstract Syntax Notation One)

A standard that defines the encoding, transmission, and encoding of data and objects that are architecture neutral.

SET Baggage

A method of appending ciphered data to a SET message.

SET CDMF Commercial Data Masking Facility

A ciphering technique based on DES that is used to transfer messages between the Acquirer Payment Gateway and the Cardholder in SET implementations.

SET Certificate Authority

A trusted party that manages the distribution of SET digital certificates, where layers of the Tree of Trust has the representation of a digital certificate.

SET Certificate Chain

A group of digital certificates used to validate a certificate in a chain.

SET Certificate Practice Statement

A group of rules that determine the suitability of certificates to particular applications and communities.

SET Certificate Renewal

An event that sees the renewal of a certificate for continued transacting purposes.

SET Consortium

An international organization that was formed when MasterCard and Visa announced SET, whose initial objective was to create an agreeable standard, and to consider STT and SEPP.

SET Digital Certificate

A means of linking an entity's identity with a public key that is carried out by a trusted party.

SET Digital Signature

A digital signature may be applied to an encrypted message. A message digest is ciphered using the sender's private key and then appended to the message, resulting in a digital signature.

SET e-Wallet

An element of a cardholder that creates the protocol and assists in acquiring and managing cardholder digital signatures.

SET Hash

An element that reduces the number of possible values using a hashing function such as the Secure Hashing Algorithm (SHA-1).

SET Idempotency

An attribute of a message that sees repetition yield a constant result.

SET Message Authentication

A process or usually subprocess that verifies that a message is received from the appropriate or legal sender.

SET Message Pair

Messages that implement the POS and certificate management in a SET implementation.

SET Message Wrapper

A top-level data structure that conveys information to message recipients.

SET Order Inquiry

A pair of set messages used to check the status of orders.

SET Out-of-Band

An activity that is not within the bounds of the SET recommendations, guidelines, and standards.

SET PKCS (Public Key Cryptography Standards)

A set of public key cryptography standards used by SET that include:

1. RSA
2. Diffie-Hellman key agrements
3. Password based encryption
4. Extended certificate syntax
5. Cryptographic message syntax
6. Private key information syntax
7. Certification request syntax

SHA-1 (Secure Hashing Algorithm)

A mechanism for reducing the number of possible values.

Shell

1. A means of entering commands using Unix and other similar operating systems. Unix shell may be considered as performing a similar task to the command line interpreter associated with DOS.
2. A term used to describe the framework of an expert system. The shell is occupied by a knowledge base that consists of IF...THEN rules that are used to solve entered problems. The knowledge (or rule) base is interpreted by an inference engine.

Shockwave

A streaming multimedia technology that uses AfterBurner compression. Its producers, Macromedia, also produce the popular Director and Authorware programs. Essentially, the technology is used to deploy Director movies over the Web, and can also do the same to applications that depend on the Lingo multimedia authoring language. Web browsers may be enabled using a Shockwave plug-in.

Shopping Basket

A metaphor used by e-commerce sites so that customers may accumulate products for purchase.

Shopping Cart

An element of an e-commerce site that permits items to be collected and purchased at a virtual checkout. Typically items are written to the client as cookies and their information is read by the e-commerce site when the final purchase is committed and the transaction is made. Another method of implementing shopping carts involves the use of forms and hidden fields, and remedies situations where the cookie function on the browser is turned off, or when the browser does not support cookies. A number of shopping cart implementations are available free, and one may be found at www.eff.org. An alternative to using a shopping cart might be to create an order form that is embedded in an HTML page using an appropriate scripting language or you may prefer to rely on the forms and templates provided by a Web development tool.

S-HTTP

A means of seamlessly integrating encryption into HTTP. It was developed by Enterprise Integration Technologies, and supports RSA, DES, triple DES, and DESX.

Signature

A means of securing transmitted matter that includes e-mail messages, and requires a Digital ID that may be purchased from many sources.

Site Server

A Microsoft solution for enhancing, deploying, and managing e-commerce Web sites.

SLIP (Serial Line Internet Protocol)

A protocol often used for serial data transmission over media that includes access technologies such as POTs.

SmartCard

A credit-card-size device that has an embedded processor chip that may store digital certificates and an e-purse, which is a stored cash value for small purchases.

SMP (Symmetric Multiprocessing)

A system that has two or more processors, but they do not operate independently with their own connected memory and I/O capabilities. Generally SMP systems are used as server variants.

A system architecture comprising multiple processors that share an interconnecting bus and memory. Systems find application as servers, and such processor designs as the Intel Pentium Pro were optimized for SMP. See Fig. S-6.

SMP systems offer processor scalability, but not in the precise processing increments associated with MPP systems. Unlike MPP systems, limitless scalability is prohibited by the interconnecting bus bandwidth that is shared.

SMTP (Simple Mail Transport Protocol)

A protocol used to transmit e-mail messages over the Internet, and across other compatible IP networks.

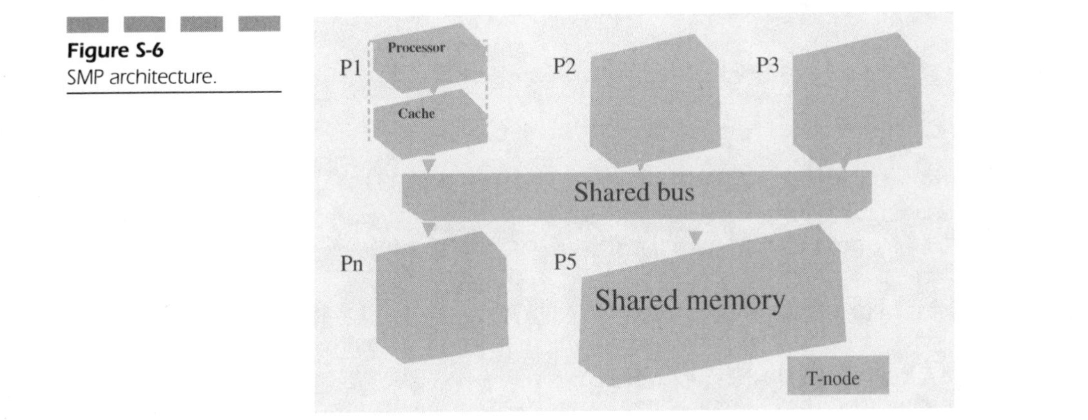

Figure S-6
SMP architecture.

Sniffing

A term used to describe the use of a sniffer program to monitor data traffic to a network or server, in order to gain access information. For instance, it may be applied to gather illegal passwords and IDs for ISP accounts, and passwords to e-mail accounts, and passwords to remote systems.

Software Distribution

A method of delivering software to users.

Software distribution may be done by downloading from the Internet. Other distribution media include floppy disk, CD-ROM, and DVD-ROM.

The first commercial CD-ROM software distribution disc from a major producer was Microsoft Office. Farallon Computing released lesser-known programs on CD-ROM prior to this, as did Microsoft itself. Also PC Sig released the world's first shareware compilation on CD-ROM in the United States.

Used as a distribution medium, DVD-ROM presents vendors with new opportunities; spare storage space may be used for program demonstrations, program documentation, training material, and advertising using DVD (MPEG-2) video.

Currently most leading software producers distribute their programs on CD-ROM, but a swing to DVD-ROM is imminent.

Sony Vaio

A popular family of notebook PCs from the Sony Corporation, each including the iLink connector, which is a synonym for FireWire or IEEE1394.

SPX/IPX

A network protocol.

SQL (Standard Query Language)

A nonprocedural language that is used to manipulate data stored in a relational DBMS.

Like other procedural languages that include Prolog, it does not have a rigidly defined series of operations to perform a function. Some 30 commands are included in the SQL specification, of which a recent version is SQL-92. However, third parties have extended it to semiproprietary variants.

The language permits the:

- Creation of table structures
- Entry, correction, and deletion of data
- Databases to be queried, so as to satisfy perhaps data requests

The SQL syntax is easily learned and understood, as it depends heavily on English words and phrases. For example, creating a database requires the statement:

```
CREATE DATABASE datawarehouse
```

SSL Web Server

A Web server the supports the Secure Sockets Layer protocol.

Stack

A contiguous series of memory locations utilized as a storage area. It is a LIFO (Last In–First Out) system in that the order in which items are dispensed is opposite of that in which they were deposited. A stack is sometimes called a push-down store. It may be used to store the return addresses from subroutines.

Streaming

1. *Streaming audio:* A method of playing audio while the audio stream is being downloaded. Streaming audio plug-ins are available for popular browsers such as Netscape Navigator. Such plug-ins are useful for tuning into radio broadcasting services on the Web.
2. *Streaming video:* A method of playing video while the video stream is being downloaded. Streaming video plug-ins are available for Netscape Navigator, while equivalent ActiveX controls operate with Microsoft Internet Explorer.

Streaming Media

A means of distributing audio, video, and multimedia in real time over the Web or similar TCP/IP network such as an intranet.

Strong Cryptosystem

A cryptosystem that is considered safe from attacks, and is difficult to crack using known techniques.

Stylus

A pen input device used to interact with, and write to, a computer or appliance. The first stylus was the light pen which was developed by the U.S. Defense SAGE project, an early warning radar system based on digital graphics technology.

A stylus might be used as a pen with a graphics tablet or bit pad. The stylus is used widely in pen computing with everything from notebook computers to PDAs. In pen computing the user simply writes directly on the screen, and data entry using normal longhand is valid.

Subnet

A method of using an IP address so that a greater number of networks may be addressed. IP addresses are designed to accommodate networks that may have between 253 and several million hosts.

In many instances, organizations wish to address a number of networks using a class C address. By creating subnets using their IP address, they may link the separate networks using a router.

Subnets are created by dividing the last octet of an IP address. The division involves reserving the most significant bits of the octet so as to provide addressing information for subnets. This may yield one of the following configurations:

- 2 subnets, each with 62 hosts
- 6 subnets, each with 30 hosts
- 14 subnets, each with 14 hosts
- 30 subnets, each with 6 hosts
- 62 subnets, each with 2 hosts

Sun Microelectronics

A division of Sun Microsystems, and manufacturer of chips that are optimized for the Java programming language, including:

- PicoJava chipset that is used in cellular phones, and computer peripherals
- MicroJava processor that is used in network devices, telecommunications hardware, and consumer games
- UltraJava that is optimized for use in 3D graphics and multimedia-related computing, much like Intel's MMX Technology

Symbolic Constant

A symbolic constant has a name, and is assigned an unchanging value. It may be used just like an integer constant. Symbolic constants improve program maintenance and updating; a single change may be made to a symbolic constant that might be used throughout a program.

A symbolic constant multiplier may be assigned the value of 10 using the C++ code:

```
#define multiplier 10
```

or,

```
const unsigned short int multiplier = 10
```

Symmetric Cryptography

A cryptosystem where the processes of encryption and decryption each require the use of a single key. Unless the recipient of the encrypted data already knows the key, it may be left to the sender to transmit its details over a secure channel.

Symmetric Cryptosystem Operation A series of processes and subprocesses that:

- Converts plaintext into ciphertext using a cipher or encryption algorithm
- Returns ciphertext to plaintext using a decryption algorithm.

Using a symmetric key and the transposition technique, the processes include:

Encryption Send the key, such as UNLOCK, for example, to the recipient using a secure channel. Arrange the key in a columnar fashion with numerals indicating their alphabetical sequence:

U 6
N 4
L 3
O 5
C 1
K 2

Arrange the plaintext, such as ATTACK VESSEL, in columns as shown below:

U 6 A V
N 4 T E
L 3 T S
O 5 A S
C 1 C E
K 2 K L

Create the ciphertext by writing the row values in sequence dictated by the numerical value, that is:

CE KL TS TE AS AV

Send the ciphertext to the recipient where a secure channel is optional.

Decryption Again the key characters are assigned numerals indicating their alphabetical sequence:

U 6
N 4
L 3
O 5
C 1
K 2

The decryption algorithm takes the ciphertext and, based on the numerical value, it creates the plaintext:

U 6 A V
N 4 T E
L 3 T S
O 5 A S
C 1 C E
K 2 K L

Miscellaneous

Sampling Rate (or Frequency)

A frequency at which an incoming analog signal is digitized. The sampling rate of an ADC influences the effectiveness of conversion.

Shannon's Theorem

A theorem that may be applied to give the maximum data transfer limit over a given medium such as an access technology:

$$I = F \log_2 (1 + S/N)$$

where F = bandwidth and S/N = signal-to-noise ratio.

Acronyms

S/I	Signal-to-interference
SA	Systems aspect
SAAL-NNI	Signaling ATM adaptation layer for network-to-network interface
SAP	Service access point
SAPI	Service access point identifier
SAR	Segmentation and reassembly
SASE	Specific application service element
SBC	Subband coding
SC	Selection combining
SCCH	Synchronization control channel
SCCP	Signaling connection control part
S-CCPCH	Secondary common control physical channel
SCH	Supplementary channel
SCI	Synchronized capsule indicator
SCO	Synchronous connection-oriented
SCP	Service control point
SCP	Signal control point
SCTP	Simple control transmission protocol
SDB	Short data bursts
SDO	Standard Development Organization
SDP	Service discovery protocol
SDU	Service data unit
SF	Spreading factor
SGCH	Signaling channel
SGSN	Serving GPRS support node
SIC	Successive interference cancellation

SID	System identification
SIG	Special interest group
SIM	Subscriber identity module
SIN	Signal-to-noise
SIR	Signal-to-interference ratio
SISO	Soft input/soft output
SIU	Serial interface unit
S-IWF	Serving system IWF
SLIC	Subscriber loop interface circuit
SM	Service management
SME	Small- to medium-sized enterprise
SMFA	System management functional area
SMG	Special Mobile Group
SML	Service management layer
SMS	Short message service
SMTP	Simple Mail Transfer Protocol
SNDC	Subnetwork dependent convergence
SNMP	Simple network management protocol
SNR	Signal-to-noise ratio
SO	Service option
SOM	Start of message
SOVA	Soft-output Viterbi algorithm
SQL	Structured Query Language
SR	Spreading rate
SRBP	Signaling radio burst protocol
SRLP	Signaling radio link protocol
SRNC	Serving RNC
SRP	Selective repeat protocol
SS	Spread spectrum
SSC	Secondary synchronization code
SSCF	Service-specific coordination function
SSCF-NNI	Service-specific coordination function for network-to-network
SSCOP	Service-specific connection-oriented protocol
SSO	Shared secret data

T

TCP/IP—Transmission Control Protocol

Transmission control protocol (TCP) is a ubiquitous transport layer protocol for IP (Internet Protocol), and is collectively referred to as TCP/IP. The segmentation and reassembly of messages is handled by IP. Selective repeat protocol (SRP): Numbered bytes (in the same packet) must be acknowledged (ACK). ACK with a sequence number "m" is an acknowledgment for all packets up to "$m - 1$".

The TCP header has 16 error detection bits for payload and header data. Error detection bits are derived by:

- Summing the 1's complements of 16bit groups in the data and in the header
- Taking the 1's complement of that sum

Data volume prior to its acknowledgment is confined to the window size (W *max*), and the TCP transmitter (Tx) does not permit more than W *max* unacknowledged packets.

ISO 8073 defines five issued classes of transport services (ISO 8073). Transport Protocol Class 4 (TP4) is aimed at unreliable networks.

IPv6

See Handovers under letter H, IPv6 under letter I.

TDMA

TDMA provides multiple access, and helps economize on radio spectrum usage. Other techniques include FDMA, CDMA, and space-division multiple

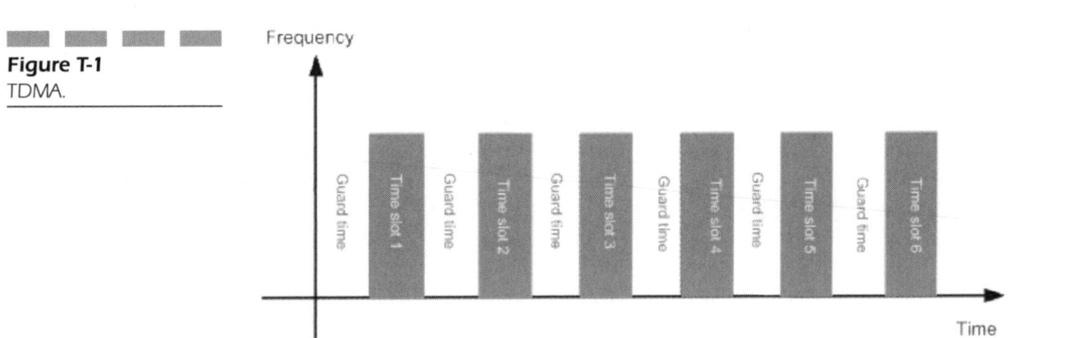

Figure T-1
TDMA.

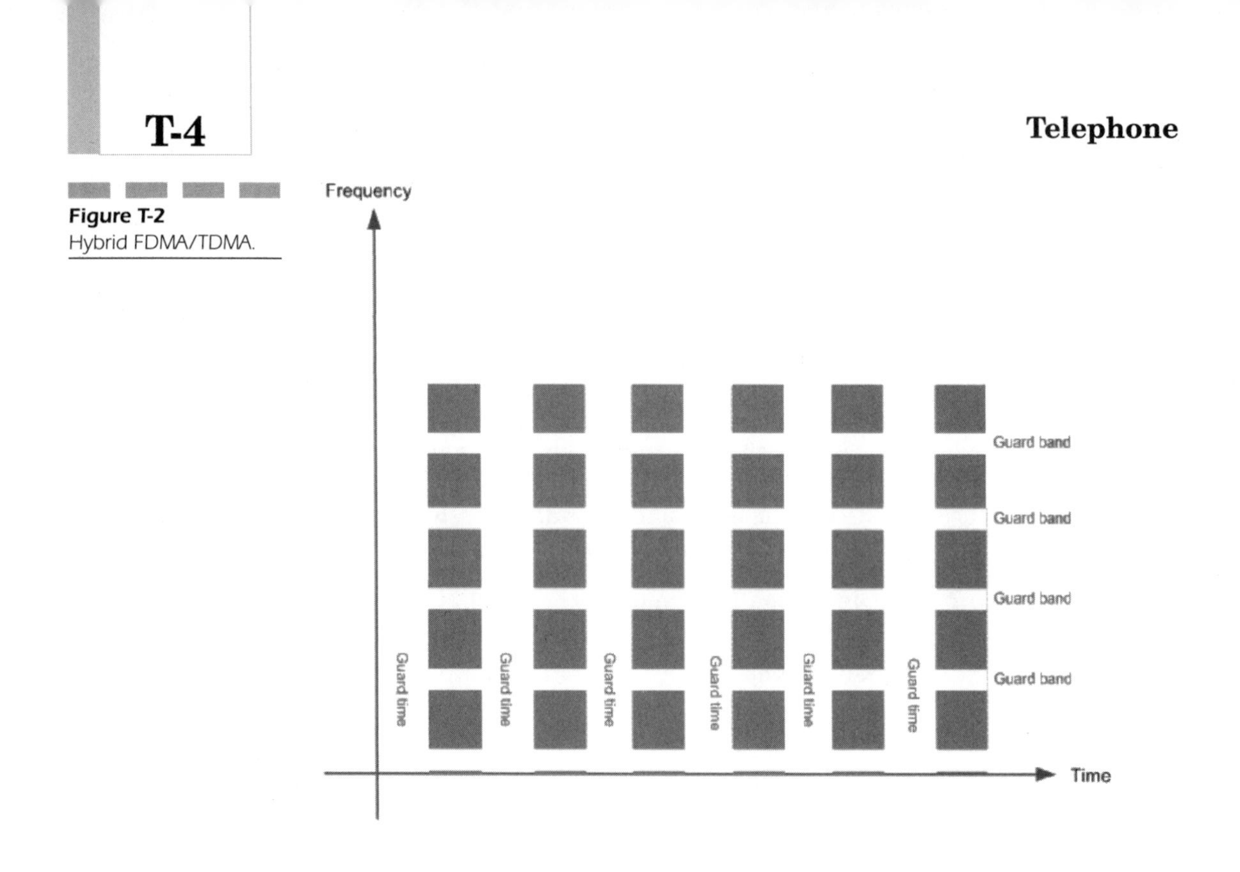

Figure T-2
Hybrid FDMA/TDMA.

access (SDMA). A TDMA system (such as GSM) allows a single user to use the entire bandwidth for short periods. See Fig. T-1.

The frequency channel is separated into time slots, which are different for the uplink and the downlink. Using GSM each frequency channel is divided into several time slots, and users are allocated one or more slots. As with many other 2G systems, GSM is a hybrid FDMA/TDMA system. See Fig. T-2.

Telephone

Alexander Graham Bell's invention of the telephone in 1876 resulted in the U.S. Patent Office granting him a patent for his technology that was described as:

> . . . the method of, and apparatus for, transmitting vocal or other sounds telegraphically. . . by causing electrical undulations, similar in form to the vibrations of the air accompanying the said vocal or other sounds.

Bell accumulated 18 patents including four for the photophone, and in 1878 he architected New Haven's telephone exchange that was the first central office switch, and eventually led to the Private-Branch Exchange (PBX). AT&T became a subsidiary of the American Bell Telephone Company, to manage and to grow American Bell.

Telephone Appliances

A telephone has transducers and a dial mechanism. Transducers that convert energy from one form to another are used in the earpiece (or speaker converting analog electrical waves into sound waves), and in the mouthpiece (microphone converting sound waves into analog electrical signals). A pulse or tone dial mechanism permits the creation of a connection with a subscriber. CS networks require a number of switches that are toggled to create a communication link between parties that is disconnected when one phone is hung up.

A CS service sees all telephones connected with a local central office that has a three-digit exchange number, and subscriber phones have a four-digit ID number. These combine to make a telephone number.

When dialing outside this local service region, customers have to dial a three-digit area code that causes the local-exchange switch to contact an appropriate interexchange switch.

Calls to other countries require a three-digit international access code and a two- or three-digit country code. The former connects the call from the local-exchange switch to interexchange switches, and to a gateway that connects with foreign gateways.

Address Signaling

Pulse signaling/dialing permits connections with analog central-office switches. Tone signaling/dialing permits connections with digital central-office switches. A raised phone creates a loop, and eventually generates a dial tone that is indicative of an idle line and available connection.

Dual-Tone Multifrequency (DTMF) Signaling–Tone Dialing Tone dialing is comparatively fast, and selects digits using two out of seven possible frequencies, where "7" sounds a 852Hz and a 1209Hz tone simultaneously, and is called Dual-Tone Multifrequency (DTMF) signaling. The central-office switch translates the tones into digits.

Pulse Dialing Rotary or pulse dialing opens and closes the loop at about 10 pulses per second, where the number of pulses represents the dialed digit. See Fig. T-3.

A local number results in a connection immediately, while those outside the locality force the central office to find a trunk that will connect it with a central office. Unlike a line circuit, or loop, a trunk circuit is shared by many users, although only one uses a trunk circuit at any given time. A circuit of 100 or more trunks may be established and exist for a call's duration. Eventually the destination local central office sends an 88V, 20Hz signal down the loop to alert the subscriber of a call waiting.

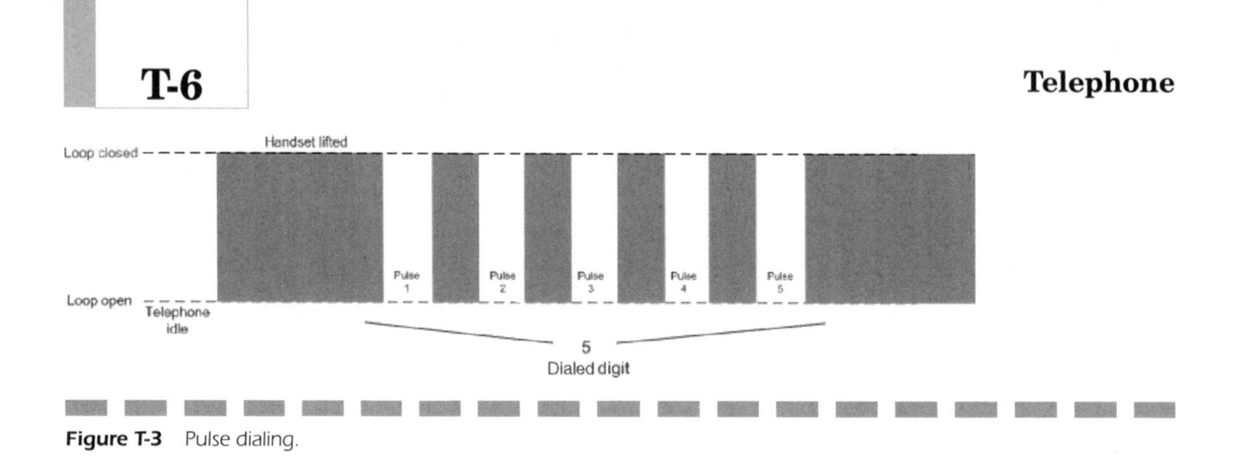

Figure T-3 Pulse dialing.

Telephone Fraud

Cloning Analog mobile phones have handshake information consisting of the Electronic Serial Number (ESN) and the Mobile Identification Number (MIN) that may be used by cloners to recode the chips of phones. Illegally cloned handsets may be sold to end-users, or set up in an area where users are time charged.

Rechipping Stolen mobile phones that are blacklisted by the operator may be rechipped with a new identity, and used until blacklisted once again.

Scanning Radio scanning equipment and a laptop computer allow a criminal to tune in to a call and steal ESNs that are broadcast at the beginning of every call. Encryption of the air interface helps combat scanning, and it is used by the North American Personal Communication Services (PCS), and by the European Global System for Mobile (GSM) communications. GSM encryption is implemented by the smart card—the Subscriber Identification Module (SIM) that holds the International Mobile-Subscriber Identity (IMSI).

Anti-Fraud Technologies

- *Authentication* is encryption-based technology, and the network requests the phone to verify itself using a mathematical equation.
- *Calling-pattern analysis* generates an alert in a service provider's fraud-management system, and is then investigated. Modern systems create subscriber profiles as references.
- *Personal Identification Numbers* (PINs) are known only to subscribers, and are required to place calls.
- *Radio-frequency fingerprinting* recognizes the unique characteristics of radio signals from cellular phones.

- *Voice verification* uses stored voice prints to confirm a legitimate mobile user. The voice-verification system may be on a public or private network, or it may be part of the Mobile Switching Center (MSC).

Towers

Towers are a key structure when deploying terrestrial mobile networks and stations, and include the inexpensive guy-wire, the more expensive monopole, and the self-supporting. The guy-wire tower has the largest footprint. The self-supporting tower may accommodate multiple carriers, and multiple users. See Figs. T-4, T-5, and T-6.

Traffic Channels (CDMA)

The traffic channel transports data using 20ms frames. The CDMA downlink transmission supports 55 traffic channels, and Walsh codes permit data modulation so as to share the same center frequency, and to be distinguishable, and provide orthogonality between channels.

The receiver sees completely separate channels, because whole periods of the Walsh functions occur in every code symbol. The Walsh codes allow 64 channels including an always-on pilot channel.

Traffic channels are assigned as requests received from the subscriber, which is informed of free traffic channels via the paging channel. See Fig. T-7.

Computers/Software

Tape Streamer

A magnetic tape storage device used to back up hard disk data as a contingency measure against data loss or data corruption resulting from system failure or an interruption of the power supply. Data recovery simply involves copying the contents of the tape streamer to a functioning hard disk.

Task Bar

A status bar used in Windows 98/NT that underlines all applications. It shows the Start button and illustrates all open applications that may be minimized or maximized. By default the Task Bar also shows the time of day.

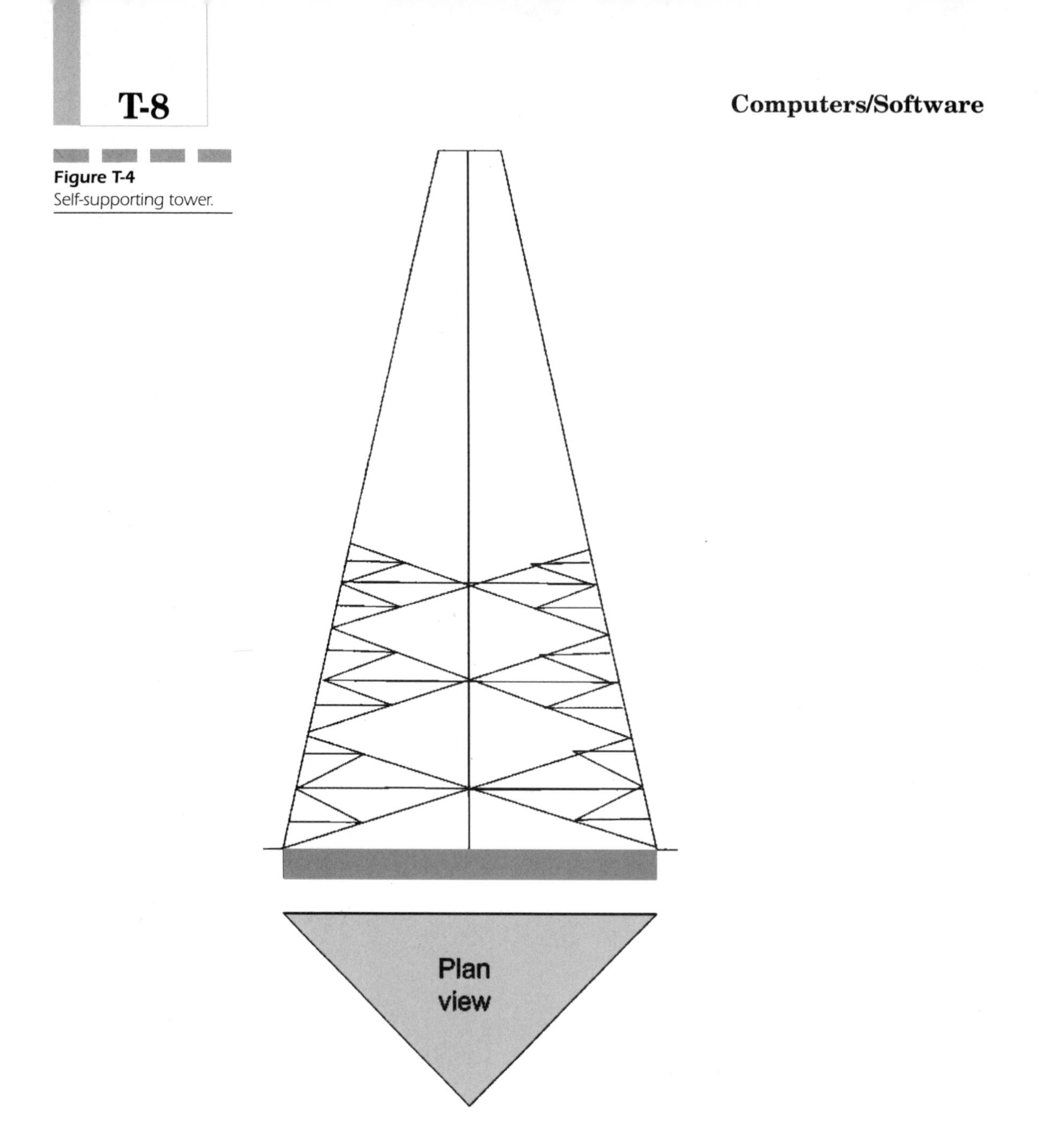

Figure T-4
Self-supporting tower.

The Windows 95/98/NT user interface centers around the task bar that provides buttons to select open applications, and it anchors the Start button that invokes the Start menu.

The Start menu has options that lead to programs as well as to submenus. Once invoked the menu system may be navigated by dragging the mouse rather than by clicking on its menu items. Programs are opened through a single mouse click.

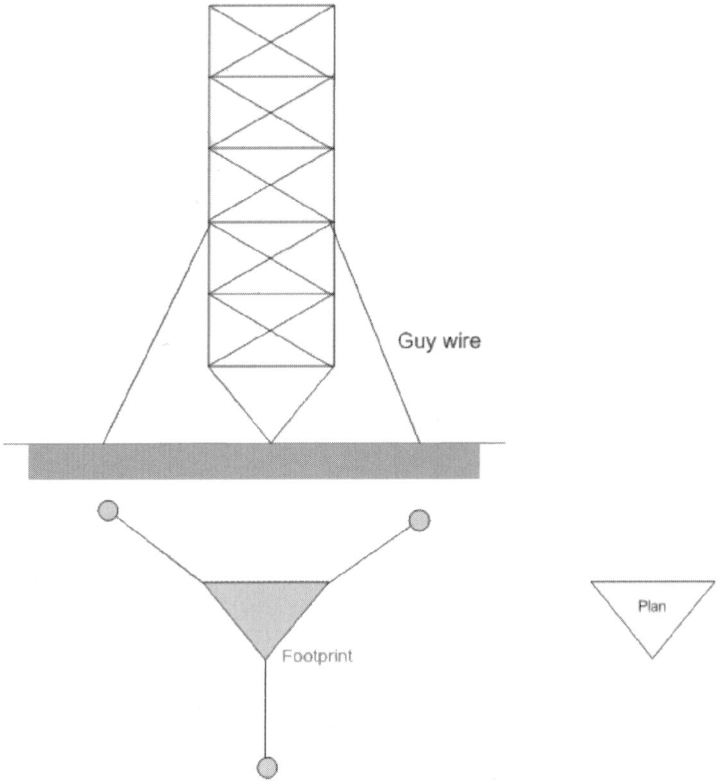

Figure T-5
Guy-wire tower.

Telnet

A connectivity mechanism that permits a client system with Internet access to operate a remote computer. The screen images shown on the remote system are also seen on the remote user's client system.

The Internet Engineering Task Force (IETF)

A publisher of specifications of Internet protocols such as TCP/IP. Further information may be obtained at www.ietf.cnri.reston.va.us.

The World Wide Web Consortium (W3C)

A publisher of specifications of Web technologies that include HTTP, HTML, and CGI. Further information may be obtained at www.w3c.org.

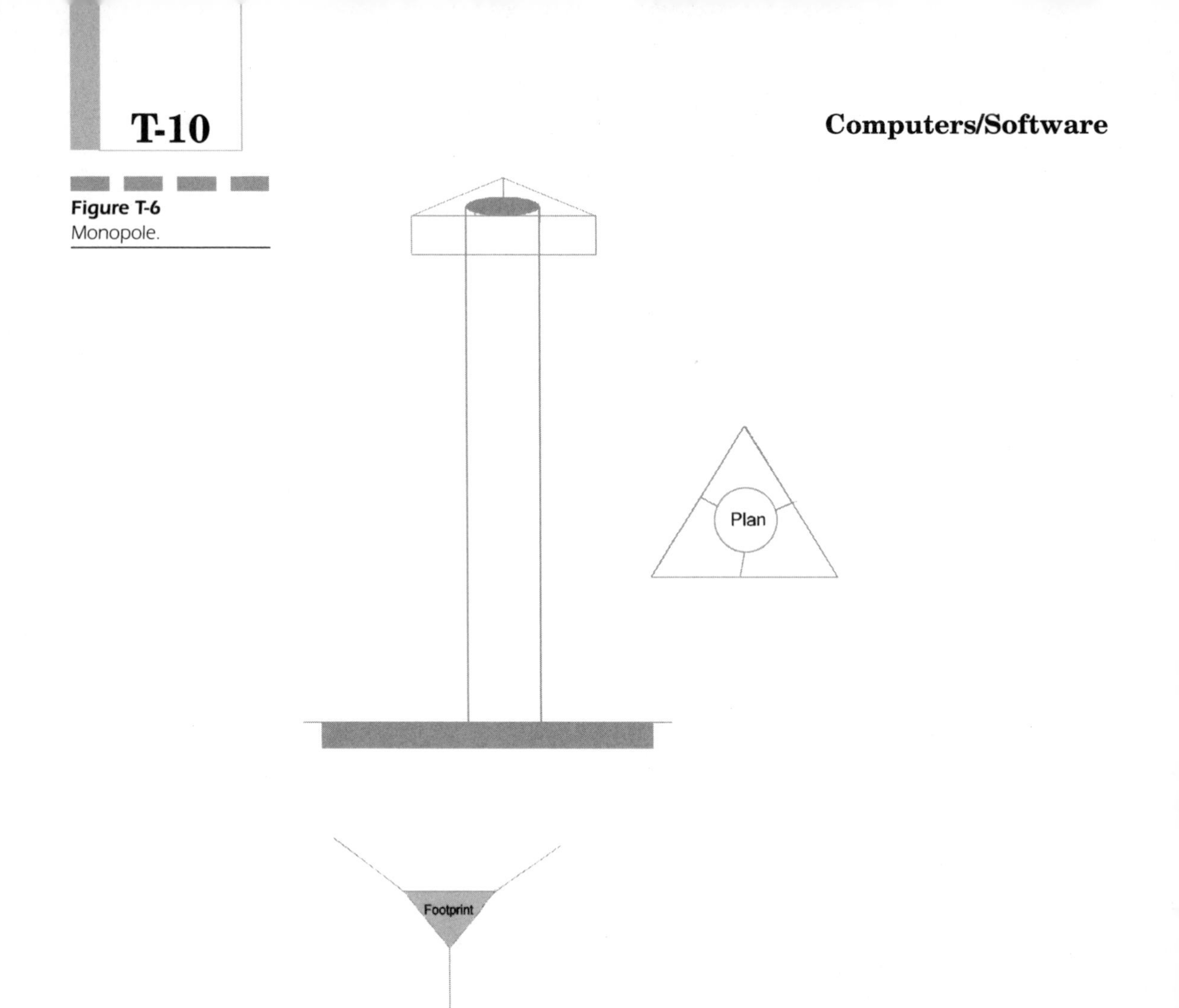

Figure T-6
Monopole.

Thin Client

A system within a client/server architecture (such as that of the Web) that features:

- Presentation that is typically in the form of a Web browser
- A portion of the application logic

Many systems connected to the Web may be described as thin clients. Thin clients require fewer hardware resources, and are therefore cheaper to deploy than fat clients.

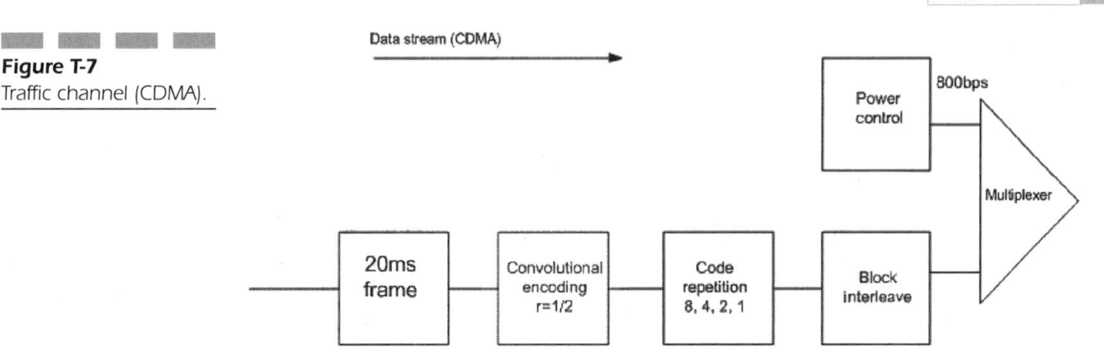

Figure T-7
Traffic channel (CDMA).

TIFF (Tag Image File Format)

An image file format maintained by the Adobe Developers Association (ADA).

Top-Down Analysis

A design approach that begins at a high level, and progresses to low-level component parts.

Touchpad

An *x-y* input device consisting of a small touch-sensitive pad or surface. It is the chosen device for notebook systems.

TP—Heavy Server

A server that runs TP monitors.

TP—Lite Server

A server that provides a portion of all the action required by full transaction processing (TP) monitors. Typically it will be able to commit changes to, and roll back changes made to operational data that are stored in an appropriate database variant.

It may be devoid of:

- Transaction coordination of multiple programs
- Resource management

Transaction

A term used to describe the data exchange and data changes that occur as the result of an interaction. The interaction might be the submission of an order form using a client browser. A transaction server is allocated the task of transaction processing (TP), and often invokes the application logic necessary to perform database interactions and manipulations. The process(es) invoked directly or indirectly by the client are collectively referred to as the transaction.

Transaction servers may include UI logic, driving the client UI, relegating the client device to little more than a dumb terminal. Typically mainframe-based transaction systems might adhere to this model. Alternatively, the UI logic, or presentation may be distributed to the client. The server consists of a TP monitor that performs transaction management and resource management. Transaction management ensures the so-called ACID properties of transactions. These are Atomicity, Consistency, Isolation, and Durability. ACID property compliance is achieved through the two-phase commit protocol. Resource management is intended to optimize the use of resources that include memory, mass storage, and processing. It may also be involved with load balancing between resources and between the software processes that may be threads.

Transaction Processing ACID Properties

Atomicity, Consistency, Isolation, and Durability (ACID) properties define the real-world requirements for transaction processing (TP). Atomicity ensures that each transaction is a single workload unit. If any subaction fails, the entire transaction is halted, and rolled back. Consistency ensures that the system is left in a stable state. If this is not possible the system is rolled back to the pretransaction state. Isolation ensures that system state changes invoked by one running transaction do not influence another running transaction. The changes must only affect other transactions, when they result from completed transactions. Durability guarantees that the system state changes of a transaction are involatile, and impervious to total or partial system failures.

Transaction Server

A transaction server is allocated the task of transaction processing (TP), and often invokes the application logic necessary to perform database interactions and manipulations. The process(es) invoked directly or indirectly by the client are collectively referred to as the transaction. Transaction servers may include UI logic, driving the client UI, relegating the client device as little more than a dumb terminal. Typically mainframe-based transaction systems might adhere to this model. Alternatively, the UI logic, or presentation may be distributed to the client. The server consists of a TP monitor that performs transaction management and resource management. Transaction management ensures the so-called ACID properties of transactions. These include Atomicity, Consistency, Isolation, and Durability. ACID property compliance is achieved through the two-phase commit protocol.

Transposition

A cryptography technique that rearranges streams of characters.

Tree of Trust

A hierarchy specified by SET used for the management of Digital Certificates including their:

- Maintenance
- Issuance
- Currency

Triple DES

An encryption technique based on a variation of the DES encryption technique. One variation sees three DES encryptions using three different keys.

Try Block

A try block is a section of code that is responsible for exception handling. Both Java and C++ support try blocks.

Two-Phase Commit

A method used in transaction processing that ensures ACID properties. It coordinates the changes made to system resources that result from transactions. It tests for their successful implementation, in which case they are committed. If not, and any one fails, they are all rolled back.

The transaction coordinator is key to the two-phase commit protocol. This queries all subordinates to verify that they are ready to commit. If the subordinates have other subordinates, these must also be queried. When all subordinates are ready to commit, the transaction coordinator records the information so as to protect it against any interruption that might be caused by a system failure.

Having received information of the readiness to commit, the transaction coordinator sends a commit command to its subordinates, and they do the same. Once the transaction coordinator has received confirmations from all subordinates, the client may be sent a transaction complete message.

Miscellaneous

T1

An AT&T designation for a digital link with a bandwidth of 1.544Mbps.

T2

An AT&T designation for a digital link with a bandwidth of 6.312Mbps.

T3

An AT&T designation for a digital link with a bandwidth of 44.736Mbps.

T50

An ITU-T designation for ASCII.

T90

An ITU-T designation for image coding that is used by Group 4 facsimile. This uses an ISDN 64kbps bearer channel.

TDM (Time Division Multiplexing)

A technique by which several different signals may be transmitted concurrently over the same physical link.

TMA (Telecommunications Managers Association)

A body whose membership is composed largely of telecommunications managers. Each year there is a TMA convention featuring state-of-the-art communications systems, techniques, and standards.

Token Ring

An IBM-developed network protocol and specification, officially named IEEE802.5.

Acronyms

TAGs	Technical Adhoc Groups
TCAP	transactions capabilities application part
TCE	traffic channel element
TCH	traffic channel
TCP	transmission control protocol
TCS	telephony control specification
TDD	time division duplexing
TDM	time division multiplexing
TDMA	time division multiple access
TE	terminal equipment
TEI	terminal equipment identity
TeleVAS	telephony value-added services
TFCI	transport-format combination indicator
TFfP	trivial FfP
THSS	time-hop spread-spectrum
TIA	Telecommunications Industry Association
TLS	transport layer security
TM	telecommunication management
TMF	Tele-Management Forum
TMN	Telecommunications Management Network
TMSI	temporary mobile subscriber identifier
TNCP	transport network control plane
TNL	transport network layer

TP4	Transport Protocol Class 4
TPC	transmit power control
TRAU	transcoder rate adapter unit
TREU	transcoder and echo canceller unit
TSG	technical specifications group
TSU	telecommunications switching unit
TTA	Telecommunications Technologies Association
TTC	Telecommunications Technology Committee
TTD	time division duplex

U

UMTS (Universal Mobile Telephone Service)

UMTS may provide global roaming and consists of orbiting satellite constellations that may integrate BTSs (Base Transmitter Stations) and BSCs (Base Switching Centers). To create this type of network, as many as 840 satellites may orbit at altitudes between 780km and 1414km. There are 840 satellites, in fact, that make up Teledesic whose Consortium is led by Bill Gates. Other mobile satellite services include Motorola's Iridium (with 66 satellites), Loran's/Qualcom's Globalstar (with 48 satellites), and TRW-Matra's Odyssey (with 10 satellites).

Aeronautical and maritime telecommunications were catalysts in the development and deployment of satellite mobile telephone services with the first maritime satellite launched in 1976. Called MARISAT, it consisted of three geostationary satellites and was used by the U.S. Navy. This later evolved into the INMARISAT (International Maritime Satellite Organization) that provide public telecommunications services to airliners.

3G UMTS user data rates that may extend to Mbps, and was shaped in part by the Third Generation Partnership Project (3GPP). 3G cellular technology greatly surpasses the multiple impacts of 2G services and, to a lesser extent, those of 2.5G. The promise of sophisticated video and multimedia wireless applications able to deliver MPEG video, and high-quality audio is attracting investment in the WASP and Mobile Telecommunications sectors, as the potentially profitable 3G services are foreseen, including:

- Video applications using MPEG standards
- Video telephony
- Videoconferencing
- Video on demand (VoD)
- Telepresence
- Surrogate services such as exploration
- Client for remote services
- Web/Internet browsing
- Client for VPN
- Client connection for teleworkers
- m-commerce—retailing, online banking, etc.
- Point of Information (POI) in real estate sector, etc.
- Point of Sale (POS) for secure purchasing
- CCTV (Closed Circuit Television) security
- UAV (Unmanned Aerial Vehicle) video communication and navigation

3G Origins

See 3G Origins in Numerals chapter, 3GPP in Numerals chapter.

UMTS Core Network Architecture

The UMTS core network (UCN) has a CS and PS network, where the former has an MSC and gateway MSC (GMSC), and the latter has a:

- Serving GPRS support node (SGSN)
- Gateway GPRS support node (GGSN)
- Domain name server (DNS)
- Dynamic host configuration protocol (DHCP) server
- Packet charging gateway
- Firewall

See Figs. U-1 and U-2.

3G-MSC

The 3G-MSC provides

- ATM/AAL2 connection with UTRAN: For transportation of user plane traffic across the I_u interface
- Call management: Handles setup messages to and from the UE
- Connection: Interconnection with networks such as PSTN and ISDN
- CS data services
  - Rate adaptation
  - Message translation for CS data services
- IN and CAMEL
- Mobility management
  - Intersystem handover
  - Attach
  - Authentication
  - HLR and SRNS relocation
- OAM agent functionality
- Short message services (SMS): Send and receive SMS data to and from the SMS-GMSC and the SMS-IWMSC
- SS7, MAP, and RANAP interfaces: The 3G-MSC terminates calls in the network, and communicates with the RNC using RANAP

Figure U-1
UMTS network.

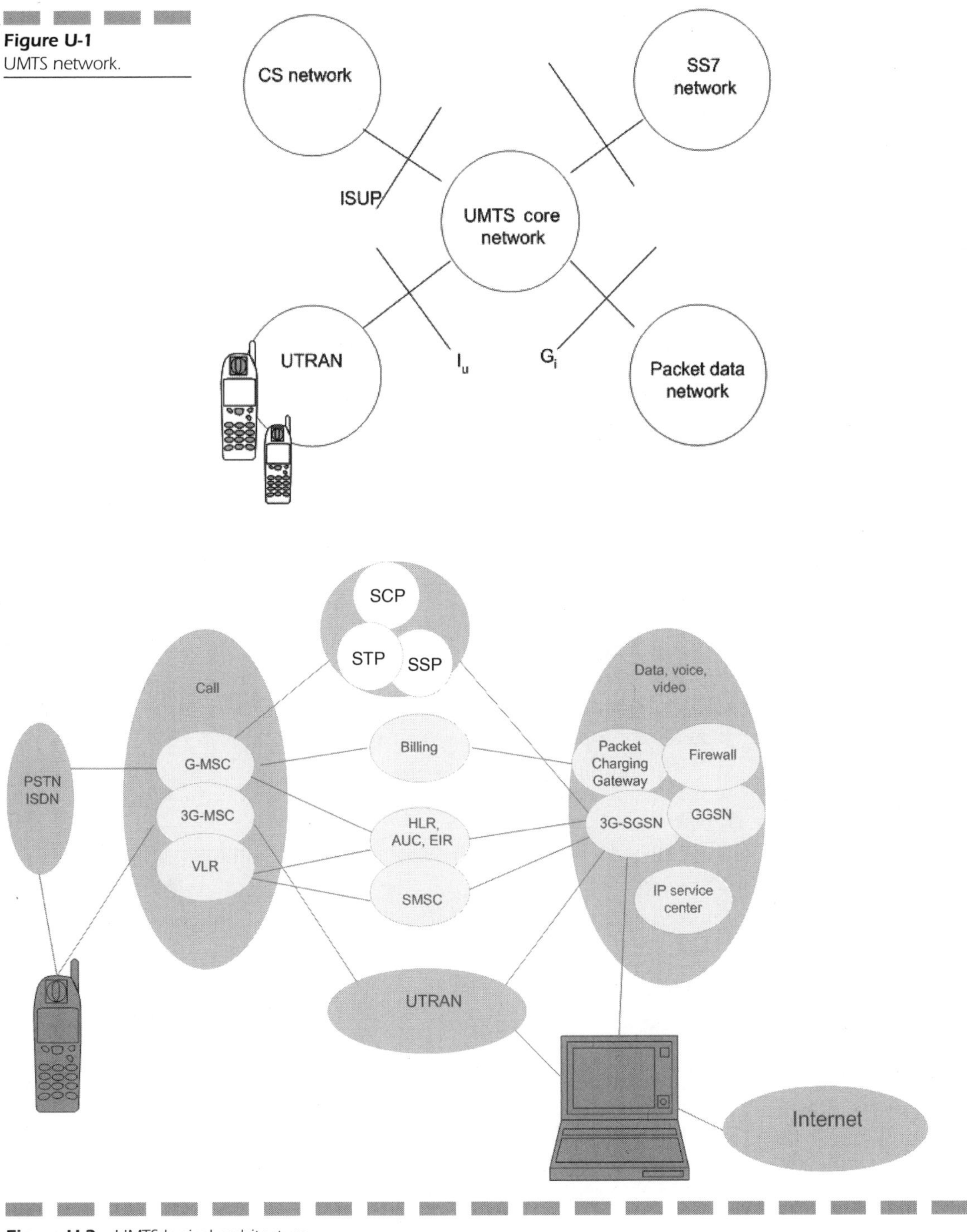

Figure U-2 UMTS logical architecture.

- Supplementary services: Handles supplementary services including call waiting
- VLR functionality: The VLR may be in 3G-MSC supporting visitor subscriber mobility, etc.

3G-SGSN

The 3G-SGSN provides

- Charging: Collects charging information
- Completion of originating or terminating sessions in the network, and with RANAP it controls/communicates with the UTRAN
- Connection with the GGSN for transporting user plane traffic using GTP
- Mobility management
  - Handles attach
  - Authentication
  - Updates HLR
  - Intersystem handover
- OAM agent functionality
- Physical connection with the UTRAN for transporting user data plane traffic over the I_u interface using GTP (GPRS tunneling protocol)
- Session management
  - Session setup messages to and from the UE
  - Session setup messages to and from the GGSN
  - Admission control
  - QoS procedures
- SMS: Send and receive SMS data to and from the SMS-GMSC and the SMS-IWMSC
- Subscriber database functionality: VLR-type database is held in the 3G-SGSN

3G-GGSN

The 3G-GGSN provides

- Access to intranets using perhaps PPP termination
- Charging: The GGSN acquires charging information
- Gateway between UMTS and data networks such as IP, X.25
- Maintenance of information location at macromobility levels, that is, SGSN

- User data screening/security
  - Subscription-based screening
  - User-controlled screening
  - Network-controlled screening
- User-level address allocation: The GGSN may allocate an address to the UE

Firewall

The backbone is firewalled to isolate it from external packet data networks, and is secured using packet filtering using ACLs access control lists.

DNS/DHCP

The DNS server translates access point names (APNs) into GGSN IP addresses, and may allow UEs to use logical names. The dynamic host configuration protocol (DHCP) server manages the allocation of IP configuration information.

UMTS Bearer Service

End-to-end network services are between two pieces of terminal equipment (TE), with a specified user QoS. A network QoS requires a defined bearer service from the source to the destination. The end-to-end service is spread across the bearer services of the TE/MT, UMTS, and the external bearer service. See Fig. U-3.

UMTS QoS

UMTS defines the QoS classes:

- Conversational where the traffic is sensitive to delay
- Streaming for real-time traffic
- Interactive for applications like Web browsing
- Background class where traffic is insensitive to latencies

Access Stratum

In the UMTS architecture, UTRAN sees the air interface isolated from the MM (mobility management) and the connection management layers. This notion takes the form of the access stratum (AS) and the nonaccess stratum (NAS). The AS includes radio access protocols that terminate in the UTRAN, while

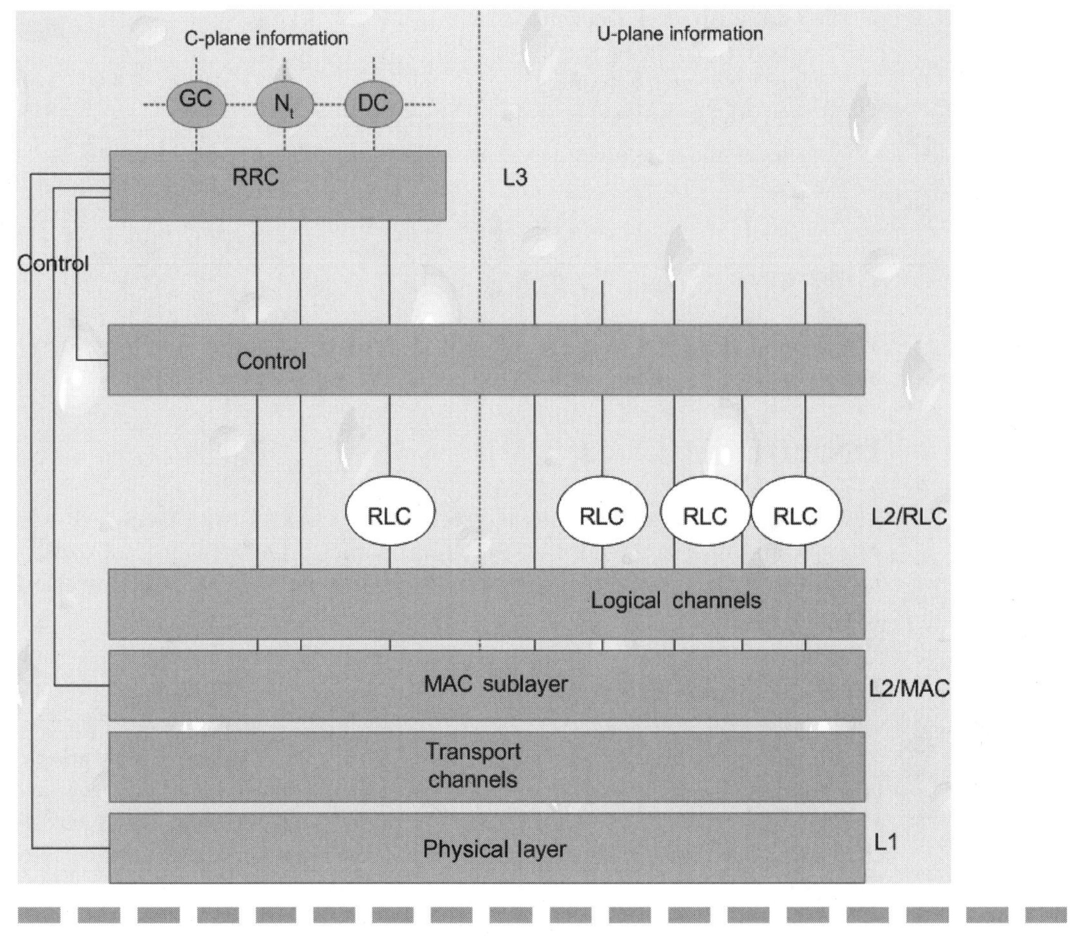

Figure U-3 UMTS bearer service layers.

the NAS has core network protocols between the UE and CN that are not terminated in the UTRAN that is transparent to the NAS.

The NAS attempts to be independent of the radio interface, as do the protocols MM, CM, GMM, and SM. These protocols are theoretically constant in terms of the radio access specification that carries them, so any 3G radio access network (RAN) should connect to any 3G CN. The GSM's MM and CM layers are almost the same as 3G NAS, and NAS layers will compare with future GSM MM and CM layers according to the phase 2+ specification.

Lower layers from the AS are different from GSM where the radio access technology (RAT) is TDMA, as opposed to CDMA for UTRAN. The protocols used therefore are also radically different, as is the packet-based GPRS protocol stack. So this reveals an obstacle in the so-called smooth migration path from 2.5G GPRS to 3G. However, the GPRS CN components can be reused when renovating the system to become a 3G solution. See Fig. U-4.

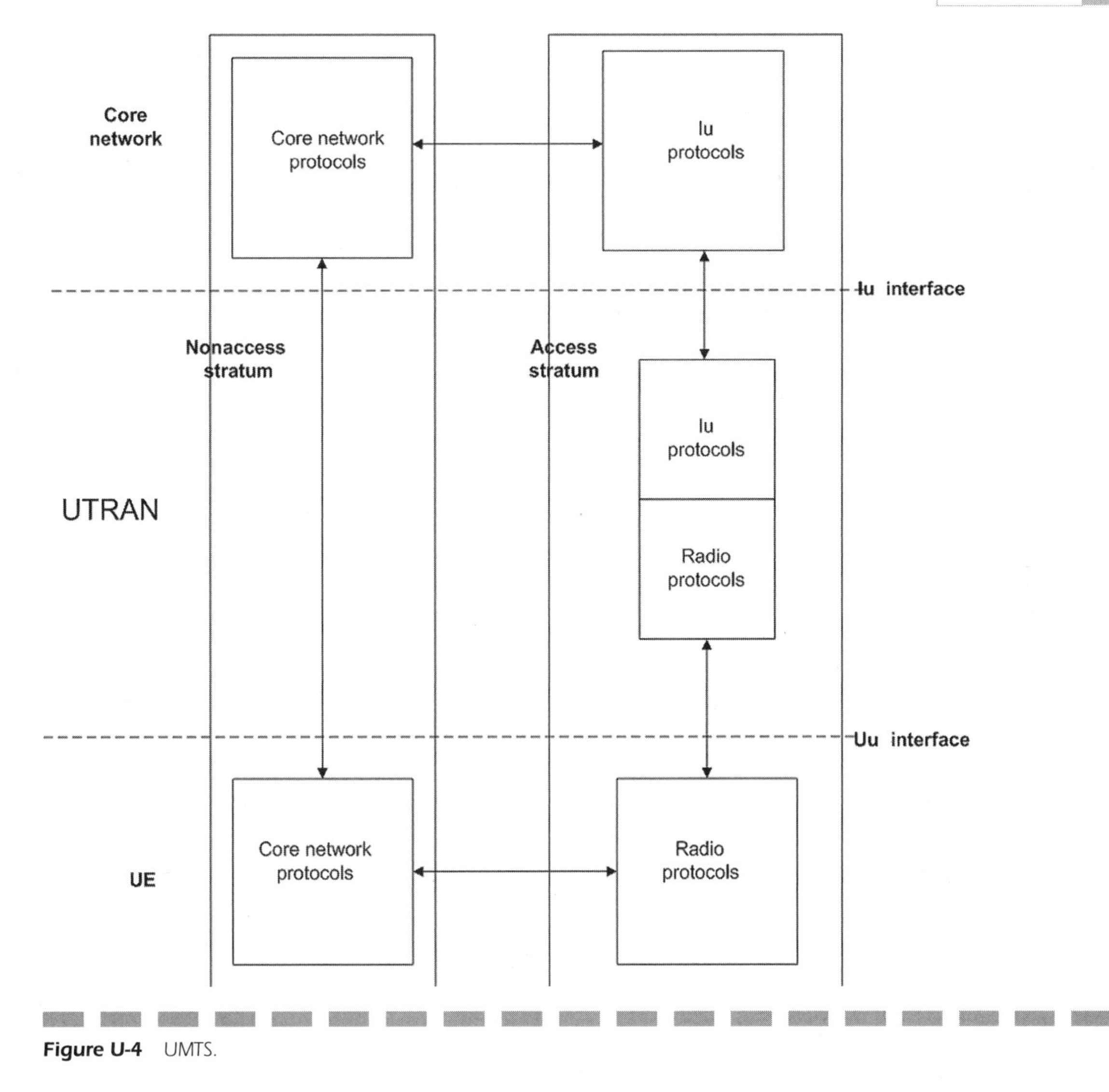

Figure U-4 UMTS.

The AS has three protocol layers that include sublayers:

Layer	**Sublayers**
Physical layer (L1)	
Data link layer (L2)	Medium Access Control (MAC) Radio link control (RLC) Broadcast/multicast control (BMC) Packet data convergence protocol (PDCP)
Network layer (L3)	Radio resource control (RRC)

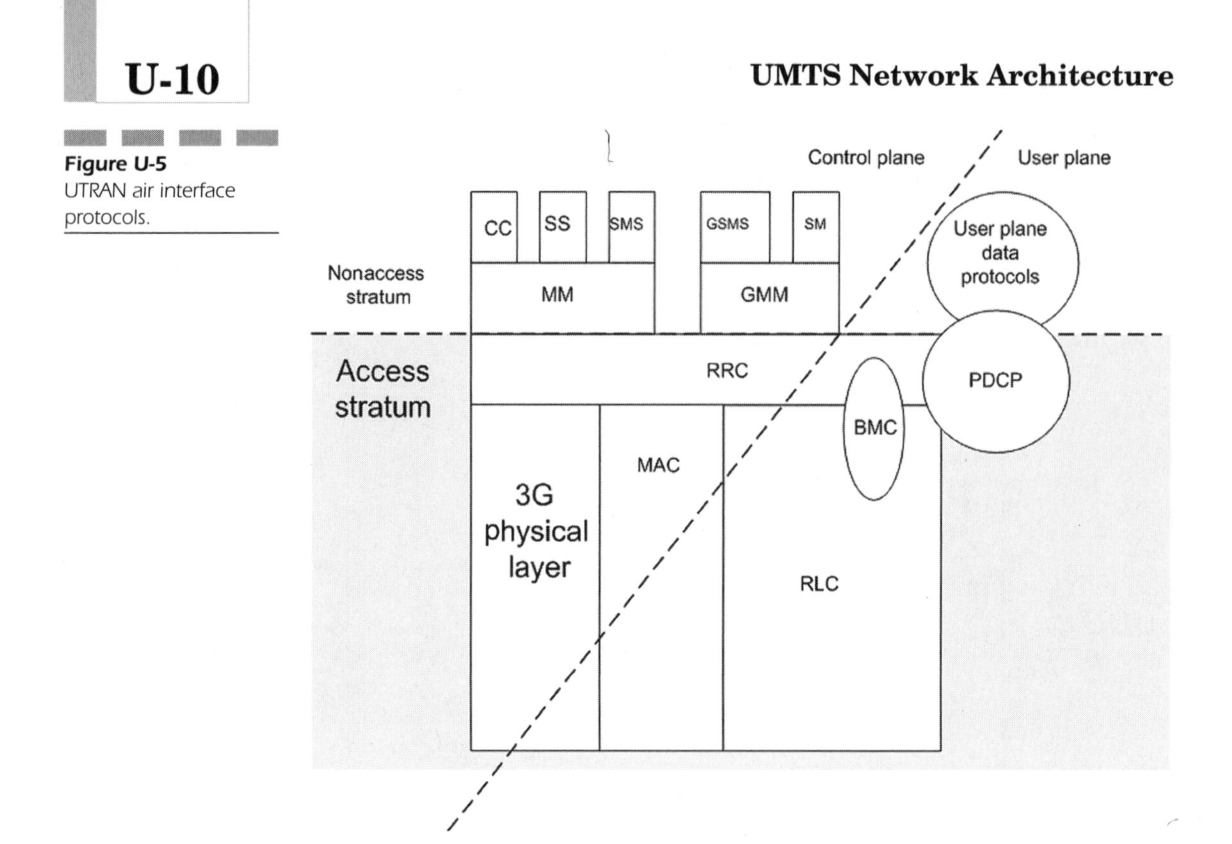

Figure U-5
UTRAN air interface protocols.

The two vertical planes include the control (C-) plane and the user (U-) plane, and both have the MAC and RLC layers, while the RRC layer is present only in the C-plane.

See Fig. U-5.

UMTS Network Architecture

See UMTS Network Architecture in Numerals chapter.

UMTS Terrestrial Radio Interface

See UMTS Terrestrial Radio Interface in Numerals chapter.

3G GGSN (Third-Generation GGSN)

A Gateway serving node for a 3G public mobile network such as UMTS.

3G QoS (Third-Generation Quality of Service)

A threshold or series of thresholds that determines the overall standard of service provided by the operator, and includes among other things:

- Latencies or delays
- Minimum bit rate guarantee that may be within a specified coverage area

3G SGSN (Third-Generation SGSN)

A server serving node for a 3G public mobile network such as UMTS.

3GPP (Third-Generation Partnership Project)

An international group of telecommunications representatives and/or entities that shaped the UMTS.

The 3GPP consisted of

- Standards organizations: ARIB (Japan), CWTS (China), ETSI (Europe), TI (USA), TTA (Korea), and TTC (Japan)
- Market representation partners: Global Mobile Suppliers Association (OSA), the OSM Association, the UMTS Forum, the Universal Wireless Communications Consortium (UWCC), and the IPv6 Forum
- Observers: TIA (USA) and TSACC (Canada)

Satellite Networks

See Mobile Satellite Service (MSS) under letter M.

Networking Approaches *See* Networking Approaches under letter N.

Computers/Software

U

A chrominance component in a video signal that comprises color information.

UI (User Interface)

A software module or program through which users interact with one or more applications.

UI Builder

A development tool used to build user interfaces. Most modern UIs are OO. Resulting UIs may be text-based, or graphics-based, as is common today. The latter naturally require the use of GUI builders that provide a means of implementing the presentation element, together with its interaction with objects, applications, and application logic.

GUIs may be built using all modern multimedia authoring tools that include Authorware, IconAuthor, and ToolBook. Programming tools such as Microsoft Visual Basic, and others included with Microsoft Visual Studio, also have the capability to construct GUIs using visual techniques. These development tools, including GUI builders, feature standard UI components or widgets that include buttons, sliders, drop-down list boxes, scroll bars, dialogues, and windows. Static GUI components might include fonts, colors, textures, patterns, etc.

The GUI will also contain containers that act as receptacles for objects or components that might be ActiveX or OLE objects. For example, using Visual Basic, a container may be used to integrate OLE objects such as the Media Player, or any compatible OLE object.

UltraJava

A chipset from Sun Microsystems that is optimized for the Java programming language. Similar to Intel MMX Technology, it is application-specific, thus optimized for 3D graphics and multimedia-related computing, including MPEG video playback. UltraJava is licensed to NEC, Samsung Electronics, LG Semicon, and Mitsubishi.

Undo

A feature provided by almost all fully specified programs. It simply cancels the last editing operation.

Uniprocessor System

A system design based on a single processor. Such serial systems might be referred to as von Neumann implementations.

Unix

See Handovers under letter H.

Unix Grep Filter

A filter that allows you to search files for specified text strings.

Uploading

A process of transferring files from a client system to a server. Usually the transfer takes place using FTP.

UPS (Uninterruptable Power Supply)

A device that prevents data loss following power supply failure or deviation.

User Block Data

A CD-ROM Mode 1 block contains 2048bytes of user data.

User Communication

A rarely used term that describes the user's interaction with a system or application.

User-Communication Device

A rarely used term that describes an input device such as a mouse, touchpad, trackball, or touch screen.

Miscellaneous

UART (Universal Asynchronous Receive Transmit)

An electronic device used for serial communications.

UDP (User Datagram Protocol)

UDP is used widely in streaming audio and video. Macromedia Shockwave Director 6.0 is among a number of leading streaming server technologies that use UDP. It does not have the reliability of TCP, and is therefore appropriate for streaming media where intensive error detection and correction are less important. Dropped packets that do not reach their destination are acceptable in streaming media. UDP, therefore, optimizes performance and makes better use of available bandwidth, because it does not insist on the retransmission of erroneous packets.

A low-level protocol, unlike Hypertext Transfer Protocol (HTTP), which is considered a high-level protocol, UDP may be exploited by multimedia streaming technologies, including ASF (Advanced Streaming Format). ASF is a container format that offers compression and protocol independence.

UPnP (Universal Plug and Play)

A technology introduced in 1999 by Microsoft as a response to SunSoft's Jini technology, and features the same discovery protocol technique used to locate registered services on networks.

URL (Universal Resource Locator)

A URL is an address of a service or Web site or Web page, and may be used by the Web browser to open specific sites and pages. For example, the Web page www.altavista.digital.com is a URL, and may be opened.

Additionally, the browser permits such URLs to be stored in a directory that might be called Favorites, or something similar. The user may then open frequently visited sites and pages, through one or two mouse clicks, depending on the browser used.

The underlying HTTP protocol implements a client/server connection for each URL that is opened by the client browser. It transmits and receives data, and carries the subsequent contents of an opened URL.

Typically, when a URL is opened, the first procedure involves finding the requested site or page on the Web. Having made an appropriate connection, the browser waits for a reply, and then downloads the ensuing page data.

Eventually, the HTTP breaks the connection with the remote server, where the requested site or page resides. This break may be carried out manually by selecting the stop button, or a similarly named button.

USB (Universal Serial Bus)

A serial interface for connecting peripheral devices.

User Authentication

A process of identifying the user of a system or program.

- The most common user authentication technique is based on tokens, such as ID names, passwords, and PIN numbers.
- User authentication can also be implemented using biometric data that may be a fingerprint, thumbprint, or retina image.

Acronyms

UBR	Unspecified Bit Rate (ATM)
UDP	User Datagram Protocol (TCP/IP)
UI	User Interface
UP	Unformatted page (Auto-Negotiation)
UPnP	Universal Plug-and-Play
USOC	Universal Service Order Code
UTMS	Universal Telephone Mobile Service
UTP	Unshielded twisted pair

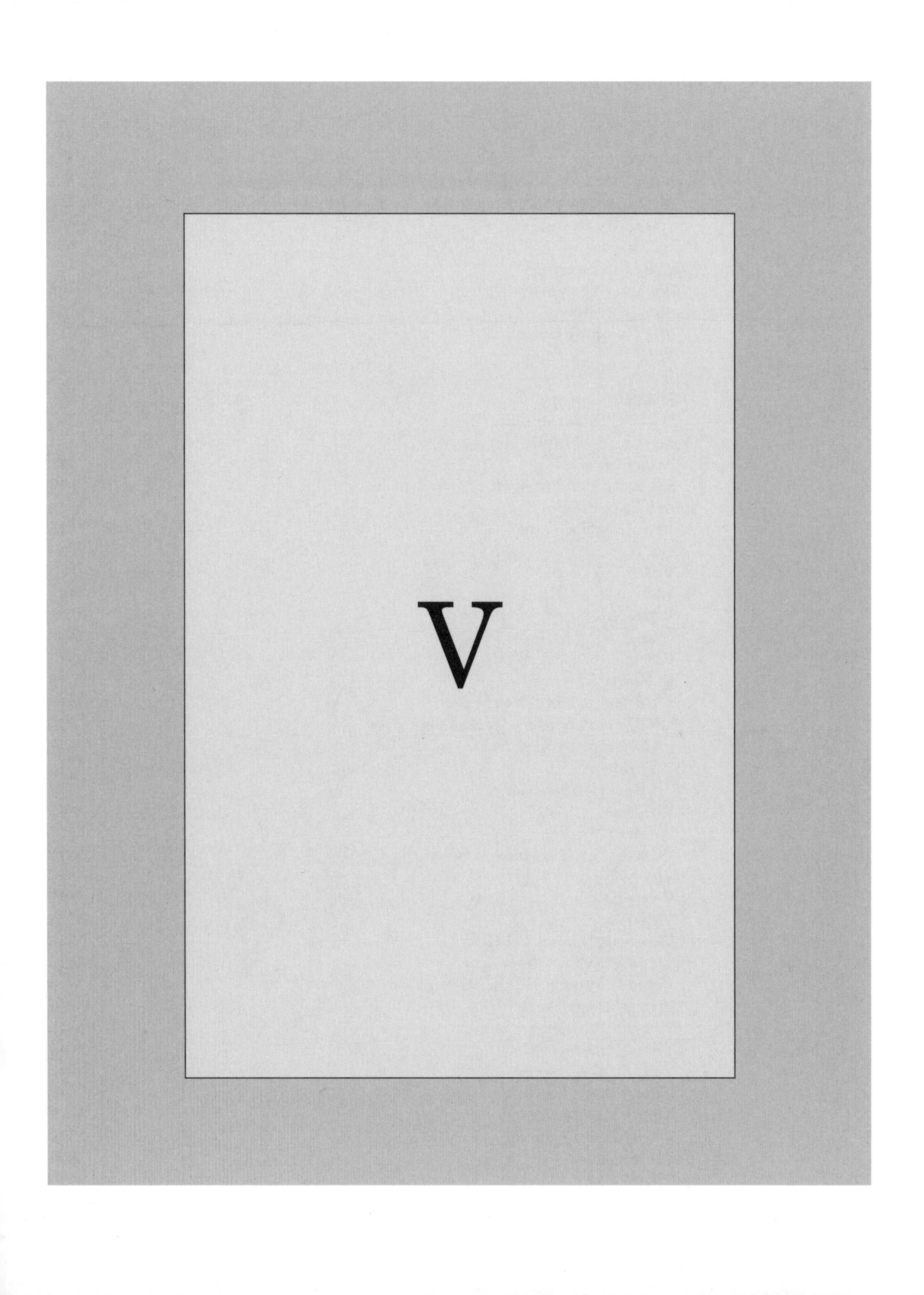

V

Video

MPEG

See MPEG under letter M.

M-JPEG (Motion-Joint Photographics Experts Group)

M-JPEG is a type of video that uses individual frames compressed according to the JPEG algorithm. It gives full frame updates as opposed to the predominantly partial frame updates of MPEG-1 video. M-JPEG video, therefore, provides random access points and lends itself to nonlinear editing.

In this respect it is more flexible on playback because applications can simply show any frozen frame of an M-JPEG sequence or play any selected frames of a sequence either backward or forward.

Other advantages of M-JPEG is that it may be compressed into other formats, including MPEG-1/2. A principal disadvantage of M-JPEG, however, is its comparatively low overall compression ratio.

Macromedia Flash Player 4 for Pocket PC

See Macromedia Flash Player 4 for Pocket PC under letter G.

Quality Settings HTML quality may be set to low, high, autolow, autohigh, or best. Quality Setting in the Publish area control anti-aliasing used in movies. Anti-aliasing needs intensive processing to smooth frames.

- Low favors playback speed and does not use anti-aliasing.
- Autolow sets Low.
- Medium uses anti-aliasing and does not smooth bitmaps, and is used most often.
- High, Auto-High, and Best map to Medium.

Generator Macromedia Flash templates may be created using Generator that may deploy Flash content. Sniffing for the browser and embedding a Macromedia Flash template allow the use of the same Macromedia Flash template to produce content for devices supporting JPG, GIF, PNG, and Macromedia Flash.

Videoconferencing

Videoconferencing permits users in remote locations to communicate in real time both visually and verbally. Systems may be divided into the categories of desktop videoconferencing using conventional desktop or notebook computers and conference room videoconferencing that typically includes appropriately large displays. Desktop videoconferencing systems include a camera, microphone, video compression/decompression hardware/software, and an interface device that connects the system to an access technology.

The interface device might be:

- A conventional modem used to connect with an ISP or intranet server, and thereafter use an Internet-based videoconferencing solution such as CU See-Me
- A cable modem that might provide high-speed Internet access via cable
- An ISDN interface that provides connection to the Internet or appropriate IP network
- A Network Interface Card (NIC) that connects to a LAN
- A wireless interface that provides connection over GSM or other mobile communications network
- Access technologies for videoconferencing include PSTN, ISDN, ADSL, cable, GSM, ATM, T1, frame relay, and proprietary wireless technologies

Point-to-point videoconferencing involves communication between two sites, while multipoint videoconferencing involves interaction among more than two sites. The latter might require a chairperson to conduct proceedings. Also the collective system might be voice activated, switching sites into a broadcasting state when the respective participant begins speaking. Internationally agreed standards relating to videoconferencing include H.320 and T.120. The former was introduced in 1990 and provides guidelines to vendors and implementers that yield appropriate levels of compatibility.

Voice over IP (VoIP)

1983 saw the ARPANET and the Internet converge on IP telephony using the so-called voice funnel that encoded voice packets. The voice funnel was a research project involving packet-based audio that could be sent between the East and West coasts over IP. Voice funnel was also used for three- and four-way conferencing.

SS7

SS7 is a signaling protocol for Voice over IP (VoIP). The SS7 protocol stack is shown in Fig. V-1, where the MTP transfers messages from the source to the destination signaling point. The ISUP (ISDN User Part) establishes phone calls, as can the SCCP that normally transports higher-layer applications like the Intelligent Network Application Part (INAP). These generally use

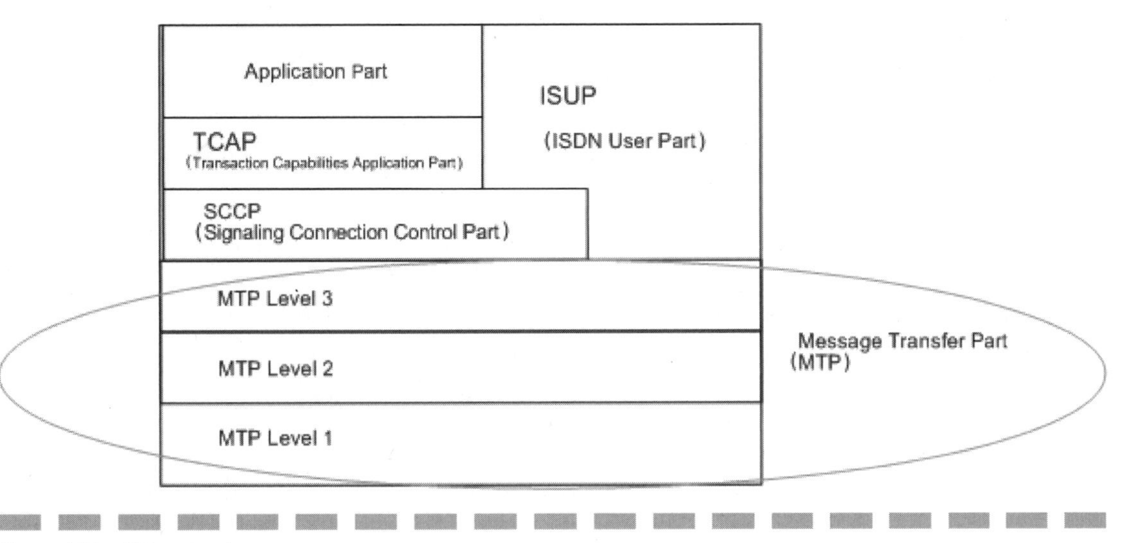

Figure V-1 SS7 protocol.

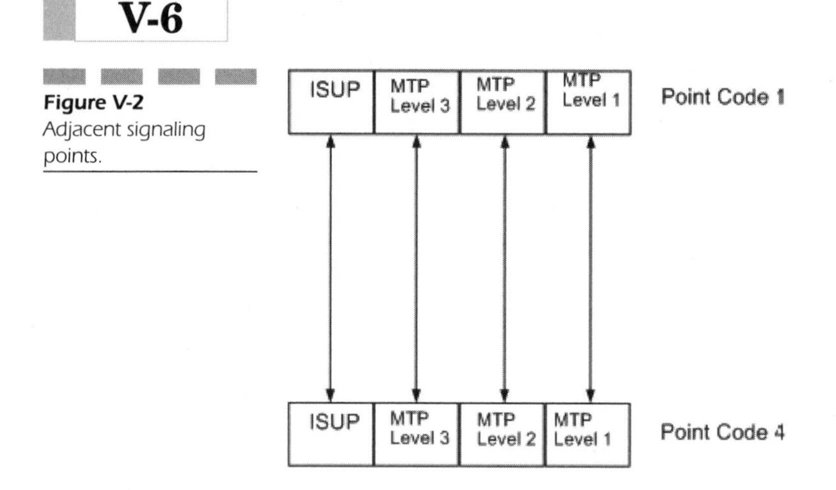

Figure V-2 Adjacent signaling points.

the services associated with the Transaction Capabilities Application Part (TCAP) that uses the SCCP services.

SCCP provides global title addressing that permits signaling between entities that may not know the target entities' point codes or signaling addresses; addresses or telephone numbers are mapped to point codes at nodes that may have initiated the message.

SS7 Communication

Communication between adjacent signaling points makes every layer a peer-to-peer link between entities that may be referred to as point code 1 and point code 4 (as shown in Fig. V-2). Figure V-3 shows nonadjacent signaling points with peer-to-peer links at layers 1 to 3, between point codes 1 and 2, 2 and 3, and 3 and 4. And at the SCCP layer, peer-to-peer links unite point codes 1 and 2 and point codes 2 and 4.

At the TCAP and Application layers, peer-to-peer links unite point codes 1 and 4, so the application at point code 1 is aware of the TCAP layer at point code 1 and the application layer at point code 4. Using the Sigtran protocol suite, it is possible to have a partial SS7 implementation with SS7-based applications at point code 4.

VoIP

Commercially desirable speech quality requires 64kbps (G.711) voice coding and CS. IP telephony places certain constraints on quality of service, so in order to remedy this situation the Real-Time Transport Protocol (RTP) has been introduced, and exists above the UDP in the protocol stack. RTP packets hold a header that provides information about their payload, including voice coding, time stamp, and the voice packet's source. The RTP Control Protocol

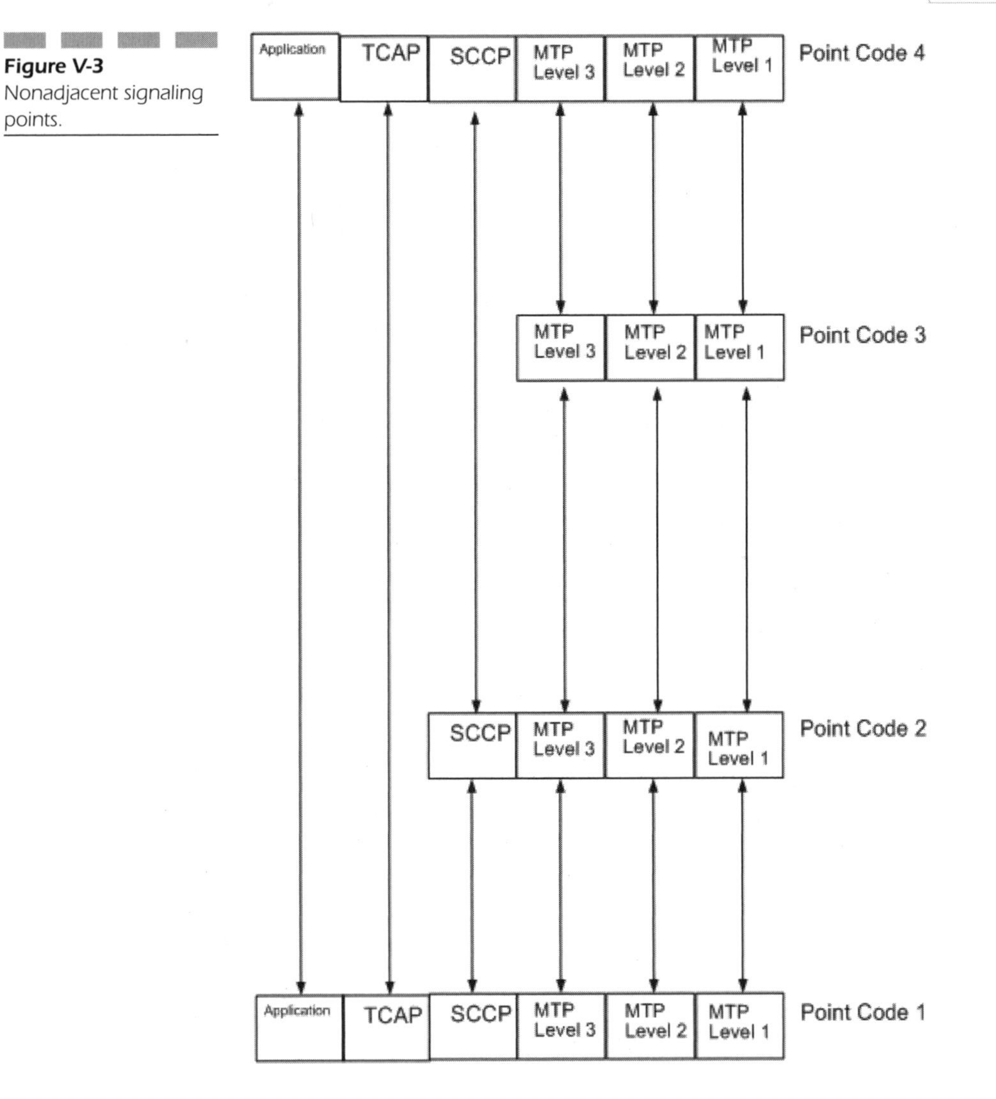

Figure V-3
Nonadjacent signaling points.

(RTCP) is without voice packets, but provides messages that are exchanged between session users. The messages hold information such as lost RTP packets, delay, and inter-arrival jitter. An RTP session opens in unison with an RTCP session, so if a UDP port number is assigned to an RTP session for packet transfer, another port number is assigned for RTCP messages.

H.323

The ITU developed the H.323 signaling protocol that is sometimes referred to as packet-based Multimedia Communications Systems for VoIP. This is

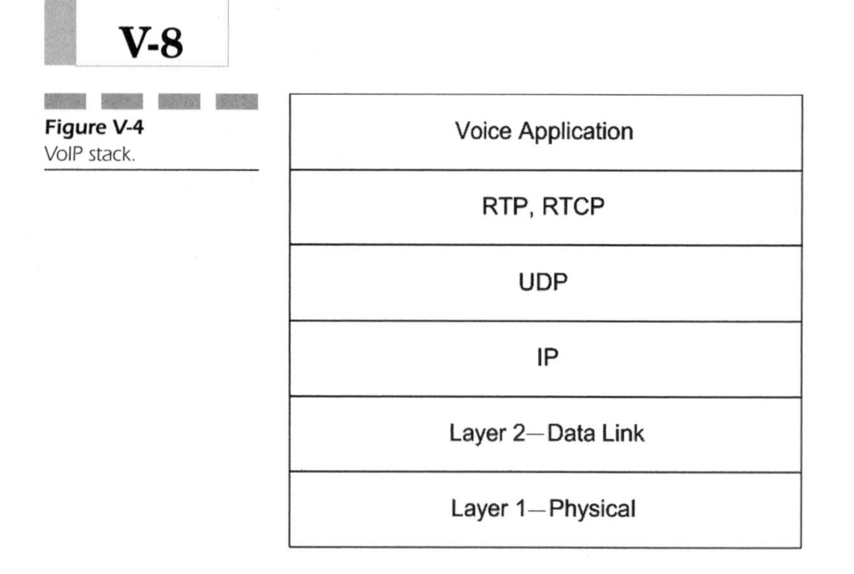

Figure V-4
VoIP stack.

currently the most popular VoIP signaling system in spite of competition from others that include the Session Initiation Protocol (SIP). Primarily, H.323 facilitates the exchange of media streams between H.323 endpoints that may be terminals, gateways, or multipoint controller units (MCUs). See Fig. V-4.

Figure V-5 shows the H.323 architecture including its terminals, gateways, gatekeepers, and (MCUs). A terminal is normally an end-user communications device with audio and perhaps video codec(s), and communicates with other endpoints. A gateway translates services between the H.323 network and another type of network, or ISDN or PSTN. The gateway supports H.323 signaling, terminates packet media, and interfaces with CS networks. The gateway may act as a H.323 terminal, and as a node in the CS network. Gatekeepers are optional and control H.323 terminals, gateways, and multipoint controllers (MCs). Gatekeepers authorize network access from endpoints.

H.323 Protocols

The H.323 protocol stack allows media exchange using RTP over UDP. The H.225.0 and H.245 protocols define exchanged messages. H.225.0 has two parts:

- A variant of ITU-T recommendation Q.931 that establishes and tears down connections between H.323 endpoints. This is call- or Q.931-signaling.

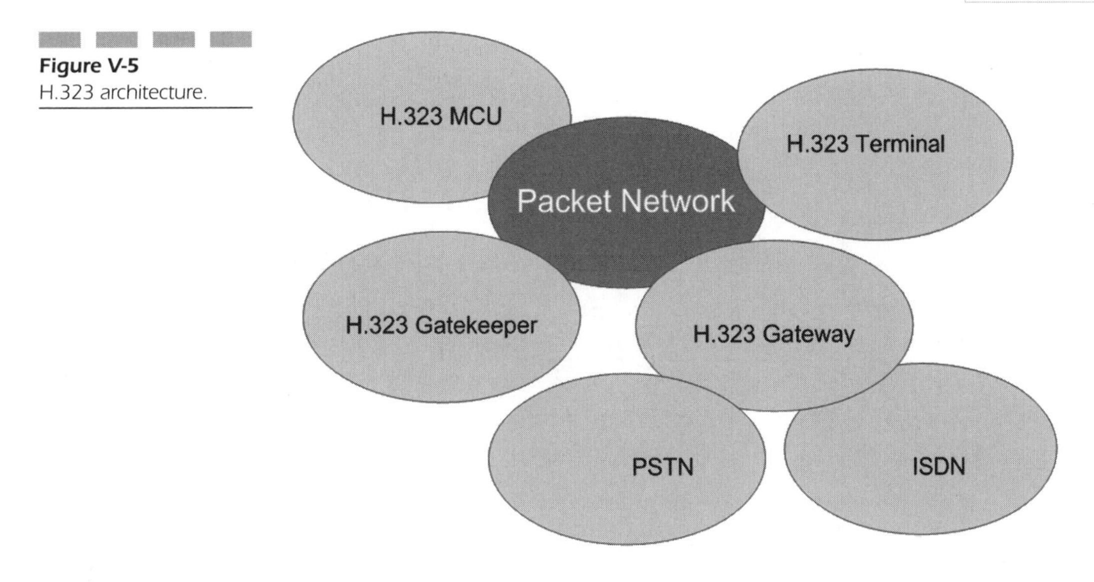

Figure V-5
H.323 architecture.

- *Registration, Admission, and Status* (RAS) signaling that enables a gatekeeper to manage the endpoints.

H.323 Call Establishment

Figure V-6 shows two H.323 endpoints establishing a VoIP call:

1. Using an Admission Request (ARQ) message, the calling terminal requests the permission to call from the gatekeeper.
2. The terminal indicates the type of call, the endpoint's own identifier, a call identifier, a call reference value, and data about the other party(ies).
3. The data about other parties includes one or more aliases and/or signaling addresses, and the bandwidth parameter.
4. The gatekeeper grants the request by replying with an Admission Confirm (ACF) message.
5. The optional callModel parameter in the ARQ indicates whether the endpoint is to send call signaling directly to the other party, or whether call signaling is passed through the gatekeeper.
6. The Setup message is the first call-signaling message to establish the call, and holds the Q.931: Protocol Discriminator, a Call ReferenceSetup, a Bearer Capability, and User-User information element.

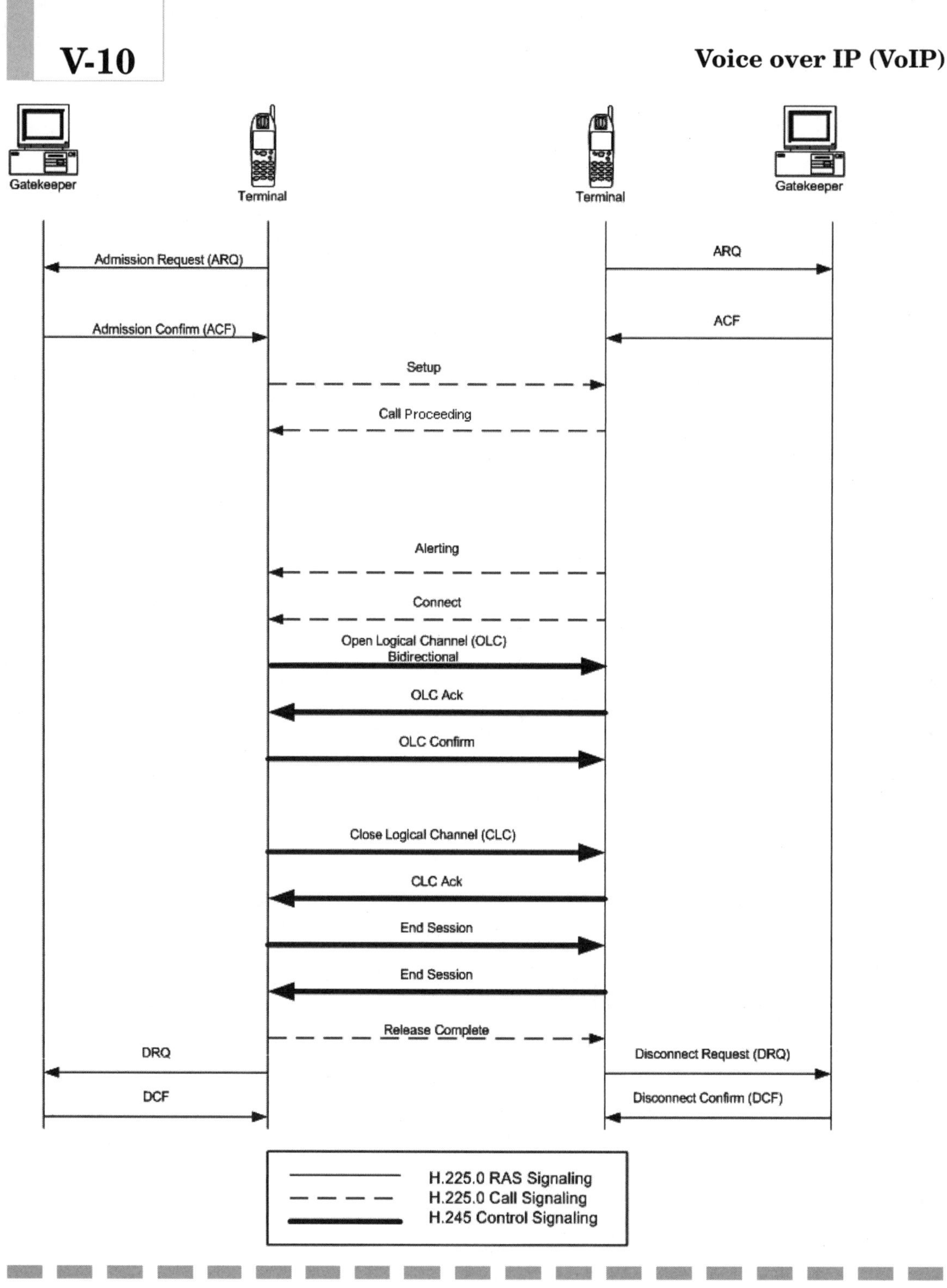

Figure V-6 H.323 calls.

7. The optional Call Proceeding message may be sent by the recipient indicating that call establishment procedures are under way.

Voice and IP Mobile Networks

Internet access configuration using a mobile IP tunnel, and Internet access configuration using voluntary L2TP tunnels and mobile IP, are two solutions for mobile IP network implementation. The mobile IP gives mobility, and forward traffic is routed between home and foreign agents over a mobile IP tunnel, and reverse traffic is routed from foreign agents to remote servers.

Internet Access Network Mobile consists of the following:

- An accounting server may interface with FA routers, collecting and storing accounting records.

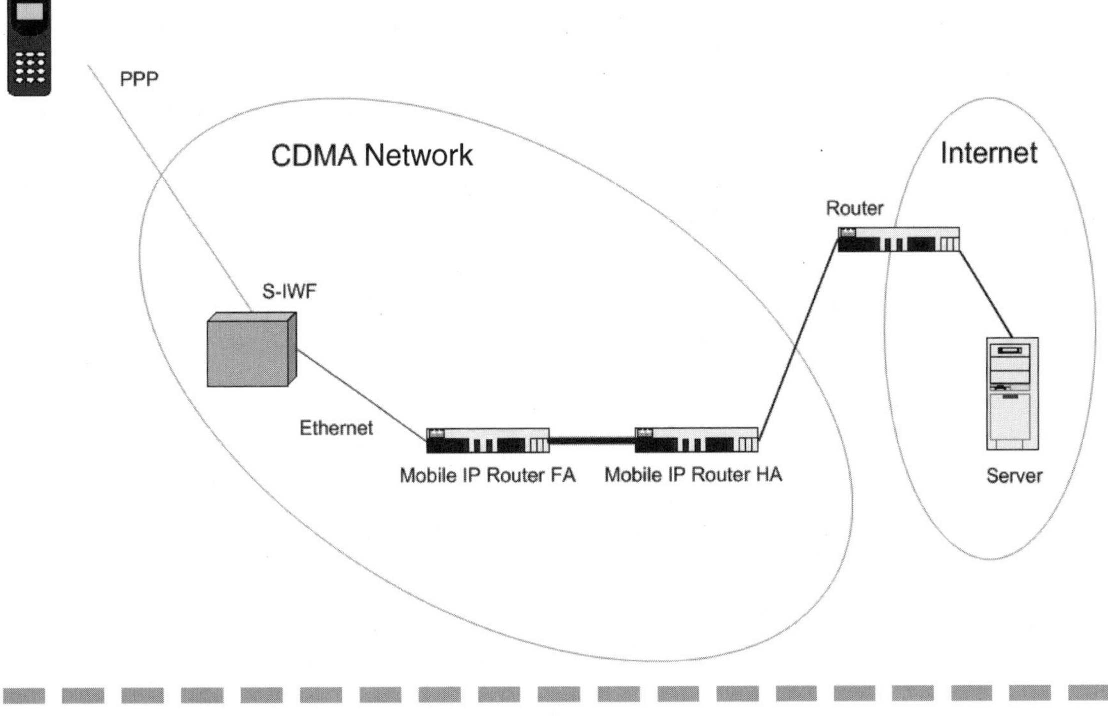

Figure V-7 Internet Access Network mobile.

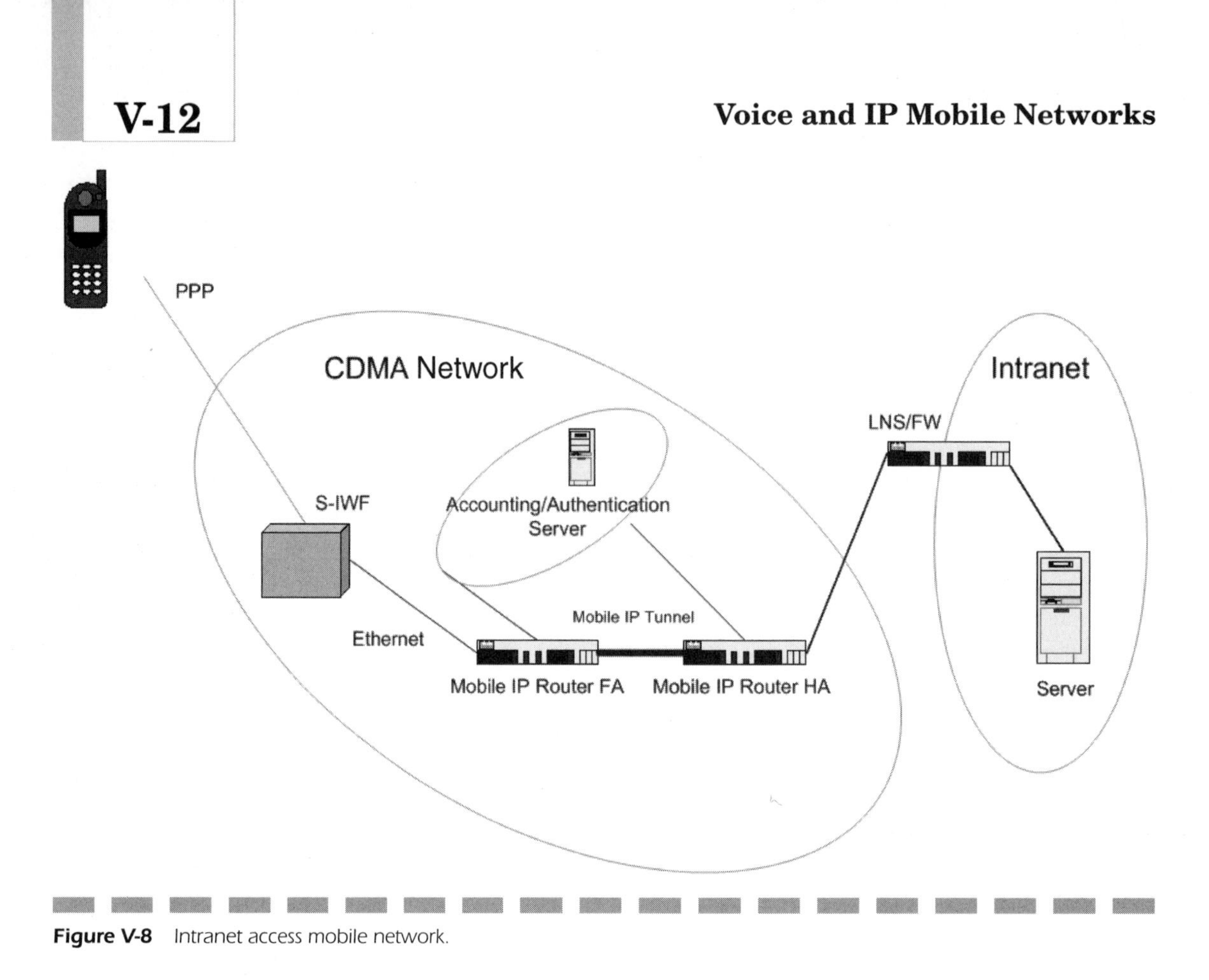

Figure V-8 Intranet access mobile network.

- Mobile IP–capable router gives the foreign agent (FA) function.
- Mobile IP–capable router gives the home agent (HA) function.
- The HA router's IP address is conveyed to the FA using the mobile–IP registration message.
- The IWF in the serving system (S-IWF) and FA router are connected using Ethernet.
- The S-IWF terminates PPP and relays received IP datagrams to the designated mobile-IP FA router on the local network.

See Figs. V-7, V-8, and V-9.

Security

Data privacy over the air is optional and can be implemented using RLP encryption (ORYX) between the mobile and packet-switching unit (PSU).

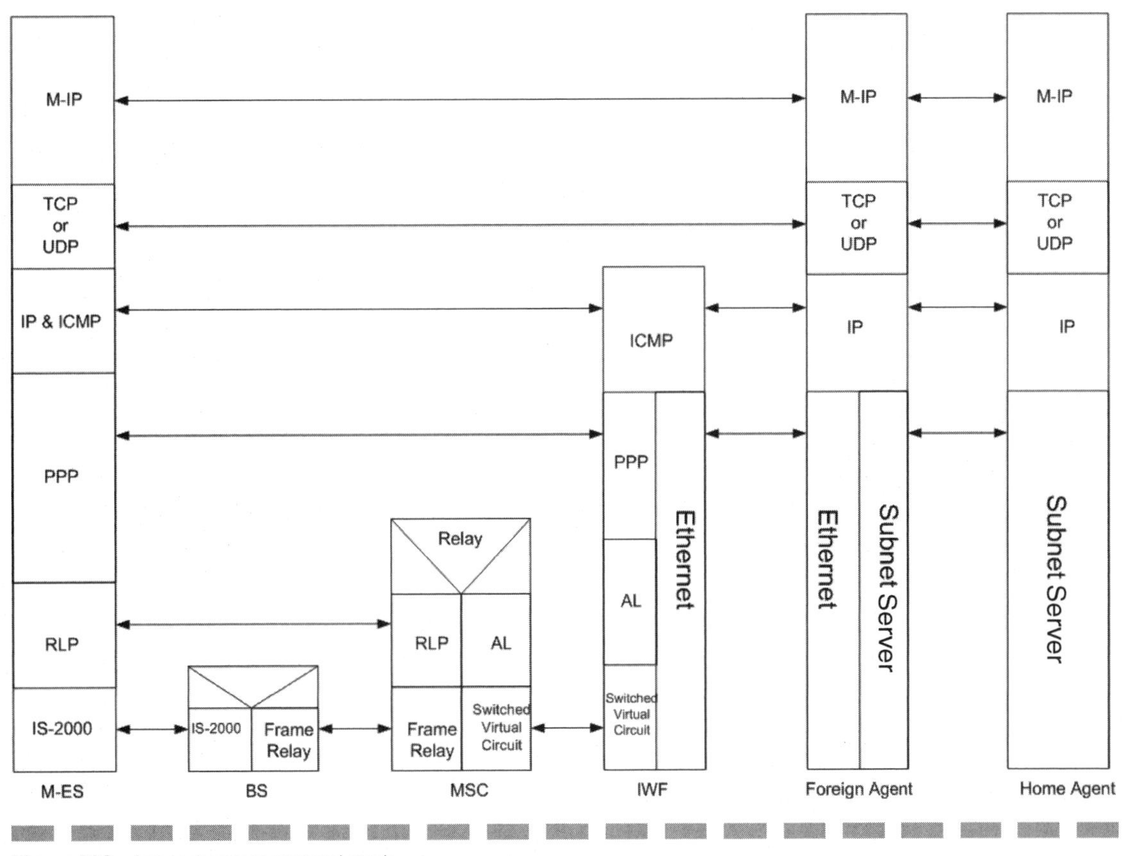

Figure V-9 Internet access protocol stack.

Three levels of authentication and authorization validation are provided: IS-41 service authorization validation (mandatory), IS-41 authentication (optional), and mobile HA authentication for mobile IP registration/reply messages (mandatory).

Security and Firewall

Authentication includes IS-41 and mobile HA during mobile IP registration. Identification in the mobile-IP registration gives antireply protection. LNS authenticates terminals using PPP authentication integrity such as challenge handshake authentication protocol (CHAP). End-to-end authentication, integrity, and confidentiality can be provided if the terminal and LNS support IP.

Voice Encoding

CDMA bypasses PCM voice digitization and uses a variable-rate voice-coding technique based on the use of a Code-Excited Linear Predictor (CELP) speech-encoding method.

Maximizing concurrent calls and optimizing the mobile network requires speech compression between the mobile and base station. The EVRC in the context of the CDMA Development Group's (CDG) implementation, provides 13kbps voice quality at 8kbps data rate. Other industry implementations include the 16kbps LD-CELP, 13kbps CELP, and ADPCM.

The EVRC speech compression algorithm bases itself on relaxed code-excited linear predictive (RCELP) coding that is based on CELP. CELP processes 20ms speech frames for coding and decoding, and for each 20ms time interval the encoder processes 160 speech samples. The volume, pitch, and rate of the voice waveform are used by the coder to represent speech at 1, 4, or 8kbps. For each 20ms speech frame, the CELP generate 10

TABLE V-1

EVRC Algorithm

EVRC Bit Allocations	Packet Type			
	Rate 1	Rate 1/2	Rate 1/8	Blank
Spectral transition indicator	1			
Line spectral pair (LSP)	28	22	8	
Pitch delay	7	7		
Delta delay	5			
Adaptive codebook gain (ACB)	9	9		
Fixed codebook shape (FCS)	105	30		
FCB gain	15	12		
Frame energy			8	
Unused	1			
Total encoded bits	171	80	16	
Mixed mode bit	1			
Frame quality indicator	12	8		
Encoder tail bits	8	8	8	8
Total bits	192	96	24	8
Rate (kbps)	9.6	4.8	1.2	0.4

linear-prediction coding filter coefficients, and with EVRC these are represented by vectors.

CELP speech coders use long-term pitch analysis to generate a 3bit pitch gain and a 7bit pitch period. The pitch analysis occurs on either four 5ms subframes, two 10ms subframes, or four 5ms subframes. This results in a number of bits per frame and gives pitch information. The EVRC algorithm categorizes speech full-rate (8.55kbps), half-rate (4kbps), and eighth-rate (0.8kbps), which occur every 20ms. See Table V-1.

Computers/Software

V.1

A standard covering power levels for telephone networks.

V.21

A standard modem speed capable of transmitting at 300bps, and receiving data at 300bps.

V.32

A standard modem speed capable of transmitting at 9600bps, and receiving data at 9600bps.

V.32terbo

An upgrade to the V.32bis standard introduced by AT&T.

V.42

An international error correction standard for modem-based communications. MNP2-4 (Microcom Networking Protocol) and LAPM (Link Access Procedure M) provide error correction.

V.90

An official designation for the 56.6kbps modem standard. 56.6kbps = 56,600bps.

VBScript

A scripting language that is a subset of Visual Basic, and may be used to deliver functionality gains to applications. VBScript may be used to enhance Web pages through the addition of:

- Event-driven objects
- ActiveX controls
- Interactive content
- Java applets

VBScript does not harness:

- OOP methodologies (1998)
- DLL calls (1998)

The VBScript syntax is similar to that of Visual Basic, and includes statements for loops, events, procedures, functions, etc.

VBScript For...Next Loop A means of repeating statements based on a true or false condition.

```
For pointer = first To last [Step step]
  ' statements
  ' statements
Next
```

VBScript While...Wend Loop

```
While condition
  'statements
  'statements
Wend
```

VBScript Procedures VBScript has Sub and Function procedures, and the latter should be used to return values (in accordance with definition of a function).

VBScript The subprocedure has the following form:

```
Sub Subroutine_Name([parameter])
  ' statements
End Sub
```

VBScript Function A function procedure is of the form:

```
Function FunctionName([parameter])
  ' statements;
End Function
```

VBXtras

An object factory that specializes in Visual Basic controls and add-ons.

Vertical Market

A specialist or niche market. Until the late 1980s multimedia was considered a vertical market.

VGA (Video Graphics Array)

An IBM PC graphics controller standard released in mid-1987 by IBM as part of its PS/2 range. As with all add-ins for PCs, graphic controllers (adapter cards) plug directly into expansion slots.

Video Capture File

A capture file is set up prior to video capture so as to optimize the rate at which digitized video may be written to a hard disk. This improves the quality level of captured video. If necessary, the target hard disk should be defragmented so that video data is written to a contiguous series of blocks thus optimizing the target hard disk performance. The specified size of the capture file should be large enough to accommodate the video sequence that is to be captured and stored. Although the capture program will usually enlarge it automatically, the possibility of complete frames being omitted or dropped during video capture is increased, as is the possibility that the capture file will become fragmented.

Video Editing

A process of editing a video file. Digital video files may be edited using programs such as VidEdit, Adobe Premier, and Asymetrix Digital Video Producer. Typical video editing operations include:

- Copying frames from point to another in a sequence
- Moving/cutting frames from one point to another in a sequence
- Copying frames from one sequence to another
- Moving frames from one sequence to another
- Deleting unwanted frames from a video sequence
- Titling video sequences
- Cropping video frames
- Altering the playback speed in terms of frames per second
- Fading colors
- Tinting colors
- Changing colors

Video for Windows

Video for Windows (VfW) permits video playback, capture, and editing. Microsoft Video for Windows, which includes the VidCap video capture program, VidEdit video editing program, BitEdit 8bit graphics editor, and PalEdit 8bit color palette editor. VidEdit provides a gateway to several video compression schemes, the variety of which depends upon the video card you have. Compression algorithms such as Intel Indeo, Microsoft RLE (Run Length Encoding), and Microsoft Video 1 can help reduce video file sizes by varying amounts. The size and quality of resultant video files may be controlled using compressors through the adjustment of compression settings. Resultant video may be added to applications through OLE (Object Linking and Embedding). It supports the AVI (Audio Video Interleaved) format and features a number of compressors including Microsoft 1, Microsoft RLE, and Intel Indeo.

Video-on-Demand (VOD)

An e-commerce implementation that permits customers to view selected purchased movies.

Video Playback fps

The playback frame rate of a video sequence.

Video Playback Frame Resolution

The frame resolution of a video sequence.

Virtual Processor

A processor that is implemented in software. It may sometimes be referred to as a virtual machine. It is a design approach used by such programming languages as Java, so that they may be system and OS independent. They may, therefore, be applied as applets in heterogeneous environments such as the Web.

Virtual Shopping

An activity where the consumer purhases items from intangible stores that are usually on the World Wide Web.

Virtual Store

An intangible store that exists on the World Wide Web.

Virtual Web Server

A Web server that is not physically implemented; rather it may exist with a number of other such virtual Web servers on the same site. Virtual Web Servers may be created using Microsoft IIS, and they may have a:

- Domain
- TCP/IP address
- Root directory

Whether or not a Web server is virtual is not known to the user.

Virus

An entity that causes a program to function incorrectly and might result in the loss or corruption of data.

VisiBroker for C++

An ORB that forms the basis for middleware implementations.

VisiBroker for Java

An ORB that forms the basis for middleware implementations.

Visual Basic

A programming language whose best-known implementation is the Microsoft Visual Basic programming tool that may be used to tackle a variety of different programming projects, including the development of:

- ActiveX controls
- Client/server applications
- Mainstream business applications
- Utilities
- Multimedia-related programs such as media players
- Leisure programs

Microsoft Visual Basic has since become VB.NET in compliance with the Microsoft .NET initiative.

Visual C++

A Microsoft C++ programming environment that may be used to create

- DLLs
- ActiveX controls
- 16bit and 32bit applications

Visual FoxPro

A Microsoft OOP database management system for creating enterprise solutions. It is supplied with the Microsoft Visual Studio.

Visual InterDev

A Microsoft development tool that is used to create Web and intranet applications. Briefly it may be used to:

- Access ODBC databases
- Script client and server Web pages

- Edit content files
- Manage multiuser Web projects

The files of an InterDev application are stored on the Web server. Files are accessed with Visual InterDev, using a local project file that points to the server and to the relevant Web. InterDev lends itself to the team collaboration environment because multiple developers can work simultaneously with files (on the Web server or in a Web). The Web project includes the Web files on the server, and the client-side, local project files. A Web Project File is available or the creation of project files that point to relevant Webs and Web servers. Visual InterDev is also supplied with the multimedia production tools:

- Image Composer that offers a sprite-based drawing environment. Each imported image becomes a sprite. A number of effects and filters are available. Support is provided for plug-ins from Adobe and Kai. It supports BMP, GIF, and TIF formats.
- Music Producer is used to create MIDI sequences.
- Media Manager empowers Windows Explorer to view media files.

Visual Java++

A Microsoft development environment for writing, compiling, and debugging Java applications and applets. Visual J++ may be used to integrate JDK packages into Java programs, and to create multithreaded Internet and intranet applications. InterDev is included in Microsoft VisualStudio 97.

Visual Programming

A programming technique where the programmer simply draws usually standard components on screen, and then attaches code to them. The code segments may be written in a line-by-line fashion or selected from a library. Many multimedia authoring tools, and modern development tools, employ visual programming techniques at various levels.

Visual SourceSafe

A useful Microsoft solution that lends itself to the team collaboration environment. Once installed, this restricts editable files to individual, authorized developers.

Visual Studio (Enterprise Edition)

A comprehensive suite of Microsoft development tools that includes:

- Visual J++
- Visual C++
- Visual Basic
- Visual InterDev
- Visual FoxPro

It also includes numerous tools, extensive documentation, and the Microsoft Developer Network (MSDN) on CD-ROM.

Visual Studio.NET

Visual Studio.NET is Microsoft's standard development environment.

VLB (Vesa Local Bus)

A standard local bus that supports compatible peripherals such as hard disk controllers, I/O cards, and graphics cards. It is internationally agreed on and backed by VESA (Video and Electronics Standards Association). It is used widely on PC designs, and gives better performance than IBM's original ISA (Industry Standard Architecture) expansion bus developed for the PC AT (Advanced Technology) in the early 1980s.

VPN (Virtual Private Network)

A network that may be built using Internet technologies, as opposed to private lines. VPNs may be LAN-to-LAN, or even extranets that include remote users that may be business partners or even customers.

VRML (Virtual Reality Modeling Language; pronounced "vermul")

A file format and a language for creating and describing objects or nodes and their behavior.

VRML extended the Open Inventor specification to include cone, cube, and cylinder primitives, along with methods for embedding hyperlinks.

Applications of VRML include:

- Multimedia presentations and titles
- Leisure software
- Virtual reality
- Web pages

Objects may be:

- Static 3-D images
- Static 2-D images
- Audio
- Multimedia
- Embedded with hyperlinks

VRML authoring tools or generators are widely available.

VRML Nodes Node properties have:

- A name that is dedicated to the class
- Parameters that offer an object's definition and have fields that contain dimensions, etc.

VRML Events The nodes are event driven, and receive and send messages such as:

- EventIn: Typically changes a property of the node
- EventOut: Sends a message from an object that might have undergone change due to an interaction with a message

Nodes interact using messages passed via ROUTE that interconnects eventOut and eventIn processes.

VRML ISO/IEC 14772

An official designation for the internationally agreed-on VRML specification.

V-standards

A set of recommendations that covers voice and data telephony. Popular V-standards cover the following full-duplex modem speeds:

V.21	300bps
V.22	1200bps
V.22bis	2400bps
V.32	9600bps
V.32bis	14,400bps
V.34	28,800bps
V.90	56,600bps

Vulnerabilities

A listing of comparative flaws in a network's defenses against illegal access.

Miscellaneous

Videotex

A service used to publish text and graphics over the PSTN. It emerged from the BT Research Laboratories in 1970s, and was launched in the form of Prestel in 1979.

Videotex uses alpha-mosaic text and graphics. The display is based on character blocks that require 7 or 8bits, and is produced from a look-up table or from a character generator.

A videotex frame consists of a matrix of such characters, consisting of 24 rows of 40 characters. Typically the frame requires 960bytes, and may be transmitted in around 6.5s over PSTN.

Using faster ISDN access technology, the transmission time is reduced.

ISDN D Channel

$$\text{Transmission time} = (960 \times 8)/16{,}000$$
$$= 480\text{ms}$$

ISDN B Channel

$$\text{Transmission time} = (960 \times 8)/64{,}000$$
$$= 120\text{ms}$$

VLAN (Virtual Local Area Network)

A network where computers may not be connected to the same physical LAN, rather they might be connected on different networks, and in remote locations. They may be configured using software and are immune to the physical location of the networked systems.

Acronyms

UWCC	Universal Wireless Communication Consortium
V	Viterbi
VAD	Voice activity detector
VASP	Value-added service provider
VC	Virtual circuit
VLR	Visitor location register
VMS	Voice message services
VPN	Virtual private network
VPU	Video provisioning unit
VRC	Variable-rate codec
VSELP	Vector sum-excited linear-predictor

W

WAP (Wireless Application Protocol)

WAP permits compatible mobile devices to access Internet sites that are usually written using WML (Wireless Markup Language). The WAP Forum (www.wapforum.org) controls development, with the *WAP* 1.0 specification released in Spring 1998. Enterprises and organizations that make WAP

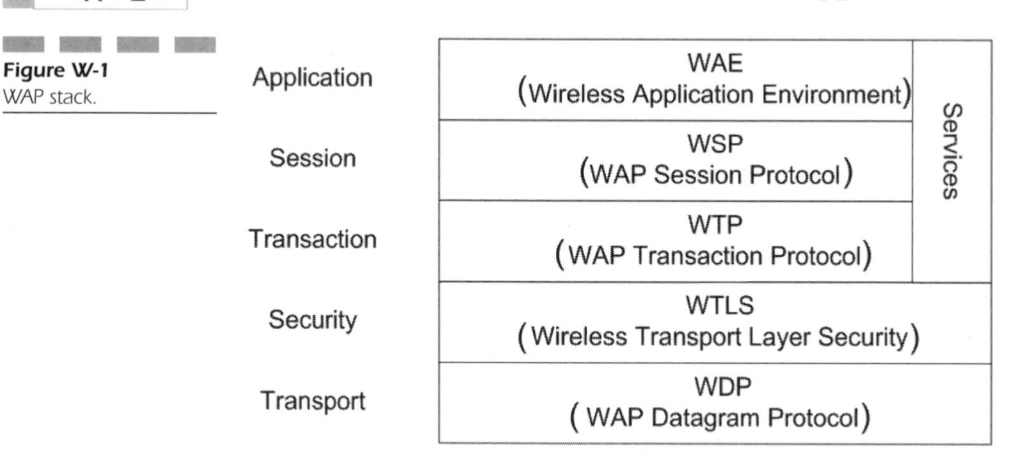

Figure W-1
WAP stack.

gateways play a key role within the WAP forum, and include Nokia and Ericsson.

The Wireless Markup Language Script (WMLScript) is a scripting language used in WAP applications. The WAP client browser requires a WMLScript virtual machine to run compiled WMLScript.

WAP operates over TDMA including D-AMPS, GSM, CDMA, and permits content using short SMS and circuit-switched cellular data (CSD). See Fig. W-1.

Wireless Transport Layer Security

The Wireless Transport Layer Security (WTLS) provides encryption capabilities for secure transactions, and is key to m-commerce applications, and operates in conjunction with the WAP Transaction Protocol (WTP) that may use TCP (reliable) and UDP (unreliable) datagram delivery. The WAP Session Protocol (WSP) is accountable for data exchange between applications, and a session has a series of request and/or response transactions that are implemented by the WTP. The WAE has WML, WMLScript, and the Wireless Telephony Application Interface (WTAI) that interfaces with network services. IP addressing is carried out using URLs.

The WAP Gateway unites the mobile network with the IP network and:

- Converts between the WAP stack to the TCP/IP stack
- Implements Domain Name Server (DNS) services
- Translates Web pages into compact encoded formats for transmission over mobile networks.
- Is sometimes implemented to convert from HTML to WML on the fly

Because there is one route from the WAP gateway over a wireless network, WTP does not support out-of-sequence packets. And to optimize bandwidth, a streamlined data transfer command set is used:

- Data Request
- ACK Reply
- Data Request
- ACK Reply
- Data Request
- ACK Reply
- Session Termination ACK

Ports

WDP uses port numbers to identify higher layers or applications such as e-mail that they may hold. Table W-1 lists such applications and/or protocols registered by the WAP Forum with the Internet Assigned Numbers Authority (IANA).

Adaptation Layer

The adaptation is a sublayer of the WDP layer, and the WDP core layer is consistent for bearer services. A different adaptation layer provides interfaces WDP with bearer services. When IP is the routing protocol, the TCP/IP UDP is adapted as the WDP. The transport of WDP over GSM's SMS may require segmented WDP datagrams.

TABLE W-1

Registered WAP Ports

Port Number	Application/Protocol
2802	WAP WTA secure connectionless session service
9200	WAP connectionless session service
9201	WAP session service
9202	WAP secure connectionless session service
9203	WAP secure session service
9204	WAPvCard
9205	WAP vCal
9207	WAP vCal.Secure

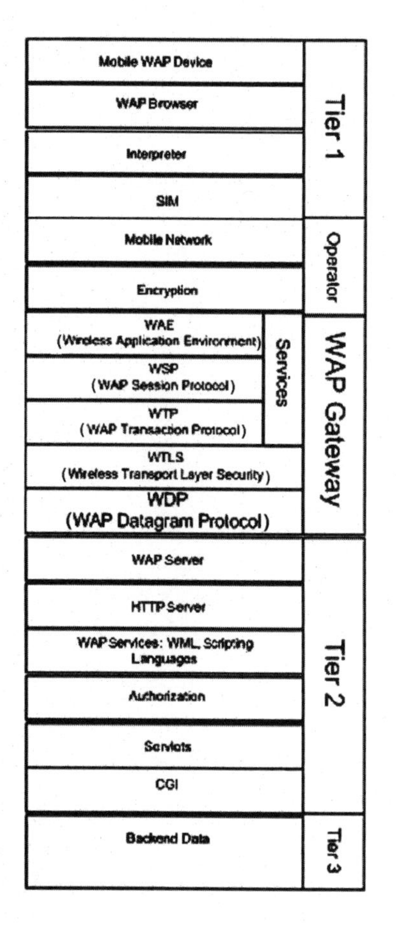

Figure W-2
WAP architecture.

WTLS

The Wireless Transport Layer Security (WTLS) helps secure WAP applications and services, and operates above the transport-layer protocol. It may authenticate users, encrypt and decrypt data, and refresh keys.

WTLS Encryption

WTLS accommodates:

- *Block-ciphering encryption* encrypts plaintext into ciphertext, and WTLS supports DES, RC5, and IDEA block ciphers.

- *Stream ciphering encryption* ex-Ors the plaintext with the output generated by a cryptic key for mobile users requiring security.
- *Digital signing* uses a one-way hash function input as a signing algorithm.
- *Public-key encryption* ensures that public-key encrypted ciphertext is decrypted using a private key.

The Wireless Application Environment

WAE is the environment that provides access to WAP Internet services, and is a gateway that couples the mobile client with the Web server. It rationalizes data from the Web server, so that the gateway is an encoder/decoder. The WAE encoding converts client requests in the WML. The client requests are transported over the WDP. The gateway also converts HTML to WML. See Fig. W-2.

WAP Gateway

A core solution that unites dissimilar networks providing a bridge between an IP network (namely the Internet) and Mobile Operator network.

Figure W-3 presents a "multiple-scenarios" high-level representation of a WAP gateway with all its surrounding software architectures and physical infrastructures. The illustration attempts to explain the basic strata beginning with the presentation layer through to the backend.

A practical implementation of WAP would obviously focus on a single descent through the shown layers and tiers that include:

Presentation:
- A handset such as GSM WAP phone, GPRS phone, or Palm
- An OS such as Palm OS or Windows CE

Mobile network:
- GSM: Vodafone, Cellnet, Orange, One2One
- GPRS

WAP gateway:
- WAP Gateway core solution: Nokia, Kannel, etc.
- OS; Windows NT, Unix, Linux

Application logic:
- Middleware CORBA, OMG NS, proprietary classes that may be written in Java, C++, and on rare occasions Objective-C.

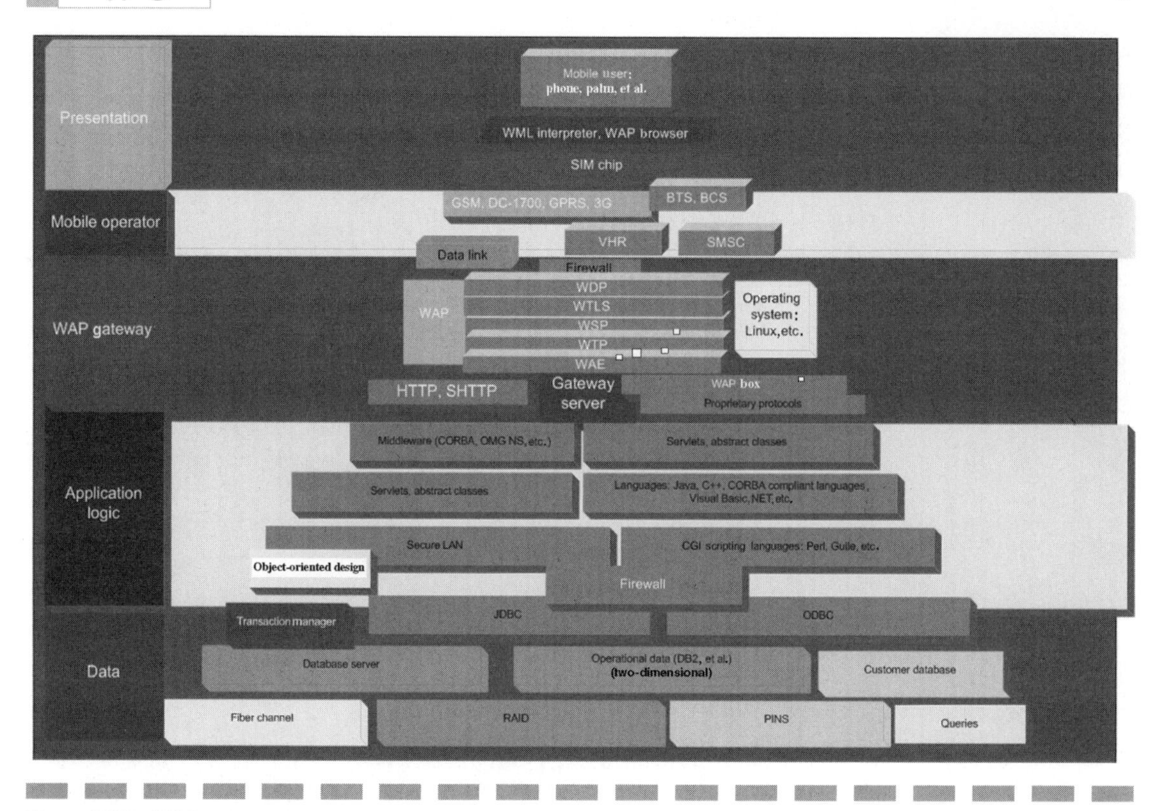

Figure W-3 WAP gateway (multiple scenarios).

- Servlets, Java classes
- CGI scripts

Data:

- Database glues: ODBC, JDBC
- Operational data

A WAP gateway core solution exhibits external interfaces, that is, what other systems it talks to, and how. Nokia and Ericsson have conventionally marketed and sold WAP gateway solutions, and open source implementations for the Linux environment include Kannel (www.kannel.org). See Fig. W-4.

The Kannel WAP gateway has up to six external interfaces:

- *SMS centers:* The SMS centers use a variety of mostly proprietary protocols (CIMD, EMI, SMPP) over TCP/IP, modem lines, or various other carriers. The gateway needs to support as many SMS center protocols as possible, and make it easy to add new ones.

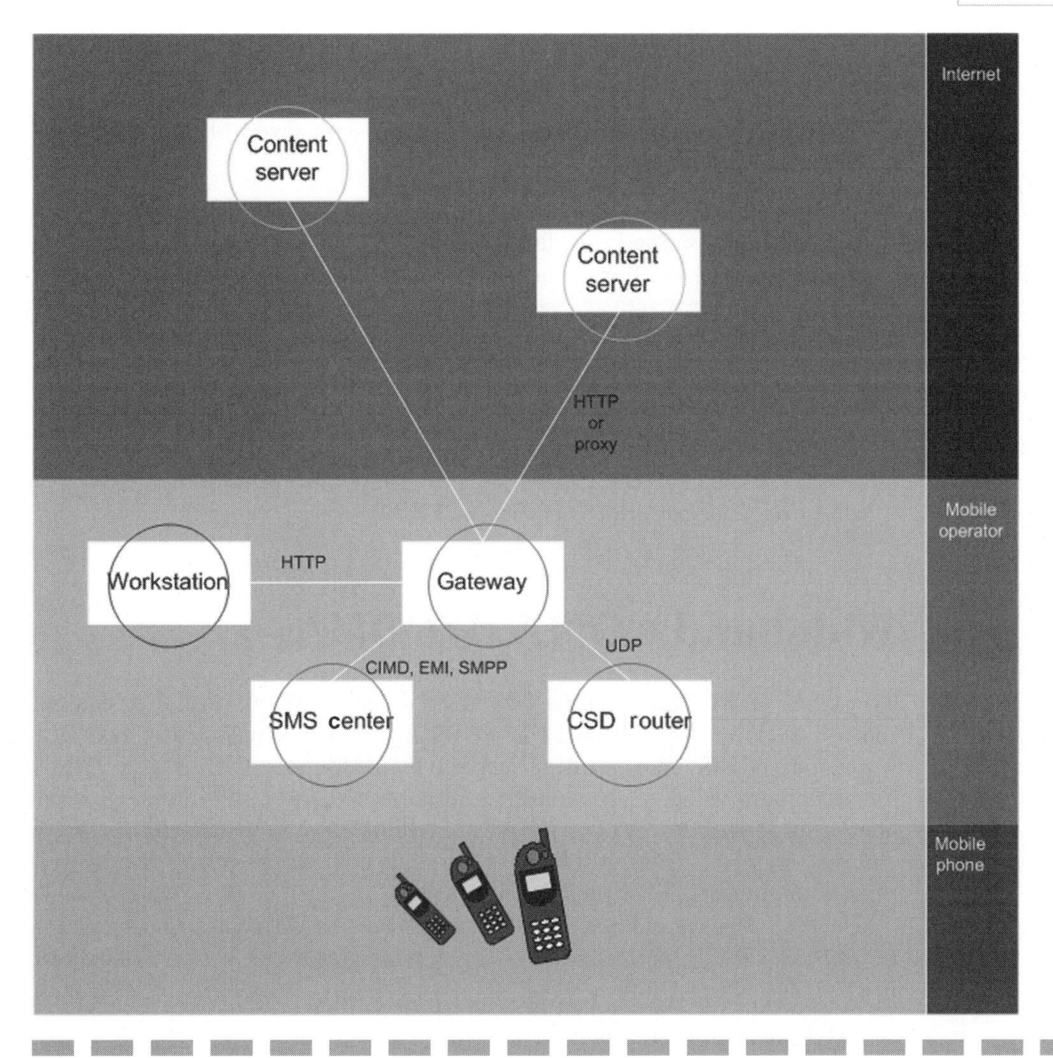

Figure W-4 External interfaces of the Kannel WAP gateway.

- *CSD routers:* Communication with phones via CSD routers is plain UDP (of the TCP/IP stack), that is, there is no special CSD protocol.
- *Configure/monitor/control workstation* (which could also be considered part of the gateway, but isn't): The c/m/c workstation uses HTTP, where the gateway works as an HTTP server, and similarly for those sending SMS via HTTP.
- *Content servers:* The content servers also use HTTP, but with them the gateway is a client. HTTP proxies also use HTTP (but a slightly different kind), and the gateway is then also a client.

- *Clients:* Clients send SMS messages via HTTP.
- *HTTP proxy.*

The WAP gateway divides the processing load on several hosts, which are of three different types:

- *Bearer box:* This host connects to the SMS centers and CSD routers, and provides a unified interface to them for the other boxes. It implements the WDP layer of the WAP stack.
- *WAP box:* These hosts run the upper layers of the WAP stack. Each session and the transactions that belong to that session are handled by the same WAP box. Sessions and transactions are not migrated between WAP boxes.
- *SMS box:* These hosts run the SMS gateway. They can't connect directly to the SMS centers, because the same SMS center connection can be used for both SMS services and WAP.

Wideband CDMA (WCDMA)

Two types of WCDMA include frequency division duplex (FDD), wideband code division multiple access (FDD/WCDMA), and time division duplex (TDD) wideband code division multiple access. WCDMA uses CDMA and TDMA for broadband usage such as audio and video. WCDMA permits more than one hundred mobile users to share a single 5MHz bandwidth radio carrier channel. By using different codes on the radio channel, multiple mobile telephones communicate on the same frequency.

The WCDMA system has user equipment (UE), a UMTS terrestrial access network (UTRAN), and a core system network. And the WCDMA network has:

- Databases and protocols
- CS and PS systems
- Gateways to public voice and data networks
- Mobile phones (or mobile stations)
- Node Bs or radio base stations
- Radio network controllers (RNCs)

WCDMA's radio channel has physical channels with 10ms frames that have 15 time-slot bursts that are assigned to users for reception and transmission. WCDMA voice communications are normally over a 5MHz carrier for the downlink and the uplink. Every carrier frequency in each cell and/or sector, has digital control channels that hold system and paging information. The digital control channels also coordinate multimedia, high-speed packet data, broadcast messaging, and fast power control.

Voice channels can be full rate where one time slot per frame is assigned to each user, so 100 users may share a radio channel. There is a variable bit-rate speech coder for adaptive multirate (AMR) speech coding, and a lower bit-rate speech coder that allows up to 200 users to share a radio channel.

FDD/WCDMA

FDD/WCDMA uses a 5MHz radio channel for the downlink and the uplink. The channels are divided into 10ms frames that have 15 time slots. Voice communication requires one or more time slots for transmitting, and one or

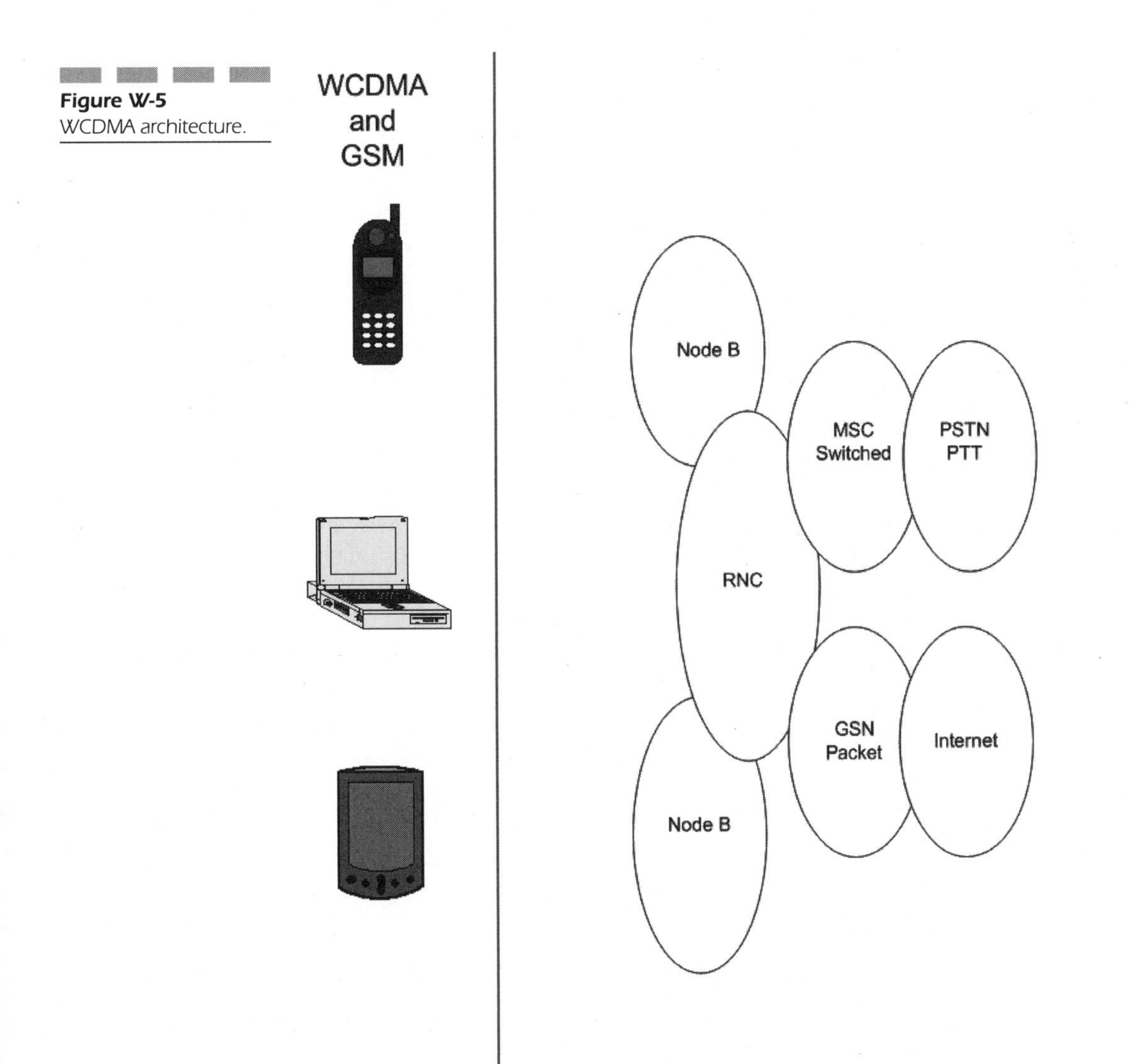

Figure W-5 WCDMA architecture.

more for receiving, and several for idle state. WCDMA mobile phones transmit on one frequency and receive on a frequency that is 190MHz higher, and it uses a code sequence to maintain a private communications link.

TDD/WCDMA

TDD/WCDMA was introduced for two-way communications on systems without two frequency bands. TDD/WCDMA uses a single 5MHz physical channel that is divided into 10ms frames that are divided into 15 time slots. It has much in common with FDD/WCDMA, including the 5MHz radio channel, frame size, and time-slot size. A comparative disadvantage of TDD is the need for guard times so as to prevent bursts of data from overlapping.

Wideband CDMA Air Interface

At the base of the stack is the physical layer, above which is layer 2 that holds the medium access control (MAC), the radio link control (RLC), the broadcast multicast control (BMC), and the Packet Data Convergence Protocol (PDCP). Layer 3 has the radio resource control (RRC), mobility management (MM), GPRS mobility management (GMM), call control (CC), supplementary services (SS), short message service (SMS), session management (SM), and GPRS short message service support (GSMS).

The physical layer has a logical interface to the MAC called PHY that offers transport channels. The Control PHY interface is between the physical layer and RRC, and it transfers control and measurement information. The UTRAN naturally operates in FDD and TDD modes, and therefore changes the functional requirements of Layer 1.

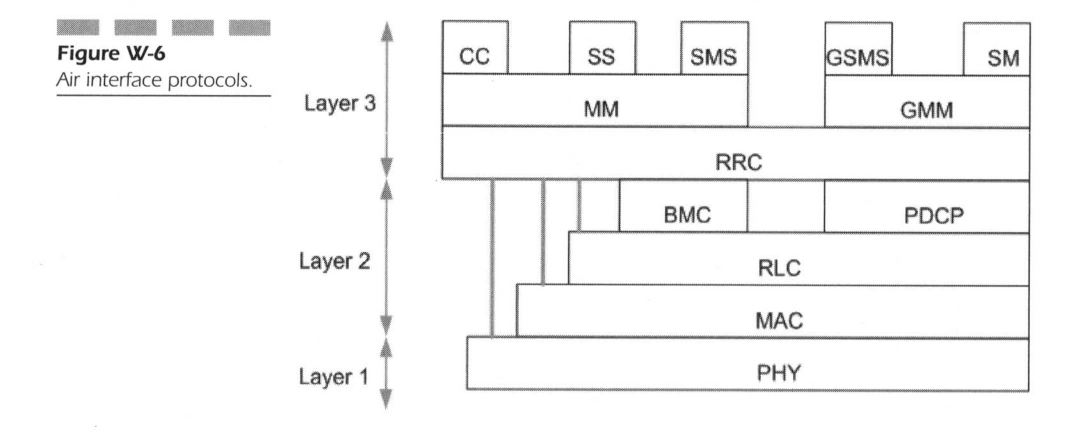

Figure W-6
Air interface protocols.

The UTRAN air interface operates at 3.84Mcps, and one 10ms radio frame is divided into 15 slots. One time slot, therefore, may transfer 2560bits. FDD spreading factors range from 4 to 256 for the uplink and from 4 to 512 for the downlink.

The physical layer is responsible for:

- Closed-loop power control
- Error detection
- FEC encoding/decoding
- Frequency and time synchronization
- Macrodiversity and soft handovers
- Mapping of CCTrCHs on physical channels
- Modulation, spreading/demodulation, and despreading
- Multiplexing of transport channels and demultiplexing
- Power weighting and combining physical channels
- Radio measurements
- Rate matching
- RF processing
- Support of uplink synchronization
- Timing advance on uplink channels

Wireless Markup Language

Wireless Markup Language (WML) is used for creating Web applications that are accessed over narrow-bandwidth media. WML has syntactic constructs that are similar to Extensible Markup Language (XML).

The Ericsson R380 is a popular WAP phone in the mass market telecommunications sector. It has a grayscale touchscreen with a resolution of 360 × 120 pixels and a 0.23 pitch, and an active screen size of almost 83 × 28 mm. Its browser includes a Browser Area, Card Title bar, and Toolbar. The browser area presents the card content and measures 310 × 100 pixels. When a card is too large for the screen, only the beginning of the card is shown at first, and a vertical scroll bar appears. Graphical components, text, and images are displayed in the top left corner, and in the same order as in the WML code.

Card Title Bar The Title bar shows which card is displayed and also browsed cards. The navigation history is reset when a loaded Card's newcontext attribute is *true*.

Toolbar The Toolbar has buttons required by the browser, and text input is done with a screenpad or character recognition screen. Three different keyboard layouts include: alpha, numeric, and national characters.

To navigate to a WAP service:

- Activate a bookmarked site using the Bookmark.
- Use the History list.
- Enter the URL in the Open Location dialog.

Design Components

Font The font used in R380 is a proportional font called Swiss A and it is used in small, normal, and big sizes.

Font Size/Height/Number of Lines

- Small/9 pixels/7.5
- Normal/10 pixels/7
- Big/12 pixels/5.5

WML The browser supports the emphasis elements big and small to change font sizes, and the elements em, strong, i, b and u.

Text Formatting

```
<wml>
<card id="first" title="Fast Burgers" newcontext="true">
<p align="center">
<br/>
<b><big><a href="#second">Welcome</a></big></b><br/>
to<br/>
Welcome to Fast Burgers.<br/><br/>
<a href="#fifth">[Contact Us]</a>
</p>
</wml>
```

Line Spacing and Line Breaks Single line spacing is used with one pixel before and after lines. Long lines are word-wrapped onto multiple lines. Text lines may include images, select lists, buttons, input fields, and hyperlinks.

WML New lines start with the br element that affects all contents in the browser, and the current alignment is used to position the line's entities.

A Line Break Example

```
<wml>
<card id="init" title="BR tag">
<p>
This sentence continues until the end of the line and
then wraps to the next line.<br/>
This phrase begins on a new line.<br/><br/>
This phrase follows two br elements.
</p>
</card>
</wml>
```

Paragraphs Paragraphs start on a new line, follow a 3-pixel gap, and have a default left alignment, but may be aligned right or centered.

WML Paragraphs are defined and aligned using the p element.

Attribute Description The align attribute may have the values: left, right, and center. The mode attribute specifies the line-wrap and makes the values *wrap* or *nowrap*.

A Paragraph Example

```
<wml>
<card id="init" title="P tag">
<p>
<b>LEFT</b><br/>
This text is left aligned, and will continue until
the end of the line and then wrap to a new line.<br/>
</p>
<p align="center">
<b>CENTER</b><br/>
This text is centered, and will continue until the
end of the line and then wrap to a new line.<br/>
</p>
<p align="right">
<b>RIGHT</b><br/>
This text is right aligned, and will continue until
the end of the line and then wrap to a new line.
```

```
</p>
</card>
</wml>
```

Indented Paragraphs Text and components may be grouped, and groups may be nested. Groups have a 20-pixel indentation and follow a 3-pixel gap.

WML A group's beginning is defined using the fieldset element.

Attribute Description The title attributes value is used as leading text, and the text following the title appears on a new line.

Fieldset

```
<card id="burgerinfo" title="What's On">
<p>
<fieldset title="Big">
Lettuce, tomato sauce, mustard, pickle, tomato
and cucumber
</fieldset>
<fieldset title="Double">
tomato sauce, mustard, pickle, tomato
and cucumber
</fieldset>
<fieldset title="Double Big">
Onions, tomato sauce, mustard, pickle, tomato
and cucumber
</fieldset>
.
.
.
.
</p>
</card>
</wml>
```

Card Title The title is defined using the title attribute in the card element.

Single Choice Lists A single choice list is shown as a drop-down listbox, showing the currently selected value in angled brackets. It is 15 pixels high, has 5 pixels of white space on either side, and adapts to the length of the text in brackets.

WML A single choice list is specified using the select element with the multiple attribute set to *No*, and each list item is specified by an option element.

A Single Choice List

```
<p>
<b>Select Burger</b>
<select>
<option>Big</option>
<option>Double</option>
<option>Double Big</option>
</select>
<a href="#burgerinfo">What's On</a>
</p>
<p>
<b>Select Meal</b>
<select>
<option>Snack</option>
<option>Medium</option>
<option>Big</option>
</select>
</p>
```

Multiple Choice Lists A multiple choice list has check boxes on separate lines that are followed by 2 pixels of white space. They may form a hierarchy of groups that have a 20-pixel indention.

WML Multiple choice lists are created using the select element with the multiple attribute set to *Yes*, and each check box item is included using the option element.

A Multiple Choice List

```
<p>
<b>Extras</b>
<select multiple="true">
<option>Fast special sauce</option>
<option>Bread crumb fries</option>
<option>Pineapple</option>
```

```
</select>
</p>
```

Buttons Buttons are displayed using the Normal Bold font, and are defined using the do element.

A Do Example

```
<p align="center">
<do type="accept" label="Continue">
<go href="#third"/>
</do><br/>
</p>
```

Three Do Elements of the Same Type

```
<p align="center">
<do type="accept" label="Continue" name="cont">
<go href="#third"/>
</do><br/>
<do type="accept" label="Contact Us" name="contact">
<go href="#fifth"/>
</do>
<do type="accept" label="Go to Start" name="start">
<go href="#first"/>
</do>
</p>.
```

Input Fields Input fields are defined using the input element that holds the shown attributes:

Attribute Description

- Type: This may have the values *text* or *password*, and if set to *password*, entered characters are shown as asterisks.
- Value: The value of the value attribute is used as default text if no preload value is defined for the input object.
- Format: Specifies an input mask for user input entries. The string has mask control characters and static text.

An Input Example

```
<p>
Name<input type="text" title="Enter name:"
  name="name"/><br/>
```

```
Address<input type="text" title="Enter address:"
name="address"/>
</p>
```

Images The browser supports images in WAP bitmap (WBMP) and also in the GIF format that have the colors: white, 0 percent black; light-gray, 25 percent black; mid-gray, 50 percent black; and black, 100 percent black. The WML img element indicates that an image is included in the text flow.

The R380 supports the following attributes:

- Alt: A text name for the image used in the placeholder.
- Src: The image's source (URI).
- Vspace: Specifies the amount of white space inserted above and below the image.
- Hspace: Specifies the amount of white space on either side of the image.
- Height: Specifies the vertical size of an image.
- Width: Specifies the horizontal size of an image.
- Align: The align attribute may have the values *top*, *middle*, and *bottom*.

An Image Example

```
<p align="center">
<img alt="baker"src="baker.gif" vspace="5" width="40"
height="30"/><br/>
<b>The burger is sizzling and being expertly prepared,
and will be delivered soon.</b><br/>
</p>
```

WLAN (Wireless LAN)

WLANs are defined by IEEE 802. with IEEE 802.11 aimed at the Physical (PHY) layer and Medium Access Control (MAC) sublayer. WLANs may replace or extend LANs, and a Basic Service Set (BSS) has more than communicating wireless nodes or stations (STAs) using a peer-to-peer architecture. A BSS has an Access Point (AP) that bridges wireless and wired LANs, and when active it prevents peer-to-peer communications. Communications to and from stations pass through the AP that is part of the wired network. This is the so-called infrastructure mode.

The Extended Service Set (ESS) is a series of overlapping BSSs that are meshed by a Distribution System (DS) such as an Ethernet LAN. See Fig. W-7.

Radio Technology

IEEE 802.11 defines PHY implemented using Direct Sequence Spread Spectrum (DSSS), or Frequency Hopped Spread Spectrum (FHSS) to conform with FCC regulations (FCC 15.247).

The Region Allocated Spectrum is as follows:

- Europe: 2.4000–2.4835GHz
- France: 2.4465–2.4835GHz
- Japan: 2.471–2.497GHz
- Spain: 2.445–2.475GHz
- United States: 2.4000–2.4835GHz

Using DSSS each information bit is combined using an XOR function with a longer Pseudo-random Numerical (PN) sequence. The resulting digital stream

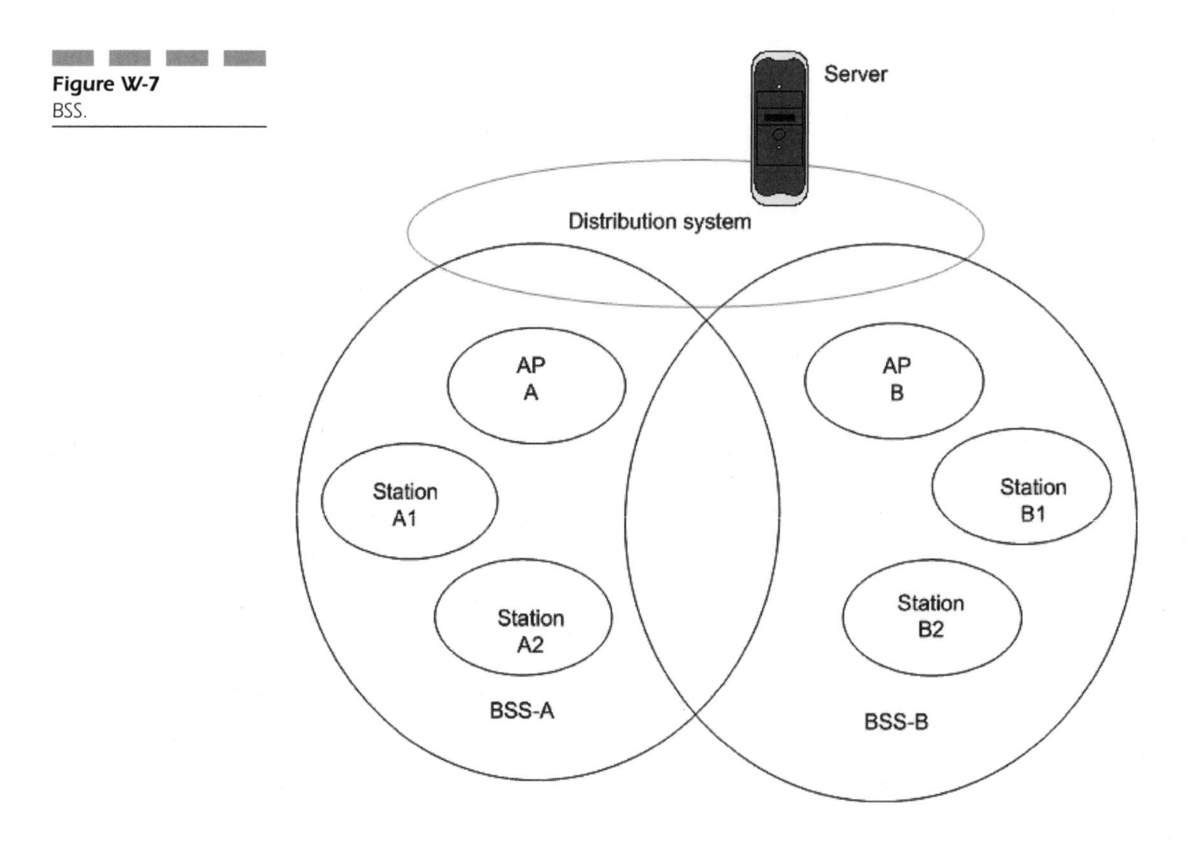

Figure W-7
BSS.

is modulated using DPSK. When receiving a DSSS signal, a matched filter correlator is used to remove the PN sequence and to create the transmitted data.

Computers/Software

W3C (The World Wide Web Consortium)

An organization dedicated to the standardization of Internet-related technologies such as HTML.

WAN (Wide Area Network)

A network of computers and interconnected LANs. Typically a WAN is spread over a greater area than a LAN.

WAV

A Microsoft standard file format for storing wave audio data. It may be used to store 8bit and 16bit wave audio at sample rates of 11.025kHz, 22.050kHz, and 44.1kHz. WAV files are compatible with all fully specified multimedia presentation programs, and multimedia authoring tools. They are also compatible with all modern Windows wave audio recorders and editors that include Sound Recorder, Creative Wave Studio, and QuickRecorder.

Wave Audio

A term often used to describe digital audio recordings usually made using an analog signal provided by a source device. Such wave audio may be distributed in real time over the Internet using Streaming server technologies, or it may be distributed using CD- and DVD-based variants. It may provide content for CD-ROM, DVD, or Web applications. Generally it may be distributed and played back using any medium that is capable of sustaining an average data transfer rate that is appropriate to the recorded wave audio quality level.

Principal parameters that drive the quality of wave audio recorded using PCM (Pulse Code Modulation) include the sampling frequency and the sample size. The wave audio quality levels that may be achieved are a function of the wave audio recording software and the sound facility on the recording system.

MPC2/3-compliant sound cards may be used to record and play wave audio in mono and in stereo at sampling rates of 11.025kHz, 22.05kHz, and 44.1kHz, using 8bit or 16bit samples. Used with appropriate software, highly specified sound cards offer higher sampling frequencies and larger sample sizes. They may make DAT quality wave audio possible that equates to 16bit samples recorded at a frequency of 48kHz. Simple sums imply that 1min of uncompressed CD-quality wave audio that amounts to 10.08Mb (10321.92kbytes), requires a DSM capable of providing an average data transfer rate of around 172.032kbytes/s. Approximate file sizes when recording 1min of 8bit stereo wave audio at different sampling rates are as follows:

11.025kHz	1.25Mb
22.050kH	2.52Mb
44.1kHz	5.04Mb
48kHz	5.49Mb

Approximate file sizes when recording 1min of 16bit stereo wave audio at different sampling rates are as follows:

11.025kHz	2.52Mb
22.050kHz	5.04Mb
44.1kHz	10.08Mb
48kHz	10.98Mb

The memory capacity consumed by a sequence is a function of quality. If it is necessary to calculate the exact memory and/or data capacity consumed, then the following simple formula may be applied:

Memory capacity required (bits)

$$= \text{sequence duration (s)} \times \text{sampling rate (Hz)} \times \text{bits per sample}$$

For example, if an 8bit sound digitizer with a sample rate of 11kHz were used to digitize a 15s sequence, then:

$$\begin{aligned}\text{Data capacity required (bits)} &= 15 \times 11{,}000 \times 8 \\ &= 1{,}320{,}000\text{bits} \\ &= 165{,}000\text{bytes} \\ &= 161.13\text{kbytes}\end{aligned}$$

Memory or disk data capacity required naturally increases linearly with increased sample rates.

Web

A global hypertext-based structure that may be navigated and browsed. It provides links to information sources and services that are termed Web sites. Tim Berners-Lee is accredited with the Web's invention, and his initial work was carried out by Berners-Lee when he was a computer scientist at the Swiss Center for Nuclear Research (CERN). Web is based on the hypertext model for information storage and retrieval. URLs are key to permitting the implantation of hypertext links and navigation schemes on the Web.

Web Cam

A Web site that features real-time video broadcasting from one or more locations. The screen updates, or the frame playback speeds, vary according to the site implementation, and may be quoted as frames per second, frames per minute, or even frames per hour. Generally Web cams provide images of locations and people from around the world and serve as entertainment, while more serious applications include CCTV-type applications or remote viewing of child care centers, for example. Web cams generally provide nonlinear broadcasting, while videoconferencing provides a bidirectional communications.

Web Page

A page that may be accessed via the Web. A Web page may include links to other pages, 2D and 3D graphics, sound bites, video, an e-mail address, and various forms for user feedback. Its underlying code or glue is HTML that may be used for formatting, as well as for holding together such components as ActiveX controls.

Web Page Description

A stream of 200 characters that exists after <BODY> tag on a Web page, and is retrieved by search engines as a description of the document. A page that may be accessed via the Web. A Web page may include links to other pages, 2D and 3D graphics, sound bites, video, an e-mail address, and various forms for user feedback. Its underlying code is HTML that may be used for formatting, as well as for holding together such components as ActiveX controls.

Web Page Title

A Web page's title is enclosed by HTML <TITLE> tags, and is used as meta data by popular search engines when retrieving Web documents, and is displayed as the document's title. A page that may be accessed via the Web. A Web page may include links to other pages, 2D and 3D graphics, sound bites, video, an e-mail address, and various forms for user feedback. Its underlying code is HTML that may be used for formatting, as well as for holding together such components as ActiveX controls.

Web Phone

See Internet Telephony under letter I.

Web Proxy

An agent that may be perceived as existing between the browser and the Internet or intranet. Typically Web proxies are used for caching Web pages in order to improve performance, hence the term *caching proxy*.

Web Security

A method of securing Web applications and their associated data from illegal unauthorized use.

Securing Web applications and their data typically involves:

- Implementing a firewall, restricting access to selected Web applications and data
- Using client-side security features of Windows NT, and security programs like Virtual Key
- Restricting access to server-side data and components that might include CGI scripts and ISAPI filters
- Monitoring system logs
- Restricting user's rights to upload files to server-side directories, so as to minimize the possibility of virus infections
- Adhering to SET guidelines
- Designing a security regime, where users require membership to access the complete site or to selected components
- Requiring site members to change their passwords
- Granting users guest rights, so that they can peruse demo Web applications and data

Web Server

An architecture that maintains the connection between the server-side processing and data, with that of the client side. The mainstay of one or more Web applications, the Web server may also implement interactions between users and server-side databases. User interaction via the browser might be processed on the client side, or on the server side. ActiveX Controls might form a basis for such client-side processing.

The Web Server interprets user requests, and implements specified tasks, such as:

- Serving HTML pages that are interpreted by the browser
- Downloading files
- Downloading Java applets
- Downloading ActiveX controls
- Interacting with server-side databases

Web servers include the Microsoft Personal Web Server that may be used for prototyping and for proving conceptual designs. With Microsoft IIS, Windows NT is used as the Web server's operating system.

Web Server Security

A set of issues that relate to securing data traffic between servers and clients so that legal usage is maintained.

Web Site

1. A physical server (or collection of such servers) and software that supports the server-side applications and data of Web applications. Users may connect with the physical or virtual Web servers contained therein, using Web addresses such as www.server.com.au. Server-side components of Web applications are numerous, including:
   - Software server components
   - ActiveX controls
   - Java applets
   - Perl scripts
2. A software solution that serves clients with a Web application.

   The application contains a page, or number of pages, and has a Web address (for example, www.testsite.com.). Such sites may be created with numerous software packages. Microsoft Publisher 98, for instance, has numerous useful wizards that guide you through the design of Web sites.

The site's interactive and media content will reside physically on the Web server, and be distributed across:

- HTML code
- Scripting languages such as JScript and VBScript
- ActiveX controls
- Java applets

Web TV

1. A Web site used in television broadcasting capacity.
2. An Internet access appliance that connects with a television. It may take the form of an STB (set top box.)

Web-Based Company

A company that uses the Web as its marketing and selling channel. Historically, such e-commerce Web sites require CGI scripts and programs in order to implement processing logic. Typically forms posted from the browser are validated in terms of credit card details, and so on, and if accepted, the customer's order is placed in the database, and processed by the vendor at an appropriate point in operations.

Wildcard

A shorthand for search strings. For example, Van Gogh *and* Amsterdam may be exchanged for Van *gh AND ?msterdam where "*" represents any series of characters and "?" replaces any single character.

WIMP (Windows, Icons, Mouse, and Pop-Up Menus)

A traditional term for the GUI environment such as Windows.

Winamp

An MP3 wave audio file player.

Windows

Windows is a GUI operating system and environment that is a proven core software technology since circa 1990, and the chosen client-side operating system for almost all client/server applications (including that of the World Wide Web). Its roots are entrenched in research carried out at the Xerox Palo Alto Research Center (PARC), and in the first commercial multitasking GUI implemented by Apple Computer for inclusion in the Apple Macintosh that was launched in 1984.

The Windows Desktop has Program icons, Active Channels, and the Taskbar that also has toolbars.

Active Desktop integrates the World Wide Web more tightly with the Windows Desktop. So-called Active Channels may be placed on the Desktop, and provide single-click access to Web sites, and the dynamic reception of information. You may also add Active Channels of your own.

The Start button provides single-click access to Start menus. These are used to open documents and applications.

The Taskbar anchors the Start button, and shows the time of day, and buttons that are used to activate open applications. It also has icons that are used to access various background applications such as Internet and network connections. There are also toolbars that provide single-click access to selected applications and features.

The founding father of the Windows concept is Douglas Engelbart, whose work was built on at Xerox PARC (Palo Alto Research Center). Microsoft's Bill Gates and Apple Computer's Steve Jobs learned of the modern Windows implementations that were developed at PARC.

This yielded the graphical user environment (GUI or "gooey") that was included in the Apple Macintosh that was launched in mid-1984. The success of the Apple Mac led Microsoft to develop a competing GUI in the form of Windows.

Work on Microsoft Windows began in earnest, with Scott MacGregor from the PARC windows initiative playing a key role. However, the early releases of Windows had little impact in a world where the PC software market was dominated largely by text-based DOS applications.

Windows was finally accepted as the de facto PC environment in the late 1980s, when version 3.0 was released.

The Microsoft Windows continuum approximates:

Windows 3.0 that supported 16bit instructions only, and featured the Program Manager that was used to organize applications and to launch them. It also featured the File Manager that was to be renamed the Windows Explorer through the launch of Windows 95. It integrated no multimedia support, because Microsoft had yet to specify the Multimedia PC-1 (MPC-1). At this time, PCs were little more than text-based appliances that offered fairly crude graphics.

Microsoft's Multimedia Extensions were launched in 1990, and could be added to Windows 3.0. These included the Media Player that is used to play audio, midi, and video files. As such, the Windows PC had become a multimedia-enabled appliance. However, it continued to be devoid of network connectivity features, and was very much a standalone implementation.

Windows 3.1 integrated the Multimedia extensions as standard.

Windows 3.11 for Workgroups included support for creating peer-to-peer Local Area Networks (LANs), in which connected computers could share their resources with other connected systems.

Windows NT (New Technology) supported 32bit instructions, and, as is the case today, was aimed at the corporate market. Its key strength is improved robustness when compared to Windows 3.1/3.11 and Windows 95/98.

Windows 95 saw the introduction of the Start menu and Taskbar that replaced the Program Manager as a means of opening applications. It also supported 32bit instructions, and was aimed at home and small office users. Networking features found in Windows 3.11 for Workgroups were also integrated in the design.

Windows 98 included new features such as the Active Desktop. Presently on the market are Windows 2000 and Windows XP.

Windows Desktop

A metaphor used by Windows, providing numerous features, which include:

- Start button that may be used to open applications and documents
- Taskbar that anchors the Start button and shows the date, as well as other important icons
- Program icons that may be double-clicked so as to open applications
- Buttons for open applications that are displayed on the Taskbar
- Time that when double-clicked invokes the Date/Time Properties
- Channel bar that provides single-click access to Web sites and information services

See Fig. W-8.

The Taskbar also serves to display numerous icons, such as those associated with connections to networks and to the Internet. These will be discussed in due course.

Windows Help System

A Windows Help system that uses Hypertext-based navigation.

Windows Media Player

A Windows program that is able to play audio, video, and midi.

Windows NT Registry

A configurable set of parameters that allow Windows NT to optimize resources for applications. The registry is stored in an initialization (INI) file, and is also used to register components that might be:

- ActiveX
- OLE
- DCOM
- COM

The regsvr32 program is used to register such components.

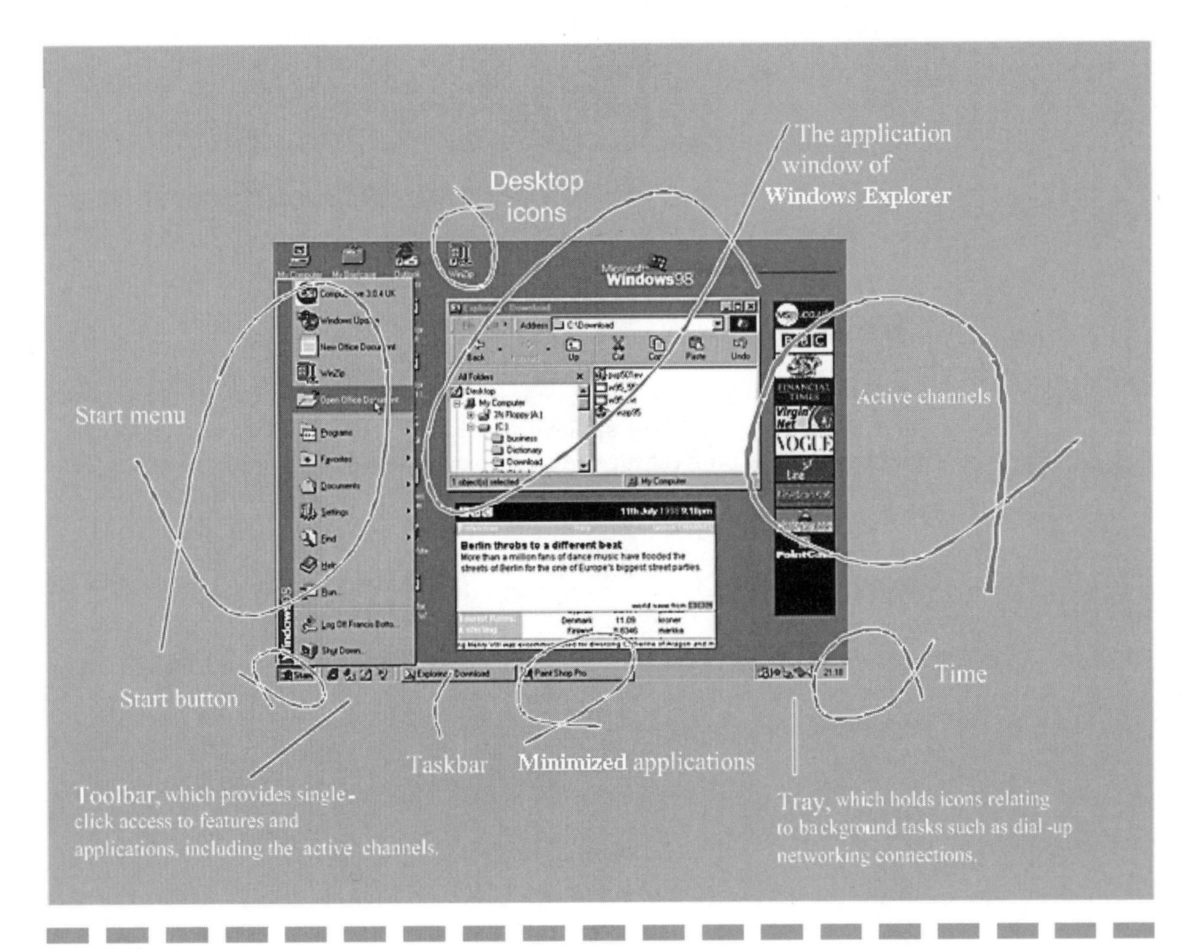

Figure W-8 Windows desktop.

Windows NT Server

A Microsoft 32bit operating system that includes the functionality of Windows NT Workstation, and an additional array of server-oriented features.

Windows NT Workstation

A Microsoft 32bit operating system that has a graphical front end. Windows NT Workstation is a complex OS, and suite of integrated applications, and includes:

- Windows Explorer that is used to browse local and remote files, open files, and launch programs
- Start menu that permits applications to be launched
- Desktop on which icons reside
- NotePad that is a simple word processor
- Network connectivity functions
- Internet connectivity functions, but has no browser (1998)
- E-mail functions

Winsock

A Windows Application Programming Interface (API) that provides input/output operations for Web applications. Its implementation takes the form of a DLL (Dynamic Link Library), and is an evolution of the Berkeley Unix sockets that provide interprocess communications both locally and over networks.

WinZip

A batch file compression/decompression utility that may be used for archiving, transmitting digital matter over narrow-bandwidth network, and accessing technologies such as analog modems.

Wizard

A software feature that guides the users through the steps required to perform a specific task. The task might be the addition of computer hardware or programs.

World Wide Web (www)

The Web is a global hypertext-based structure that may be navigated and browsed. It provides links to information sources and services that are termed Web sites. Tim Berners-Lee is accredited with the Web's invention, and is currently the Director of W3C (World Wide Web Consortium). Berners-Lee carried out the initial work when he was a computer scientist at the Swiss Centre for Nuclear Research (CERN).

A key facet of the Internet, the Web is based on the hypertext model for information storage and retrieval. Universal Resource Locators (URLs) or Web addresses are key to permitting the implantation of hypertext links and navigation schemes on the Web. It can support mixed media, including video.

It was released in 1992 by CERN. Its origins are in hypertext, hypermedia, and multimedia models and concepts. A Web page may include links to other pages, 2D and 3D graphics, sound bites, video, an e-mail address, and various forms for user feedback. Its underlying code or glue is HTML that may be used for formatting, as well as for holding together such components as ActiveX controls.

Wrappering

A process used to migrate a conventional program structure to that of an object. The program is renovated in terms of the addition of an object interface. Thereafter it may be stimulated as any other object.

www.netcraft.co.uk/

A Web site that may be visited to gain information about Web servers.

Miscellaneous

Wired Equivalent Privacy

A security protocol specified in IEEE Wireless Fidelity standard that gives wireless local area networks (WLANs) security and privacy. Data encryption secures inks between clients and access points. Other security measures include password protection, end-to-end encryption, virtual private networks (http://searchNetworking.techtarget.com/sDefinition/0,,sid7_gci213324,00.html), and authentication.

Acronyms

WACS	Wireless access communication system
WAE	Wireless application environment
WANU	Wireless access network unit
WAP	Wireless application protocol
WARC	World Administration Radio Conference
WARC-92	World Administration Conference 1992
WASU	Wireless access-subscriber unit
WCC	WML communication control
W-CDMA	Wideband CDMA
WCMP	Wireless control message protocol
WDP	Wireless datagram protocol
WECA	Wireless Ethernet Compatibility Alliance
WEP	Wired equivalent privacy
WG	Working Group
WIMS	Wireless multimedia service
WIN	Wireless intelligent networking
WLAN	wireless LAN
WLT	Wireless line transceiver
WML	Wireless markup language
WPA	Weak pilot alert
WP-CDMA	Wideband packet CDMA
WPKT	WLL packet control
WPOS	wireless point-of-sale
WRRC	WLL radio resource control
WS	Work station
WSP	Wireless session protocol
WTA	Wireless telephony application
WTAI	Wireless telephony application interface
WTLS	Wireless transport layer security
WTP	Wireless transaction protocol

INDEX

Index